Tucker

Fundamentals of Fiber Orientation

Charles L. Tucker III

Fundamentals of Fiber Orientation

Description, Measurement and Prediction

HANSER
Hanser Publishers, Munich

The Author:
Charles L. Tucker III, Urbana, Illinois, USA

MIX
Paper from
responsible sources
FSC
www.fsc.org
FSC® C083411

Distributed by:
Carl Hanser Verlag
Postfach 86 04 20, 81631 Munich, Germany
Fax: +49 (89) 98 48 09
www.hanserpublications.com
www.hanser-fachbuch.de

Editor: Mark Smith
Production Management: Cornelia Speckmaier
Cover concept: Marc Müller-Bremer, www.rebranding.de, Munich
Cover image: © Charles L. Tucker III
Cover design: Max Kostopoulos
Typesetting: Charles L. Tucker III
Printed and bound by CPI books GmbH, Leck
Printed in Germany

ISBN: 978-1-56990-875-4
E-Book ISBN: 978-1-56990-876-1

To my mentors

Nam P. Suh
K. K. Wang
Frederick A. Leckie

Preface

In 1977 an engineer from General Motors asked me a question. I did not know it at the time, but that question would change my life.

I was a Ph.D. student at MIT and General Motors was a sponsor of my research, which was about adding short glass fibers to reaction injection molded polyurethanes. I was studying the impingement mixing process, and I also explored the trade-offs between adding more and longer fibers (to increase stiffness and decrease thermal expansion) or using fewer and shorter fibers (to make mixing easier).

Researchers at General Motors were making plaques by hand and measuring mechanical properties. The question they asked was, "When we cut out samples in the mold filling direction we get one set of properties, but when we cut samples perpendicular to that we get different values. Which ones should we use?"

At the time I understood that they had flow-induced fiber orientation. However, we had no way to quantify orientation effects, much less predict what they might be in a molded part. Still, the question was a good one and it stayed with me.

A few years later, as a new assistant professor, I decided to work on flow-induced orientation. It would be several years before I had a meaningful answer to the original question, and many more years before fiber orientation software became available. The models and concepts that underly those software packages are the subject of this book.

In my own work on fiber orientation I was helped by many people. I have dedicated this book to three of them:

Nam P. Suh was my Ph.D. advisor at MIT, and later became President of the Korea Advanced Institute of Science and Technology. He taught me that I should choose research problems because they were important, and that working with industry is a great way to know which problems actually matter. He also taught me to believe in my own ideas, and to be willing to follow them even if they were unconventional. Professor Suh shaped my approach to research in a powerful way, and I am grateful to count him as a mentor and friend.

K. K. Wang, professor at Cornell University and founder of the Cornell Injection Molding Project (CIMP), set an example for doing research that was both industrially relevant and scientifically rigorous—an example I have tried to follow. In my early career he advised me on how to navigate the challenges of university-industry collaborations, and he has supported my work in all its stages. He and his CIMP colleagues made seminal contributions to computer-aided engineering for injection molding, and that field would not be what it is today without him. All of us who use injection molding simulations owe K. K. a debt of gratitude.

Frederick A. Leckie was my colleague at the University of Illinois, and he is the person responsible for orientation tensors. Fred's office was next to mine, and when my students and I were solving fiber orientation distribution functions in 2-D, Fred said, "You should use these tensors that Turan Onat and I are using for microscopic damage in metals." At first I was unable to grasp the idea. However, one day Fred came in and said, "I was at home sick yesterday so I made some notes about how you could use tensors for your fibers." He handed me one sheet of paper with equations and sketches, and from that I was able to understand the concept. My paper with Suresh Advani [AT87] was the eventual result. Orientation tensors have since become the standard way to predict, measure or even talk about fiber orientation, but without Fred's generosity and persistence we might not have them today. Users of the DIGIMAT software will enjoy knowing that one of its principal creators, Prof. Issam Doghri, did his post-doctoral studies with Prof. Leckie and benefitted in similar ways.

These mentors set a great example for me, and I tried to follow that example with my own graduate students and post-docs. The ones who worked on fiber orientation or mold filling, and who thus contributed to this book, are: Ahmed Adb El-Rahman, Suresh Advani, Alexander Bakharev, Randy Bay, Peter Brown, Carolyn Wright Camacho, Che-Yang Chen, Robert Davis, Scott Davis, Francisco Folgar, William Haverty, Henry Huynh, William Jackson, Tim Konicek, Craig Korte, Ching-Chih Lee, Erwin Liang, Peter Loughlin, Mark Lovrich, Tim Osswald, Jay Phelps, Chad Sandstrom, Douglas Smith, Gregorio Vélez-García, Brent VerWeyst, Jin Wang, Eric Wetzel, Yeoung-Eun Yoo and Martin Zentner. Thank you for your hard work and your patience with my many requests. You have put your own stamp on this field, and I am proud of each one of you.

I had the chance to collaborate with a number of colleagues from other universities, always benefitting from their insights and ideas. From Purdue University I worked with Anthony Favaloro, Michael Sangid and Bhisham Sharma; from Universidade Federal de São Carlos, Brazil, Joaquim Cintra; from the University of Illinois at Urbana-Champaign, Daniel Tortorelli; from the University of Minho, Portugal, Júlio Viana and António Cunha; and from Virginia Tech, Donald Baird and Peter Wapperom. Thank you, my friends.

Following the pattern of my student years, I maintained industrial collaborations throughout my research career. In addition to financial sponsorship, these connections were a source of valuable insights, vital information and vibrant collaborations. These sponsors and collaborators include Autodesk, Inc. (Franco Costa, Xiaoshi Jin, Peter Kennedy and Jin Wang); Delphi Automotive Systems (John O'Gara and Michael Wyzgoski); Deere & Co.; Dow Chemical Company (Robert McGee and Selim Yalvaç); DuPont de Nemours, Inc. (Albert Hirsch); Ford Motor Co. (Carl Johnson, Yang Li and Danielle Zeng); GenCorp (Jose Castro, Richard Griffith and James Stevenson); General Electric Corp. (Vijay Stokes and Gerry Trantina); General Motors Corp. (Martin Barone, David Caulk and Peter Foss); Moldex3D (Srikar Vallury and Anthony Yang); Navistar International; Oak Ridge National Laboratory (Barbara Frame and Vlastimil Kunc); Owens Corning Fiberglas Corp. (Robert Beck, Douglas Denton, Allan Isham and Donald Sage); Pacific Northwest National Laboratory (Leo Fifield, James Holbery, Ba Nghiep Nguyen and Mark Smith); Sundstrand Corp.; and Toyota North America (Umesh Gandhi). Additional sponsors of my research include the American Chemistry Council, the National Science Foundation, and the United States Department of Energy. I am indebted to all of them for their support.

I would not have written this book without the generosity of some dear friends and colleagues. Professors Tim Osswald and Lih-Sheng (Tom) Turng invited me to speak at the Polymer Symposium at the University of Wisconsin-Madison in 2015. There I met Dr. Mark Smith from Hanser, and we began discussing this book. At first I was hesitant to start the project. Then, in 2018, Prof. R. Byron Pipes invited me to give a series of lectures on orientation to the students and staff of the Composites Manufacturing and Simulation Center at Purdue University. Developing materials for 14 hours of lectures and interacting with the students at CMSC convinced me that a text on orientation would be valuable, and those materials became the backbone of this book. It took Mark Smith another two years to get me to the point of writing, but his patience and persistence paid off. Tim, Tom, Byron and Mark, my thanks to all of you.

The images from CAE software in Chapt. 1 were created by Dr. Jin Wang of Autodesk, Inc., who also provided important suggestions for the text. I am grateful for his help and for the support of Dr. Franco Costa.

Doctors Armin Kech, Susanne Kugler and Thomas Riedel of Robert Bosch GmbH generously provided the micro-CT data and images in Chapt. 3, sharpened my understanding of micro-CT measurements, and also provided direct fiber simulation images and data for Chapt. 9. They are at the forefront of technology in fiber orientation, and I greatly appreciate their assistance.

I am also grateful to my colleagues Bill Mischo and Mary Schlembach of the Grainger Engineering Library at the University of Illinois at Urbana-Champaign for cheerfully locating hard-to-find references, and for maintaining a world-class resource for scholars. The team at `le-tex publishing services` resolved many LaTeX issues for me, and I thank them for making that part of the work easy.

My wife, Laurie Tucker, is truly the better part of me. An author in her own right, she understood what it would take for me to write this book, and she was unwavering in her support. Thank you, my love. If we have another 47 years together it will not be enough.

As I write this, we have reached the 100th anniversary of George Barker Jeffery's paper to the Royal Society of London, *The motion of ellipsoidal particles immersed in a viscous fluid* [Jef22]. This paper is the foundation of all fiber orientation analysis. Jeffery was a mathematician who was interested in exact solutions to physical problems, and he probably gave little thought to practical applications of his work. A biographical memoir written shortly after Jeffery's death in 1957 [Tit58] gives no special mention to this paper. However, citations to it began to rise in the 1960s, and it now receives more than 100 new citations each year. It is undoubtedly Jeffery's most important work. We can be grateful to Prof. Jeffery for solving a significant problem, once and for all. In addition to his mathematical research, Jeffery devoted much of his professional life to education and to the training of teachers. I like to think that he would be pleased with the impact of his work on two counts: first, that it has been useful to so many people, and second, that it has underpinned the training of many bright and capable students. We can all hope for such a legacy.

Charles L. Tucker III

Urbana, Illinois
January 2022

Contents

1 Introduction

1.1 Discontinuous Fiber Composites

A composite is any material with two or more constituents that retain their identities in the final material. Everyday examples range from plywood, which is layers of wood veneer glued together, to concrete, a mixture of gravel and sand bonded by Portland cement.

Many composites contain fibers, often glass or carbon, embedded in a polymer matrix. When the material must be as strong, stiff and light as possible, and when high manufacturing cost can be tolerated, the fibers are continuous and are either aligned in thin layers much like plywood, or woven into a fabric. Continuous fiber composites are widely used in aircraft, wind-turbine blades, boat hulls, and in sporting goods such as tennis rackets, bicycles and golf clubs.

When manufacturing costs must be low or when the final component has a complex shape, the fibers are cut or chopped into shorter lengths to create a *discontinuous fiber composite* (DFC). The fibers and matrix polymer are combined when the matrix is in a liquid state, then formed into final shape by processes like injection or compression molding. Discontinuous fiber composites are used in business machines, power tools and many other products. In automobiles they are used for body panels, structural components and under-the-hood components. Since the 1970s the automotive industry has been the largest user of fiber reinforced composites [GGOS20], and almost all automotive composites have discontinuous fibers. Discontinuous fiber composites are the focus of this book.

Many discontinuous fiber composites have thermoplastic matrices and are processed by injection molding. These are often subdivided into short fiber/thermoplastics (SFTs) and long fiber/thermoplastics (LFTs). Pellets of the short fiber materials are prepared by mixing the fibers and matrix in a compounding extruder, and average fiber lengths are less than 1 mm. Long fiber materials are prepared by impregnating a continuous fiber tow with the matrix resin, usually by an extrusion process, then cutting the coated fibers to a length of approximately 12 mm to make pellets. Pellets of either type are processed in a conventional injection molding machine, where they are re-melted in the plasticizing screw and injected into the mold. For LFT materials this substantially reduces the fiber length, but final lengths greater than 1 mm are possible with careful processing. We examine both SFTs and LFTs in detail in this book.

Another important type of discontinuous fiber composite is sheet molding compound (SMC), which has a thermosetting polymer matrix with chopped fibers. The molding compound is initially produced by combining chopped fiber bundles, 25 to 50 mm long, with a liquid resin, particulate filler and other chemicals at room temperature. The resin is allowed to thicken until the sheet is can be handled but is still soft. Layers of this material are cut and stacked in a specified shape, called the charge, then placed into a hot compression mold. Closing the mold squeezes the charge, causing it to flow and fill the mold cavity, while heat from the mold initiates a curing reaction that hardens the polymer. After one to two minutes, the mold is opened and the solid part is removed.

There are many other types of discontinuous fiber composites. In direct in-line processing of long fiber/thermoplastic composites (D-LFTs) the polymer is melted in a compounding extruder, and continuous glass fibers are added at a port partway down the extruder barrel. The hot fiber/resin mixture is extruded from a die, cut into a strand, and placed immediately into a mold where it is compression molded to final shape. This process can preserve fiber lengths as long as 60 mm in the finished part.

With glass mat thermoplastic (GMT) composites, a mat of reinforcing fibers is impregnated with a polymer resin and solidified into an initial sheet form. The fibers may be continuous, woven or chopped. In a subsequent step, blanks cut from the sheets are re-heated, placed into a compression mold and formed into final shape.

Discontinuous fiber composites are also used with additive manufacturing or 3-D printing. This is most commonly done via fused-filament fabrication (FFF) or fused deposition modeling (FDM), in which a bead of molten polymer is extruded from a heated die and laid down on the bed of the 3-D printer, where it solidifies. By moving the die and/or the printer bed, beads of polymer are laid down in rows in the desired pattern. Subsequent layers are deposited on top to create the final shape. By adding discontinuous fibers to the feed material, a fiber reinforced material can be printed [RHJS18, MGB+18, OBW+20].

Some composites are reinforced with disk-shaped particles such as glass flakes or ceramic platelets. These materials are often processed by injection or compression molding. Because the particles can orient under flow, these materials have the same connection between processing, microstructure and properties as discontinuous fiber composites. The orientation and mechanical property models in this book can be applied to these composites as well.

■ 1.2 Processing, Microstructure and Properties

While discontinuous fiber composites offer an attractive combination of mechanical properties, light weight and low cost, designing with them presents a challenge. This is because the properties of a DFC component depend on how the final part is made.

Figure 1.1 shows an example. The two boxes in this figure were injection molded from the same glass-fiber/polyamide SFT material using the same mold. However, one box is straight and the other is badly warped. The difference is the location of the gate where the molten fiber/polymer mixture first enters the mold cavity. The box on the left uses a film gate across the near end, while the warped box on the right uses a small tab gate partway down the left side. The two gate locations produce different mold filling patterns, which create different patterns of fiber orientation in the two moldings.

For both moldings, the material is above room temperature when it solidifies, then cools to room temperature after it is removed from the mold. The fibers reduce the thermal contraction of the polymer matrix along their length, but have only a small effect in other directions. If the fibers have a preferential orientation, then as the material cools it will contract less in the fiber direction but more in the perpendicular direction. In the box on the right in Fig. 1.1, the direction of fiber orientation is very different from one location to another, and this causes

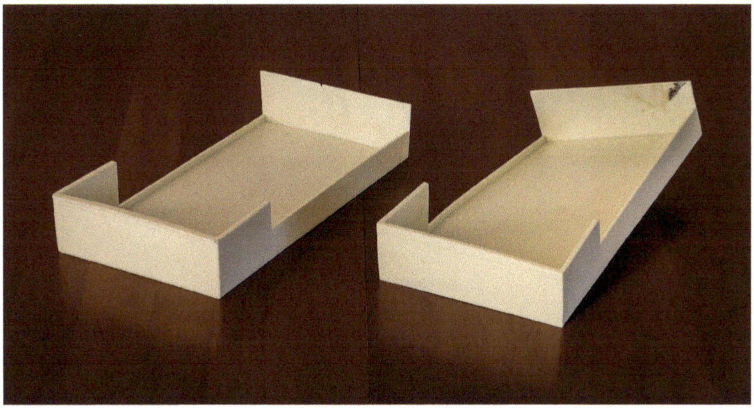

Figure 1.1 Two injection molded boxes. Photo by the author.

the box to warp as it cools. In the box on the left, the fibers are oriented in nearly the same direction at all locations, so this box does not warp. Processing has affected the properties of the material in these boxes.

A second example appears in Fig. 1.2, which shows stress-strain curves for samples of sheet molding compound. All of the curves are for the same batch of material and were molded to the same final thickness in the same rectangular plaque mold. However, some samples were made by cutting the charge to nearly fit the mold, while other samples used charges with smaller area but more layers. The charges were all designed to flow in only one direction as they were compressed, as in Fig. 4.28(a). Also, the tensile specimens were cut from the plaques so that some were loaded parallel to the flow direction (0°), while some were loaded perpendicular to the flow direction (90°).

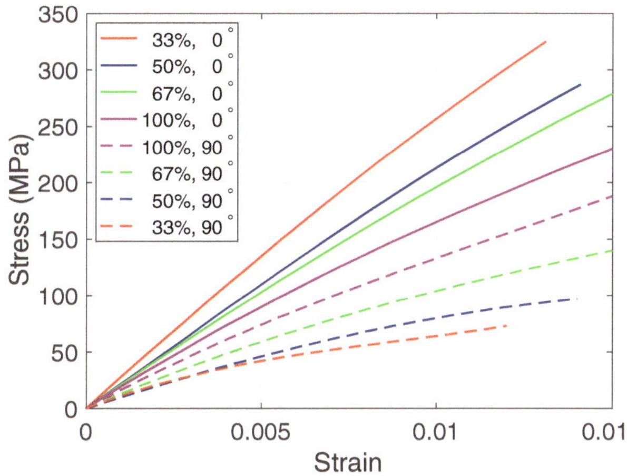

Figure 1.2 Tensile stress-strain curves for samples of sheet molding compound with 65% by weight of glass fibers. The legend gives the percentage of the mold area initially covered by the charge and the direction of the tensile test relative to the flow. Data from [CT84].

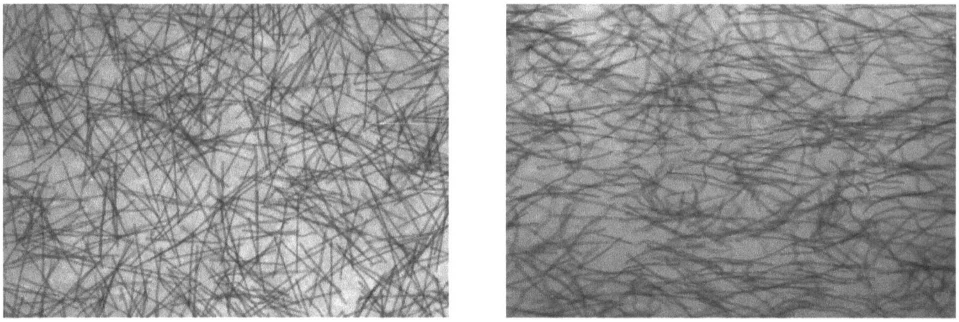

(a) 100% charge (b) 33% charge

Figure 1.3 X-ray images of two of the samples of sheet molding compound from Fig. 1.2. The flow direction is horizontal and the fibers are 25 mm long [Che84].

These variations in processing and testing produce the wide range of properties seen in Fig. 1.2. The strongest specimen has an ultimate tensile strength of 325 MPa, five times greater than the strength of the weakest specimen, 73 MPa. The elastic moduli show similar large variations.

These property variations follow a pattern. For each initial charge shape, the material is stronger and stiffer parallel to the flow, weaker and less stiff perpendicular to the flow. The smaller the charge, the greater the difference between the two directions.

These differences in mechanical properties arise because of differences in fiber orientation. Figure 1.3 shows X-ray images from two of the plaques. In these samples, some of the fibers were made from a glass with a high lead content and are opaque to X-rays, so they appear as dark fibers in the images. The fibers in the 100% charge sample are almost randomly oriented, while fibers in the 33% charge sample have a strong horizontal (flow-direction) orientation. Fibers reinforce the polymer matrix primarily along their axes. This makes the 33% charge sample much stronger and stiffer than the 100% charge sample in the flow direction, but weaker and less stiff perpendicular to flow. The other two charges have intermediate degrees of orientation and intermediate properties. Again, processing has a major effect on properties.

Discontinuous fiber composites present a striking contrast to materials like steel and aluminum, whose elastic moduli can be looked up in any materials handbook and are the same in any direction. The properties of a discontinuous fiber composite depend on the amount of flow, on the loading direction, and may vary from one location to another within a part.

How can one design with these materials? If we use only the minimum properties that can occur in a molded part, the design will be overly conservative and the part will be heavy and expensive. To use DFCs effectively, a designer must have good information about the actual properties at each location in the component. This can depend on processing conditions and part geometry, as well as on the intrinsic properties of the fiber and the matrix.

In these examples the connection between properties and processing is fiber orientation. If we can predict how the fibers are oriented in a molded part and relate that orientation to the mechanical properties at each location, we can determine how processing will affect the properties. We can then design components that use discontinuous fiber composites effectively. The required calculations are complex, but computer-aided engineering software is available to do most of the work.

1.3 Computer-Aided Engineering Workflow

The most widely used computer-aided engineering tools for DFCs are for injection molding. Many of these software packages can also simulate compression molding and can treat both thermoplastic and thermosetting matrices.

With these software tools, each analysis starts from the final part geometry, represented by a CAD solid model. This determines the shape of the mold cavity. An initial choice for the material is specified, and the analyst adds some manufacturing-related information, such as runners and gates for an injection mold, or the charge shape and location for compression molding. The analysis then proceeds in three major steps:

1. A **mold filling simulation** calculates the velocity, pressure and temperature of the polymer/fiber mixture as a function of time during filling, at all locations in the mold. Fiber orientation predictions are integrated into this calculation, and also proceed step by step as the mold fills. The fiber length distribution may be calculated, and the simulation may extend into the packing and cooling phases of the process. Output from this step includes the fiber orientation (and possibly fiber length distribution) at each location in the final part.
2. **Mechanical property predictions** use the fiber and matrix properties and the microstructural data from the first step to find direction-dependent properties (stiffness, thermal expansion, etc.) at each location in the final part.
3. A **structural analysis** is performed on the solid part using the finite element method. This calculation incorporates the mechanical properties from the previous step on an element-by-element basis. Multiple load cases may be analyzed to determine dimensional stability of the part (warping and shrinkage), deflections under specified loads, local stresses and loads at failure.

Figures 1.4 and 1.5 illustrate these steps for a water pump housing. The initial CAD model is shown in Fig. 1.4(a). In this figure, gates have been added, and are indicated by the yellow cones. A finite element mesh is then generated, dividing this geometry into small, three-dimensional elements. A portion of this mesh is shown in Fig. 1.4(b). The mold filling analysis can then proceed. Figure 1.4(c) shows a portion of the filling pattern from this analysis. At this instant in time the colored portion of the mold has been filled, and the colors indicate where the flow front was at previous times.

A portion of the predicted fiber orientation pattern is shown in Fig. 1.5(a). The colored lines show the main directions of orientation, with the red and yellow lines indicating highly aligned fibers, and blue or green lines indicating more random orientations. A cutting plane has been used to provide a view across the part thickness, and orientation does vary across the thickness.

This orientation information is converted into an elastic modulus value for each element in the mesh, as shown in Fig. 1.5(c). Each element has direction-dependent properties, and the figure shows the tensile modulus in the main fiber direction from Fig. 1.5(a). This modulus information is used, together with an orientation-dependent thermal expansion for each element, to calculate the size and shape of the part after it has cooled to room temperature. Deflections from this calculation are shown in Fig. 1.5(c), with the deflections amplified 10× to make them more visible. The base of the part is slightly curved, and the large hole in the center is no longer perfectly round.

All of these analyses may be done in a single integrated software package, as in the example above, or separate packages may be used with information flowing between them.

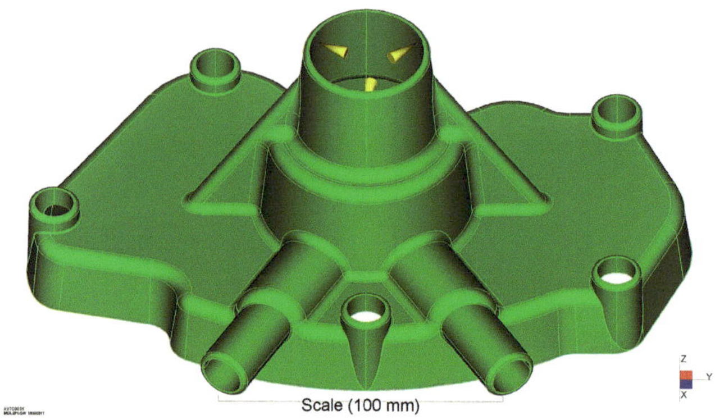

(a) CAD model

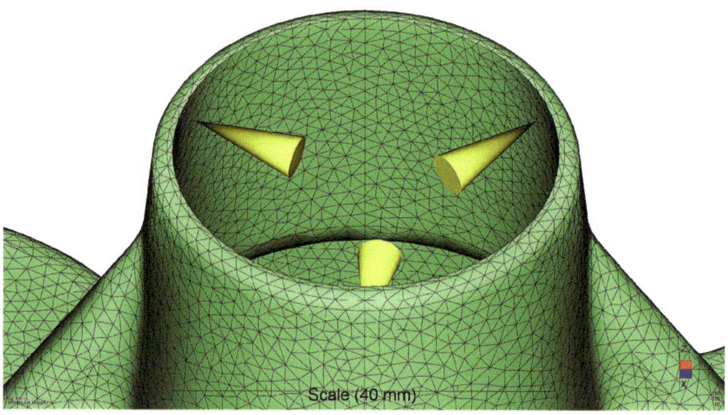

(b) Finite element mesh near the gates

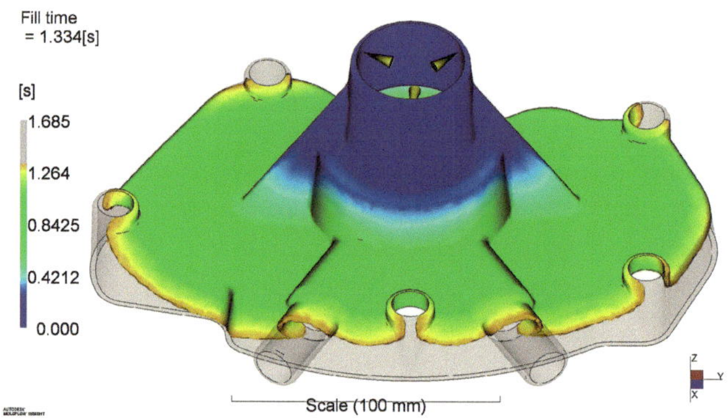

(c) Mold filling pattern

Figure 1.4 Initial steps in the CAE workflow for an injection molded, fiber reinforced part. Images provided by Autodesk, Inc.

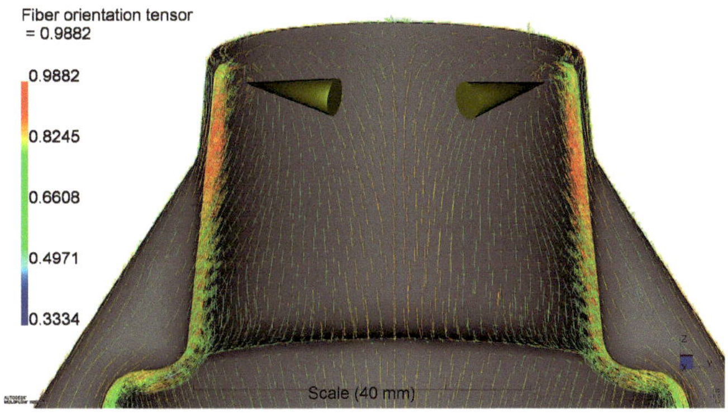

(a) Fiber orientation near the gates (cutaway view)

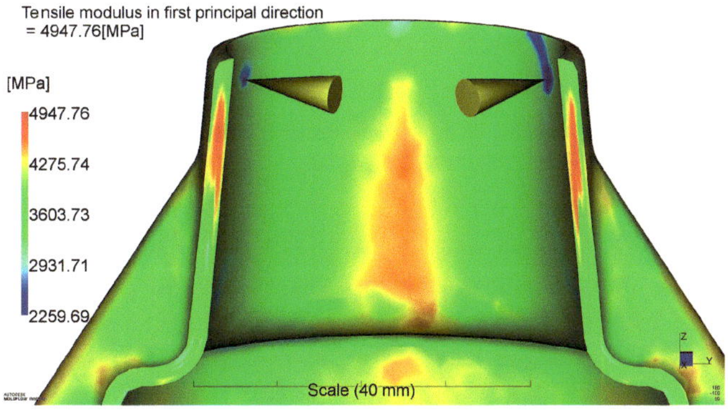

(b) Tensile modulus in the main orientation direction

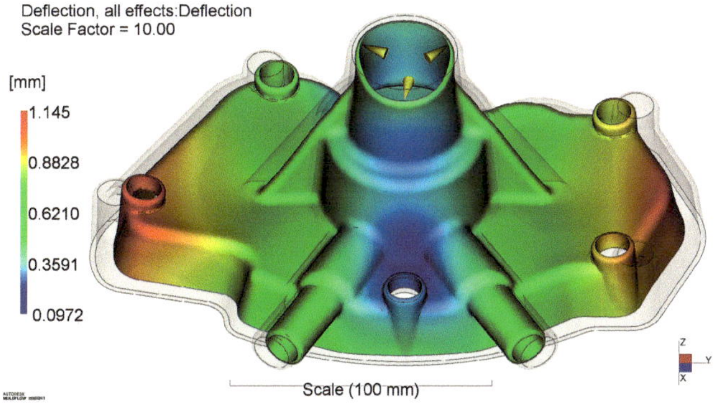

(c) Deflection after cooling

Figure 1.5 Additional steps in the CAE workflow for an injection molded, fiber reinforced part. Images provided by Autodesk, Inc.

At any step, detailed results are viewed and interpreted. If the results do not meet design requirements, one can revise the processing conditions, material choice or part design, then start the analysis again. This cycle of analysis, interpretation and revision continues until a suitable combination of part design, material and processing conditions has been found.

1.4 Scope and Organization of the Book

This book is written for users of CAE software who want to know how fiber orientation calculations work and how to use their software effectively. The primary focus is on calculation of fiber orientation and fiber length within a mold filling simulation. We also discuss the simplest mechanical property calculations—linear elastic stiffness and thermal expansion—to show how fiber orientation is connected to properties. The goal is to cover all the models that are used in commercial software, explain their underlying concepts, and show when to use each option and how to set the parameters for each model.

Another key audience is graduate students who are beginning research in this field. For these readers the book provides an organized presentation of the state of the art as of this writing, with selected references to recent research.

Flow and heat transfer models for mold filling are not discussed, as that information is available elsewhere [Ken95, ZTF11]. Similarly, we do not discuss finite element structural analysis.

While it is easy to see that fiber orientation affects properties, predicting those properties quantitatively takes some effort. An essential step is to choose the right variables to describe the microstructure. The best microstructural variables meet the following requirements:

- They have a precise mathematical definition.
- They can be measured experimentally.
- The ways they change during processing can be modeled using physical principles.
- Their effect on mechanical properties can be quantified, also using physical principles.

All of these requirements are met by the fiber orientation and length variables used in CAE software, and this book is designed around them:

- Chapter 2 defines the variables used to describe fiber orientation and fiber length. Chief among these is the second-order orientation tensor. This is the principal tool used for fiber orientation, and without it practical CAE tools for discontinuous fiber composites would not exist. Fiber length distributions and fiber length averages are also presented.
- Chapter 3 discusses how fiber orientation and fiber length are measured. For orientation we cover both traditional measurements on planar sections and modern micro-CT measurements. Methods for fiber length include burning off the matrix and micro-CT.
- Chapter 4 discusses the orientation of single fibers and Jeffery's equation. It also reviews basic types of deformation, and shows how two simple rules can qualitatively explain the orientation patterns in almost all situations.

- Chapter 5 explains how the orientation statistics for groups of fibers are predicted, and examines the models that are used in CAE software to make quantitative orientation predictions.
- Fibers can also alter the flow, and ways to model this are discussed in Chapt. 6. This requires modeling the rheology of fiber suspensions.
- Fiber length modeling is presented in Chapt. 7. These models are especially useful for LFT materials, to help preserve as much fiber length as possible in the final part.
- The connection between fiber orientation and mechanical properties is discussed in Chapt. 8. The focus here is on linear elastic behavior and thermal expansion. Property models for nonlinear stress-strain behavior, damage and failure also exist, but those topics are outside the scope of the book.
- Chapter 9 provides an overview of two advanced models from recent research. One, shear-induced fiber migration, causes the fiber volume fraction to vary from point to point. Models of this type are now appearing in commercial software. The other, direct simulation of multiple, interacting fibers, has the potential to affect how fiber orientation is modeled in the future.

Readers are assumed to be familiar with fluid mechanics and material behavior at the level of a bachelor's degree in engineering, as well as calculus of multiple variables, differential equations and matrix algebra.

Tensors are the natural mathematical language for fiber orientation modeling and are used extensively here. Readers who are less familiar with tensors will find the discussion of orientation tensors and the examples in Section 2.3 helpful. Appendix A summarizes the types of tensor notation used in this book and shows how to perform mathematical operations involving tensors. In practice, all of these calculations are done numerically by a computer, and Appendix A shows how vector and tensor operations are carried out in MATLAB.

The cited references are either foundational papers or representative of recent research. No attempt has been made to comprehensively review the literature, and there are many excellent papers on fiber orientation and related topics that are not cited here.

Key ideas and equations appear in shaded boxes, and each chapter ends with a summary.

1.5 Accompanying Software

For readers who want to explore fiber orientation and mechanical property models in more detail, a set of MATLAB functions is available at *https://github.com/charlestucker3/Fiber-Orientation-Tools*. Most of the calculations in this book were performed using this toolkit, and its capabilities are mentioned throughout the text. Documentation at that website gives more information. Fiber orientation models in Python are provided by Meyer [Mey21].

2 Describing Fiber Orientation and Fiber Length

The first step in dealing with fiber orientation is to describe it mathematically. We begin with single fibers, then discuss statistical distributions of groups of fibers. For this, there are two main options: probability density functions (also called fiber orientation distribution functions) and orientation tensors. The tensor description is a key tool both for predicting orientation and for presenting experimental data, and it plays a central role in the chapters that follow.

Fiber length plays a role in properties as well. Here we discuss fiber length distributions and different average values of fiber length.

2.1 Single Fiber Orientation

We primarily consider fibers that are straight and rigid (i.e., they do not bend by any noticeable amount). Both glass and carbon fibers have circular cross-sections, so each fiber is a circular cylinder. The orientation of any fiber is then described by the orientation of its symmetry axis. The angles θ and ϕ, shown in Fig. 2.1, provide one way to do this.

An alternate description, which we will use extensively, is a unit vector **p** that points along the fiber axis. This vector is also shown in Fig. 2.1. The components of this vector are p_1, p_2 and p_3, and they are related to the angles θ and ϕ according to

$$\begin{Bmatrix} p_1 \\ p_2 \\ p_3 \end{Bmatrix} = \begin{Bmatrix} \sin\theta\cos\phi \\ \sin\theta\sin\phi \\ \cos\theta \end{Bmatrix} \tag{2.1}$$

How can **p** have three components when only two angles are needed to describe the orientation of a fiber? **p** also has unit length, which requires that

$$\mathbf{p}\cdot\mathbf{p} = 1 \qquad \text{or} \qquad p_1^2 + p_2^2 + p_3^2 = 1 \tag{2.2}$$

Thus, only two components of **p** are independent.

While the vector **p** has a head and a tail, a fiber does not. As Fig. 2.2 shows, for any fiber there are two equally valid ways to choose **p**. If the unit vector $\mathbf{p}_1$ describes a fiber, then so does the vector $\mathbf{p}_2 = -\mathbf{p}_1$. Similarly, if a fiber has angles θ_1 and ϕ_1, we could also describe it as having angles (θ_2, ϕ_2) where $\theta_2 = \pi - \theta_1$ and $\phi_2 = \phi_1 + \pi$. All equations that deal with fiber orientation must respect these relationships.

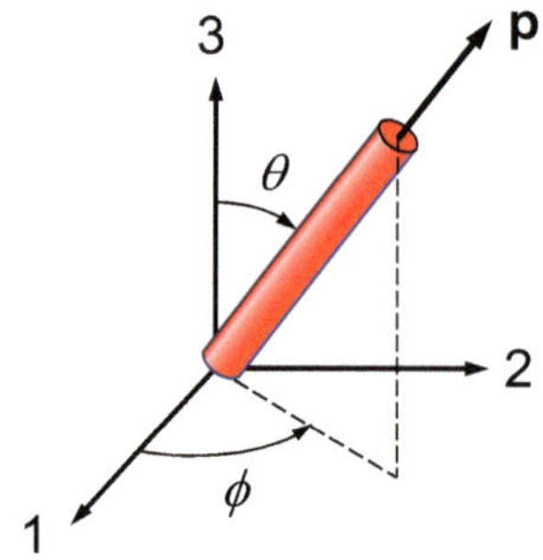

Figure 2.1 The angles θ and ϕ and the unit vector $\mathbf{p}$ that define the orientation of a single, rigid fiber with respect to Cartesian coordinate axes.

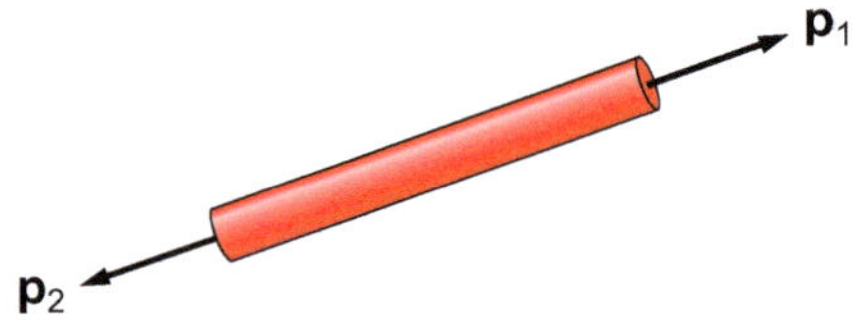

Figure 2.2 The orientation of a single fiber is described by either $\mathbf{p}_1$ or $\mathbf{p}_2$, where $\mathbf{p}_2 = -\mathbf{p}_1$.

2.2 Distributions of Fiber Orientation

An injection molded short fiber composite can easily have 10,000 fibers per mm^3, so for realistic composites we must consider many fibers. Describing the orientation of each individual fiber would be a huge task, and is usually unnecessary. Instead, one describes the statistics of the fiber orientation. This is most easily understood when all fibers lie in a plane, so we will treat that case first, then generalize to three-dimensional fiber orientation.

2.2.1 Planar Orientation

Consider a composite where all fibers lie in the 1–2 plane. This is a good approximation for compression molded sheet molding compound (SMC), where the fibers are long compared to the final part thickness. For planar orientation all fibers have $\theta = \pi/2$, and each fiber's orientation is described by a single angle ϕ. Similarly, the orientation vector $\mathbf{p}$ always has $p_3 = 0$, so p_1 and p_2 describe an individual fiber and $p_1^2 + p_2^2 = 1$.

Figure 1.3 showed two X-ray images of SMC samples. The orientations appear nearly random in the sample on the left, and have a strong horizontal alignment in the sample on the right. (The fibers are also not straight, an issue we will ignore for now and return to in Chapt. 9.)

One can measure the angle ϕ for many fibers and collect the data in a histogram, like those in Fig. 2.3. Note that the data has been restricted to $-90° < \phi \leq 90°$. This is a typical way to treat experimental data for planar orientation.

In Fig. 2.3 the histogram on the left indicates a mostly random orientation, while the histogram on the right shows a strong tendency for fibers to have ϕ close to zero. This corresponds to our

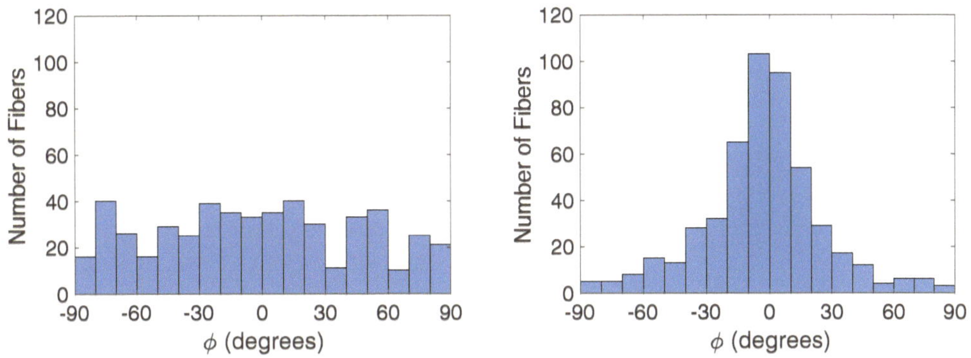

(a) 100% charge (b) 33% charge

Figure 2.3 Histograms of orientation angle ϕ for 500 fibers measured from each of the samples in Fig. 1.3 [CT84].

visual impressions of the X-ray images. Looking closely, the histogram on the left does show a slight preference for the 0° direction compared to the 90° direction, a feature that is not easily detected by looking at the X-ray image.

A histogram is a discrete approximation to the underlying probability density function for orientation. For planar orientation states we will use $\psi_\phi(\phi)$ to denote this function, which is also called the *orientation distribution function*. This function is defined such that the probability of any fiber having a ϕ value between some angle ϕ^* and $\phi^* + \mathrm{d}\phi$ is

$$P(\phi^* \leq \phi < \phi^* + \mathrm{d}\phi) = \psi_\phi(\phi^*)\,\mathrm{d}\phi \tag{2.3}$$

The function ψ_ϕ must satisfy some requirements for any orientation state. First, every fiber must have some orientation, so the integral of ψ_ϕ over all values of ϕ must equal unity:

$$\int_0^{2\pi} \psi_\phi(\phi)\,\mathrm{d}\phi = 1 \tag{2.4}$$

We will refer to this as the *normalization* requirement. Also, because ϕ is equivalent to $\phi + \pi$, the orientation distribution function must be *periodic* with period π:

$$\psi_\phi(\phi + \pi) = \psi_\phi(\phi) \tag{2.5}$$

If the fibers are randomly oriented in the plane, then all values of ϕ are equally likely and $\psi_\phi(\phi)$ is constant. The normalization requirement (2.4) then tells us that

$$\psi_\phi(\phi) = \frac{1}{2\pi} \qquad \text{random-in-plane orientation} \tag{2.6}$$

2.2.1.1 Matching ψ_ϕ to Data

In later sections we will write equations for $\psi_\phi(\phi)$ and develop models to predict it, so we will want to be able to compare those predictions to experimental data like that in Fig. 2.3. This is done by scaling the histogram data on the vertical axis. Let ϕ_i, $i = 1, \ldots, n$, be the angles at the centers of the histogram bins, where n is the number of bins. Also assume that the bins span a total angle of π. Let $\Delta\phi = \phi_{i+1} - \phi_i$ be the bin width, which is the same for all bins, and let N_i be the number of fibers assigned to the bin with ϕ_i. We know that the distribution function at the center of bin i should be proportional to N_i,

$$\psi_\phi(\phi_i) = kN_i \tag{2.7}$$

where k is some constant of proportionality to be determined. We want the distribution function to satisfy the normalization condition, Eqn. (2.4), such that

$$\int_0^{2\pi} \psi_\phi(\phi)\,\mathrm{d}\phi \approx 2\sum_{i=1}^{n} \psi_\phi(\phi_i)\,\Delta\phi = 1 \tag{2.8}$$

The factor of two appears in the second expression because the ϕ_i values span an angular range of π, but the normalization requirement is written over the full range of 2π. Combining Eqns. (2.7) and (2.8) gives the value of k and allows us to write the correspondence between histogram data and the orientation distribution function:

$$\psi_\phi(\phi_i) = \frac{N_i}{2\Delta\phi \sum_{j=1}^{n} N_j} \tag{2.9}$$

While we often plot data with ϕ in degrees, the value of $\Delta\phi$ in this equation must be in radians.

2.2.2 Three-Dimensional Orientation

For three-dimensional orientation the set of all possible orientations (θ, ϕ) or $\mathbf{p}$ is the surface of a unit sphere, as shown in Fig. 2.4. We will call this surface the *orientation space.*

The orientation distribution function for 3-D orientation, written as $\psi(\theta, \phi)$ or $\psi(\mathbf{p})$, is defined such that the probability of a fiber lying between θ^* and $\theta^* + \mathrm{d}\theta$ and between ϕ^* and $\phi^* + \mathrm{d}\phi$ is

$$P(\theta^* \le \theta < \theta^* + \mathrm{d}\theta, \phi^* \le \phi < \phi^* + \mathrm{d}\phi) = \psi(\theta^*, \phi^*) \sin\theta^*\,\mathrm{d}\theta\,\mathrm{d}\phi \tag{2.10}$$

The factor of $\sin\theta^*$ appears, as in any problem using spherical coordinates, because an increment $\mathrm{d}\theta\,\mathrm{d}\phi$ covers different amounts of area at different values of θ^*.

The integral of any function $f(\theta, \phi)$, or equivalently $f(\mathbf{p})$, over the orientation space can be written in either of two equivalent ways:

$$\int_{\theta=0}^{\pi} \int_{\phi=0}^{2\pi} f(\theta, \phi)\,\sin\theta\,\mathrm{d}\theta\,\mathrm{d}\phi = \oint f(\mathbf{p})\,\mathrm{d}\mathbf{p} \tag{2.11}$$

In the latter expression, $\mathrm{d}\mathbf{p}$ is a small solid angle corresponding to the small area on the surface of the sphere $\sin(\theta)\,\mathrm{d}\theta\,\mathrm{d}\phi$, as shown in Fig. 2.4. The $\mathbf{p}$ notation is compact and will be useful in describing concepts, while the (θ, ϕ) notation is useful for deriving results algebraically or for numerical calculations.

The normalization requirement for 3-D orientation is

$$\oint \psi(\mathbf{p})\,\mathrm{d}\mathbf{p} = 1 \quad \text{or} \quad \int_{\theta=0}^{\pi} \int_{\phi=0}^{2\pi} \psi(\theta, \phi)\,\sin\theta\,\mathrm{d}\theta\,\mathrm{d}\phi = 1 \tag{2.12}$$

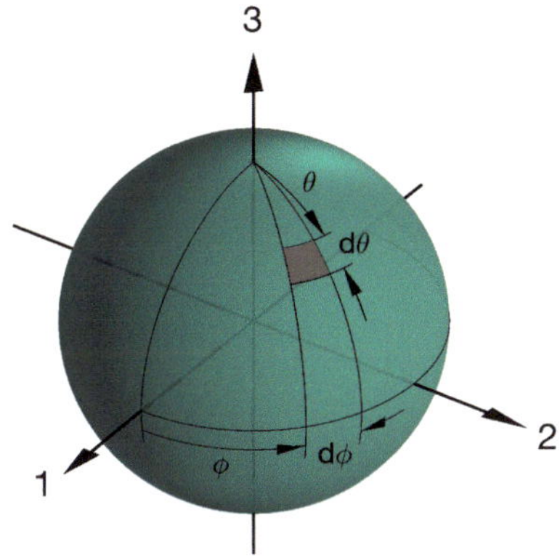

Figure 2.4 The unit-sphere orientation space for three-dimensional orientation. The gray shaded region has area $\sin(\theta)\,\mathrm{d}\theta\,\mathrm{d}\phi = \mathrm{d}\mathbf{p}$.

If the fibers are oriented randomly in space then all directions $\mathbf{p}$ are equally likely and $\psi(\mathbf{p})$ is constant. The surface area of a sphere with unit radius is 4π, so the normalization requirement (2.12) tells us that

$$\psi(\mathbf{p}) = \frac{1}{4\pi} \qquad \text{3-D random orientation} \tag{2.13}$$

We saw in Section 2.1 that any fiber can be represented by two different $\mathbf{p}$ vectors or by two different combinations of (θ, ϕ). This produces the *periodicity* requirement for 3-D orientation, which is

$$\psi(\mathbf{p}) = \psi(-\mathbf{p}) \qquad \text{or} \qquad \psi(\theta, \phi) = \psi(\pi - \theta, \phi + \pi) \tag{2.14}$$

Any point on the unit sphere must have the same value of ψ as the polar opposite point. If we know $\psi(\mathbf{p})$ on a hemisphere of the orientation space then we have complete information about ψ.

In principle, $\psi(\mathbf{p})$ can take on any form that satisfies the normalization and periodicity conditions. In practice, the distribution functions for discontinuous fiber composites vary smoothly with $\mathbf{p}$ and are reasonably well approximated by certain mathematical forms[1]. Figure 2.5 shows the orientation distribution functions for three examples:

(a) The fibers are strongly oriented along the 1 axis, and the orientation is symmetric about that axis. This is the type of alignment that occurs in a converging nozzle, where the flow is in the 1 direction.

(b) Many fibers lie close to the 1–2 plane, but the orientation is not perfectly planar. The orientation is symmetric about the 3 axis, with no preferred direction in the 1–2 plane. A biaxial squeezing flow, as seen in some compression moldings, would create this type of alignment by squeezing along the 3 axis.

(c) There is strong alignment in the flow (1) direction, with some spread in the crossflow (2) direction but less spread in the thickness (3) direction. The direction of greatest orienta-

[1] These forms are discussed in Section 2.4.

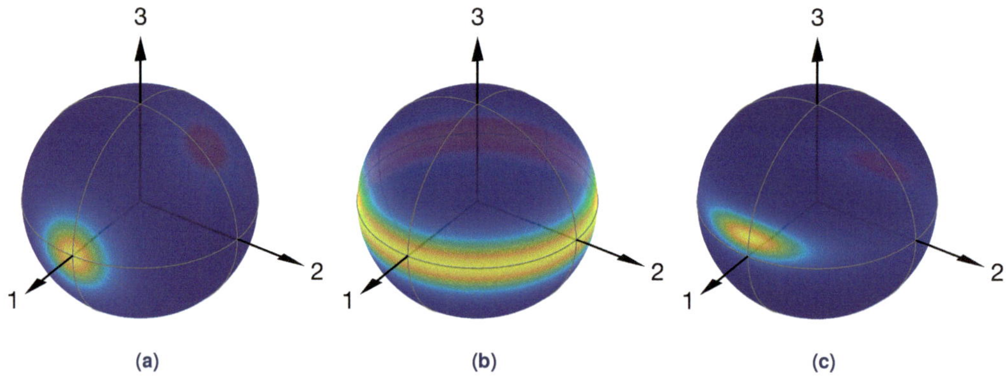

Figure 2.5 Examples of 3-D orientation distribution functions. Lowest and highest values of ψ are indicated by dark blue and yellow, respectively. (a) Axisymmetric alignment in the 1 direction. (b) Nearly planar alignment, uniformly distributed in the 1–2 plane. (c) Orientation similar to the shell layer in an injection molded composite. The corresponding orientation tensors appear in Eqns. (2.40–2.42).

tion is also tilted out of the 1–2 plane. This orientation state is similar to the shell layer in an injection molded composite. It is created by simple shear flow in the 1 direction with velocity changing in the 3 direction.

In all three examples in Fig. 2.5 the pattern on the front of the sphere is mirrored on the back of the sphere, as required by the periodicity condition.

2.3 Orientation Tensors

2.3.1 Introduction

The distribution function $\psi(\mathbf{p})$ contains a great deal of information, and even if we can calculate or measure the full function, interpreting it is not a simple task. It would be very helpful to have some kind of average, perhaps analogous to the mean or the standard deviation for scalar data, to help describe and interpret orientation distributions.

The average value of $\mathbf{p}$ is not helpful in this regard. The sign of $\mathbf{p}$ is arbitrary and, because of periodicity, each $\mathbf{p}$ should be cancelled by a corresponding $-\mathbf{p}$, so the average value of $\mathbf{p}$ is always equal to zero.

We could deal with the sign of $\mathbf{p}$ by averaging $\mathbf{p}$ squared, but there are several ways to multiply a vector by itself. The dot product is not helpful because $\mathbf{p}\cdot\mathbf{p} = 1$ for every $\mathbf{p}$, so an average of dot products will aways equal unity. The cross product is also not helpful, since $\mathbf{p}\times\mathbf{p} = 0$ for any $\mathbf{p}$, so an average of the cross products will aways equal zero.

A way to multiply $\mathbf{p}$ with itself that *is* useful here is the *dyadic product*. We will write this product as $\mathbf{pp}$, without an intervening symbol. This way of multiplying two vectors produces a tensor whose ij component is the product $p_i p_j$, for all combinations of i and j. In matrix form the components of **pp** are

$$\mathbf{pp} = \begin{bmatrix} p_1 p_1 & p_1 p_2 & p_1 p_3 \\ p_2 p_1 & p_2 p_2 & p_2 p_3 \\ p_3 p_1 & p_3 p_2 & p_3 p_3 \end{bmatrix} \tag{2.15}$$

In index notation the dyadic product of **p** with itself is $p_i p_j$.

The average of the dyadic product **pp** is very useful. We will give this quantity its own symbol **A**, and call it the *orientation tensor*, or more precisely the second-order orientation tensor [AT87]. Using angle brackets $\langle\,\rangle$ to represent the average, the second-order orientation tensor is

$$\mathbf{A} \equiv \langle \mathbf{pp} \rangle \quad \text{or} \quad A_{ij} \equiv \langle p_i p_j \rangle \tag{2.16}$$

The components of **A** written in matrix form are

$$\begin{bmatrix} A_{11} & A_{12} & A_{13} \\ A_{21} & A_{22} & A_{23} \\ A_{31} & A_{32} & A_{33} \end{bmatrix} = \begin{bmatrix} \langle p_1 p_1 \rangle & \langle p_1 p_2 \rangle & \langle p_1 p_3 \rangle \\ \langle p_2 p_1 \rangle & \langle p_2 p_2 \rangle & \langle p_2 p_3 \rangle \\ \langle p_3 p_1 \rangle & \langle p_3 p_2 \rangle & \langle p_3 p_3 \rangle \end{bmatrix} \tag{2.17}$$

When Eqn. (2.1) is used to express the components of **p** in terms of θ and ϕ, the components of the orientation tensor are:

$$\begin{bmatrix} A_{11} & A_{12} & A_{13} \\ A_{21} & A_{22} & A_{23} \\ A_{31} & A_{32} & A_{33} \end{bmatrix} = \begin{bmatrix} \langle \sin^2\theta \cos^2\phi \rangle & \langle \sin^2\theta \cos\phi \sin\phi \rangle & \langle \cos\theta \sin\theta \cos\phi \rangle \\ \langle \sin^2\theta \cos\phi \sin\phi \rangle & \langle \sin^2\theta \sin^2\phi \rangle & \langle \cos\theta \sin\theta \sin\phi \rangle \\ \langle \cos\theta \sin\theta \cos\phi \rangle & \langle \cos\theta \sin\theta \sin\phi \rangle & \langle \cos^2\theta \rangle \end{bmatrix} \tag{2.18}$$

How should one take the average here? Suppose that we wish to describe a group of N fibers and we know the orientation of each fiber. Let $(\mathbf{p})_k$ represent the orientation of the kth fiber, $k = 1$ to N. The average of **pp** is then computed as the average over all fibers:

$$\mathbf{A} = \langle \mathbf{pp} \rangle = \frac{1}{N} \sum_{k=1}^{N} (\mathbf{pp})_k \quad \text{or} \quad A_{ij} = \langle p_i p_j \rangle = \frac{1}{N} \sum_{k=1}^{N} (p_i p_j)_k \tag{2.19}$$

That is, we take the dyadic product of each $(\mathbf{p})_k$ with itself, then average the products for all fibers.

With Eqn. (2.19) we can compute the orientation tensor for any group of fibers. The function p2A in the accompanying software does this calculation.

Example 2.3.1 Orientation tensor for aligned fibers

What is the orientation tensor when all fibers are aligned in the 1 direction as in Fig. 2.6(a)?

All the fibers have

$$\mathbf{p} = \begin{Bmatrix} 1 \\ 0 \\ 0 \end{Bmatrix} \tag{2.20}$$

so $p_1 p_1 = 1$ and all other $p_i p_j$ products equal zero. The dyadic product is then

$$\mathbf{pp} = \begin{bmatrix} 1 & 0 & 0 \\ 0 & 0 & 0 \\ 0 & 0 & 0 \end{bmatrix} \tag{2.21}$$

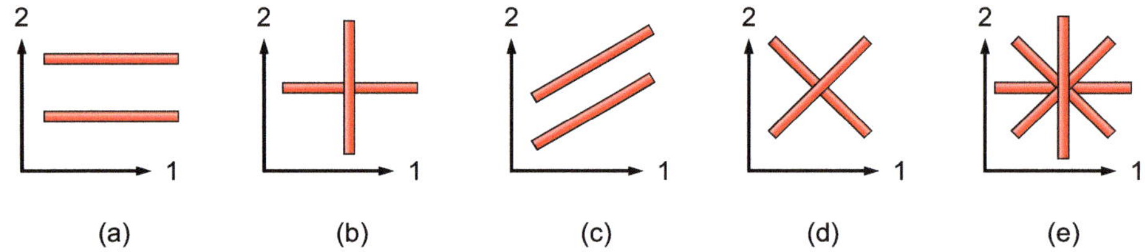

Figure 2.6 Fiber arrangements for Examples 2.3.1–2.3.10. All fibers lie in the 1–2 plane.

The tensor **pp** is the same for every fiber, so the average ⟨**pp**⟩ equals the **pp** tensor for one fiber. This gives the orientation tensor as

$$\mathbf{A} = \begin{bmatrix} 1 & 0 & 0 \\ 0 & 0 & 0 \\ 0 & 0 & 0 \end{bmatrix} \tag{2.22}$$

■

Example 2.3.2 Orientation tensor, fibers at $0°$ and $90°$

Suppose that half the fibers are aligned in the 1 direction and half the fibers are aligned in the 2 direction, all in the 1–2 plane, as in Fig. 2.6(b). What is the orientation tensor?

The fibers aligned in the 1 direction have the **p** vector and **pp** tensor from the previous example, Eqns. (2.20) and (2.21). The fibers aligned with the 2 axis have

$$\mathbf{p} = \begin{Bmatrix} 0 \\ 1 \\ 0 \end{Bmatrix} \quad \text{and} \quad \mathbf{pp} = \begin{bmatrix} 0 & 0 & 0 \\ 0 & 1 & 0 \\ 0 & 0 & 0 \end{bmatrix} \tag{2.23}$$

The orientation tensor is the average of the **pp** tensors from the two cases,

$$\mathbf{A} = \frac{1}{2}\begin{bmatrix} 1 & 0 & 0 \\ 0 & 0 & 0 \\ 0 & 0 & 0 \end{bmatrix} + \frac{1}{2}\begin{bmatrix} 0 & 0 & 0 \\ 0 & 1 & 0 \\ 0 & 0 & 0 \end{bmatrix} \tag{2.24}$$

so for this arrangement of fibers the orientation tensor is

$$\mathbf{A} = \begin{bmatrix} 0.5 & 0 & 0 \\ 0 & 0.5 & 0 \\ 0 & 0 & 0 \end{bmatrix} \tag{2.25}$$

■

Equations (2.22) and (2.25) suggest that

Each diagonal component of the orientation tensor can be interpreted as the fraction of fibers oriented along the corresponding coordinate axis.

This rule is a useful starting point for understanding orientation tensors.

What about the off-diagonal components? Here is an example to explore those.

Example 2.3.3 Orientation tensor, all fibers at 30°

What is the orientation tensor when all fibers are oriented in the 1–2 plane at 30° to the 1 axis, as in Fig. 2.6(c)?

All of these fibers have $\theta = \pi/2$ and $\phi = 30^\circ = \pi/6$. Using Eqn. (2.1), the orientation vector for all fibers is

$$\mathbf{p} = \begin{Bmatrix} \cos(\pi/6) \\ \sin(\pi/6) \\ 0 \end{Bmatrix} = \begin{Bmatrix} 0.8660 \\ 0.5000 \\ 0 \end{Bmatrix} \tag{2.26}$$

Since all the fibers have the same orientation $\mathbf{p}$, the orientation tensor equals the $\mathbf{pp}$ tensor for one fiber. Multiplying out all the combinations of the components of $\mathbf{p}$, we find that the orientation tensor is

$$\mathbf{A} = \begin{bmatrix} 0.7500 & 0.4330 & 0 \\ 0.4330 & 0.2500 & 0 \\ 0 & 0 & 0 \end{bmatrix} \tag{2.27}$$

■

The diagonal components tell us that the fibers are oriented more in the 1 direction than in the 2 direction, and there is no orientation in the 3 direction. However, now the off-diagonal components A_{12} and A_{21} are non-zero. We know that in this example the main direction of orientation is between the 1 and 2 axes, which suggests the following interpretation:

> Orientation tensors have non-zero off-diagonal components when some principal directions of orientation are not aligned with the coordinate axes.

We will be more precise about this in Section 2.3.4.

Here are two more examples that illustrate a different feature of orientation tensors.

Example 2.3.4 Orientation tensor, fibers at $\pm 45^\circ$

What is the orientation tensor when half the fibers are oriented at +45° to the 1 axis and half are at −45°, with all fibers in the 1–2 plane? See Fig. 2.6(d).

All fibers have $\theta = \pi/2$, with half of the fibers having $\phi = 45^\circ = \pi/4$ and half having $\phi = -45^\circ = -\pi/4$. Using Eqn. (2.1), the two orientation vectors are

$$(\mathbf{p})_1 = \begin{Bmatrix} \cos(\pi/4) \\ \sin(\pi/4) \\ 0 \end{Bmatrix} = \begin{Bmatrix} \sqrt{2}/2 \\ \sqrt{2}/2 \\ 0 \end{Bmatrix} \qquad (\mathbf{p})_2 = \begin{Bmatrix} \cos(-\pi/4) \\ \sin(-\pi/4) \\ 0 \end{Bmatrix} = \begin{Bmatrix} \sqrt{2}/2 \\ -\sqrt{2}/2 \\ 0 \end{Bmatrix} \tag{2.28}$$

The two dyadic products are

$$(\mathbf{p})_1(\mathbf{p})_1 = \begin{bmatrix} 0.5 & 0.5 & 0 \\ 0.5 & 0.5 & 0 \\ 0 & 0 & 0 \end{bmatrix} \qquad (\mathbf{p})_2(\mathbf{p})_2 = \begin{bmatrix} 0.5 & -0.5 & 0 \\ -0.5 & 0.5 & 0 \\ 0 & 0 & 0 \end{bmatrix} \tag{2.29}$$

Averaging these two results gives the orientation tensor,

$$\mathbf{A} = \begin{bmatrix} 0.5 & 0 & 0 \\ 0 & 0.5 & 0 \\ 0 & 0 & 0 \end{bmatrix} \tag{2.30}$$

This is identical to the orientation tensor for fibers at 0° and 90°, Eqn. (2.25). ■

Example 2.3.5 Second-order tensor, fibers at $0°$, $90°$ and $\pm 45°$

What is the second-order tensor when the fibers are evenly divided between 0°, 90° and ±45° to the 1 axis, all in the 1–2 plane, as in Fig. 2.6(e)?

This orientation state is a 50:50 combination of the results from Examples 2.3.2 and 2.3.4, so the orientation tensor here will be the average of the tensors for those two cases. However, both of those cases had the same orientation tensor, Eqns. (2.25) and (2.30), so we can immediately write

$$\mathbf{A} = \begin{bmatrix} 0.5 & 0 & 0 \\ 0 & 0.5 & 0 \\ 0 & 0 & 0 \end{bmatrix} \tag{2.31}$$

■

Examples 2.3.2, 2.3.4 and 2.3.5 show that multiple arrangements of fibers can have the same orientation tensor.

2.3.2 Relationship to the Distribution Function

To relate the orientation tensor to the orientation distribution function, we need an average that only uses information about the orientation state from $\psi(\mathbf{p})$. We define the *orientation average* of any function $f(\mathbf{p})$ as the integral of f over all orientations $\mathbf{p}$, weighted by the distribution function $\psi(\mathbf{p})$:

$$\langle f(\mathbf{p})\rangle \equiv \oint f(\mathbf{p})\,\psi(\mathbf{p})\,\mathrm{d}\mathbf{p} \equiv \int_{\theta=0}^{\pi}\int_{\phi=0}^{2\pi} f(\theta,\phi)\,\psi(\theta,\phi)\,\sin\theta\,\mathrm{d}\theta\,\mathrm{d}\phi \tag{2.32}$$

This is the formal definition of the orientation averaging operator $\langle\,\rangle$, and readers familiar with probability theory will recognize $\langle f\rangle$ as the expected value of f. Using this form of the orientation average, the formal definition of the second-order orientation tensor is

$$\mathbf{A} \equiv \langle\mathbf{pp}\rangle \equiv \oint \mathbf{pp}\,\psi(\mathbf{p})\,\mathrm{d}\mathbf{p} \quad \text{or} \quad A_{ij} \equiv \langle p_i p_j\rangle \equiv \int_{\theta=0}^{\pi}\int_{\phi=0}^{2\pi} p_i p_j\,\psi(\theta,\phi)\,\sin\theta\,\mathrm{d}\theta\,\mathrm{d}\phi \tag{2.33}$$

This can be used to compute the orientation tensor corresponding to any $\psi(\mathbf{p})$. Equation (2.33) is the formal definition of $\mathbf{A}$ and, strictly speaking, Eqn. (2.19) is an estimate for $\mathbf{A}$ using an unbiased sample of fibers.

We can now combine this result with the insight from Examples 2.3.2, 2.3.4 and 2.3.5 to say that

Each orientation distribution function corresponds to one and only one second-order orientation tensor, but multiple distribution functions can produce the same second-order tensor.

Here is an example using Eqn. (2.33) to go from $\psi(\mathbf{p})$ to $\mathbf{A}$.

Example 2.3.6 Orientation tensor, 3-D random orientation

What is the orientation tensor when the fibers are oriented randomly in all directions?

From Eqn. (2.13) we know that $\psi(\mathbf{p}) = 1/4\pi$. Using this in the (θ, ϕ) version of Eqn. (2.33) we can write

$$A_{ij} = \frac{1}{4\pi} \int_{\theta=0}^{\pi} \int_{\phi=0}^{2\pi} p_i p_j \sin\theta \, d\theta \, d\phi \tag{2.34}$$

To find each individual component, substitute expressions for p_i and p_j from Eqn. (2.1) and work out the integrals. Here we do the A_{33} component as an example. Since $p_3 = \cos\theta$, we have

$$A_{33} = \frac{1}{4\pi} \int_{\theta=0}^{\pi} \int_{\phi=0}^{2\pi} \cos^2\theta \sin\theta \, d\theta \, d\phi \tag{2.35}$$

Working out the details,

$$A_{33} = \frac{1}{4\pi} \int_{\phi=0}^{2\pi} d\phi \int_{\theta=0}^{\pi} \cos^2\theta \sin\theta \, d\theta = \left(\frac{1}{4\pi}\right)(2\pi)\left(\frac{2}{3}\right) = \frac{1}{3} \tag{2.36}$$

Following through with the remaining components gives the orientation tensor for 3-D random orientation:

$$\mathbf{A} = \begin{bmatrix} 1/3 & 0 & 0 \\ 0 & 1/3 & 0 \\ 0 & 0 & 1/3 \end{bmatrix} \tag{2.37}$$

■

Not surprisingly, all three diagonal components are equal, since the fibers are evenly distributed among the three coordinate directions. The off-diagonal components are all zero, since there are no preferred directions of orientation in a 3-D random state.

Equation (2.37) is an important result, and we will refer to 3-D random orientation as the *isotropic* orientation state. In tensor notation this is often written as

$$\mathbf{A} = \frac{\mathbf{I}}{3} \quad \text{or} \quad A_{ij} = \frac{1}{3}\delta_{ij} \qquad \text{isotropic orientation} \tag{2.38}$$

Here $\mathbf{I}$ is the second-order *identity tensor* and δ_{ij} is the *Kronecker delta*[2],

$$\mathbf{I} \equiv \begin{bmatrix} 1 & 0 & 0 \\ 0 & 1 & 0 \\ 0 & 0 & 1 \end{bmatrix} \quad \text{and} \quad \delta_{ij} \equiv \begin{cases} 1 \text{ when } i = j \\ 0 \text{ when } i \neq j \end{cases} \tag{2.39}$$

[2] While δ_{ij} is the ij component of $\mathbf{I}$, when using index notation it is standard to write δ_{ij} rather than I_{ij}.

As further examples, the three distribution functions in Fig. 2.5 have orientation tensors of

$$\mathbf{A} = \begin{bmatrix} 0.60 & 0 & 0 \\ 0 & 0.20 & 0 \\ 0 & 0 & 0.20 \end{bmatrix} \quad \text{axisymmetric about the 1 axis} \tag{2.40}$$

$$\mathbf{A} = \begin{bmatrix} 0.45 & 0 & 0 \\ 0 & 0.45 & 0 \\ 0 & 0 & 0.10 \end{bmatrix} \quad \text{nearly planar, uniform in the 1–2 plane} \tag{2.41}$$

$$\mathbf{A} = \begin{bmatrix} 0.60 & 0 & 0.10 \\ 0 & 0.30 & 0 \\ 0.10 & 0 & 0.10 \end{bmatrix} \quad \text{typical for a shell layer} \tag{2.42}$$

2.3.3 Properties of the Orientation Tensor

The orientation tensor condenses important information about the distribution function into an array of nine numbers. In fact, it only takes five numbers to specify the second-order orientation tensor. Here is why.

First, in each of the examples above the orientation tensor is symmetric about its main diagonal. This is true in general since, from the definition, $\langle p_i p_j \rangle = \langle p_j p_i \rangle$ for any i and j, so that

$$\mathbf{A} = \mathbf{A}^T \quad \text{or} \quad A_{ij} = A_{ji} \tag{2.43}$$

where $\mathbf{A}^T$ is the transpose of $\mathbf{A}$. We say that $\mathbf{A}$ is a symmetric tensor. The symmetry of $\mathbf{A}$ means that three of the off-diagonal components are duplicates of other off-diagonal components, leaving six independent components. For example, if A_{11}, A_{22}, A_{33}, A_{23}, A_{31} and A_{12} are specified, then we can easily fill in $A_{32} = A_{23}$, $A_{13} = A_{31}$ and $A_{21} = A_{12}$.

In addition, the sum of the diagonal entries from upper left to lower right (i.e., the *trace* of the tensor, $A_{11} + A_{22} + A_{33}$) always equals unity. Again this comes from the definition and the fact that $p_1^2 + p_2^2 + p_3^2 = 1$ for any fiber. This requirement is usually written as

$$\operatorname{tr}\mathbf{A} = 1 \quad \text{or} \quad A_{ii} = 1 \tag{2.44}$$

In this equation $\operatorname{tr}\mathbf{A}$ denotes the trace of $\mathbf{A}$. The second expression uses the *summation convention,* in which a repeated index (in this case ii) implies a sum over all values of that index. In other words,

$$A_{ii} \quad \text{means} \quad \sum_{i=1}^{3} A_{ii} = A_{11} + A_{22} + A_{33} \tag{2.45}$$

Equation (2.44) relies on $\psi(\mathbf{p})$ being properly normalized, so it is often referred to as the *normalization condition* for the orientation tensor.

If we want to condense the second-order orientation tensor to the minimum amount of information, we can remove one diagonal component from the previous list (say, A_{33}) and recover it from the other two diagonal components using Eqn. (2.44) (e.g., $A_{33} = 1 - A_{11} - A_{22}$).

2.3.4 Principal Directions and Principal Values

Look again at the orientation distribution functions in Fig. 2.5. Are there main directions of orientation for each case? The fibers in Fig. 2.5(a) are strongly aligned around the 1 axis, and in Fig. 2.5(b) the fibers lie close to the 1–2 plane. The fibers in Fig. 2.5(c) are clustered around an axis in the 1–3 plane that is close to the 1 direction, and 1–3 is a plane of symmetry. Each distribution function has one or more major directions of orientation. We will call these the *principal directions* or *principal axes* of orientation.

The second-order orientation tensor allows us to find the principal directions for any orientation state. In addition, each principal direction has an associated *principal value* of orientation. The principal values tell us the extent to which the fibers are random or aligned, regardless of the directions of that alignment. The principal directions tell us how the distribution is situated in space, regardless of the degree of alignment. This separation will be useful in a number of places.

The principal values of a tensor are also its *eigenvalues*, and the principal directions are its *eigenvectors*. These are usually calculated numerically, using software that finds the eigenvalues and eigenvectors of a matrix (e.g., [PTVF92]). We will say more about the properties of eigenvalues and eigenvectors in a moment, but first here is an example.

Example 2.3.7 Computing principal values and directions

Using MATLAB, find the principal values and principal directions for the distribution function in Fig. 2.5(c). Use $\mathbf{e}_1$, $\mathbf{e}_2$ and $\mathbf{e}_3$ to denote unit vectors along the principal directions (the eigenvectors), and λ_1, λ_2 and λ_3 to denote the corresponding principal values (the eigenvalues).

The orientation tensor for this distribution, given in Eqn. (2.42), is

$$\mathbf{A} = \begin{bmatrix} 0.60 & 0.00 & 0.10 \\ 0.00 & 0.30 & 0.00 \\ 0.10 & 0.00 & 0.10 \end{bmatrix} \tag{2.46}$$

The MATLAB function that finds eigenvalues and eigenvectors is `eig`. If we enter $\mathbf{A}$ as a 3×3 matrix and give the command `[Q Lambda] = eig(A)`, MATLAB returns two matrices:

$$\mathbf{Q} = \begin{bmatrix} 0.9820 & 0.0000 & -0.1891 \\ 0.0000 & 1.0000 & 0.0000 \\ 0.1891 & 0.0000 & 0.9820 \end{bmatrix} \tag{2.47}$$

$$\mathbf{\Lambda} = \begin{bmatrix} 0.6193 & 0.0000 & 0.0000 \\ 0.0000 & 0.3000 & 0.0000 \\ 0.0000 & 0.0000 & 0.0807 \end{bmatrix} \tag{2.48}$$

The columns of $\mathbf{Q}$ are the components of the principal-direction vectors $\mathbf{e}_1$, $\mathbf{e}_2$ and $\mathbf{e}_3$, and the diagonal entries in $\mathbf{\Lambda}$ are the principal values λ_1, λ_2 and λ_3. $\mathbf{\Lambda}$ will always be diagonal, i.e., its off-diagonal components will always be zero. Using this, the three prin-

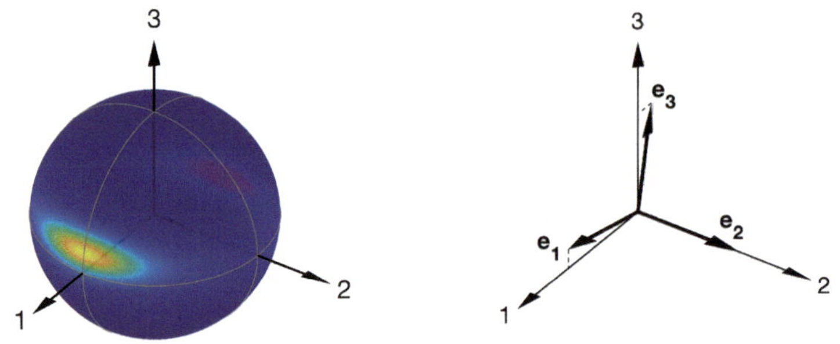

Figure 2.7 The orientation distribution function (left) and principal directions (right) for Example 2.3.7.

cipal directions and their corresponding principal values are

$$\mathbf{e}_1 = \begin{Bmatrix} 0.9820 \\ 0.0000 \\ 0.1891 \end{Bmatrix} \qquad \lambda_1 = 0.6193 \tag{2.49}$$

$$\mathbf{e}_2 = \begin{Bmatrix} 0.0000 \\ 1.0000 \\ 0.0000 \end{Bmatrix} \qquad \lambda_2 = 0.3000 \tag{2.50}$$

$$\mathbf{e}_3 = \begin{Bmatrix} -0.1891 \\ 0.0000 \\ 0.9820 \end{Bmatrix} \qquad \lambda_3 = 0.0807 \tag{2.51}$$

These principal directions are shown together with the distribution function in Fig. 2.7. As expected, $\mathbf{e}_1$ is close to the 1 direction but tilted slightly toward the 3 axis, and $\mathbf{e}_2$ is parallel to the 2 axis. The third principal direction $\mathbf{e}_3$ is perpendicular to both $\mathbf{e}_1$ and $\mathbf{e}_2$.

The second eigenvalue λ_2 equals A_{22}, and this will always be the case when an eigenvector is directed along a coordinate axis. The largest eigenvalue λ_1 is a bit larger than the largest diagonal component A_{11}, and the smallest eigenvalue λ_3 is a bit smaller than the smallest diagonal component A_{33}. This is also expected when some off-diagonal components are non-zero. ■

This calculation gives precise values for the principal directions and the principal values, using only the second-order orientation tensor as input. If a different orientation distribution function had the same second-order tensor, it would have the same principal directions and principal values.

2.3.4.1 Properties of Eigenvalues and Eigenvectors

We said that principal values and principal directions of orientation are the eigenvalues and eigenvectors of the second-order orientation tensor. Here we show what that means, and list some important mathematical properties of eigenvalues and eigenvectors.

Multiplying some vector $\mathbf{v}$ by a tensor, say $\mathbf{A}$, gives a new vector[3] $\mathbf{u} = \mathbf{A}\cdot\mathbf{v}$. In general, $\mathbf{u}$ will point in a different direction than $\mathbf{v}$. However, for a given tensor there are some special vectors where the vector resulting from this multiplication has the *same direction* as the original vector, though it may have a different magnitude. These special vectors are called *eigenvectors*. A vector $\mathbf{e}$ is an eigenvector of $\mathbf{A}$ if

$$\mathbf{A}\cdot\mathbf{e} = \lambda\mathbf{e} \tag{2.52}$$

The product of $\mathbf{A}$ times $\mathbf{e}$ is the original vector $\mathbf{e}$ scaled by a factor of λ, which is the *eigenvalue* corresponding to $\mathbf{e}$. Since λ is a scalar, $\lambda\mathbf{e}$ and $\mathbf{e}$ point in the same direction.

A symmetric second-order tensor like $\mathbf{A}$ has three eigenvalues $(\lambda_1, \lambda_2, \lambda_3)$ and three eigenvectors $(\mathbf{e}_1, \mathbf{e}_2, \mathbf{e}_3)$. These have the following properties:

(a) The eigenvalues $(\lambda_1, \lambda_2, \lambda_3)$ are all real numbers. (This only requires that the tensor be symmetric.)

(b) Eigenvalues and eigenvectors are paired with one another. Thus, $\mathbf{A}\cdot\mathbf{e}_1 = \lambda_1\mathbf{e}_1$ and $\mathbf{A}\cdot\mathbf{e}_2 = \lambda_2\mathbf{e}_2$, but $\mathbf{A}\cdot\mathbf{e}_1 \neq \lambda_2\mathbf{e}_1$ (unless $\lambda_2 = \lambda_1$).

(c) If $\mathbf{e}_i$ is an eigenvector then $k\mathbf{e}_i$ is also an eigenvector, where k is any constant. This means that the important information in an eigenvector is its direction, and the magnitude of an eigenvector is arbitrary. It is common to give each eigenvector unit length, and we follow that convention here. MATLAB and other software tools typically calculate eigenvectors with unit length as well.

(d) As a corollary to (c), if $\mathbf{e}_i$ is an eigenvector then $-\mathbf{e}_i$ is also an eigenvector. We can reverse the direction of an eigenvector without changing its properties.

(e) The eigenvectors are orthogonal to one another. If each eigenvector has unit length then

$$\mathbf{e}_i\cdot\mathbf{e}_j = \delta_{ij} \tag{2.53}$$

That is, the dot product of any eigenvector with itself is unity, while the dot product of two different eigenvectors is zero.

(f) Using (e), the three eigenvectors $\mathbf{e}_1$, $\mathbf{e}_2$ and $\mathbf{e}_3$ are the basis vectors for a Cartesian coordinate system. We will refer to this system as the *principal axes* of the tensor.

The eigenvalues of an orientation tensor $\mathbf{A}$ have some additional properties, but before listing those we need to discuss coordinate transformation.

2.3.4.2 Relationship to Coordinate Transformation

The components of any tensor are tied to a coordinate system, and the same tensor will have different component values in some other coordinate system. We can calculate tensor components in any coordinate system if we know the components in some other coordinate system and the relative orientation of the two systems. This calculation, which is called *coordinate transformation*, provides more insight into the tensors $\mathbf{Q}$ and $\mathbf{\Lambda}$ from Example 2.3.7.

Let $\hat{\boldsymbol{\delta}}_1$, $\hat{\boldsymbol{\delta}}_2$ and $\hat{\boldsymbol{\delta}}_3$ be the unit vectors directed along the 1, 2 and 3 coordinate axes, respectively. Also, let $\bar{\mathbf{A}}$ be the orientation tensor expressed in the principal axes, i.e., the $\mathbf{e}_1$, $\mathbf{e}_2$, $\mathbf{e}_3$ coordinate

[3] See Eqn. (A.27).

system. By writing Eqn. (2.52) once for each eigenvalue/eigenvector pair and summing the equations, one can show that $\mathbf{A}$ and $\bar{\mathbf{A}}$ are related by

$$\mathbf{A} = \mathbf{Q} \cdot \bar{\mathbf{A}} \cdot \mathbf{Q}^T \qquad \text{or} \qquad A_{ij} = Q_{ik} \bar{A}_{kl} Q_{jl} \tag{2.54}$$

This is the standard equation for transforming a tensor from one coordinate system to another, and $\mathbf{Q}$ is the *transformation matrix* that relates the two coordinate systems. The components of $\mathbf{Q}$ are given by

$$Q_{ik} = \hat{\boldsymbol{\delta}}_i \cdot \mathbf{e}_k \tag{2.55}$$

That is, each component of $\mathbf{Q}$ is the dot product of one unit vector from the 1, 2, 3 coordinate axes and one unit vector from the principal axes. Since the vectors have unit length, Q_{ik} equals the cosine of the angle between $\hat{\boldsymbol{\delta}}_i$ and $\mathbf{e}_k$. Equation (2.55) is consistent with the interpretation of $\mathbf{Q}$ in Example 2.3.7: each column of $\mathbf{Q}$ contains the components of one eigenvector in the 1, 2, 3 coordinate system.

Because the eigenvectors are orthogonal, Eqn. (2.53), $\mathbf{Q}$ is an *orthogonal matrix*. This means that its inverse is equal to its transpose,

$$\mathbf{Q} \cdot \mathbf{Q}^T = \mathbf{I} \qquad \text{or} \qquad Q_{ik} Q_{jk} = \delta_{ij} \tag{2.56}$$

Returning to the eigenvalues, one can also show (again using Eqn. (2.52)) that

$$\bar{\mathbf{A}} = \begin{bmatrix} \bar{A}_{11} & 0 & 0 \\ 0 & \bar{A}_{22} & 0 \\ 0 & 0 & \bar{A}_{33} \end{bmatrix} = \begin{bmatrix} \lambda_1 & 0 & 0 \\ 0 & \lambda_2 & 0 \\ 0 & 0 & \lambda_3 \end{bmatrix} = \boldsymbol{\Lambda} \tag{2.57}$$

That is, the tensor in the principal axes $\bar{\mathbf{A}}$ is diagonal, and its diagonal elements are the eigenvalues. Thus, the $\boldsymbol{\Lambda}$ matrix returned by the MATLAB function `eig` is $\bar{\mathbf{A}}$, the principal-axis version of the tensor.

Since the eigenvalues are the diagonal components of $\bar{\mathbf{A}}$, we can apply the trace property of $\mathbf{A}$, Eqn. (2.44), and show that the three eigenvalues of any second-order orientation tensor always sum to unity:

$$\lambda_1 + \lambda_2 + \lambda_3 = 1 \tag{2.58}$$

Also, since each λ_i is a diagonal component of $\bar{\mathbf{A}}$, it cannot be negative, nor can it be greater than unity:

$$0 \le \lambda_i \le 1 \tag{2.59}$$

These restrictions apply only to the eigenvalues of the second-order orientation tensor.

Finally, Eqn. (2.54) can be manipulated to express the second-order orientation tensor explicitly in terms of its eigenvalues and eigenvectors,

$$\mathbf{A} = \lambda_1 \mathbf{e}_1 \mathbf{e}_1 + \lambda_2 \mathbf{e}_2 \mathbf{e}_2 + \lambda_3 \mathbf{e}_3 \mathbf{e}_3 \tag{2.60}$$

This equation applies in any coordinate system; one simply uses the components of the eigenvectors in the coordinate system of interest. Note that $\mathbf{e}_1\mathbf{e}_1$, $\mathbf{e}_2\mathbf{e}_2$ and $\mathbf{e}_3\mathbf{e}_3$ are dyadic products, just like $\mathbf{pp}$.

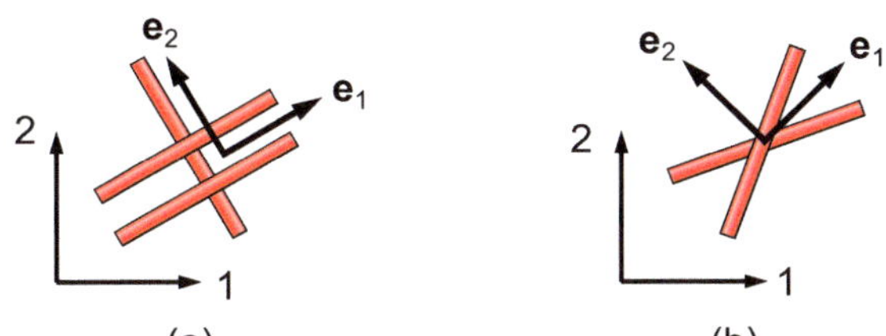

Figure 2.8 The principal directions or eigenvectors for two simple fiber orientation states. All fibers lie in the 1–2 plane so $\mathbf{e}_3$ is parallel to the 3 axis.

2.3.4.3 Indeterminate Eigenvectors

Figure 2.8 shows two simple arrangements of fibers and their eigenvectors. In both cases all of the fibers lie in the 1–2 plane, so $\mathbf{e}_3$ points along the 3 axis and $\lambda_3 = 0$.

(a) In Fig. 2.8(a) two-thirds of the fibers are aligned in one direction and one-third are perpendicular. Here the eigenvectors $\mathbf{e}_1$ and $\mathbf{e}_2$ match the two fiber directions.

(b) The fibers in Fig. 2.8(b) are at $\pm 25°$ to a symmetry axis. In this case $\mathbf{e}_1$ lies along the symmetry axis, and $\mathbf{e}_2$ is perpendicular. In this example, no fibers are actually oriented in either the $\mathbf{e}_1$ or the $\mathbf{e}_2$ directions.

For both of these examples the eigenvectors are well defined. This is always the case when all three eigenvalues are distinct (i.e., $\lambda_1 \neq \lambda_2 \neq \lambda_3$).

The direction of the principal axes becomes less clear when two eigenvalues are equal. The fiber configurations in Fig. 2.6(b) and (d) provide a good example. The $0°/90°$ arrangement in Fig. 2.6(b) and the $\pm 45°$ arrangement in Fig. 2.6(d) are identical, except for a rotation of $45°$ about the 3 axis. Examples 2.3.2 and 2.3.4 showed that the second-order tensor for both orientation states is

$$\mathbf{A} = \begin{bmatrix} 0.5 & 0 & 0 \\ 0 & 0.5 & 0 \\ 0 & 0 & 0 \end{bmatrix} \tag{2.61}$$

Since the orientation tensor is diagonal, the 1, 2, 3 axes must be in a principal axis system, and the eigenvalues must be $\lambda_1 = \lambda_2 = 0.5$ and $\lambda_3 = 0$. Yet the orientation states in the two figures are different. How can this be?

Whenever two eigenvalues are equal, the corresponding eigenvectors are not uniquely determined. In this example, $\mathbf{e}_1$ or $\mathbf{e}_2$ could lie anywhere in the 1–2 plane and still be eigenvectors of this orientation tensor, as long as $\mathbf{e}_1$ and $\mathbf{e}_2$ are perpendicular to one another. Since the fibers are evenly distributed in the 1–2 plane (at least as far as the second-order orientation tensor goes), any orthogonal pair of unit vectors in the 1–2 plane can serve as the first two eigenvectors for Eqn. (2.61). The third eigenvector $\mathbf{e}_3$ is uniquely determined and is parallel to the 3 axis.

Two other examples of indeterminate eigenvectors are provided by Figs. 2.5(a) and (b). The corresponding orientation tensors are given in Eqns. (2.40) and (2.41).

In Fig. 2.5(a) the fibers are strongly aligned around the 1 axis, with a spread that is symmetric around that axis. $\mathbf{A}$ in Eqn. (2.40) is diagonal, so the 1, 2, 3 coordinate axes are suitable eigenvectors, and the eigenvalues are $\lambda_1 = 0.60$ and $\lambda_2 = \lambda_3 = 0.20$. Because $\lambda_2 = \lambda_3$, the eigenvectors $\mathbf{e}_2$ and $\mathbf{e}_3$ are not uniquely determined. As a pair, they could be rotated by any amount around the 1 axis and still be eigenvectors of this tensor.

In Fig. 2.5(b) the fibers are nearly planar in the 1–2 plane, with orientations uniformly distributed around the 3 axis. Equation (2.41) shows that again the 1, 2, 3 axes are suitable eigenvectors, and the eigenvalues are $\lambda_1 = \lambda_2 = 0.45$ and $\lambda_3 = 0.10$. Now, with $\lambda_1 = \lambda_2$, the eigenvectors $\mathbf{e}_1$ and $\mathbf{e}_2$ are not uniquely determined, and any pair of orthogonal unit vectors that both lie in the 1–2 plane could serve just as well. Because λ_3 is distinct, $\mathbf{e}_3$ is uniquely determined and must be parallel to the 3 axis.

Indeterminate eigenvectors do not present any difficulties in practice, and software algorithms for finding eigenvalues and eigenvectors will handle such cases gracefully, producing eigenvector/eigenvalue pairs that satisfy Eqn. (2.52) with all eigenvectors orthogonal. As a software user, be aware that the eigenvectors produced in these cases will be legal, but not unique.

When two eigenvalues are close to one another, one should interpret eigenvector information carefully. Experimentally measured orientation tensors always have some uncertainty, which carries over into the eigenvalues. Similarly, numerical predictions always contain some noise due to discretization and round-off errors. If two eigenvalues are nearly equal, a small change in the tensor components can produce a large change in the eigenvectors. In such cases one should interpret the data accordingly. For example, if λ_1 and λ_2 are nearly equal, then one might say that the orientation is nearly random in the plane defined by $\mathbf{e}_1$ and $\mathbf{e}_2$, but give no importance to the directions given by $\mathbf{e}_1$ or $\mathbf{e}_2$. In this case the value of $\mathbf{e}_3$ is still meaningful.

When all three principal values are equal we must have $\lambda_1 = \lambda_2 = \lambda_3 = 1/3$, because of Eqn. (2.58). This is the isotropic state, Eqn. (2.38), and all three of the principal directions are indeterminate. Again this presents no practical problem because, for the isotropic distribution, there is no preferred direction of orientation.

As before, when all three eigenvalues are close to one another in experimental or numerical data, then one should say that the orientation state is nearly isotropic, and place little importance on the directions of the eigenvectors.

2.3.4.4 Numerical Details

Up to this point we have glossed over some details that are important when doing numerical calculations. For Example 2.3.7, MATLAB did not produce exactly the $\mathbf{Q}$ and $\mathbf{\Lambda}$ matrices given by Eqns. (2.47) and (2.48). Instead, for

$$\mathbf{A} = \begin{bmatrix} 0.60 & 0.00 & 0.10 \\ 0.00 & 0.30 & 0.00 \\ 0.10 & 0.00 & 0.10 \end{bmatrix} \tag{2.62}$$

typing `[Q Lambda] = eig(A)` into MATLAB actually gave

$$\mathbf{Q} = \begin{bmatrix} -0.1891 & 0.0000 & -0.9820 \\ -0.0000 & 1.0000 & 0.0000 \\ 0.9820 & 0.0000 & -0.1891 \end{bmatrix} \tag{2.63}$$

$$\mathbf{\Lambda} = \begin{bmatrix} 0.0807 & 0.0000 & 0.0000 \\ 0.0000 & 0.3000 & 0.0000 \\ 0.0000 & 0.0000 & 0.6193 \end{bmatrix} \tag{2.64}$$

The eigenvectors and eigenvalues have been returned in a different order than in Eqns. (2.47) and (2.48). The numbering of the eigenvalues and their corresponding eigenvectors is arbitrary, and they can be re-numbered in any convenient way. With MATLAB, as with many software tools, it is up to the user to sort the eigenvalues and their eigenvectors into some order, if desired.

In the examples above, we re-ordered the eigenvalues and eigenvectors so that

$$\lambda_1 \geq \lambda_2 \geq \lambda_3 \tag{2.65}$$

This is convenient, and we will continue this convention for the remainder of the book. The function `eigsort` in the accompanying software returns eigenvalues and eigenvectors sorted according to Eqn. (2.65).

Comparing the third column of Eqn. (2.63) with $\mathbf{e}_1$ in Eqn. (2.49), we see that the two vectors point in opposite directions. Here we took advantage of property (d) from Section 2.3.4.1 to change the sign of $\mathbf{e}_1$ so that it was pointing close to the $\hat{\boldsymbol{\delta}}_1$ direction. This made for a cleaner example, but in large numerical calculations one uses the eigenvectors as calculated.

Finally, note that $Q_{21} = -0.0000$, when we expect the 2-direction component of that eigenvector to be zero. This is a case where a quantity that should be zero comes out not quite equal to zero, because of round-off errors. The actual value of Q_{21} in this case is -5.5×10^{-17}. Since these calculations are done in double precision, values less than 10^{-15} are effectively zero, so this value is too.

2.3.5 The Fourth-Order Orientation Tensor

While the second-order orientation tensor **A** is the average of the (dyadic) product of two **p**s, we will also need an orientation tensor that is the product of four **p**s. This is the *fourth-order orientation tensor*, and its definition is

$$\mathbb{A} \equiv \langle \mathbf{pppp} \rangle \quad \text{or} \quad \mathbb{A}_{ijkl} \equiv \langle p_i p_j p_k p_l \rangle \tag{2.66}$$

Second-order tensors (of which stress and strain are familiar examples) have two indices and, in general, $3^2 = 9$ components. Fourth-order tensors have four indices, so in general they have $3^4 = 81$ components. Readers who have studied the mechanics of composite materials will have encountered another fourth-order tensor, the stiffness tensor $\mathbb{C}$, which describes the linear elastic stress-strain behavior of an anisotropic material.

The fourth-order orientation tensor is symmetric with respect to any pair of indices. This means that we can exchange any two indices and the value of the tensor component will be unchanged. For example, $\mathbb{A}_{1231} = \mathbb{A}_{2131}$. There are 24 ways to permute four indices, so the full symmetry of the $\mathbb{A}$ tensor means that

$$\mathbb{A}_{ijkl} = \mathbb{A}_{jikl} = \mathbb{A}_{ijlk} = \mathbb{A}_{klij} = \mathbb{A}_{kjil} = \mathbb{A}_{kilj} = \quad \ldots \quad \text{(24 total terms)} \tag{2.67}$$

As a comparison, the stiffness tensor, which we will use in Chapt. 8, has *only* the following symmetries:

$$\mathbb{C}_{ijkl} = \mathbb{C}_{jikl} = \mathbb{C}_{ijlk} = \mathbb{C}_{klij} \tag{2.68}$$

So, for example, $\mathbb{C}_{1123} = \mathbb{C}_{1132} = \mathbb{C}_{2311}$, but in general $\mathbb{C}_{1123} \neq \mathbb{C}_{1213}$. Any tensor with the symmetries of Eqn. (2.68) is said to have the *symmetries of an elasticity tensor.* The fourth-order orientation tensor $\mathbb{A}$ has the symmetries of an elasticity tensor, plus many more.

The fourth-order orientation tensor also has a property related to the normalization property of **A**. Recall that, using the summation convention, $\mathbb{A}_{ijkk}$ is a concise way of writing $\sum_{k=1}^{3} \mathbb{A}_{ijkk}$. Performing this operation on any two indices of the fourth-order orientation tensor gives the second-order orientation tensor:

$$\mathbb{A}_{ijkk} = A_{ij} \tag{2.69}$$

Thus, the fourth-order tensor contains all the information in the second-order tensor.

Because of its many symmetries and because of Eqn. (2.69), only 14 components of $\mathbb{A}$ are independent. Showing this will be much easier using contracted notation, which we discuss next.

2.3.5.1 Contracted Notation

We can write the components of a second-order tensor as a 3×3 matrix, but for a fourth-order tensor we would need to write a matrix that is 3×3 × 3×3. This is not easy to do! A convenient way to display fourth-order tensors is *contracted notation.* This is also useful for programming calculations that involve fourth-order tensors.

If a second-order tensor is symmetric, we can specify the tensor using the three diagonal components and three independent off-diagonal components[4]. These components are conveniently written as a 6×1 column vector. For the orientation tensor this column-vector form is

$$\{\mathbf{A}\} = \begin{Bmatrix} A_{11} \\ A_{22} \\ A_{33} \\ A_{23} \\ A_{31} \\ A_{12} \end{Bmatrix} = \begin{Bmatrix} A_{1} \\ A_{2} \\ A_{3} \\ A_{4} \\ A_{5} \\ A_{6} \end{Bmatrix} \tag{2.70}$$

Here we have used curly braces, {**A**}, to indicate that we are writing a column vector of the components of **A**. The ordering of the tensor components in the middle expression is standard, and in the last expression we have numbered the contracted components by their column-vector indices. We can convert between the tensor indices A_{ij} and the contracted-notation indices A_m using

$$A_{ij} \leftrightarrow A_m \tag{2.71}$$

with ij and m related as in Table 2.1. For MATLAB computations, the `tens2vec` function in the accompanying software transforms a tensor from 3×3 matrix form to 6×1 column-vector form, and the `vec2tens` function does the reverse.

In this book, second-order tensors are denoted by capital roman letters[5], so a capital letter with two indices (e.g., A_{12}) is a tensor component and a capital letter with one index (e.g., A_6) is a contracted component.

4 We want a scheme that works for any symmetric tensor, so we will not use the normalization property of **A**.

5 Exceptions to this rule include the total stress tensor $\boldsymbol{\sigma}$, the extra-stress tensor $\boldsymbol{\tau}$ and the strain tensor $\boldsymbol{\varepsilon}$.

Table 2.1 Correspondence between tensor indices and contracted-notation indices

Tensor indices *ij*	**11**	**22**	**33**	**23 or 32**	**31 or 13**	**12 or 21**
Contracted index *m*	1	2	3	4	5	6

For fourth-order tensors we consider tensors that are symmetric with respect to their first two indices, $\mathbb{A}_{ijkl} = \mathbb{A}_{jikl}$, and their last two indices, $\mathbb{A}_{ijkl} = \mathbb{A}_{ijlk}$. Then we can contract the first two indices using Table 2.1 and contract the second-two indices in the same way. Now each tensor component $\mathbb{A}_{ijkl}$ corresponds to a contracted component $\mathbb{A}_{mn}$, where ij contracts to m and kl contracts to n:

$$\mathbb{A}_{ijkl} \leftrightarrow \mathbb{A}_{mn} \tag{2.72}$$

The resulting contracted fourth-order tensor can be written as a 6×6 matrix:

$$\begin{aligned}[\mathbb{A}] &= \begin{bmatrix} \mathbb{A}_{1111} & \mathbb{A}_{1122} & \mathbb{A}_{1133} & \mathbb{A}_{1123} & \mathbb{A}_{1131} & \mathbb{A}_{1112} \\ \mathbb{A}_{2211} & \mathbb{A}_{2222} & \mathbb{A}_{2233} & \mathbb{A}_{2223} & \mathbb{A}_{2231} & \mathbb{A}_{2212} \\ \mathbb{A}_{3311} & \mathbb{A}_{3322} & \mathbb{A}_{3333} & \mathbb{A}_{3323} & \mathbb{A}_{3331} & \mathbb{A}_{3312} \\ \mathbb{A}_{2311} & \mathbb{A}_{2322} & \mathbb{A}_{2333} & \mathbb{A}_{2323} & \mathbb{A}_{2331} & \mathbb{A}_{2312} \\ \mathbb{A}_{3111} & \mathbb{A}_{3122} & \mathbb{A}_{3133} & \mathbb{A}_{3123} & \mathbb{A}_{3131} & \mathbb{A}_{3112} \\ \mathbb{A}_{1211} & \mathbb{A}_{1222} & \mathbb{A}_{1233} & \mathbb{A}_{1223} & \mathbb{A}_{1231} & \mathbb{A}_{1212} \end{bmatrix} \\ &= \begin{bmatrix} \mathbb{A}_{11} & \mathbb{A}_{12} & \mathbb{A}_{13} & \mathbb{A}_{14} & \mathbb{A}_{15} & \mathbb{A}_{16} \\ \mathbb{A}_{21} & \mathbb{A}_{22} & \mathbb{A}_{23} & \mathbb{A}_{24} & \mathbb{A}_{25} & \mathbb{A}_{26} \\ \mathbb{A}_{31} & \mathbb{A}_{32} & \mathbb{A}_{33} & \mathbb{A}_{34} & \mathbb{A}_{35} & \mathbb{A}_{36} \\ \mathbb{A}_{41} & \mathbb{A}_{42} & \mathbb{A}_{43} & \mathbb{A}_{44} & \mathbb{A}_{45} & \mathbb{A}_{46} \\ \mathbb{A}_{51} & \mathbb{A}_{52} & \mathbb{A}_{53} & \mathbb{A}_{54} & \mathbb{A}_{55} & \mathbb{A}_{56} \\ \mathbb{A}_{61} & \mathbb{A}_{62} & \mathbb{A}_{63} & \mathbb{A}_{64} & \mathbb{A}_{65} & \mathbb{A}_{66} \end{bmatrix}\end{aligned} \tag{2.73}$$

The square brackets, $[\mathbb{A}]$, indicate that we are writing a matrix of the contracted components of the tensor $\mathbb{A}$. All fourth-order tensors in this book are denoted by capital letters in the blackboard bold font, like $\mathbb{A}$ and $\mathbb{C}$. Any symbol in this font with four indices is a tensor component, while a symbol in this font with two indices is a contracted component.

While we have used the second- and fourth-order orientation tensors as examples, contracted notation can be used with any symmetric second-order tensor, such as stress or strain, and with any fourth-order tensor having the requisite symmetries, such as the stiffness tensor $\mathbb{C}$.

We can now determine how many of the components of $\mathbb{A}$ are independent. The symmetries used for contracted notation have already reduced the number of independent components from 81 to 36. Then, since the fourth-order tensor also allows the first pair of indices to be exchanged with the second pair, $\mathbb{A}_{ijkl} = \mathbb{A}_{klij}$, the contracted components must have $\mathbb{A}_{mn} = \mathbb{A}_{nm}$. Thus, the matrix of components in Eqn. (2.73) is symmetric. Because each element below the main diagonal duplicates an element above the main diagonal, this reduces the number of independent components to 21. (The stiffness tensor $\mathbb{C}$ has this symmetry as well, so in general it has 21 independent components.)

Looking at the components on and above the main diagonal, there are still many duplicated components. For example, $\mathbb{A}_{1122} = \mathbb{A}_{1212}$, and $\mathbb{A}_{2331} = \mathbb{A}_{3312}$. In fact, all of the components in the last three rows are duplicates of components from the first three rows. However, none of the components in the first three rows (on or above the main diagonal) are duplicates of one another. This leaves 15 independent components on the first three rows that are on or above the main diagonal, and we have used all of the symmetries of $\mathbb{A}$.

Finally, we can use normalization to eliminate one more component. Combining Eqn. (2.69) with the normalization property of the second-order tensor, Eqn. (2.44), gives

$$\mathbb{A}_{iikk} = A_{ii} = 1 \tag{2.74}$$

Writing this out explicitly,

$$\begin{aligned} &\mathbb{A}_{1111} + \mathbb{A}_{1122} + \mathbb{A}_{1133} \\ &+\mathbb{A}_{2211} + \mathbb{A}_{2222} + \mathbb{A}_{2233} \\ &+\mathbb{A}_{3311} + \mathbb{A}_{3322} + \mathbb{A}_{3333} = 1 \end{aligned} \tag{2.75}$$

Thus, the fourth-order orientation tensor has 14 independent components.

In contracted notation, Eqn. (2.75) reads

$$\begin{aligned} &\mathbb{A}_{11} + \mathbb{A}_{12} + \mathbb{A}_{13} \\ &+\mathbb{A}_{21} + \mathbb{A}_{22} + \mathbb{A}_{23} \\ &+\mathbb{A}_{31} + \mathbb{A}_{32} + \mathbb{A}_{33} = 1 \end{aligned} \tag{2.76}$$

These components occupy the 3×3 sub-matrix in the upper left-hand corner of $[\mathbb{A}]$, and normalization of the fourth-order tensor requires these nine components to sum to unity.

2.3.5.2 Examples

Example 2.3.8 Fourth-order tensor, all fibers aligned

Find the fourth-order tensor when all fibers are aligned with the 1 axis, Fig. 2.6(a).

For the fourth-order tensor a good approach is to use tensor indices $ijkl$ and refer to Eqn. (2.73) for which components to calculate and their positions in the 6×6 contracted matrix.

This example follows on from Example 2.3.1. From Eqn. (2.20) $p_2 = p_3 = 0$, so any component of $p_i p_j p_k p_l$ with a 2 or 3 among its indices will equal zero. The only combination that is non-zero is $p_1 p_1 p_1 p_1$, which equals unity. Writing all the components in contracted form gives

$$[\mathbb{A}] = \begin{bmatrix} 1 & 0 & 0 & 0 & 0 & 0 \\ 0 & 0 & 0 & 0 & 0 & 0 \\ 0 & 0 & 0 & 0 & 0 & 0 \\ 0 & 0 & 0 & 0 & 0 & 0 \\ 0 & 0 & 0 & 0 & 0 & 0 \\ 0 & 0 & 0 & 0 & 0 & 0 \end{bmatrix} \tag{2.77}$$

■

Example 2.3.9 Fourth-order tensor, fibers at 0° and 90°

Find the fourth-order tensor when half the fibers are aligned with the 1 axis and half the fibers are aligned with the 2 axis, Fig. 2.6(b).

In the previous example, Eqn. (2.77) equals **pppp** for a fiber aligned with the 1 axis. Using the same argument as the previous example, a fiber aligned with the 2 axis has $p_2 p_2 p_2 p_2 = 1$ and all other $p_i p_j p_k p_l = 0$. Averaging these two **pppp** matrices gives the result,

$$[\mathbb{A}] = \begin{bmatrix} 0.5 & 0 & 0 & 0 & 0 & 0 \\ 0 & 0.5 & 0 & 0 & 0 & 0 \\ 0 & 0 & 0 & 0 & 0 & 0 \\ 0 & 0 & 0 & 0 & 0 & 0 \\ 0 & 0 & 0 & 0 & 0 & 0 \\ 0 & 0 & 0 & 0 & 0 & 0 \end{bmatrix} \tag{2.78}$$

■

Example 2.3.10 Fourth-order tensor, fibers at ±45°

Find the fourth-order tensor when half the fibers are aligned at 45° to the 1 axis and half the fibers are aligned at −45°, all in the 1–2 plane, as in Fig. 2.6(d).

Since $p_3 = 0$ for all fibers, any component $p_i p_j p_k p_l$ that has i, j, k or l equal to 3 will itself be equal to zero. From Eqn. (2.73) this leaves five components in the first three rows, on or above the main diagonal, whose indices contain only 1 and 2. Equation (2.28) gives the **p** vectors for this case. To simplify the next steps use $c = \cos(\pi/4)$ and $s = \sin(\pi/4)$, and note that $\cos(-\pi/4) = c$ and $\sin(-\pi/4) = -s$. Now the five components of interest are

$$\mathbb{A}_{1111} = \frac{1}{2}\left(c^4 + c^4\right) = 0.25 \tag{2.79}$$

$$\mathbb{A}_{2222} = \frac{1}{2}\left(s^4 + (-s)^4\right) = 0.25 \tag{2.80}$$

$$\mathbb{A}_{1122} = \frac{1}{2}\left(c^2 s^2 + c^2(-s)^2\right) = 0.25 \tag{2.81}$$

$$\mathbb{A}_{1112} = \frac{1}{2}\left(c^3 s + c^3(-s)\right) = 0 \tag{2.82}$$

$$\mathbb{A}_{2212} = \frac{1}{2}\left(cs^3 + c(-s)^3\right) = 0 \tag{2.83}$$

Placing these into the matrix and filling in the remaining components using symmetry, we find the fourth-order orientation tensor:

$$[\mathbb{A}] = \begin{bmatrix} 0.25 & 0.25 & 0 & 0 & 0 & 0 \\ 0.25 & 0.25 & 0 & 0 & 0 & 0 \\ 0 & 0 & 0 & 0 & 0 & 0 \\ 0 & 0 & 0 & 0 & 0 & 0 \\ 0 & 0 & 0 & 0 & 0 & 0 \\ 0 & 0 & 0 & 0 & 0 & 0.25 \end{bmatrix} \tag{2.84}$$

While this arrangement of fibers had the same second-order orientation tensor as the previous example (see Examples 2.3.2 and 2.3.4), the fourth-order tensors are different.

■

Example 2.3.11 Fourth-order tensor, fibers at 0°, 90° and ±45°

Find the fourth-order tensor when the fibers are evenly divided between 0°, 90° and ±45° to the 1 axis, all in the 1–2 plane. See Fig. 2.6(e).

This type of orientation is frequently used in laminates of continuous fiber composites, and is called a *quasi-isotropic* layup. Since the set of orientations is a 50:50 combination of Example 2.3.9 and Example 2.3.10, we can average the fourth-order tensors from those two examples, Eqns. (2.78) and (2.84). The result is

$$[\mathbb{A}] = \begin{bmatrix} 0.375 & 0.125 & 0 & 0 & 0 & 0 \\ 0.125 & 0.375 & 0 & 0 & 0 & 0 \\ 0 & 0 & 0 & 0 & 0 & 0 \\ 0 & 0 & 0 & 0 & 0 & 0 \\ 0 & 0 & 0 & 0 & 0 & 0 \\ 0 & 0 & 0 & 0 & 0 & 0.125 \end{bmatrix} \tag{2.85}$$

This set of orientations has the same second-order tensor as the two preceding examples, but the fourth-order tensor is different for all three cases. ■

A random-in-plane orientation has the same **A** as Examples 2.3.9–2.3.11 (e.g., Eqn. (2.25)) and the fourth-order tensor given by Eqn. (2.85). Thus, while fourth-order tensors give more detailed information about orientation than second-order tensors, there are still aspects of the orientation distribution function that fourth-order tensors cannot discriminate.

Equation (2.38) gave the second-order orientation tensor for an isotropic (3-D random) orientation. The corresponding fourth-order tensor can be derived following the pattern of Example 2.3.6. The result is

$$[\mathbb{A}] = \begin{bmatrix} 1/5 & 1/15 & 1/15 & 0 & 0 & 0 \\ 1/15 & 1/5 & 1/15 & 0 & 0 & 0 \\ 1/15 & 1/15 & 1/5 & 0 & 0 & 0 \\ 0 & 0 & 0 & 1/15 & 0 & 0 \\ 0 & 0 & 0 & 0 & 1/15 & 0 \\ 0 & 0 & 0 & 0 & 0 & 1/15 \end{bmatrix} \quad \text{isotropic orientation} \tag{2.86}$$

■ 2.4 Reconstructing Distribution Functions

Sometimes we have only orientation tensor information but need the full orientation distribution function. Computing a distribution function from tensor information is called *reconstructing* the distribution function.

Multiple distribution functions can have the same orientation tensor, so there is no unique way to reconstruct a distribution function. There are, however, some options. In this section we discuss series approximations (Fourier series for planar orientation and spherical harmonics for 3-D orientation) and a form that is particular to fiber orientation called the Jeffery distribution. The series approximations have limited usefulness, but they illuminate the relationships between orientation tensors of various orders. The Jeffery distribution is a better choice for fiber orientation distributions, and is the recommended option.

2.4.1 Planar Orientation

First, consider planar orientation states. Suppose that we know $\mathbf{A}$, and possibly $\mathbb{A}$, and wish to reconstruct a function $\widetilde{\psi}_\phi(\phi)$ that approximates the actual distribution function $\psi_\phi(\phi)$. We will require that $\widetilde{\psi}_\phi$ be properly normalized and periodic, Eqns. (2.4) and (2.5). In addition, orientation tensors calculated from $\widetilde{\psi}_\phi$ must match the original $\mathbf{A}$, and also $\mathbb{A}$ if we used it in forming $\widetilde{\psi}_\phi$.

2.4.1.1 Fourier Series

Since any planar distribution function is periodic in ϕ, a Fourier series representation seems like a good option. We know that ψ_ϕ has period π, but we will start with the general Fourier series for a function of period 2π. The Nth order Fourier approximation is then

$$\widetilde{\psi}_\phi(\phi) = \frac{a_0}{2} + \sum_{n=1}^{N} \left[a_n \cos(n\phi) + b_n \sin(n\phi)\right] \tag{2.87}$$

and the coefficients are calculated according to

$$a_n = \frac{1}{\pi}\int_0^{2\pi} \psi_\phi(\phi)\cos(n\phi)\, \mathrm{d}\phi \tag{2.88}$$

$$b_n = \frac{1}{\pi}\int_0^{2\pi} \psi_\phi(\phi)\sin(n\phi)\, \mathrm{d}\phi \tag{2.89}$$

It appears that we need the unknown distribution function ψ_ϕ to find the approximation $\widetilde{\psi}_\phi$; however, this is not necessarily the case. To see this, let us work out a_2 in detail. From Eqn. (2.88),

$$a_2 = \frac{1}{\pi}\int_0^{2\pi} \psi_\phi(\phi)\cos(2\phi)\, d\phi = \frac{1}{\pi}\int_0^{2\pi} \psi_\phi(\phi)\left(\cos^2\phi - \sin^2\phi\right) \mathrm{d}\phi \tag{2.90}$$

For planar orientation $\cos\phi = p_1$ and $\sin\phi = p_2$, so we can write

$$a_2 = \frac{1}{\pi}\int_0^{2\pi} \psi_\phi(\phi)\, p_1 p_1\, \mathrm{d}\phi - \frac{1}{\pi}\int_0^{2\pi} \psi_\phi(\phi)\, p_2 p_2\, \mathrm{d}\phi \tag{2.91}$$

We now have two integrals of ψ_ϕ over the full range of ϕ, weighted by $p_1 p_1$ and $p_2 p_2$ respectively. These are the orientation tensor components A_{11} and A_{22}, so we obtain

$$a_2 = \frac{1}{\pi}(A_{11} - A_{22}) \tag{2.92}$$

The Fourier coefficient a_2 is completely determined by A_{11} and A_{22}, without knowing anything more about the actual distribution ψ_ϕ. This is a useful result!

Repeating the exercise for the other coefficients shows that the odd-order coefficients ($n = 1,3,5,\ldots$) are all zero, because ψ_ϕ actually has period π. The even-order coefficients up to

$n = 4$ are

$$a_0 = \frac{1}{\pi} \tag{2.93}$$

$$a_2 = \frac{1}{\pi}(A_{11} - A_{22}) \tag{2.94}$$

$$b_2 = \frac{2}{\pi}A_{12} \tag{2.95}$$

$$a_4 = \frac{1}{\pi}(\mathbb{A}_{1111} + \mathbb{A}_{2222} - 6\mathbb{A}_{1122}) \tag{2.96}$$

$$b_4 = \frac{4}{\pi}(\mathbb{A}_{1112} - \mathbb{A}_{1222}) \tag{2.97}$$

The second-order coefficients a_2 and b_2 are determined by the second-order orientation tensor, and the fourth-order coefficients a_4 and b_4 are determined by the fourth-order orientation tensor.

Evaluating higher-order terms would show that the sixth-order coefficients are determined by the sixth-order orientation tensor, the eighth-order coefficients by the eighth-order orientation tensor, and so on. Just as there are an infinite number of terms in a Fourier series, there are an infinite number of orientation tensors, in all even orders[6]. For example, the sixth-order orientation tensor is the average of the product of six **p**s, and has six indices:

$$\mathscr{A} \equiv \langle \mathbf{pppppp} \rangle \qquad \text{or} \qquad \mathscr{A}_{ijklmn} \equiv \langle p_i p_j p_k p_l p_m p_n \rangle \tag{2.98}$$

These higher-order tensors are directly related to the higher-order terms in a series representation of the distribution function, so each successive higher-order tensor provides more detailed information about the distribution function.

While we will see the sixth-order tensor again briefly in Section 5.4, in practice only the second- and fourth-order tensors are available, so we cannot make use of higher-order terms in the Fourier series for reconstructing the distribution function.

Substituting the coefficients from Eqns. (2.93–2.97) into Eqn. (2.87) gives a practical approximation for ψ_ϕ using Fourier series:

$$\begin{aligned}\widetilde{\psi}_\phi(\phi) = \frac{1}{\pi}\bigg[&\frac{1}{2} + (A_{11} - A_{22})\cos(2\phi) + 2A_{12}\sin(2\phi)\\ &+ (\mathbb{A}_{1111} + \mathbb{A}_{2222} - 6\mathbb{A}_{1122})\cos(4\phi) + 4(\mathbb{A}_{1112} - \mathbb{A}_{1222})\sin(4\phi)\bigg]\end{aligned} \tag{2.99}$$

We will refer to $\widetilde{\psi}_\phi$ using only the first line of Eqn. (2.99) as the second-order (or $N = 2$) approximation, and $\widetilde{\psi}_\phi$ using both lines as the fourth-order (or $N = 4$) approximation.

2.4.1.2 Jeffery Distribution, Planar Orientation

Fiber orientation distribution functions are not arbitrary; they arise from the physical process of flow-induced orientation. This suggests that distribution functions should have typical shapes, and that these shapes could be deduced from the mathematics of flow-induced orientation. This analysis has been carried out for fibers that follow Jeffery's equation, which

[6] The odd-order tensors, $\langle \mathbf{p} \rangle$, $\langle \mathbf{ppp} \rangle$, etc. are all zero, again owing to the periodicity of the distribution function. This is true for both planar and 3-D orientation.

governs the orientation of isolated, single fibers. The analysis that leads to this distribution function is discussed in Section 5.4.2.3. For this section we take the Jeffery distribution as given and explore its properties.

For planar orientation, the Jeffery distribution can be found directly from the second-order tensor **A**. One first finds the eigenvalues and eigenvectors of **A**. Let λ_1 be the largest eigenvalue, and $\mathbf{e}_1$ the corresponding eigenvector. For any planar orientation state the largest eigenvalue λ_1 fully characterizes the degree of alignment, since $\lambda_1 + \lambda_2 + \lambda_3 = 1$ and for planar orientation $\lambda_3 = 0$.

Next, calculate $\hat{\phi}$, the orientation angle of $\mathbf{e}_1$, such that

$$\mathbf{e}_1 = \begin{Bmatrix} \cos\hat{\phi} \\ \sin\hat{\phi} \\ 0 \end{Bmatrix} \tag{2.100}$$

Also calculate an auxiliary variable F,

$$F = \sqrt{\frac{\lambda_1}{1-\lambda_1}} \tag{2.101}$$

Now the Jeffery distribution for planar orientation is

$$\tilde{\psi}_\phi(\phi) = \frac{F^2}{2\pi\left[(1-F^4)\cos^2(\phi-\hat{\phi}) + F^4\right]} \tag{2.102}$$

We did not use the fourth-order tensor to construct the Jeffery distribution, but using Eqn. (2.102) we can compute $\mathbb{A}$. This means that the Jeffery distribution can also be used to approximate the fourth-order tensor $\mathbb{A}$ in terms of the second-order tensor **A**. This particular approximation is called the *natural closure*, and we examine it in Section 5.4.2.3.

2.4.1.3 Comparison of Planar Orientation Reconstructions

How useful are these approximations to ψ_ϕ? Figure 2.9 shows four distribution functions measured from the SMC compression moldings discussed in Section 1.2 [CT84]. The orientation states range from nearly random in plane ($\lambda_1 = 0.55$) to strongly aligned ($\lambda_1 = 0.82$).

In each case, **A** and $\mathbb{A}$ were calculated from the data, then used in Eqn. (2.99) to generate second-order and fourth-order Fourier approximations. The second-order tensor was also used to calculate the Jeffery distribution for each case. The reconstructed distribution functions are plotted as lines in the figure, and the original data was scaled using Eqn. (2.9) so that it can be compared directly.

When the fibers are only slightly aligned, as in Fig. 2.9(a), all approximations do a good job of representing the data, and there is little difference between them. If the distribution function were exactly random in plane ($\lambda_1 = 0.5$) then all three approximations would give the exact result, $\psi_\phi = 1/2\pi$.

As the fibers become more aligned, as in Figs. 2.9(b) and (c), the second-order Fourier approximation becomes less accurate, and cannot match the sharpness of the peak in orientation near $\phi = 0$. The fourth-order approximation does a better job, and is quite good for Fig. 2.9(b); however, for Fig. 2.9(c) it still cannot quite capture the highest and lowest parts of the distribution.

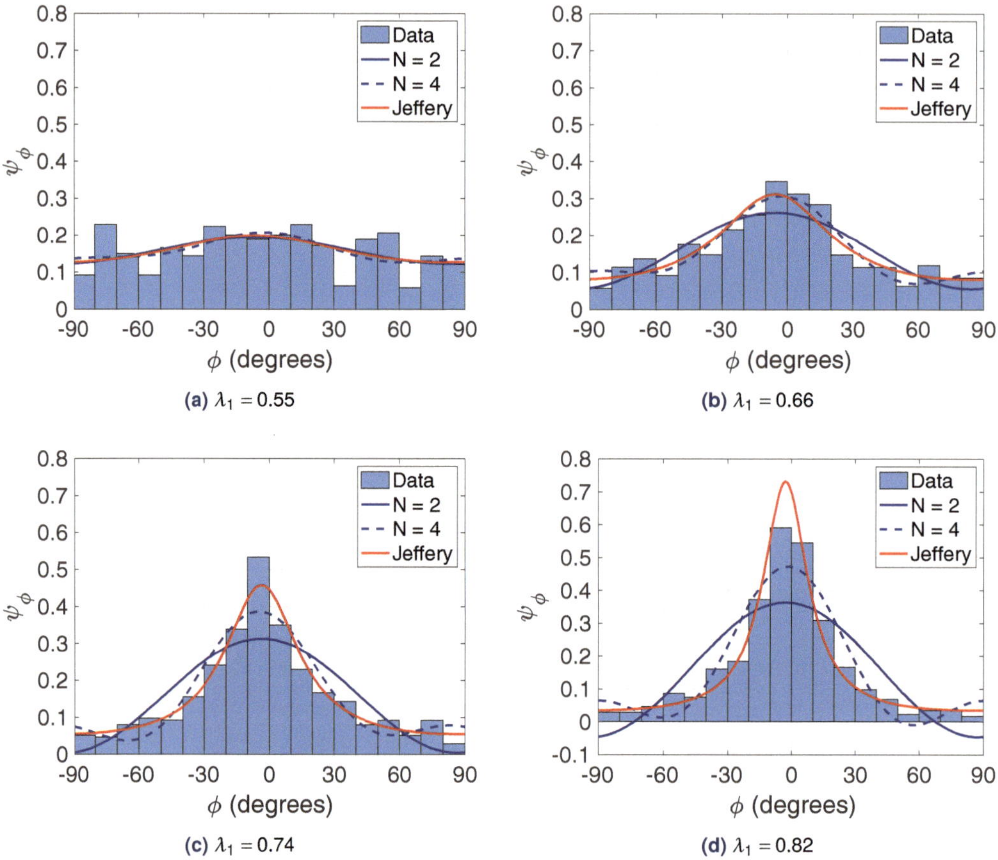

Figure 2.9 Planar orientation distribution functions and reconstructions using Fourier series approximations of second order ($N = 2$) and fourth order ($N = 4$) as well as the Jeffery distribution. Data from [CT84]. Distributions (a) and (d) are the same as in Fig. 2.3. λ_1 is the largest eigenvalue of the second-order orientation tensor of the experimental data.

The Jeffery distribution captures the sharpness of the peak quite well in both Figs. 2.9(b) and (c), and does a better job of representing the distribution function across the entire range of ϕ.

In Fig. 2.9(d) the fibers are strongly aligned, and neither Fourier approximation is particularly accurate. Worse yet, the second-order Fourier approximation goes negative near $\phi = 90°$ and $\phi = -90°$. Any probability distribution function must have $\psi_\phi(\phi) \geq 0$ for all ϕ since there is no such thing as a negative probability, so this particular approximation is physically impossible. Using such an approximation in any subsequent calculation could produce further unrealistic results. The fourth-order Fourier approximation also dips below zero near $\phi = 60°$. The Jeffery distribution always has positive values, it never exhibits the "wiggles" of the fourth-order Fourier approximation, and it reproduces this distribution function quite well over the full range of ϕ.

This example illustrate the strengths and limitations of Fourier series reconstructions. The Fourier series approach works well for slightly aligned fibers, but as the fibers become more aligned the low-order approximations do a poor job of representing ψ_ϕ, and can be negative for some values of ϕ. The second-order approximation will have negative values when $\lambda_1 > 3/4$. Fourier representations at orders well above $N = 4$ can be more accurate, but as noted before one rarely has the information needed to calculate the higher-order coefficients.

In contrast, the Jeffery distribution never exhibits negative values, and it matches the shapes of the experimental distribution functions well for all values of ϕ over a wide range of alignment λ_1. For this reason we recommend the Jeffery distribution, Eqn. (2.102), when reconstructing ψ_ϕ.

2.4.2 3-D Orientation

The 3-D analog of the Fourier series approach is to approximate $\psi(\mathbf{p})$ using spherical harmonic functions. As a tool for reconstructing approximate distribution functions, spherical harmonics have the same limitations as Fourier series: the low-order approximations based on $\mathbf{A}$ and $\mathbb{A}$ lose accuracy as the fibers become aligned, and $\tilde{\psi}(\mathbf{p})$ can be negative for some values of $\mathbf{p}$. For that reason we will not show the details here. Expressions for the series approximation of 3-D distribution functions in terms of orientation tensors up to fourth order are given by Onat and Leckie [Ona84, OL88] (see also [AT87]), while Jack and Smith add expressions for the sixth-order [JS04] and eighth-order [JS08] terms.

A better choice is the Jeffery distribution for 3-D orientation. This was first derived by Dinh and Armstrong [DA84], and it has the form

$$\tilde{\psi}(\mathbf{p}) = \frac{1}{4\pi\left(\mathbf{p}\cdot\mathbf{B}^{-1}\cdot\mathbf{p}\right)^{3/2}} \tag{2.103}$$

Here, $\mathbf{B}^{-1}$ is a symmetric, second-order tensor called the Piola tensor that can be calculated from the deformation gradient tensor $\mathbf{F}$. We will meet these tensors in Section 4.1.2. Any $\mathbf{B}^{-1}$ generates a distribution function $\tilde{\psi}(\mathbf{p})$ that corresponds to some $\mathbf{A}$, so using Eqn. (2.103) for reconstruction requires that we find the $\mathbf{B}^{-1}$ for our known $\mathbf{A}$. The functions in Fig. 2.5 were all generated this way, and are examples of the Jeffery distribution.

Equations giving $\mathbf{A}$ for any $\mathbf{B}^{-1}$ are worked out in detail by Montgomery-Smith et al. [MSHJS11][7]. The accompanying software[8] provides a function `A2F` that implements these equations, and an inverse function `F2A` that can be used to find the $\mathbf{B}^{-1}$ for any $\mathbf{A}$. Section 5.4.2.3 discusses the origins of the Jeffery distribution.

Just as we saw in planar orientation, the 3-D Jeffery distribution varies smoothly with $\mathbf{p}$, is never negative, and provides a good match for experimental orientation distributions over the full range of orientation states. Thus, it is the recommended form for recovering distribution functions in 3-D.

---

[7] Montgomery-Smith's $\mathbf{B}$ is our $\mathbf{B}^{-1}$.

[8] *https://github.com/charlestucker3/Fiber-Orientation-Tools*

2.5 Fiber Length

A single fiber is conveniently described as a circular cylinder of length ℓ and diameter d_f. However, in practice most composites have a variety of fiber lengths, so we need a fiber length distribution function. This is normally written using a set of discrete fiber lengths.

Choose some convenient increment of fiber length $\Delta\ell$, and let $\ell_i = i\,\Delta\ell$, where $i = 1$ to n and n is large enough that ℓ_n is greater than or equal to the length of the longest fiber in the composite.

Let N_i be the number of fibers having length ℓ_i. This could be the number of fibers in a measured sample, or the expected number of fibers per unit volume at some point in a composite. The set of numbers N_i, $i = 1$ to n, is the *number-based fiber length distribution.* If we measured the lengths of all the fibers in a sample and plotted the data as a histogram with bins of width $\Delta\ell$ centered on the ℓ_i values, this would be a plot of the number-based distribution function.

An alternative description is a *weight-based fiber length distribution,* where each value w_i is proportional to the weight of fibers of length ℓ_i. Glass and carbon fibers normally have a nearly uniform diameter d_f, so the weight of each fiber is proportional to its length. The relationship between the number-based and weight-based distributions is then

$$w_i = kN_i\ell_i \tag{2.104}$$

where k is a normalization factor. A convenient choice for k is

$$k = \left[\sum_{i=1}^{n} N_i\ell_i\right]^{-1} \qquad \text{which gives} \qquad \sum_{i=1}^{n} w_i = 1 \tag{2.105}$$

Average fiber lengths are helpful for concisely describing a distribution of orientations, and for comparing one distribution to another. There are two distinct ways to compute an average fiber length: the *number average length* L_n and the *weight average length* L_w:

$$L_n = \frac{\sum_{i=1}^{n} \ell_i N_i}{\sum_{i=1}^{n} N_i} \tag{2.106}$$

$$L_w = \frac{\sum_{i=1}^{n} \ell_i w_i}{\sum_{i=1}^{n} w_i} = \frac{\sum_{i=1}^{n} \ell_i^2 N_i}{\sum_{i=1}^{n} \ell_i N_i} \tag{2.107}$$

Note that L_n and L_w are unchanged if every N_i or every w_i is multiplied by a constant.

If, instead of histogram data, one simply has the lengths ℓ_k of N fibers, $k = 1$ to N, the average lengths are

$$L_n = \frac{1}{N}\sum_{k=1}^{N} \ell_k \tag{2.108}$$

$$L_w = \sum_{k=1}^{N} \ell_k^2 \Big/ \sum_{k=1}^{N} \ell_k \tag{2.109}$$

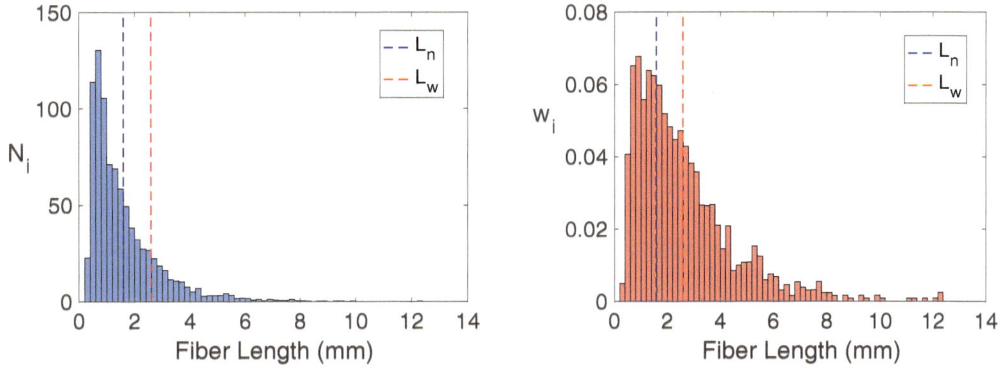

(a) Number based **(b)** Weight based

Figure 2.10 Experimental fiber length data for an injection molded PP sample reinforced with 40% by weight of long glass fibers. Average lengths are $L_n = 1.59\,\text{mm}$ and $L_w = 2.59\,\text{mm}$. Data from [Phe09].

It is always true that $L_w \geq L_n$. The two averages will be equal only if all fibers have the same length, so the difference between L_n and L_w and the ratio L_w/L_n are indicators of the breadth of the fiber length distribution.

Figure 2.10 shows the fiber length distributions for a long fiber/polypropylene material that has been injection molded. Notice that the number-based distribution is dominated by fibers shorter than 2 mm (of which there are many), while fibers longer than 8 mm are barely visible. In contrast, the presence of the longer fibers is more apparent in the weight-based distribution, which also gives a more balanced picture of the importance of the shorter fibers. Number-based and weight-based data contain the same information, and one can easily convert between the two using Eqn. (2.104). However, plots of the weight-based distribution give a better picture of the fibers that are present in a sample.

Models that predict fiber length, which we shall examine in Chapt. 7, work directly with the fiber length distribution function, either N_i or w_i. While L_n and L_w are useful for describing fiber length, they are not used directly in predictive models, and there is no need to recover the distribution function from the average lengths.

2.6 Number-Average and Weight-Average Orientation

The following question arises when measuring fiber orientation: Suppose we know the orientation $(\mathbf{p})_k$ and the length ℓ_k for each of N fibers, $k = 1$ to N, and not all the lengths are the same. How should we compute an orientation tensor?

One option is to ignore the length information and take an average of the $(\mathbf{pp})_k$ values, weighting all fibers equally. By analogy with L_n, this is a *number-average orientation tensor.* We will denote this as $\mathbf{A}^{(n)}$, and its components are

$$A_{ij}^{(n)} = \frac{1}{N} \sum_{k=1}^{N} (p_i p_j)_k \tag{2.110}$$

A second option is to weight each fiber by its length, so that a longer fiber contributes more to the average than a shorter fiber. This is the analog of L_w, a *weight-average orientation tensor* that we will denote as $\mathbf{A}^{(w)}$. Its components are

$$A_{ij}^{(w)} = \sum_{k=1}^{N} (p_i p_j)_k \, \ell_k \Big/ \sum_{k=1}^{N} \ell_k \tag{2.111}$$

The situation where fibers of different length have different orientation states can be described using a joint distribution function $\Psi(\mathbf{p}, \ell)$, defined such that the probability of a fiber having an orientation between $\mathbf{p}^*$ and $\mathbf{p}^* + \mathrm{d}\mathbf{p}$ and a length between ℓ^* and $\ell^* + \mathrm{d}\ell$ equals $\Psi(\mathbf{p}^*, \ell^*)\, \mathrm{d}\mathbf{p}\, \mathrm{d}\ell$. This function is normalized in the usual way,

$$\int_{\ell=0}^{\infty} \oint_{\mathbf{p}} \Psi(\mathbf{p}, \ell)\, \mathrm{d}\mathbf{p}\, \mathrm{d}\ell = 1 \tag{2.112}$$

Just as L_n and L_w give partial information about the detailed fiber length distribution, $\mathbf{A}^{(n)}$ and $\mathbf{A}^{(w)}$ give partial information about the joint orientation and length distribution function $\Psi(\mathbf{p}, \ell)$.

It is common to assume that the orientation and length distributions are *separable*, such that

$$\Psi(\mathbf{p}, \ell) = \psi(\mathbf{p}) N(\ell) \tag{2.113}$$

where $N(\ell)$ is a continuous version of the fiber length distribution function, normalized so that $\int_0^{\infty} N(\ell)\, \mathrm{d}\ell = 1$. If Eqn. (2.113) is true, the fibers at each length ℓ all have the same orientation distribution function $\psi(\mathbf{p})$, and the fibers at each orientation $\mathbf{p}$ all have the same length distribution function $N(\ell)$. Then $\mathbf{A}^{(n)} = \mathbf{A}^{(w)}$ and there is no difference between the number-average and weight-average orientation tensors.

The assumption that length and orientation are separable is convenient, and until recently there was no way to obtain data to test this assumption. However, recent micro-CT measurements show that the shorter fibers in an injection molded composite tend to be less aligned than the longer fibers [HRW$^+$19, JAS19]. In that case, $\mathbf{A}^{(n)}$ and $\mathbf{A}^{(w)}$ are different. This is currently a research issue, and computer-aided engineering software calculates a single orientation tensor at each point in the part. When interpreting measurement data, we will see in Chapt. 3 that the result of most orientation measurements is a weight-average tensor $\mathbf{A}^{(w)}$. This is a useful outcome, since $\mathbf{A}^{(w)}$ counts each fiber in the orientation tensor in proportion to its volume in the composite, and is a better choice for mechanical property predictions than $\mathbf{A}^{(n)}$.

2.7 Chapter Summary

The orientation of a single fiber can be described by a pair of angles (θ, ϕ) or by a unit vector $\mathbf{p}$.

The orientation of a group of fibers can be described by an orientation distribution function $\psi(\mathbf{p})$ or by one or more orientation tensors. The second-order orientation tensor $\mathbf{A} = \langle \mathbf{pp} \rangle$ is the most common tensor description. Each diagonal component of $\mathbf{A}$ can be interpreted as the fraction of fibers oriented along the corresponding coordinate axis. Off-diagonal components that are non-zero indicate that a principal direction of orientation is not aligned with the coordinate axes.

The second-order orientation tensor can be decomposed into eigenvalues and eigenvectors. The eigenvectors are the principal axes of orientation, and the corresponding eigenvalues show what fraction of fibers are oriented along each principal axis.

The fourth-order orientation tensor $\mathbb{A} = \langle \mathbf{pppp} \rangle$ gives additional information about the orientation state. Any orientation distribution function has a unique set of orientation tensors, but neither the second- nor the fourth-order tensor specify a unique distribution function.

There are several way to reconstruct an approximate distribution function from orientation tensor information. The best choice is the Jeffery distribution, which has both planar and 3-D versions.

Many composites have a distribution of fiber lengths. This can be described using either a number-based or a weight-based distribution, and the two descriptions are interchangeable. The weight-based distribution is recommended for displaying data. The number-average fiber length L_n and the weight-average length L_w are convenient indicators of fiber length, though they only provide partial information about the fiber length distribution. When all fibers have the same length, $L_w = L_n$. Otherwise, the ratio L_w / L_n indicates the breadth of the distribution.

When fibers have a distribution of lengths, one can define a number-based orientation tensor $\mathbf{A}^{(n)}$ and a weight-based tensor $\mathbf{A}^{(w)}$. Most experimental measurements produce $\mathbf{A}^{(w)}$.

3 Measuring Fiber Orientation and Length

Now that we have well-defined ways to describe fiber orientation and length, the next step is to measure these quantities experimentally. This chapter looks at measurement techniques, covering both the mathematical principles and the practical aspects of acquiring and interpreting data.

Fiber orientation is measured on solid samples in one of two ways: by cutting and polishing a planar section and examining the section under a microscope, or by using high-resolution X-ray computed tomography (micro-CT) to image the fibers in three dimensions. Both approaches are well established, and we discuss them in Sections 3.1 and 3.2.

There are also two general approaches to measuring fiber length: a fully destructive sampling method where the matrix is burned off so the fibers can be separated for measurement, and a non-destructive method using micro-CT. Both approaches are discussed in Section 3.3.

3.1 Planar Section Measurements of Orientation

The earliest quantitative measurements of fiber orientation in short fiber reinforced plastics were made from polished sections of solid samples [FF85, Kon87, FE88]. Figure 3.1 shows an example section. Glass and carbon fibers are circular cylinders, so the intersection of a fiber and a section plane is an ellipse. By measuring the geometry of each ellipse one can deduce the orientation of each fiber. Averaging the data for many fibers then yields an estimate for the orientation tensor or the orientation distribution function. This method is often referred to as the *method of ellipses.*

3.1.1 Mathematical Principles

The basic measurement principle is illustrated in Fig. 3.2. Let the section plane be the 1–2 plane, so the 3 axis is normal to the plane. For each ellipse determine the principal axes in the 1–2 plane and the major and minor diameters M and m. The orientation angle ϕ is the in-plane angle between the 1 axis and the principal axis corresponding to M, while the angle θ between the fiber axis and the 3 axis is found from

$$\cos\theta = \frac{m}{M} \tag{3.1}$$

This procedure is repeated for every fiber in the sample.

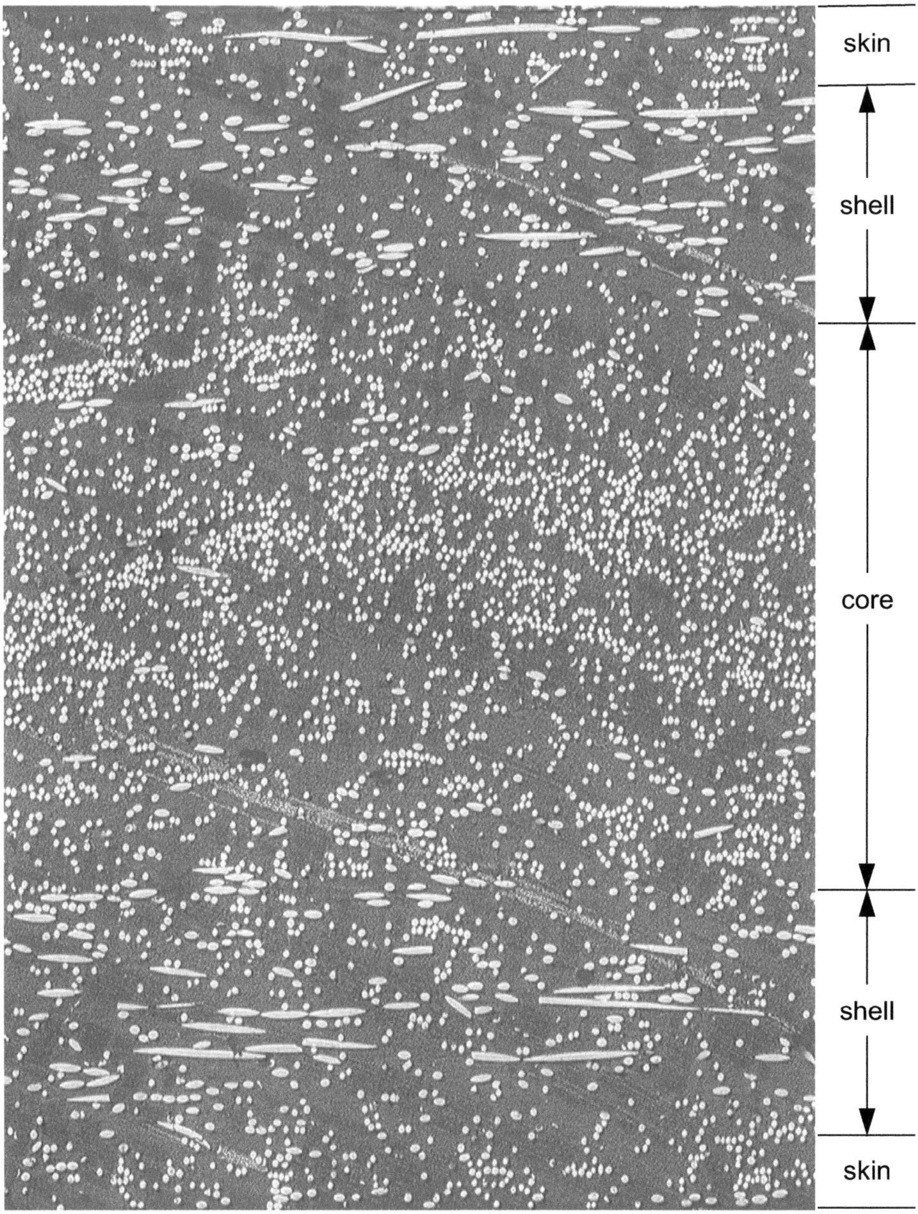

Figure 3.1 Polished cross-section from an injection molded PP plaque with 40% by weight long glass fibers [Phe09]. The vertical dimension of the figure is the entire 3 mm thickness of the molding and the filling direction is from left to right.

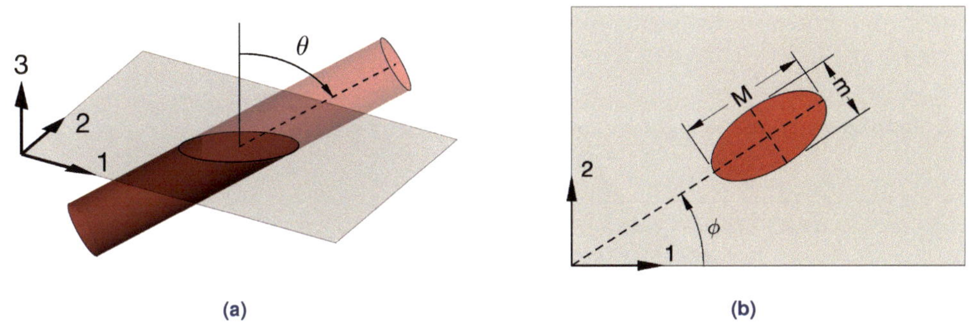

Figure 3.2 The intersection of a cylindrical fiber and a plane is an ellipse, whose in-plane orientation ϕ, major diameter M and minor diameter m give the fiber's orientation in three dimensions.

3.1.1.1 Correcting for Sample Bias

Next, we want to find the orientation tensor for the sample. Number the fibers from $k = 1$ to N. The orientation vector $(\mathbf{p})_k$ for the kth fiber is easily found from the angles (θ_k, ϕ_k) using Eqn. (2.1). One might think that the orientation tensor is found by simply averaging $(\mathbf{pp})_k$ over all the fibers on the section, as in Eqn. (2.19), but this would be incorrect.

As Fig. 3.3 shows, a fiber oriented nearly parallel to the section plane (θ_k close to $\pi/2$) is less likely to intersect a section plane than a fiber nearly normal to the plane (θ_k close to zero). As a result, the set of fibers that appear on any section are a biased sample: fibers that are nearly normal to the section are overrepresented, and fibers nearly parallel to the section plane are underrepresented. Similarly, a longer fiber is more likely to appear on the section plane than a shorter fiber with the same θ. Because of these biases, a simple average over the fibers appearing on a section plane does not approximate an average of fibers in the sample volume. Fortunately, we can correct for this bias.

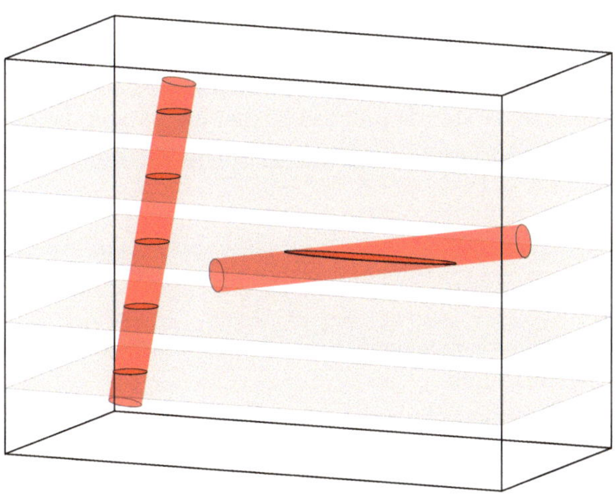

Figure 3.3 A fiber is more likely to appear on a section plane if has a greater projected length in the direction normal to the plane.

If a fiber is present in the sample volume, the probability that it will appear on the section plane is proportional to its length projected normal to the plane. We can obtain an unbiased average from planar section data if we divide each fiber's contribution by its projected length, and normalize accordingly.

If all fibers have the same length ℓ then this unbiased average is

$$A_{ij} = \frac{\sum_{k=1}^{N} (p_i p_j)_k / (\ell \cos\theta_k)}{\sum_{k=1}^{N} 1/(\ell \cos\theta_k)} \tag{3.2}$$

The ℓs in the numerator and denominator cancel, and we can write this as

$$A_{ij} = \frac{\sum_{k=1}^{N} (p_i p_j)_k F_k}{\sum_{k=1}^{N} F_k} \tag{3.3}$$

where F_k is a weighting factor, given in this case by

$$F_k = \frac{1}{\cos\theta_k} \tag{3.4}$$

Equation (3.3) is the basic equation for computing orientation tensors from planar section data, though we will see in a moment that F_k may need some modification.

Fourth-order components can be computed in the same way:

$$\mathbb{A}_{ijmn} = \frac{\sum_{k=1}^{N} (p_i p_j p_m p_n)_k F_k}{\sum_{k=1}^{N} F_k} \tag{3.5}$$

What about fibers that have different lengths? If we include individual fiber lengths ℓ_k in Eqn. (3.2) then we are computing a number-average orientation tensor $\mathbf{A}^{(n)}$ as in Eqn. (2.110),

$$A_{ij}^{(n)} = \frac{\sum_{k=1}^{N} (p_i p_j)_k / (\ell_k \cos\theta_k)}{\sum_{k=1}^{N} 1/(\ell_k \cos\theta_k)} \tag{3.6}$$

While we know θ_k for each fiber, there is no way to determine the individual fiber lengths ℓ_k from planar section data, so we cannot proceed further with this equation.

However, if we compute the weight-average orientation tensor $\mathbf{A}^{(w)}$ from Eqn. (2.111), the unbiased average is

$$A_{ij}^{(w)} = \frac{\sum_{k=1}^{N} (p_i p_j)_k \ell_k / (\ell_k \cos\theta_k)}{\sum_{k=1}^{N} \ell_k / (\ell_k \cos\theta_k)} = \frac{\sum_{k=1}^{N} (p_i p_j)_k (1/\cos\theta_k)}{\sum_{k=1}^{N} (1/\cos\theta_k)} \tag{3.7}$$

The individual lengths ℓ_k have cancelled in both the numerator and denominator, and we are back to Eqn. (3.3). This shows that

Analysis of planar section data using Eqn. (3.3) produces a weight-average orientation tensor.

Returning to the weighting function F_k, some fibers will cross the section plane close to their ends and appear as partial ellipses. Figure 3.1 contains many examples. The choice of the weighting function F_k depends on how these fibers are treated in the analysis.

Bay [BT92c] included all objects half an ellipse or larger when computing orientation tensors. This means that a fiber will be counted if the section plane intersects its symmetry axis on or between the endpoints of its axis. The projected length of this axis normal to the section plane is $\ell \cos\theta_k$, so Eqn. (3.4) is the corresponding weighting function.

While this function has been used in a number of studies, F_k becomes infinite as $\theta_k \to 90°$ and the analysis can fail. In reality, as θ_k approaches 90° the diameter of the fiber influences its projected length normal to the section plane. A detailed analysis of this geometry can be done, but the relevant angle range is just a few degrees for typical fiber aspect ratios, and θ_k is rarely measured with enough precision for that analysis to be useful. Instead, Bay [BT92c] recommends defining a cutoff angle θ_c using

$$\cos\theta_c = \frac{d_f}{\ell} \tag{3.8}$$

and truncating Eqn. (3.4) at that value. The final tensor components are not sensitive to the fiber length ℓ in Eqn. (3.8), and either L_n or L_w (if known) can be used in place of ℓ, or a typical fiber length for the material can be used.

The final weighting function is then

$$F_k = \begin{cases} 1/\cos\theta_k & \text{for} \quad 0 \le \theta_k \le \theta_c \\ 1/\cos\theta_c & \text{for} \quad \theta_c < \theta \le \pi/2 \end{cases} \tag{3.9}$$

This form has been widely used, and is recommended for most purposes.

If every fiber that appears on the cross section is included in the average, even if it is less than half an ellipse, then the weighting function should be [Kon87, ZBL97]

$$F_k = \frac{1}{\cos\theta_k + (d_f/\ell)\sin\theta_k} \tag{3.10}$$

This function also considers fibers whose θ value is close enough to $\pi/2$ that the section plane passes through both ends, so the fibers appear like rectangles. Again, an average value is used for ℓ. This form has no singularities, though the θ_k and ϕ_k values for small portions of an ellipse may have large measurement errors.

In the accompanying software, the function `thetaphi2A` implements Eqn. (3.3) with either Eqn. (3.9) or (3.10) as the weighting function.

Here is an example that illustrates the importance of the weighting function.

Example 3.1.1 Orientation tensor for fibers on a planar section

Figure 3.4 shows a set of ellipses on a planar section, and the θ and ϕ values are given in the table below. Before doing any calculations, estimate the values of A_{11}, A_{22} and A_{33} based on your visual impression of the figure. Then calculate the second-order orientation tensor. For simplicity, use Eqn. (3.4) for the weighting function. Also calculate the orientation tensor without a weighting function. How accurate is your visual impression, and what errors are introduced by not weighting the data?

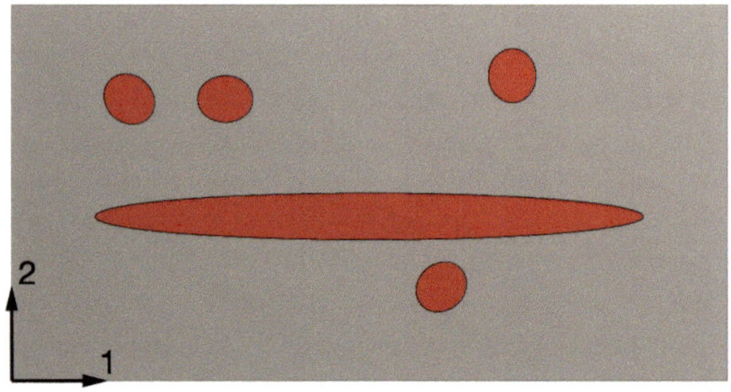

Figure 3.4 Fibers on a section plane for Example 3.1.1.

θ (degrees)	ϕ (degrees)
30	0
30	45
30	90
30	135
85	0

To solve this, compute the components of $(\mathbf{p})_k$ for each fiber from Eqn. (2.1), use $F_k = 1/\cos\theta_k$ for the weighting function, and calculate the tensor components using Eqn. (3.3).

Writing out the solution for A_{33} in detail, we have

$$A_{33} = \frac{4\cos^2(30°)/\cos(30°) + \cos^2(85°)/\cos(85°)}{4/\cos(30°) + 1/\cos(85°)} = 0.221 \tag{3.11}$$

To calculate the full set of tensor components, it is worth writing a spreadsheet or short MATLAB script. The correct answer for this data set is

$$\mathbf{A} = \begin{bmatrix} 0.743 & 0.000 & 0.093 \\ 0.000 & 0.036 & 0.075 \\ 0.093 & 0.075 & 0.221 \end{bmatrix} \tag{3.12}$$

If, instead, we average each fiber's contribution without a weighting function (which is an incorrect method), the calculation for A_{33} is

$$A_{33} = \frac{4\cos^2(30°) + \cos^2(85°)}{5} = 0.602 \tag{3.13}$$

and the full set of (incorrect) tensor components is

$$\mathbf{A} = \begin{bmatrix} 0.298 & 0.000 & 0.104 \\ 0.000 & 0.100 & 0.209 \\ 0.104 & 0.209 & 0.602 \end{bmatrix} \tag{3.14}$$

In this case, the un-weighted data suggest that the primary orientation direction is close to the 3 axis, when in fact the fibers are strongly oriented in the 1 direction. This points out how important it is to use the weighting factor F_k in Eqn. (3.3) when analyzing planar section data.

Most readers' impressions of Fig. 3.4 are closer to the un-weighted results (3.14) than to the correct, weighted results (3.12). The human visual system tends to assign equal weighting to the objects on a planar section. However, any slender ellipse that appears on a section, like the fiber with $\theta = 85^\circ$ in this example, is important, because it implies the presence of many more fibers with similar orientations that are not on the section. If a section contains a noticeable fraction of fibers with long slender ellipses, then the fibers are strongly aligned in that direction. ■

3.1.1.2 Ambiguous Angles and Ambiguous Tensor Components

A fundamental limitation of the method of ellipses is that two different orientations can produce the same ellipse on the measurement plane. Figure 3.5 shows an example. If we restrict θ to the range $[0, \pi/2]$ then we cannot distinguish between ϕ and $\phi + \pi$ based on the ellipse alone. These two possible orientations are quite different, except for the special cases of $\theta = 0$ and $\theta = \pi/2$.

This ambiguity affects only some of the off-diagonal components. To see which ones, first note that $\cos(\phi + \pi) = -\cos\phi$ and $\sin(\phi + \pi) = -\sin\phi$. Now consider A_{11}. This is the average of $\sin^2\theta\cos^2\phi$, and $\cos^2\phi$ does not change value if we change angles from ϕ to $\phi + \pi$. Thus, the contribution of each fiber to A_{11} does not depend on which of the two possible ϕ values is used for that fiber, and A_{11} is determined without ambiguity. In contrast, A_{13} is the average of $\cos\theta\sin\theta\cos\phi$, and this changes sign if we change angles from ϕ to $\phi + \pi$, so A_{13} is affected by the ambiguity.

Write the orientation tensor in terms of θ and ϕ as in Eqn. (2.18):

$$\begin{bmatrix} A_{11} & A_{12} & \underline{A_{13}} \\ A_{21} & A_{22} & \underline{A_{23}} \\ \underline{A_{31}} & \underline{A_{32}} & A_{33} \end{bmatrix} = \begin{bmatrix} \langle\sin^2\theta\cos^2\phi\rangle & \langle\sin^2\theta\cos\phi\sin\phi\rangle & \underline{\langle\cos\theta\sin\theta\cos\phi\rangle} \\ \langle\sin^2\theta\cos\phi\sin\phi\rangle & \langle\sin^2\theta\sin^2\phi\rangle & \underline{\langle\cos\theta\sin\theta\sin\phi\rangle} \\ \underline{\langle\cos\theta\sin\theta\cos\phi\rangle} & \underline{\langle\cos\theta\sin\theta\sin\phi\rangle} & \langle\cos^2\theta\rangle \end{bmatrix} \tag{3.15}$$

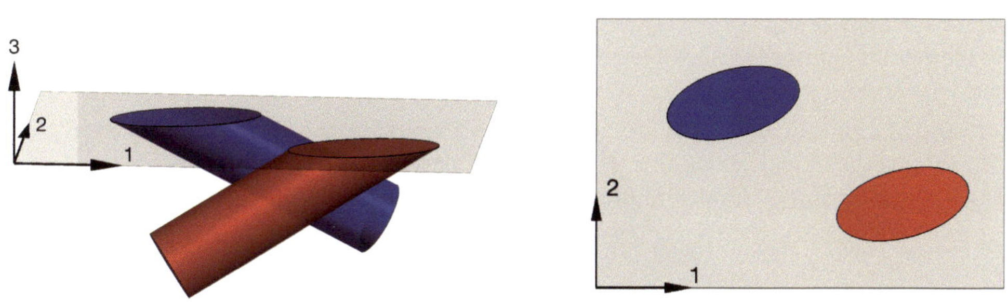

(a) (b)

Figure 3.5 Ambiguity in the method of ellipses. The red and blue fibers have ϕ values that differ by π but their ellipses on the section plane are identical.

The diagonal entries A_{11}, A_{22} and A_{33}, as well as $A_{12} = A_{21}$, are determined unambiguously. The underlined components, $A_{13} = A_{31}$ and $A_{23} = A_{32}$, contain either $\sin\phi$ or $\cos\phi$ to the first power, so they are affected by the ambiguity in ϕ.

Note that A_{13} and A_{23} are ambiguous components when the 1–2 plane is the section plane. Different off-diagonal components will be ambiguous if the coordinate axes are re-labelled.

There are several options for dealing with this ambiguity.

Take advantage of symmetry. In some cases the section plane can be chosen to take advantage of the symmetry of the part geometry and processing flow, so that the ambiguous components should be zero by symmetry.

Compute upper and lower bounds. One can compute rigorous bounds on the ambiguous components. The upper bound is found by replacing $\cos\phi_k$ with $|\cos\phi_k|$ and $\sin\phi_k$ with $|\sin\phi_k|$ for every fiber when calculating an ambiguous component. The lower bound is found by using $-|\cos\phi_k|$ and $-|\sin\phi_k|$. The true values of A_{13} and A_{23} for fibers on the section must lie between these bounds. However, the bounds are not always tight enough to provide useful information.

For Example 3.1.1, the bounds are $-0.137 \le A_{13} \le 0.137$ and $-0.075 \le A_{23} \le 0.075$. In this example the actual value of $A_{13} = 0.093$ falls between the bounds, while the actual value of $A_{23} = 0.075$ equals the upper bound.

Acquire additional information. If one can obtain images from two closely spaced section planes and match the ellipses for each fiber on the two planes, ϕ can be measured without ambiguity [DCA97, ZPB01]. The second plane can be generated by re-polishing the sample, then aligning the two images by matching the ellipses of fibers with θ close to zero. Second-plane data can also be acquired without re-polishing the sample using confocal scanning laser microscopy, which can focus a short distance below the surface of the sample [CAD95, EC01]. Another option is to etch away a few extra microns of matrix material in the final stages of sample preparation, leaving the fibers slightly exposed. This reveals the "shadow" of the solid portion of each fiber beneath the section plane and resolves the ambiguity [VGWB$^+$12].

Analyze multiple sections. If three mutually perpendicular sections are analyzed, one can determine all of the off-diagonal tensor components without ambiguity. The different sections are normally taken from separate but nominally identical samples.

3.1.1.3 Systematic Errors in θ

The method of ellipses is also subject to a systematic error in determining θ, one that affects fibers with small θ values.

Let m and M represent the actual minor and major diameters of an ellipse on the section plane, and let θ_a be the actual orientation of the fiber as determined from Eqn. (3.1). If there is an error in measuring m by some amount e_m, then the measured angle θ_m will satisfy

$$\cos\theta_m = \frac{m - e_m}{M} = \cos\theta_a \left[1 - \frac{e_m}{m}\right] \tag{3.16}$$

Figure 3.6 compares the measured and actual values of θ for different error sizes (e_m/m). As θ_a approaches zero, the measured value θ_m asymptotes to a non-zero value that depends on the error size. Even a 1% error in m will give $\theta_m = 8°$ when $\theta_a = 0$. It is simply hard to measure small values of θ accurately using the method of ellipses.

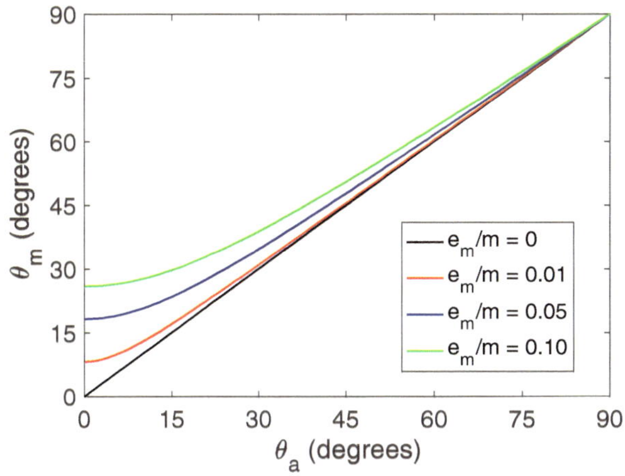

Figure 3.6 Measured angle θ_m as a function of the actual angle θ_a according to Eqn. (3.16) for the method of ellipses. e_m is the error in measuring the minor diameter m.

The error e_m is controlled in part by the resolution of the digital image used to analyze the section. It is tempting to assume that e_m equals the pixel size. However, O'Connell and Duckett [OD91] argue that this is overly pessimistic, and that e_m is closer to one-tenth the pixel size.

This systematic error in θ can usually be managed by taking high-resolution images, and by careful choice of the section plane. A resolution that gives at least 20 pixels over the fiber diameter is recommended. Resolutions higher than about 33 pixels over the fiber diameter do not give improved results for θ_m [ECV+01]. This may be because the fibers themselves are not perfectly circular. Carbon fibers are very slightly elliptical in cross-section, such that a fiber with $\theta_a = 0$ and perfectly measured major and minor diameters could have θ_m in the range of 0 to 8° [OD91].

Happily, the impact of this error on the orientation tensor is not severe, since we only require values of $\cos\theta$ and $\sin\theta$, not θ itself. Also, ellipses with small θ values receive a small weighting factor when computing the orientation tensor, which further reduces the error's effect.

It is possible to apply a correction to each θ_m value to bring it closer to θ_a. This can be based on careful assessment of the errors in θ [ECV+01] or can be as simple as choosing a θ_m value below which θ_a will be set to zero [Phe09]. However, few researchers have done this.

If the fibers are highly aligned, a section plane whose normal is inclined relative to the fiber orientation direction can greatly improve measurement accuracy [Yur87]. This option is especially useful for aligned, continuous fiber composites.

3.1.2 Practical Aspects of Measurement

3.1.2.1 Sample Preparation and Imaging

To obtain reliable data from a planar section, the sample must be polished carefully, and high-resolution images obtained.

Protocols for preparing and polishing samples are given by Velez et al. [VGWK+12] and Sharma et al. [SNN+18]. The goal is to obtain crisp images of the fiber ellipses, and to avoid damaging the fibers during sample preparation. One must also maintain alignment of the sample as it is mounted, polished and imaged.

Once the sample has been polished, it must be further treated to enhance the contrast between the fibers and the matrix. Etching the sample in an oxygen plasma [MLG99, VGWK+12] is a standard method for increasing contrast, and it works with a variety of polymers. At this step, additional etching can be done to further expose the fibers and allow the ambiguity in ϕ to be resolved [VGWB+12].

Most workers acquire images of the sample using an optical microscope. To obtain high resolution for each ellipse while covering a total area that can be millimeters on each side, one must acquire multiple, overlapping images and stitch them together to form an image for analysis. Figure 3.1 was formed this way. The full image, covering 3×10 mm, was assembled from approximately 1,000 individual images.

The acquisition of multiple, overlapping images is best done with a motorized, computer-controlled microscope stage. Autofocusing is important, since the sample is not perfectly planar but each image should be in focus.

The programmed movement of the microscope stage is not accurate enough to align neighboring images. Instead, neighboring images are aligned in software by matching the overlapping portions of the images. A variety of software tools are available to perform this image stitching function.

Once suitable images have been acquired and stitched together, the next step is to analyze them.

3.1.2.2 Image Analysis

Digital image analysis provides a good way to process thousands of fibers quickly and consistently, and is the standard technique for implementing the method of ellipses. There is no commercial software package for method-of-ellipses measurement of fiber orientation, so most researchers have assembled their own software for this step.

Image analysis starts with a grayscale image, as in Fig. 3.7(a). A *thresholding* operation is used to convert this to an initial binary image, as in Fig. 3.7(b), where each pixel is either 1 (white) or 0 (black). Figure 3.7(b) uses a single threshold, but more sophisticated thresholding operations may give better results [SNN+18].

Next the binary image is *filtered* to remove as many white pixels that are not fibers as possible. These pixels can arise from noise, polishing scratches, or small pieces of broken fiber in the sample. It is best not to remove objects that are partial ellipses, since these can provide valuable information. A variety of automatic filtering algorithms and parameters have been used [ECV+01, VGWB+12, SNN+18].

The filtered image is then *segmented*. This operation ensures that the pixels for each fiber are separated from all other fibers by at least one line of black pixels. A sophisticated segmenting algorithm that handles many cases well is described by [SNN+18]. Unfortunately, no segmentation algorithm is perfect: some long ellipses can be over-segmented into shorter pieces, while tightly clustered fibers may not be separated from one another. If high-quality data is required, some manual touch-up of the segments is usually needed. This is also an opportunity to remove any artifacts that were not caught during the filtering step.

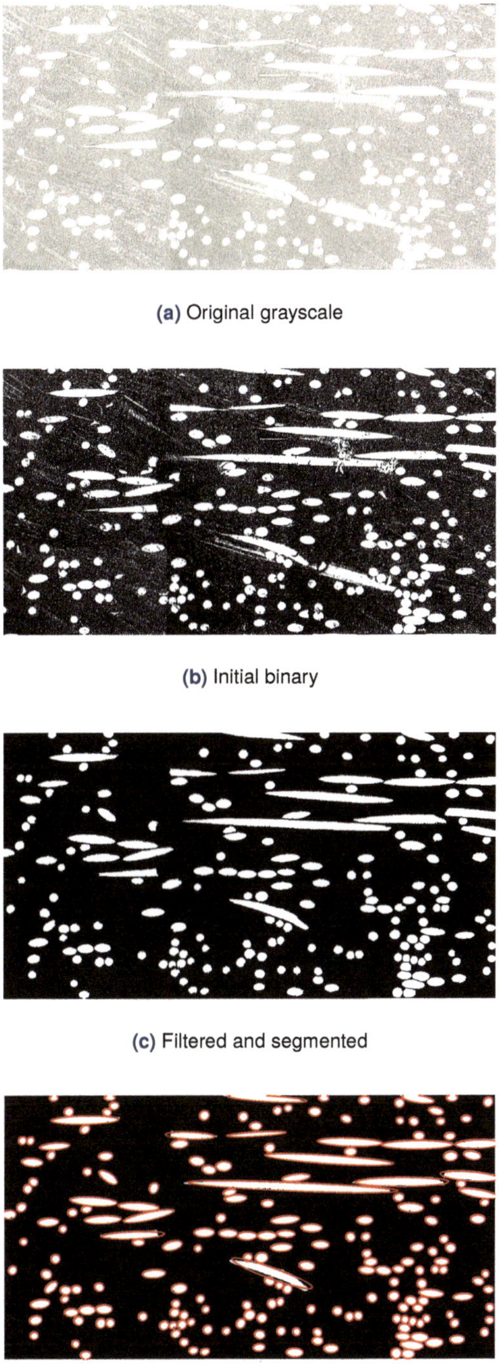

(a) Original grayscale

(b) Initial binary

(c) Filtered and segmented

(d) Fitted ellipses

Figure 3.7 Image processing steps for the method of ellipses. This is the lower left corner of Fig. 3.1. Images (a–c) from Phelps [Phe09].

The result of the filtering and segmentation steps is an image like Fig. 3.7(c). Now, standard image analysis functions can identify the pixels belonging to each fiber.

The final step in image analysis is to determine the principal-axis direction ϕ and principal diameters M and m for each ellipse. This is done by fitting an ellipse to each connected group of white pixels in the image. In image processing, these groups of pixels are called *objects*. Figure 3.7(d) shows a set of ellipses fitted to the objects in Fig. 3.7(c).

Ellipses are fitted using one of two algorithms:

Second moments: All of the pixels in the object are used to compute the centroid of the object and its moments of inertia. The centroid and moments define a unique ellipse, from which M, m and ϕ are determined [ECV⁺01]. The MATLAB function `regionprops` is one of many image processing tools that performs this calculation.

The second-moment algorithm produces a result for any object, whether it is close to an ellipse or not. Results for nearly elliptical objects are quite accurate, but the calculated orientations for non-elliptical objects are less accurate. The ellipses in Fig. 3.7(d) were found using the second moment technique. A few fits to non-elliptical objects can be seen there. For fiber orientation measurement it is common practice to use these ellipses, even if the object on the section plane is not perfectly elliptical.

Boundary fitting: The pixels on the boundary of the object are identified, and a least-squares calculation is used to find the best-fitting ellipse [Mle99, ECV⁺01]. If a detailed filtering operation is used to eliminate boundary pixels that are not part of a smooth ellipse (as when the section plane passes through the end of a fiber) then the full underlying ellipse shape can be determined with good accuracy.

Eberhardt et al. [ECV⁺01] compared the two methods, using confocal scanning laser microscopy to obtain an accurate reference orientation for each fiber, and found similar accuracy from the two algorithms.

With either method, some additional filtering may be performed afterward to eliminate objects whose elliptical fits are not plausible shapes for a fiber. For example, one might eliminate objects whose measured m is much smaller than the fiber diameter.

The result of this step is a set of (θ_k, ϕ_k) values for each fiber in the sample. The orientation tensor is then calculated using Eqn. (3.3) and a suitable weighting function.

3.1.2.3 Sample Size and Orientation

To obtain the most useful data, the size and orientation of the sample should also be chosen carefully.

The angles (θ, ϕ) of an individual fiber are determined most accurately when the fiber lies close to the plane of the section (i.e., when θ is close to 90°). For injection molded composites the fibers tend to be nearly parallel to the mold surfaces, which suggests using a "top view" section plane that is parallel to the mold surfaces. However, fiber orientation in injection molded parts varies strongly across the part thickness, and a good measurement should capture this. To avoid the work of polishing many planes across the part thickness, most injection molded parts are measured using a "side view," where one dimension of the section plane is the thickness direction and the other dimension is the mold filling direction (if that is known). Figure 3.1 is such an image.

For these side-view sections, one typically images the entire thickness of the molding along a width of one to ten millimeters. Then the image is "sliced" computationally across the thickness, and the fibers in each slice are used to compute the orientation tensor at that position across the thickness. Each tensor component is then plotted as a function of the thickness-direction coordinate.

Nine or 11 slices across the thickness is the minimum number that is useful, and many measurements use 19 or 21 slices. An odd number of slices is preferred, to obtain a data point exactly on the midplane.

The width of the image determines the number of fibers per slice. Wider images give more fibers, which reduces sampling errors. However, the width should not be so great that the orientation changes from one side of the sample to the other. For injection molded composites an image width that gives several hundred fibers per slice is common, and a few thousand fibers per slice is not unusual. The number of fibers per slice will vary with the orientation state; alignment normal to the section plane will give a larger number of fibers, while alignment parallel to the section plane will give fewer fibers. Velez et al. [VGWB⁺12] suggest that useful data for small areas can be obtained when $\sum(1/\cos\theta_k) \geq 100$. This metric accounts for the fact that slices where most θ values are close to zero will contain many more ellipses than slices where θ is close to 90°.

When the fibers are grouped in bundles, as in some types of sheet molding compound, a larger sample area may be needed so that the measurement is not dominated by a few bundles.

3.1.3 Example: Core/Shell Structures in Injection Molded Composites

Let us look now at Figure 3.1 in more detail. This is an injection molded plaque of polypropylene with 40% by weight of long glass fibers. The mold filling direction is horizontal, the part thickness direction is vertical, and the image shows the full 3 mm thickness of the part.

Visually, the image suggests a layered structure, with two *shell* layers surrounding a central *core* layer. Fibers in the shell are aligned in the flow direction, and the core has a crossflow alignment. This sample also has thinner *skin* layers, at the upper and lower mold surfaces, that are less aligned than the shells. The approximate locations of these layers are suggested in the figure.

This core/shell structure is typical for injection molded parts, for reasons that will be discussed in Section 4.3. The shell layers are always aligned in the flow direction, while the core can have a crossflow, random or flow-direction alignment, depending on the flow geometry. Skin layers are not always present (e.g., Fig. 5.16), and are more likely to develop when the mold filling speed is slow.

To analyze the orientation, the full height of this image is divided into 21 slices, as shown in Fig. 3.8(a). This gives approximately 600 fibers per slice. The orientation is determined for each slice, and the results are shown in Fig. 3.8(b). For this figure we have transformed to a global coordinate system where the 1 axis is the flow direction, the 2 axis is the crossflow direction, and the 3 axis is the thickness direction. $x_3 = 0$ denotes the midplane of the part, and the total part thickness is $2h$. This will be our standard coordinate system for molded parts.

While it seems easy to visually identify the boundaries between the core and shell in Fig. 3.1, the quantitative data shows that these layers are not distinct. Instead, the orientation changes

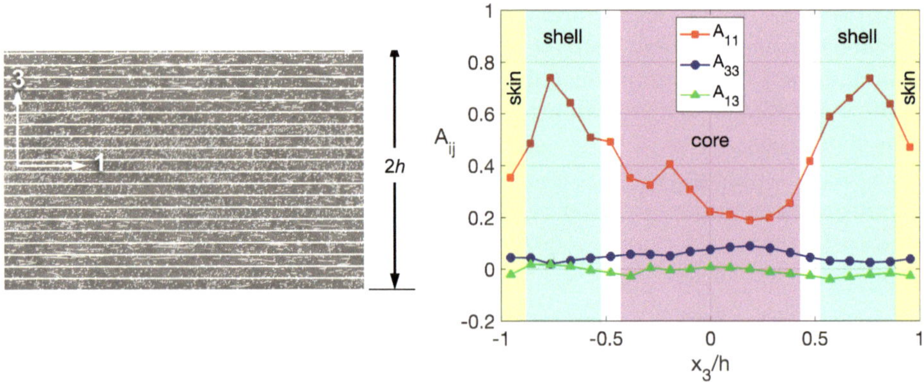

(a) Slices and global coordinate axes (b) Measured orientation

Figure 3.8 Analysis of the orientation in Fig. 3.1. Injection molded plaque with 40% by weight of long glass fibers in PP. Sample image and data from Phelps [Phe09].

gradually across the thickness, from a crossflow alignment at the midplane ($x_3/h = 0$) to flow-direction alignment around $x_3/h = \pm 0.75$. This is another way in which visual interpretations of polished sections can be misleading.

The core, shell and skin regions for this sample are indicated approximately on Fig. 3.8(b).

The thickness-direction orientation, described by A_{33}, is close to zero in this part, so all the fibers lie close to the flow/crossflow plane. A_{13}, which is the unambiguous off-diagonal component in this measurement, is also small. Near-planar orientation is also typical for injection molded parts.

We have not graphed data for A_{22} here. In principle one can determine A_{22} from the plots, since $A_{11} + A_{22} + A_{33} = 1$. For injection moldings, where A_{33} is small, the graph of A_{22} vs. x_3 looks like the A_{11} graph flipped vertically around $A_{ij} = 0.5$.

The concept of core, shell and skin layers provides a convenient way to talk about the structure of injection molded parts. However, in real parts these layers are not distinct. Both measurements and predictions show that the fiber orientation varies smoothly across the thickness of an injection molded part.

3.1.4 Measurement Accuracy

It is not easy to place rigorous error estimates on planar section data. As with any measurement, one needs to estimate the magnitudes of systematic errors, measurement errors and sampling errors.

Systematic errors are those that bias the average results of the measurement. Mis-aligning the section plane relative to the molded part is one example. Sharma et al. [SNN$^+$18] provide quantitative estimates of the systematic errors associated with image resolution, polishing errors, focus errors and stitching errors. These are errors that one tries to minimize by using best practices throughout the measurement process.

Every measurement has some resolution limit, and *measurement errors* come from these sources. For the method of ellipses, one can estimate the error in determining θ and ϕ, and algebraically work out how these errors propagate into the tensor components through Eqn. (3.3). See [ECV+01] for details. The result is an estimate for the standard deviations of each tensor component. However, sampling errors are typically much larger than measurement errors, so these estimates by themselves are rarely useful.

Sampling errors occur because we can only measure a finite number of fibers, rather than the entire population. The larger the sample size, the smaller the sampling error. One can estimate the sampling error *if* the orientation of each fiber on the section plane is independent of the orientations of all other fibers on the section [BT92c, Mle99, ECV+01]. Unfortunately this is not true in practice. Instead, the orientation of each fiber is correlated with those of its neighbors, and we do not have a way to quantify this correlation.

A good way to get a sense of the variability in orientation data is to compare multiple measurements on nominally identical samples. Figure 3.9 shows an example. All of the data curves in this figure come from a single sample that was injection molded from polypropylene with 50% by weight of long carbon fibers [SNN+18]. The part is a center-gated disk, 3.2 mm thick. The polished section, shown in Fig. 3.10(a), is divided into 21 slices across the thickness direction and 8 columns along the flow direction. Each column of slices produces one of the reddish lines in Fig. 3.9, and each data point on those lines is based on approximately 900 fibers. The large blue circles in the figure use the full width of the sample in the flow direction, so each blue data point is based on approximately 7,200 fibers.

Overall, A_{11} has the usual core/shell/skin structure, and the small values of A_{33} show that the fibers are oriented primarily in the flow/crossflow plane. The core has a strong crossflow alignment ($A_{11} < 0.5$), which is normal for a center-gated disk. The core is also quite thick and the shell layers are not strongly aligned in the flow direction, which is typical of long fiber injection molding compounds. A_{11} decreases near the upper and lower mold surfaces, so this part has skin layers.

With 900 fibers per data point one would expect fairly consistent curves of A_{11} vs. x_3. However, this is not the case. As Fig. 3.9(a) shows, there are considerable variations in $A_{11}(x_3)$ from column to column, and these variations appear more or less random. This reflects sampling errors, which are large here because of the patchy nature of orientation in this sample. A line connecting the blue circles, which combine the data from all the columns, would be smoother, representing the average of the reddish curves. Still, there are places like $x_3/h = 0.3$ where data from a few columns pushes a blue point out of line. The data for A_{33} is more consistent because the value of A_{33} is small across the entire sample, but a few of the reddish curves still show large deviations.

Figure 3.10 provides some additional insights into this particular data set. Portions of fiber bundles that survived the injection process are visible in the original image in Fig. 3.10(a). Where these intersect the measurement plane, the local orientation is strongly influenced by the bundle orientation. This effect is quantified in Figs. 3.10(b) and (c), which show the measured orientation in each small area of the image. Each column of rectangles in these figures corresponds to one of the reddish lines in Fig. 3.9, and each small rectangle corresponds to one of the data points. While the general core/shell structure is consistent across the columns, there is a lot of variability from point to point. Long fiber materials like this sample tend to have more of this type of variability, though the effect is still present in short fiber molding compounds.

While we cannot write a formula for the sampling error as a function of the number of fibers per data point, information like Fig. 3.9 should be helpful when interpreting orientation data.

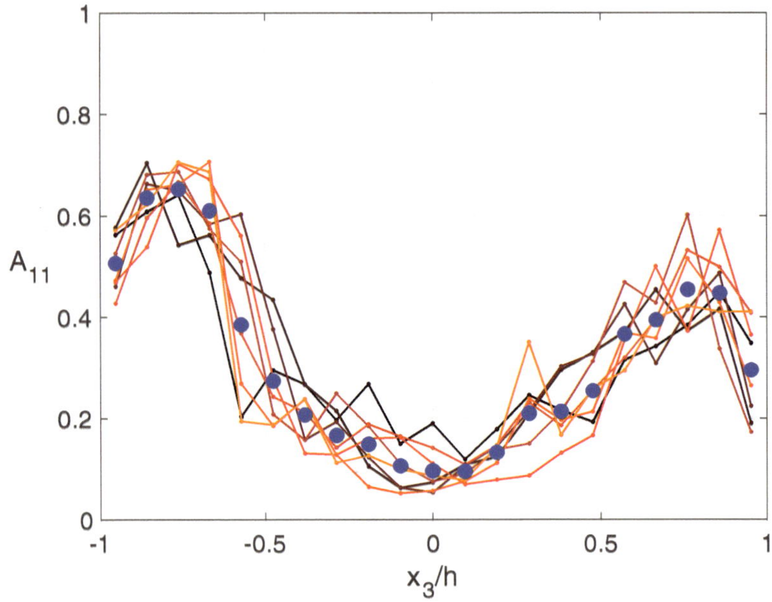

(a)

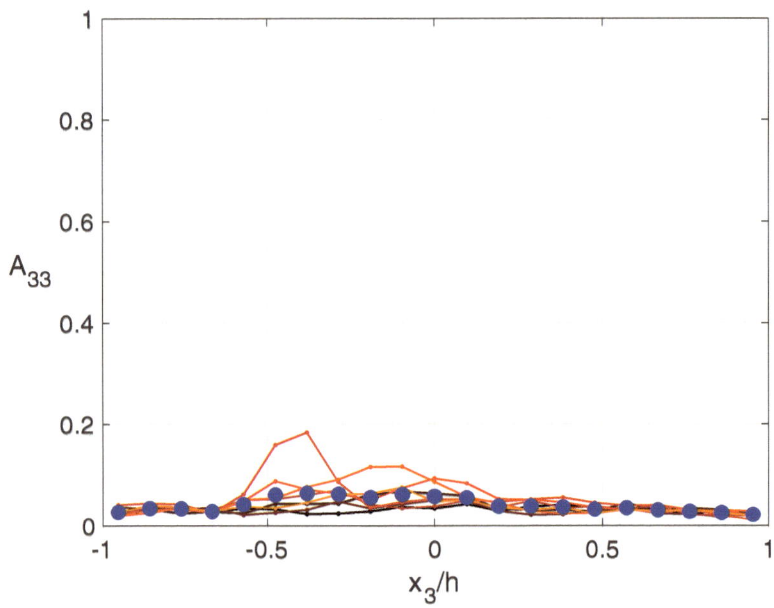

(b)

Figure 3.9 Orientation measurements on the PP/long carbon fiber sample in Fig. 3.10. The blue circles represent data across the entire width of the image in the flow (x_1) direction. Thin reddish lines are measurements for different portions of the sample width, corresponding to the columns in Fig. 3.10(b) and (c). The reddish lines use approximately 900 fibers per data point.

(a) Original image, thresholded and segmented

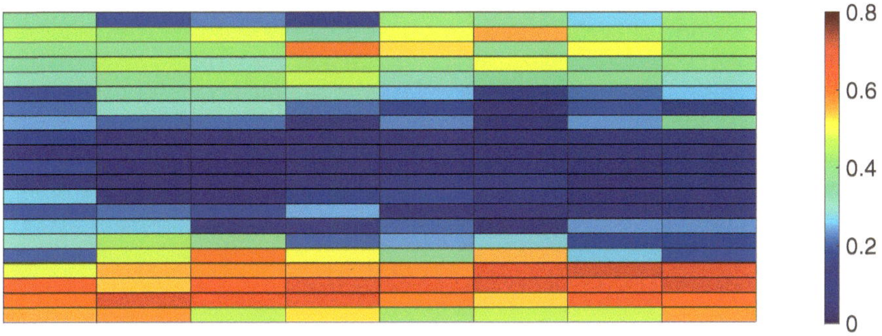

(b) Local A_{11} values

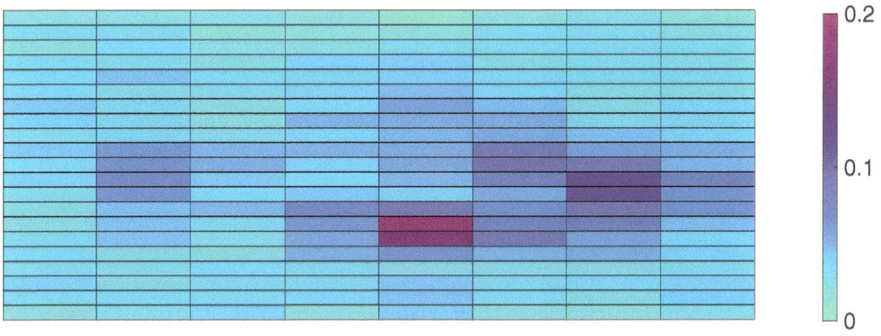

(c) Local A_{33} values

Figure 3.10 Cross-section analyzed in Fig. 3.9. Polypropylene with 50% by weight of long carbon fibers. Colors in (b) and (c) indicate the tensor component values measured in the corresponding areas of (a). The flow direction x_1 is horizontal and the thickness direction x_3 is vertical. Image (a) from Sharma [SNN+18].

3.2 Micro-CT Measurements of Orientation

A second approach to measuring fiber orientation is to use high-resolution X-ray computed tomography, or micro-CT. This produces a 3-D image of the sample with enough detail to resolve the individual fibers. The main steps are to choose a sample size, scan the sample to acquire a 3-D image, and analyze the image to determine fiber orientation. We will treat these steps in turn.

3.2.1 Image Acquisition and Sample Size

The main components of a micro-CT scanner are shown in Fig. 3.11. A small X-ray source illuminates the sample, projecting its X-ray image onto the detector. The brightness of each point on this image is determined by the absorption and scattering of X-rays within the sample, along a line from the source to the detector. The sample is mounted on a rotating stage, and projections of the sample are captured and recorded for many different rotation angles. Rotating the sample through 360° at increments as small as 0.12° is typical. Scanning a sample can take from half an hour to a few hours in a laboratory micro-CT machine. The radiation of a synchrotron is much more intense that in a lab scanner, and a synchrotron scan can be completed in minutes or even seconds.

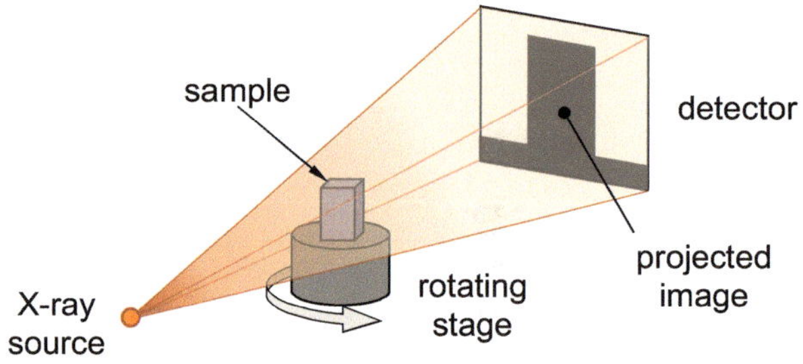

Figure 3.11 The main components of a micro-CT scanner.

The recorded images are processed in a computer, and the X-ray absorption distribution on each horizontal slice of the sample is reconstructed from the corresponding line of pixels on the detector (e.g., [Buz08, Her09, Hse09, KS01]). These slices are then stacked to assemble a 3-D digital image of the sample. Software for controlling the scanner, acquiring the images and reconstructing the images is normally provided with the scanner.

Figure 3.12 shows a scan of a short fiber composite processed by injection molding. The material is polybutylene-terephthalate (PBT) with 30% by weight of short glass fibers. The sample is approximately 3×2×2 mm in size, and the x_3 direction is the full thickness of the molding. This sample is taken from a gear window housing (point 2 in [KLC+20b, KDS+20]). The crowded nature of this composite, where every fiber is close to many neighboring fibers, is apparent. A close look at the figure also suggests a core/shell structure to the orientation, particularly looking at the plane that is normal to the x_2 axis. We will see quantitative data for this sample in Fig. 3.14.

Figure 3.12 Micro-CT image of an injection molded composite. The sample is approximately 3×2×2 mm, and is PBT with 30% by weight short glass fibers. Image by Dr. Thomas Riedel, Robert Bosch GmbH.

The X-ray attenuation characteristics of a material depend primarily on density and atomic number. Glass fibers have more than twice the density of most polymer matrices, so glass fibers can be readily imaged using absorption-based micro-CT. However, carbon fibers are close in density to most polymers, so they are difficult to image clearly using absorption. One option for micro-CT of carbon-fiber composites is to use a phase contrast method [ROCM20], which requires particular capabilities from the scanning device.

While micro-CT is often described as being non-destructive, it is usually necessary to cut a small piece from the molded part for analysis. This is because of the interrelationship between sample size, voxel size and scanner resolution.

From Fig. 3.11, we can see that moving the stage closer to the X-ray source increases the magnification and reduces the physical size of the voxels. Conversely, the sample can be viewed at lower magnification, and the physical size of the voxels made larger, by moving the sample farther from the source. Micro-CT scanners typically provide this adjustment. However, there are two important limits on the magnification that can be used.

First, any given micro-CT scanner has a maximum *resolution*, which controls the smallest feature it can focus on. Resolution is expressed either as the maximum number of lines per millimeter or the minimum line spacing that can be resolved at any magnification. Scanner resolution is determined by factors such as the focal spot size of the X-ray source, detector element size, scanner geometry, number of projections and the reconstruction algorithm [Buz08, Hse09]. The best current micro-CT scanners can resolve line spacings less that 1 μm.

Second, the micro-CT sample can only be reconstructed correctly if the entire sample width is in view at all rotation angles. This means that the projection of the maximum width of the sample cannot be larger than the width of the detector. The detector has a finite number of pixels, which determines the voxel size in the final digital image. This relationship is

$$\text{voxel size} \geq \frac{\text{sample width}}{\text{number of pixels across the detector}} \tag{3.17}$$

Some practitioners recommend using only the central 50% of the detector width, and thus half the number of pixels across the detector, to reduce artifacts and obtain the most accurate reconstruction [dBGlR17].

If the scanner resolution is less than or equal to the voxel size, then the voxel size determines the resolution of the final image. If the scanner resolution is larger than the voxel size, only features the size of the scanner resolution and larger will be resolved.

The in-focus voxel size needed to measure fiber orientation depends on the image analysis method, but for injection molded composites the voxels are normally much smaller than the fiber diameter. Most glass fibers have a diameter in the range of 10 to 20 μm, so one often needs voxels as small as 2 μm on each edge.

Suppose that we want to scan a composite with 2 μm voxels, the scanner resolution is 1 μm, the detector has 2,048 × 2,048 pixels and we use the full width of the detector. Then the sample width must be less than 4.1 mm in any direction perpendicular to the rotation axis.

For injection molded composites, micro-CT samples are commonly cut to 2 × 2 to 3 × 3 mm in cross-section. This still includes tens of thousands of fibers in the final image. If the stage can be moved vertically, which is normally the case, then the third dimension of the sample can be greater than the lateral dimensions. If larger voxels are acceptable, the physical dimensions of the sample can be increased accordingly.

While it is routine to cut small pieces from a molding for analysis, some scanner/reconstruction-software combinations are capable of *region of interest* (RoI) scanning [BDD⁺19]. With RoI CT, only a portion of a larger molding is imaged and analyzed, and it may not be necessary to cut up the molding for analysis.

Micro-CT has also been used to track individual fibers in a flowing suspension [LOD⁺16]. A viscous fluid with a few fibers was placed in a specially designed squeeze flow rheometer. A synchrotron provided the X-rays, allowing images to be captured in real time as the rheometer was rotated in front of the beam.

3.2.2 Image Analysis

Once the scans have been acquired and the 3-D image reconstructed, the next step is to analyze the image. There are two main schemes for analyzing micro-CT images for fiber orientation. One assigns an orientation vector **p** to each fiber voxel, while the other identifies all the voxels belonging to each individual fiber and determines a single **p** for the whole fiber. Each scheme has its advantages, and either one can provide accurate orientation data if used properly. We explain these two approaches, then compare their accuracy and capabilities. We also discuss options for analyzing samples where the fibers are organized into bundles or platelets.

3.2.2.1 Voxel Orientations

A micro-CT image is a three-dimensional array of numbers that give the intensity of X-ray absorption at each voxel. Let $I(\mathbf{x})$ be this intensity array, where $\mathbf{x} = [x_1, x_2, x_3]^T$ denotes position within the image. The task of image processing is to compute an orientation tensor from I.

A widely used approach is to segment the image into fiber voxels and matrix voxels, and assign an orientation vector $\mathbf{p}(\mathbf{x})$ to each fiber voxel. The orientation tensor for the sample is then computed as the average over all of the fiber voxels,

$$\mathbf{A} = \frac{1}{n_f} \sum_{\mathbf{x} \in \text{fibers}} \mathbf{p}(\mathbf{x})\mathbf{p}(\mathbf{x}) \tag{3.18}$$

where n_f is the number of voxels that belong to fibers. A similar expression can be used to find the fourth-order tensor. A micro-CT analysis treats the entire volume of the sample, so there is no sampling bias and no need for a weighting function. Note that a voxel-orientation scheme does not identify individual fibers, only fiber voxels.

There are two popular options for determining voxel orientation. Both use spatial derivatives of the image intensity $I(\mathbf{x})$.

The first option, the **structure tensor**, is formed using the first derivatives of I. In its basic form, the structure tensor **S** has components

$$S_{ij} = \frac{\partial I}{\partial x_i}\frac{\partial I}{\partial x_j} \quad \text{or} \quad \mathbf{S} = \begin{bmatrix} \left(\frac{\partial I}{\partial x_1}\right)^2 & \frac{\partial I}{\partial x_1}\frac{\partial I}{\partial x_2} & \frac{\partial I}{\partial x_1}\frac{\partial I}{\partial x_3} \\ \frac{\partial I}{\partial x_2}\frac{\partial I}{\partial x_1} & \left(\frac{\partial I}{\partial x_2}\right)^2 & \frac{\partial I}{\partial x_2}\frac{\partial I}{\partial x_3} \\ \frac{\partial I}{\partial x_3}\frac{\partial I}{\partial x_1} & \frac{\partial I}{\partial x_3}\frac{\partial I}{\partial x_2} & \left(\frac{\partial I}{\partial x_3}\right)^2 \end{bmatrix} \tag{3.19}$$

The derivatives of I can be evaluated using a multi-point finite difference formula [SLW15] or using Gaussian-derivative filters [DSF$^+$18]. The basic tensor in Eqn. (3.19) is then averaged over a small window centered on each voxel to assign a final structure tensor to each voxel.

To determine the local fiber direction, an eigenvalue/eigenvector analysis is performed on the final structure tensor at each voxel. The local orientation **p** is chosen as the eigenvector of **S** associated with the eigenvalue having the *smallest* absolute value. The smallest eigenvalue is used because the fiber axis should be the direction along which I varies the least. This operation is performed for each voxel in the image, then the fiber voxels are identified and Eqn. (3.18) is used to compute **A**.

The second option, the **Hessian matrix**, is formed using the second derivatives of I. The image is first smoothed so that the edges of the fibers are gray but the core of each fiber is white. This

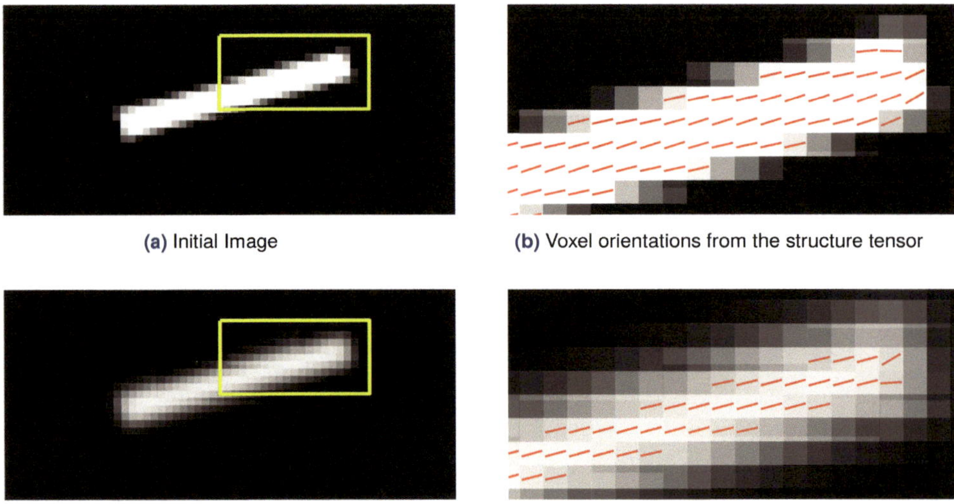

(a) Initial Image

(b) Voxel orientations from the structure tensor

(c) Image filtered for Hessian analysis

(d) Voxel orientations from the Hessian matrix

Figure 3.13 Two-dimensional example of image processing for voxel-based orientation analysis of micro-CT data.

imparts a curvature to $I(\mathbf{x})$ in the directions normal to the fiber axis, leaving a ridge of white voxels along the axis. The Hessian matrix $\mathbf{H}$ for each voxel is then computed as

$$H_{ij} = \frac{\partial^2 I}{\partial x_i \partial x_j} \qquad \text{or} \qquad \mathbf{H} = \begin{bmatrix} \frac{\partial^2 I}{\partial x_1^2} & \frac{\partial^2 I}{\partial x_1 \partial x_2} & \frac{\partial^2 I}{\partial x_1 \partial x_3} \\ \frac{\partial^2 I}{\partial x_2 \partial x_1} & \frac{\partial^2 I}{\partial x_2^2} & \frac{\partial^2 I}{\partial x_2 \partial x_3} \\ \frac{\partial^2 I}{\partial x_3 \partial x_1} & \frac{\partial^2 I}{\partial x_3 \partial x_2} & \frac{\partial^2 I}{\partial x_3^2} \end{bmatrix} \tag{3.20}$$

As before, the eigenvalues and eigenvectors of $\mathbf{H}$ are found, and $\mathbf{p}$ for each voxel is chosen as the eigenvector associated with the eigenvalue having the smallest magnitude. Because of the initial smoothing, the magnitude of the second derivative of I is large in any direction normal to the fiber axis, but small along the fiber axis. Equation (3.18) is then used to compute the orientation tensor.

Figure 3.13 shows a 2-D example that illustrates these schemes. The initial image in Fig. 3.13(a) contains a single fiber. The resolution of this image is fairly coarse, with 4 pixels across the fiber diameter. A close-up of the orientation vectors found using the structure tensor is shown in Fig. 3.13(b), and there is a unique $\mathbf{p}$ vector for each pixel that is bright enough to be considered part of the fiber. These vectors follow the overall direction of the fiber well, with some small variations at the edges and the tip.

For the Hessian matrix scheme, the initial image is first smoothed using a Gaussian filter, producing Fig. 3.13(c). Hessian analysis then gives the orientation vectors shown in Fig. 3.13(d). In this case the threshold for identifying fiber pixels emphasizes the core of the fiber, where the Hessian $\mathbf{p}$ vectors are most accurate. Again the overall direction of the fiber is accurately captured, with a few variations at the fiber tip.

Because this example has a single fiber, the corresponding orientation tensor should have a largest eigenvalue of $\lambda_1 = 1$, and the corresponding eigenvector should be oriented along the fiber axis. The structure tensor and Hessian matrix approaches are both quite accurate here, giving $\lambda_1 > 0.99$ and getting the orientation of the eigenvector correct within 0.3°. Both schemes work well in this example, despite the fairly coarse resolution of the original image. Of course real micro-CT data is three-dimensional, with much larger images.

Both the structure tensor and the Hessian matrix approaches have internal parameters that must be chosen appropriately to obtain good orientation data. These include the size of the template used to compute derivatives, the size of the averaging window for the structure tensor, and the size of the initial smoothing window for the Hessian matrix. These parameters must be tuned to match the image resolution and the fiber diameter, though selecting good values is not difficult.

Other image processing schemes (anisotropic Gaussian filters, main axis of inertia) have also been used to find local voxel orientations. A comparison of these methods shows that the structure tensor and Hessian matrix are just as accurate as other schemes and are faster to compute [WSEO08, PDB⁺18]. For this reason, either **S** or **H** is typically available in commercial software packages for analyzing micro-CT fiber images.

For either method, one must decide which voxels belong to the fibers and which belong to the matrix. A basic approach is to use a thresholding operation based on the voxel intensity I to identify fiber voxels. However, somewhat better results are obtained if one also uses the structure tensor or Hessian values to assist with the thresholding. In that case, fiber voxels are those that have a high intensity I and a large difference between the largest and smallest eigenvalues of **S** or **H**. This focuses the orientation results on voxels with a strong directionality, omitting small chips of the fiber material and other locations where $\mathbf{p}$ is not well defined.

Since longer fibers will have more voxels than shorter fibers, the orientation tensor measured by a voxel-based approach is a weight-average orientation, as in Eqn. (2.111). It is not possible to determine a number-average orientation with a voxel-based approach.

3.2.2.2 Fiber Reconstruction

The other main approach to micro-CT orientation analysis is to reconstruct each individual fiber. This was the first approach used with micro-CT to measure fiber orientation in injection molded composites [EC02, SKK⁺11], and this capability is now present in some commercial software for analyzing micro-CT images, e.g. [KMK⁺20].

The initial image processing steps are to segregate the voxels that belong to the fibers, then skeletonize each fiber into a string of single, connected voxels along its core. Voxel orientation methods like the Hessian matrix may be used as part of the algorithm, which can improve the segregation of fiber voxels [SKK⁺11]. Some fibers may incorrectly be broken into multiple segments in this process, so these segments must be identified and rejoined. There are also locations where two or more fibers nearly touch, and these fibers must be separated. It can be difficult to process all of these cases accurately with an automated scheme, so some researchers have developed tools for manually touching up the segmentation [HS19]. There is no widely accepted algorithm for fiber reconstruction, and readers interested in this subject should consult the literature [EC02, SKK⁺11, SJKK15, AS18, HRW⁺19].

Individual fibers can only be reconstructed from a high-resolution image. Voxels 2 μm or less are typically used for glass fibers, resulting in at least six voxels across the fiber diameter. This can limit the overall size of the sample. Image processing times for fiber reconstruction are also typically much longer than processing times for voxel orientation schemes.

Once the fibers have been reconstructed, it is straightforward to determine the orientation vector **p** and length ℓ of each fiber. For short fiber composites the fibers tend to be quite straight, so an end-to-end vector is used to find **p**. If the fibers have significant curvature, then one might divide each fiber into segments that are shorter than the minimum radius of curvature and form an orientation tensor using the segment orientations [GSKS16].

Because pieces of fiber material as small as a few microns will be detected, it is customary to discard fibers shorter than a certain length, say $\ell < 2d_f$, from both the orientation and length analyses. Otherwise the results can be skewed by the large number of very small "fibers."

Since the length of each fiber is known, both the number-average and weight-average orientation tensors, Eqns. (2.110) and (2.111), can be determined with fiber reconstruction. While the literature is often unclear about which orientation tensor is being reported, the weight-average orientation should be used for most purposes.

A weight-average orientation is produced by both planar section and voxel orientation measurements, so it is the correct quantity for comparing different measurement methods. Also, the mechanical properties of the solid composite are affected by different fiber lengths in proportion to their weight fraction rather than their number, so the weight-average orientation should be more representative of final properties than the number-average orientation.

3.2.2.3 Example Data

Quantitative orientation data from the scan in Fig. 3.12 is shown in Fig. 3.14. The scan used a 2.4 μm voxel size and the orientation was determined with a voxel-based method.

The 3 axis is the thickness direction in this sample. The sample has the familiar core/shell structure, with the fibers lying primarily in the 1–2 plane (A_{33}, A_{23} and A_{31} are all small).

The shell orientation is primarily in the 1 direction, though A_{12} is non-zero, so the principal direction of orientation lies between the 1 and 2 axes. This is a key advantage of micro-CT measurements as compared to planar section data: all of the tensor components are determined without ambiguity. This allows the directionality of the orientation to be determined clearly,

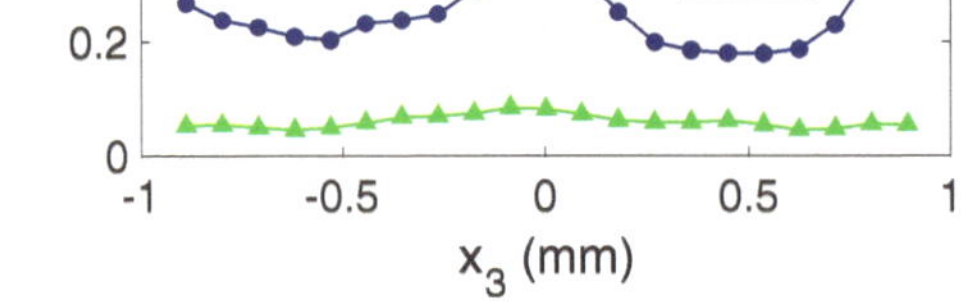

(a) Diagonal components

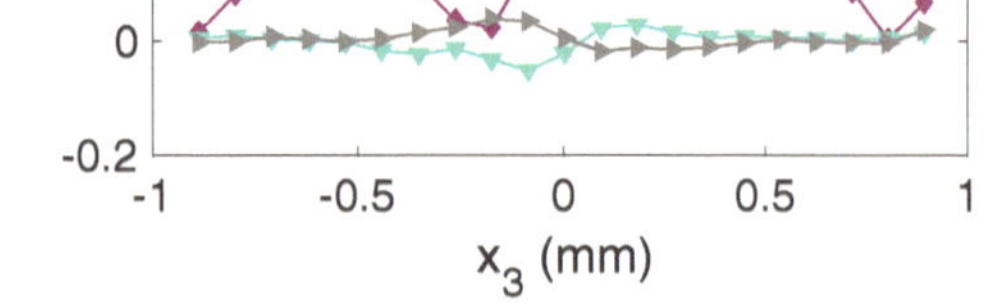

(b) Off-diagonal components

Figure 3.14 Fiber orientation measured from the micro-CT image in Fig. 3.12. PBT with 30% by weight short glass fibers. Data from Dr. Thomas Riedel, Robert Bosch GmbH.

and is especially valuable with complex part geometries, where the local flow direction is not known ahead of time. One can easily do an eigenvalue/eigenvector analysis at each measurement point and determine the principal values of orientation and the principal directions at each value of x_3/h.

3.2.2.4 Comparison of Image Analysis Schemes

A careful comparison of voxel-orientation and fiber-identification methods [KMK+20] shows that the two methods produce similar, accurate results for high-resolution CT scans (1.4 μm or 3.2 μm voxels for 17 μm glass fibers).

For lower-resolution scans, the fiber reconstruction algorithm cannot separate the fibers and cannot produce a meaningful result. In contrast, the voxel-orientation method continues to produce useful results for voxels the size of the fiber diameter or larger [BDD+19, KMK+20]. The ability to provide meaningful measurements at lower scan resolution is a key advantage of the voxel-orientation approach, and can be used either to speed up the scanning process or to analyze larger samples.

Decreasing the scan resolution does affect the accuracy of voxel-based results. Compared to scans using a 3.2 μm voxel, lower-resolution scans properly capture the main direction of orientation and the spatial variations (e.g., the core/shell structure), but give measured orientations that are significantly less aligned than the high-resolution scan data [BDD+19]. For injection molded composites, one should only use lower-resolution scans (say, voxels larger than about 3 μm for 10 μm diameter fibers) when some imprecision in the tensor components is acceptable.

3.2.2.5 Samples with Fiber Bundles or Platelets

Composites in which the fibers are organized into bundles or platelets present a special challenge for micro-CT measurement. Within each bundle or platelet, a few hundred to a few thousand fibers have nearly the same orientation. A micro-CT measurement that detects each individual fiber will provide information about a few bundles or platelets, but will not necessarily give useful information about the overall fiber orientation in the material.

For this type of composite, researchers have had success using micro-CT measurements that resolve the bundles or platelets but do not resolve individual fibers. These measurements use a voxel-orientation scheme for image analysis, exactly as described above, except that now the voxel size and analysis parameters are tuned to the bundle diameter or the platelet thickness.

Good results have been reported for SMC, which has fiber bundles, using a voxel size of 17.3 μm [SBG+17]. Pre-preg platelet composites have been analyzed using voxels of 50 μm [DSF+18], which is approximately half of the platelet thickness. In that case, not only was the fiber direction determined using the smallest eigenvalue of the structure tensor, the platelet normal was also determined using the largest eigenvalue of the structure tensor.

As discussed above, these measurements tend to capture the main direction and spatial variations of orientation accurately, but may underestimate the anisotropy of the orientation distribution function.

3.3 Length Measurement

There are two approaches to measure the fiber length distribution: fully destructive sampling, where the matrix is burned off to reveal the fibers; and micro-CT measurement, where the fibers remain in place in the scanned sample. Each method has its advantages and disadvantages.

3.3.1 Destructive Sampling

The fully destructive method uses the following steps:

Cut an initial sample out of the molded component. In most cases the full thickness of the molding is analyzed. The remaining dimensions of this sample should be at least twice as long as the longest fibers that could be present. For example, if a long fiber injection molding compound has pellets with 12 mm long fibers, then the initial sample should be at least 24 mm × 24 mm. This ensures that none of the fibers that will be measured are shortened when the sample is cut.

Heat the sample to pyrolyze the polymer matrix, while constraining the sample. Pyrolysis is typically done at temperatures from 500 to 650 °C and requires 1 to 2 hours. Glass-fiber composites can be pyrolyzed in air, but carbon fibers will oxidize in air at these temperatures and should be pyrolyzed in a nitrogen atmosphere [SKF$^+$18]. Constraining the sample in a metal box or aluminum foil keeps the fiber network in place for the next step. Pyrolysis should remove the matrix, leaving only the fibers.

Down-select a subset of the fibers for analysis. An initial sample for pyrolysis has too many fibers for a practical analysis. Historically, researchers have used tweezers or some other means to select a subset of fibers for analysis, but this can introduce errors large enough to make the measurements meaningless [GBY$^+$18]. The recommended approach, developed by Kunc and co-workers [KFN$^+$07], is to use a syringe and hypodermic needle to inject a small plug of liquid epoxy resin across the fiber mass, then allow the epoxy to solidify and trap some of the fibers. The needle is inserted all the way through the sample thickness, then the syringe plunger is depressed as the needle is withdrawn from the sample. The resulting cylinder of epoxy has a diameter of 1.5 to 2.5 mm.

Once the epoxy has set, the sample is released from the holder and all fibers that are not adhered to the epoxy are removed. This can be done by gently shaking the sample and/or applying a short burst of low-pressure air. The epoxy and its attached fibers now look something like Fig. 3.15. The cylinder and attached fibers are then subjected to a second pyrolysis. This removes the epoxy, so that only the down-selected fibers remain.

Disperse the fibers on a substrate suitable for scanning. The down-selected fibers need to be separated and spread out so that individual fibers can be seen. This must be done without breaking any of the fibers. Placing the fibers in water and briefly sonicating the water, dispersing the fibers in a turbulent air flow, and using an electrical corona from a high-frequency generator (for carbon fibers) have all been used. The fibers are then spread onto one or more Petri dishes, glass plates or sheets of paper [KFN$^+$07, GBY$^+$18, SKF$^+$18]. Ideally the fibers should be well separated, with few contacts between individual fibers.

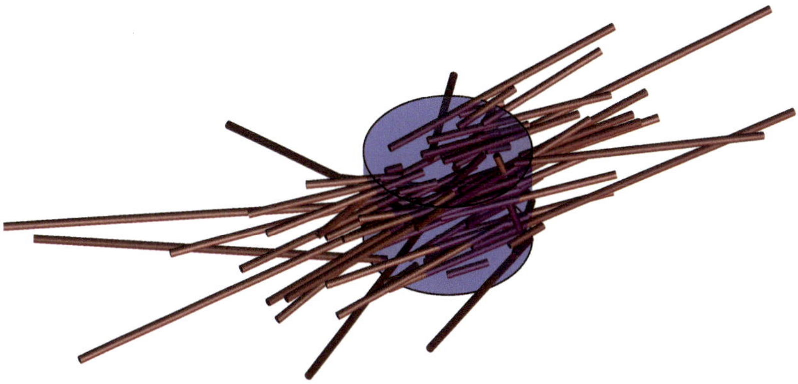

Figure 3.15 Schematic of down-selected fibers (red) captured by a cylinder of epoxy resin (blue).

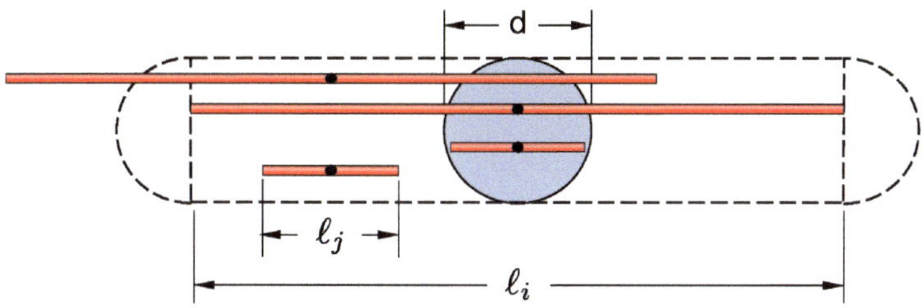

Figure 3.16 Basis of the down-sampling correction for measuring the fiber length distribution, Eqn. (3.21). The blue circle is the epoxy cylinder, seen from the top.

Scan the fibers and measure the length of each fiber. Modern flat-bed scanners work well for acquiring images of the dispersed fibers. Scans at 2400 dpi are suitable for glass fibers, while carbon fibers have a smaller diameter and should be scanned at 4800 dpi. Standard image analysis software can automatically determine the lengths of isolated fibers, but will often fail when two or more fibers intersect. Some researchers have used software to manually measure the length of each fiber (e.g., by clicking on the ends of each fiber), though an automated scheme that deals with intersecting fibers has been reported [GBY$^+$18].

Analyze the data, correcting for sampling bias. The down-selection procedure described above is more likely to capture longer fibers than shorter fibers, and this must be accounted for in the data analysis.

The issue is illustrated in Fig. 3.16. This shows a top view of the epoxy cylinder, and for simplicity we assume that the fibers lie in the plane normal to the cylinder axis. (This is normally the case for injection molded samples.) The figure shows two long fibers with length ℓ_i and two shorter fibers with length ℓ_j. Both long fibers are captured by the epoxy cylinder, but only one of the shorter fibers is captured, despite the fact that its centroid is the same distance from the cylinder as one of the longer fibers.

To derive the correction factor, let the diameter of the epoxy cylinder be d and its height h. Also, let $\hat{N}_i$ be the number of fibers of length ℓ_i captured by the epoxy cylinder in the down-selection procedure, and let N_i be the number of fibers of length ℓ_i per unit volume of the composite. $\hat{N}_i$ is the initial measurement data and we want to find the corrected values N_i.

From Fig. 3.16, a fiber of length ℓ_i will intersect the epoxy cylinder if its centroid lies anywhere within the dotted oval in the figure, at any place along the height of the cylinder. The volume of this space is $[\ell_i d + \pi d^2/4]h$. The true number of fibers of this length per unit volume in the sample is proportional to $\hat{N}_i$ divided by the volume: $N_i \propto \hat{N}_i/[\ell_i d + \pi d^2/4]h$. The values of d and h will be the same for every fiber length ℓ_i. Also, the N_i values can all be multiplied by a constant without changing the fiber length distribution. Thus, we can simplify this equation and calculate the corrected fiber length distribution as

$$N_i = \frac{\hat{N}_i}{1 + (4\ell_i/\pi d)} \tag{3.21}$$

Once this correction has been applied, the equations in Section 2.5 can be used to calculate the number-average and weight-average fiber lengths and the weight-based length distribution.

The fiber length data in Fig. 2.10 was obtained using this procedure. In that figure, which is from a long fiber injection molding material, the longest fibers are more than 12 mm long but the average fiber lengths are much shorter, $L_n = 1.59\,\text{mm}$ and $L_2 = 2.59\,\text{mm}$. The lengths of 2,000 fibers were measured in this data set.

A substantial number of fibers must be included in the down-selected sample to obtain an accurate result. Goris et al. [GBY$^+$18] recommend at least 2,000 fibers per measurement. They also recommend repeating the measurement five times when highly accurate data is needed. Sharma et al. [SKF$^+$18] also used 2,000 fibers per measurement, though they found that L_n and L_w could be determined within 5% error with samples of 800 fibers.

For injection molded samples, the resulting fiber length distribution is an average value across the thickness of the initial sample. However, models suggest that the fiber length should vary across the part thickness. It is possible to get separate measurements for the core and shell layers, using the total thickness on one sample and grinding away a shell layer on a second sample [BO20]. These measurements confirm that long fiber injection molding materials have longer fibers in the core and somewhat shorter fibers in the shell.

Here is an example that illustrates the importance of correcting for sampling bias.

Example 3.3.1 Analysis of experimental fiber length data

A fiber length measurement is performed as described above, and the epoxy cylinder for down-selection is 2 mm in diameter. The measurement finds 10 fibers that are 1 mm long, 2 fibers that are 11 mm long and no other fibers. Calculate a normalized weight-based distribution, the number-average fiber length and the weight-average fiber length, correcting the data for sampling bias. Also calculate the same quantities without the correction and compare the results.

Denote the two fiber lengths as $\ell_1 = 1\,\text{mm}$ and $\ell_{11} = 11\,\text{mm}$. The raw measurement data is $\hat{N}_1 = 10$ and $\hat{N}_{11} = 2$, with all other $\hat{N}_i = 0$.

To correct the data for sample bias we use Eqn. (3.21), and find

$$N_1 = \frac{\hat{N}_1}{1 + (4\ell_1/\pi d)} = \frac{10}{1 + (4 \times 1/2\pi)} = 6.110 \tag{3.22}$$

$$N_{11} = \frac{\hat{N}_{11}}{1 + (4\ell_{11}/\pi d)} = \frac{2}{1 + (4 \times 11/2\pi)} = 0.250 \tag{3.23}$$

The correct number-average and weight-average fiber lengths are then

$$L_n = \frac{\sum \ell_i N_i}{\sum N_i} = \frac{1 \times 6.110 + 11 \times 0.250}{6.110 + 0.250} = 1.39\,\text{mm} \tag{3.24}$$

$$L_w = \frac{\sum \ell_i^2 N_i}{\sum \ell_i N_i} = \frac{1^2 \times 6.110 + 11^2 \times 0.250}{1 \times 6.110 + 11 \times 0.250} = 4.10\,\text{mm} \tag{3.25}$$

For the weight-based distribution, use the normalization factor from Eqn. (2.105) in Eqn. (2.104). This gives

$$w_1 = \frac{\ell_1 N_1}{\sum \ell_i N_i} = \frac{1 \times 6.110}{1 \times 6.110 + 11 \times 0.250} = 0.690 \tag{3.26}$$

$$w_{11} = \frac{\ell_{11} N_{11}}{\sum \ell_i N_i} = \frac{11 \times 0.250}{1 \times 6.110 + 11 \times 0.250} = 0.310 \tag{3.27}$$

The shorter fibers are almost 70% of the weight of the fiber length distribution, while the longer fibers are just over 30% by weight.

If we ignore the correction for sample bias and use $\hat{N}_i$ instead of N_i, the results are quite different. Using hats to denote these (incorrect) results, the average lengths are

$$\hat{L}_n = \frac{\sum \ell_i \hat{N}_i}{\sum \hat{N}_i} = \frac{1 \times 10 + 11 \times 2}{10 + 2} = 2.76\,\text{mm} \tag{3.28}$$

$$\hat{L}_w = \frac{\sum \ell_i^2 \hat{N}_i}{\sum \ell_i \hat{N}_i} = \frac{1^2 \times 10 + 11^2 \times 2}{1 \times 10 + 11 \times 2} = 7.88\,\text{mm} \tag{3.29}$$

The down-selection procedure is biased towards longer fibers, so without the correction the average fiber lengths are substantially over-estimated.

Similarly, the (incorrect) weight-based fiber length distribution is

$$\hat{w}_1 = \frac{\ell_1 \hat{N}_1}{\sum \ell_i \hat{N}_i} = \frac{1 \times 10}{1 \times 10 + 11 \times 2} = 0.313 \tag{3.30}$$

$$\hat{w}_{11} = \frac{\ell_{11} \hat{N}_{11}}{\sum \ell_i \hat{N}_i} = \frac{11 \times 2}{1 \times 10 + 11 \times 2} = 0.688 \tag{3.31}$$

This is a complete reversal of the correct distribution, suggesting erroneously that the short fibers are just over 30% of the weight and the long fibers are nearly 70% of the weight.

While this example is made up, it shows how important the correction in Eqn. (3.21) is when measuring fiber length.

As a more realistic example, if the correction of Eqn. (3.21) had not been applied to the data in Fig. 2.10, the measured average lengths would have been $\hat{L}_n = 2.14\,\text{mm}$ and $\hat{L}_w = 3.49\,\text{mm}$, compared to the correct values of $L_n = 1.59\,\text{mm}$ and $L_w = 2.50\,\text{mm}$. ■

3.3.2 Micro-CT Data for Length

An alternate approach to measuring fiber length is to use a micro-CT scan [SKK+11, SJKK15]. This is only possible when the image analysis scheme reconstructs each fiber, as described in Section 3.2.2.2.

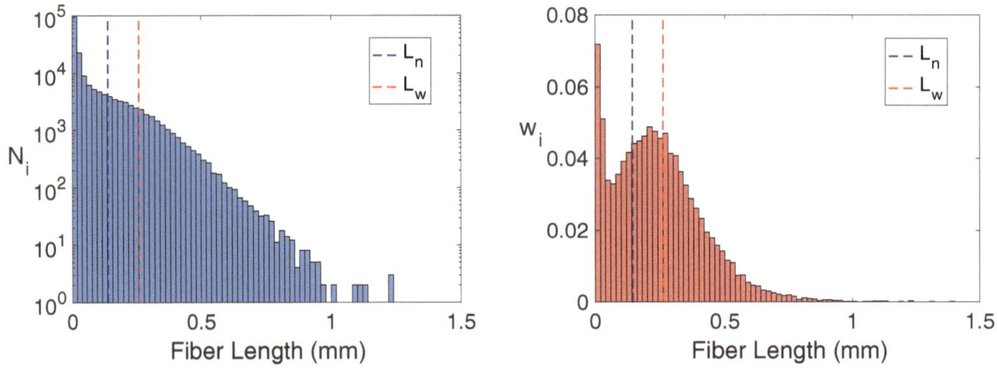

(a) Number based

(b) Weight based

Figure 3.17 Micro-CT fiber length data from an injection molded PP composite with 30% by weight short glass fibers [HAXS19]. Average lengths are $L_n = 0.143$ mm and $L_w = 0.261$ mm.

Figure 3.17 shows fiber length data obtained from a micro-CT scan [HAXS19]. The material is polypropylene reinforced with 30% by weight of short glass fibers, injection molded, and the scanned volume is a cylinder 2.4 mm in diameter and 2.5 mm tall.

Because the micro-CT scan detects every fiber in the sample, the sampling-bias correction of Eqn. (3.21) is not required. It is important to set a minimum length below which objects are excluded from the fiber length distribution, and a minimum length of $2d_f$ is common. Figure 3.17 follows this practice, and contains data from 85,677 individual fibers. This is every fiber in the scanned volume whose length is greater than 20 μm, compared to the fiber diameter of 10 μm. An additional 95,201 objects shorter than 20 μm were detected by the CT scan, but excluded from the analysis.

With this large number of fibers, the histograms in Fig. 3.17 are quite smooth. Note that the number-based distribution N_i is plotted on a logarithmic scale, to make the presence of the longer fibers more apparent. The weight-based distribution in Fig. 3.17(b) provides a good visual sense of the fiber length distribution. With average lengths of $L_n = 0.143$ mm and $L_w = 0.261$ mm this is definitely a short fiber composite. These average lengths are approximately one-tenth the average lengths of the long fiber injection molded composite in Fig. 2.10.

Micro-CT measurements for fiber length are useful for composites with very short fibers, but may not be suitable for materials like long fiber injection molding compounds, which have average lengths of a few millimeters and maximum fiber lengths of 12 mm or more. A fiber reconstruction scheme requires a high-resolution scan, which can limit the size of the scanned sample to a few millimeters in cross-section. This could prevent the full length of longer fibers from appearing in the scanned volume, biasing the measurement toward shorter fibers. For long fiber materials, the destructive procedure of Section 3.3.1 may be the best way to accurately measure the fiber length distribution.

3.4 Chapter Summary

Fiber orientation and fiber length can both be measured experimentally, using multiple techniques.

Fiber orientation is measured either from optical microscopy on polished planar sections, or from a micro-CT scan of a 3-D sample. The planar section method is always destructive, and a micro-CT scan may also require a small sample to be cut from the molded part.

For the planar section method, the 3-D orientation of each fiber is determined from its elliptical footprint on the section plane. The fibers on a planar section are a biased sample, but a weighting function corrects for this bias when computing the orientation tensor. All three diagonal components of the second-order orientation tensor and one off-diagonal component are determined unambiguously; the other off-diagonal components can only be bounded. While sample preparation and analysis are time-consuming, planar section measurements are a standard against which other methods are judged.

Micro-CT analyses either reconstruct each fiber in the sample, or determine a local orientation vector for each fiber voxel without identifying individual fibers. Both schemes provide accurate orientation data as long as the voxel size is small compared to the fiber diameter. All components of the orientation tensor are determined without ambiguity. The voxel orientation scheme can also provide orientation data using lower-resolution scans. This reduces quantitative accuracy, but can be useful when the fibers are arranged in bundles or platelets, or when one wants a general idea of the orientation in a large sample.

With either method, results are usually reported in terms of orientation tensor components. Accuracy is similar for planar section and micro-CT measurements, provided each measurement is done properly.

Fiber length is measured either by burning off the matrix and dispersing the residual fibers, or directly from a micro-CT scan.

For the burn-off method, a controlled down-selection process is used to choose a reasonable number of fibers, usually a few thousand, for length measurement. After burn-off, a plug of liquid epoxy is injected into the fiber network and allowed to solidify. Any fibers not captured by the epoxy are discarded. A second burn-off removes the epoxy, then the down-selected fibers are dispersed, placed on a scanner and imaged for length measurement. This produces a biased sample, but the bias is controlled and the results can be corrected for this.

If a micro-CT scan is analyzed to reconstruct each fiber, then the fiber length distribution can readily be obtained. This scheme is most suitable for composites where the longest fibers are not much longer than a millimeter, so that cutting the sample to the size required for a high-resolution scan does not bias the length distribution.

Fiber length results are usually reported using the number-average length L_n and the weight-average length L_w. If the full fiber length distribution is reported, the weight-based distribution w_i is preferred, as this conveys how the fiber mass is distributed across different lengths.

4 Flow Orientation of Single Fibers

We are now ready to examine how flow affects fiber orientation. This chapter looks at the motion of a single fiber (i.e., a fiber without any neighbors), laying the foundation for discussing the more realistic case of closely spaced groups of fibers in Chapt. 5. We first examine descriptions of fluid motion (flow kinematics), then turn to the centerpiece of the chapter: Jeffery's equation. This remarkable analytical result is the starting point for all models of flow-induced fiber orientation. In addition to its theoretical importance, Jeffery's equation gives us two rules that qualitatively explain *all* fiber orientation patterns observed in molded parts. We present those two rules, and illustrate their application in injection and compression molding.

4.1 Flow Kinematics

4.1.1 Deformation Rate and Vorticity

Consider the flow of a fluid that, for the moment, contains no fibers or other particles. The motion of the fluid is described by giving the velocity $\mathbf{v}$ of fluid particles as a function of position $\mathbf{x}$ and time t:

$$\mathbf{v} = \mathbf{v}(\mathbf{x}, t) \qquad \text{or} \qquad v_i = v_i(x_j, t) \tag{4.1}$$

A mold filling simulation uses the equations of fluid mechanics to calculate the velocity field at each time step during filling[1]. For now, we will treat the velocity field as known, either because it is imposed in some rheological measurement device or because it is calculated as part of a molding simulation.

If all the fluid in the neighborhood of a given point moves with the same velocity, then a fiber at that point will simply move along with the fluid. For fibers to change orientation, the fluid velocity must change with position. The first measure of this change is the set of derivatives of velocity with respect to position, $\partial v_1/\partial x_1$, $\partial v_1/\partial x_2$, $\partial v_2/\partial x_1$, etc. The full set of these derivatives is a 3×3 matrix called the *velocity gradient tensor* $\mathbf{L}$:

$$\mathbf{L} = \begin{bmatrix} \frac{\partial v_1}{\partial x_1} & \frac{\partial v_1}{\partial x_2} & \frac{\partial v_1}{\partial x_3} \\ \frac{\partial v_2}{\partial x_1} & \frac{\partial v_2}{\partial x_2} & \frac{\partial v_2}{\partial x_3} \\ \frac{\partial v_3}{\partial x_1} & \frac{\partial v_3}{\partial x_2} & \frac{\partial v_3}{\partial x_3} \end{bmatrix} \tag{4.2}$$

[1] See Section 6.4 for the equations that govern the velocity field.

We can write this in index notation as

$$L_{ij} = \frac{\partial v_i}{\partial x_j} \tag{4.3}$$

The velocity gradient tensor **L** is not necessarily symmetric. However, any tensor can be written as the sum of a symmetric part and an anti-symmetric part as follows:

$$\mathbf{L} = \mathbf{D} + \mathbf{W} \qquad \text{where} \qquad \mathbf{D} = \frac{1}{2}(\mathbf{L} + \mathbf{L}^T) \qquad \text{and} \qquad \mathbf{W} = \frac{1}{2}(\mathbf{L} - \mathbf{L}^T) \tag{4.4}$$

For any tensor **L**, the tensor **D** is symmetric ($D_{ij} = D_{ji}$) and the tensor **W** is anti-symmetric ($W_{ij} = -W_{ji}$).

This way of decomposing **L** is important. **D** is the *rate-of-deformation tensor*, and it describes how the fluid is changing shape; **W** is the *vorticity tensor*, and it describes rigid-body rotation of the fluid.

The components of the rate-of-deformation tensor **D** in index notation are

$$D_{ij} = \frac{1}{2}\left(\frac{\partial v_i}{\partial x_j} + \frac{\partial v_j}{\partial x_i}\right) \tag{4.5}$$

Writing out the details,

$$\mathbf{D} = \begin{bmatrix} \frac{\partial v_1}{\partial x_1} & \frac{1}{2}\left(\frac{\partial v_1}{\partial x_2} + \frac{\partial v_2}{\partial x_1}\right) & \frac{1}{2}\left(\frac{\partial v_1}{\partial x_3} + \frac{\partial v_3}{\partial x_1}\right) \\ \frac{1}{2}\left(\frac{\partial v_2}{\partial x_1} + \frac{\partial v_1}{\partial x_2}\right) & \frac{\partial v_2}{\partial x_2} & \frac{1}{2}\left(\frac{\partial v_2}{\partial x_3} + \frac{\partial v_3}{\partial x_2}\right) \\ \frac{1}{2}\left(\frac{\partial v_3}{\partial x_1} + \frac{\partial v_1}{\partial x_3}\right) & \frac{1}{2}\left(\frac{\partial v_3}{\partial x_2} + \frac{\partial v_2}{\partial x_3}\right) & \frac{\partial v_3}{\partial x_3} \end{bmatrix} \tag{4.6}$$

If the fluid density is constant, then conservation of mass requires that

$$\nabla \cdot \mathbf{v} = 0 \qquad \text{or} \qquad \frac{\partial v_1}{\partial x_1} + \frac{\partial v_2}{\partial x_2} + \frac{\partial v_3}{\partial x_3} = 0 \tag{4.7}$$

Comparing this to the diagonal entries in Eqn. (4.6) shows that an equivalent statement is

$$\operatorname{tr}\mathbf{D} = 0 \tag{4.8}$$

Polymer processing flows take place at nearly constant density, so we will routinely assume that $\operatorname{tr}\mathbf{D}$ is zero.

The components of the vorticity tensor **W** in index notation are

$$W_{ij} = \frac{1}{2}\left(\frac{\partial v_i}{\partial x_j} - \frac{\partial v_j}{\partial x_i}\right) \tag{4.9}$$

Writing out the details we have

$$\mathbf{W} = \begin{bmatrix} 0 & \frac{1}{2}\left(\frac{\partial v_1}{\partial x_2} - \frac{\partial v_2}{\partial x_1}\right) & \frac{1}{2}\left(\frac{\partial v_1}{\partial x_3} - \frac{\partial v_3}{\partial x_1}\right) \\ \frac{1}{2}\left(\frac{\partial v_2}{\partial x_1} - \frac{\partial v_1}{\partial x_2}\right) & 0 & \frac{1}{2}\left(\frac{\partial v_2}{\partial x_3} - \frac{\partial v_3}{\partial x_2}\right) \\ \frac{1}{2}\left(\frac{\partial v_3}{\partial x_1} - \frac{\partial v_1}{\partial x_3}\right) & \frac{1}{2}\left(\frac{\partial v_3}{\partial x_2} - \frac{\partial v_2}{\partial x_3}\right) & 0 \end{bmatrix} \tag{4.10}$$

Because of its anti-symmetry, **W** always has zeros on the main diagonal.

4.1.1.1 Invariants and the Scalar Strain Rate

The rate of deformation near a point in a flowing fluid is fully described by the tensor **D**, but it is often useful to have a scalar measure of deformation rate. How can we form a useful scalar measure of a tensor?

We want a measure we can calculate from the tensor components of **D** that will not change if we transform those tensor components to some other coordinate system, say using Eqn. (2.54). Any scalar that has this property is called an *invariant* of the tensor, because it does not change value under coordinate rotation.

A vector has one invariant: its magnitude. If we know the components v_1, v_2, v_3 of the vector **v**, we can compute its magnitude $|\mathbf{v}| = \sqrt{\mathbf{v}\cdot\mathbf{v}} = \left(v_1^2 + v_2^2 + v_3^2\right)^{1/2}$. This will have the same value no matter what coordinate system we use to express **v**.

A second-order tensor has three invariants, and there are several ways to choose them. One convenient set of invariants is [BAH87]

$$I_D = \mathrm{tr}(\mathbf{D}) = D_{ii} \tag{4.11}$$

$$II_D = \mathrm{tr}(\mathbf{D}^2) = D_{ij}D_{ji} \tag{4.12}$$

$$III_D = \mathrm{tr}(\mathbf{D}^3) = D_{ij}D_{jk}D_{ki} \tag{4.13}$$

These are called the first, second and third invariants, respectively, and the subscript denotes the tensor whose invariant is begin computed.

We cannot use the first invariant I_D to form a scalar measure of the rate of deformation, because $\mathrm{tr}\,\mathbf{D} = 0$ for a fluid with constant density. The third invariant III_D is also not useful because it equals zero for any planar flow. We can form a useful measure from the second invariant, Eqn. (4.12). We would like our measure to have the same units as the components of **D**, so the scalar magnitude of the rate of deformation $\dot{\gamma}$ is defined as

$$\dot{\gamma} \equiv \sqrt{2II_D} = \sqrt{2D_{ij}D_{ji}} \tag{4.14}$$

While $\dot{\gamma}$ is properly called the *scalar strain rate*, it is often simply called the strain rate. For any flow, $\dot{\gamma}$ provides a measure of how fast the fluid is deforming.

Equations (4.11–4.13) are not the only possible choices for tensor invariants. Another commonly used set is [Mal69]

$$I_1 = \mathrm{tr}\,\mathbf{D} \tag{4.15}$$

$$I_2 = \frac{1}{2}\left[(\mathrm{tr}\,\mathbf{D})^2 - \mathrm{tr}\left(\mathbf{D}^2\right)\right] \tag{4.16}$$

$$I_3 = \det \mathbf{D} \tag{4.17}$$

These are also referred to as the first, second and third invariants. In addition, the three eigenvalues of a tensor do not change under coordinate transformation, so they provide yet another option for the invariants. We will use both the eigenvalues of the orientation tensor and the invariants of **A** in Eqns. (4.15–4.17) to form closure approximations in Section 5.4.2.

4.1.1.2 Basic Flows

To provide some understanding of the rate-of-deformation and vorticity tensors, and to define some basic flows that we will study later, here are examples showing the velocity fields and the

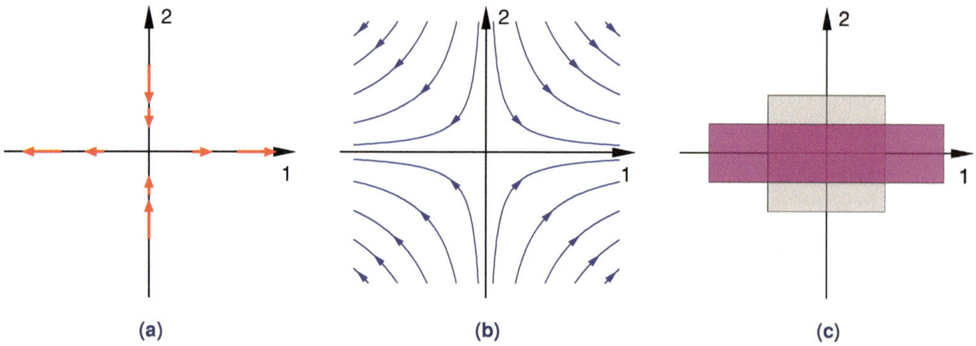

Figure 4.1 Planar extensional flow in the 1–2 plane for $\dot{\varepsilon}_P > 0$, showing (a) velocity vectors on the axes, (b) streamlines and (c) an initial fluid element (gray) and its shape at a later time (color).

tensors **L**, **D** and **W** for some important basic flows. We will study how fibers orient in these flows in later sections.

Example 4.1.1 Planar extension

Find the velocity gradient, rate-of-deformation and vorticity tensors for planar extensional flow, where the velocity field is

$$v_1 = \dot{\varepsilon}_P \, x_1 \tag{4.18}$$

$$v_2 = -\dot{\varepsilon}_P \, x_2 \tag{4.19}$$

$$v_3 = 0 \tag{4.20}$$

and $\dot{\varepsilon}_P$ is the elongation rate.

Figure 4.1(a) shows velocity vectors along the x_1 and x_2 axes: the fluid is stretching in the 1 direction and contracting in the 2 direction. The velocity is zero at the origin. The full velocity field produces the streamlines shown in Fig. 4.1(b). As shown in Fig. 4.1(c), in this flow a square of material stretches in the 1 direction and contracts in the 2 direction.

Using either Eqn. (4.2) or Eqn. (4.3), the velocity gradient tensor for this flow is

$$\mathbf{L} = \begin{bmatrix} \dot{\varepsilon}_P & 0 & 0 \\ 0 & -\dot{\varepsilon}_P & 0 \\ 0 & 0 & 0 \end{bmatrix} \tag{4.21}$$

Then, from Eqn. (4.4) or Eqns. (4.5) and (4.9), the rate-of-deformation and vorticity tensors are

$$\mathbf{D} = \begin{bmatrix} \dot{\varepsilon}_P & 0 & 0 \\ 0 & -\dot{\varepsilon}_P & 0 \\ 0 & 0 & 0 \end{bmatrix} \quad \text{and} \quad \mathbf{W} = \begin{bmatrix} 0 & 0 & 0 \\ 0 & 0 & 0 \\ 0 & 0 & 0 \end{bmatrix} \tag{4.22}$$

If $\dot{\varepsilon}_P > 0$ then the fluid is stretching along the 1 axis, which corresponds to $D_{11} > 0$. If the fluid is not changing volume then it must also be contracting along the 2 axis, and $D_{22} < 0$. Thus, the diagonal components of the rate-of-deformation tensor indicate stretching (if positive) or contraction (if negative) along the corresponding axes. Recall that $\operatorname{tr}\mathbf{D} = 0$ to maintain constant volume, and that is true in this case.

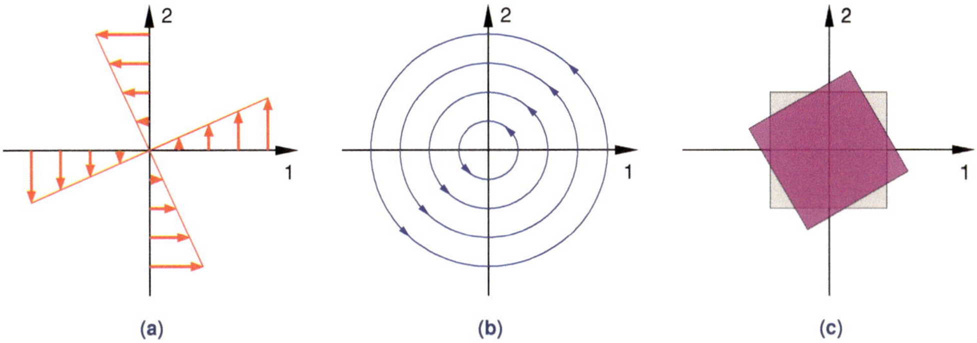

Figure 4.2 Rigid-body rotation in the 1–2 plane for $\Omega > 0$, showing (a) velocity vectors on the axes, (b) streamlines and (c) an initial fluid element (gray) and its shape at a later time (color).

Using Eqn. (4.14) the scalar strain rate is

$$\dot{\gamma} = 2\dot{\varepsilon}_P \tag{4.23}$$

All components of the vorticity tensor are zero, matching our observation that there is no rigid-body rotation in this flow. ■

Example 4.1.2 Rigid-body rotation

Find the **L**, **D** and **W** tensors for a rigid-body rotation in the 1–2 plane. The velocity field is

$$v_1 = -\Omega x_2 \tag{4.24}$$

$$v_2 = \Omega x_1 \tag{4.25}$$

$$v_3 = 0 \tag{4.26}$$

where Ω is the angular velocity.

Figure 4.2 shows the velocity vectors, streamlines and the motion of a fluid element for $\Omega > 0$. The streamlines are circular, and any fluid element rotates about the x_3 axis without changing its shape.

Equation (4.2) or Eqn. (4.3) gives the velocity gradient tensor as

$$\mathbf{L} = \begin{bmatrix} 0 & -\Omega & 0 \\ \Omega & 0 & 0 \\ 0 & 0 & 0 \end{bmatrix} \tag{4.27}$$

Equation (4.4) or Eqns. (4.5) and (4.9) then give the rate-of-deformation and vorticity tensors as

$$\mathbf{D} = \begin{bmatrix} 0 & 0 & 0 \\ 0 & 0 & 0 \\ 0 & 0 & 0 \end{bmatrix} \quad \text{and} \quad \mathbf{W} = \begin{bmatrix} 0 & -\Omega & 0 \\ \Omega & 0 & 0 \\ 0 & 0 & 0 \end{bmatrix} \tag{4.28}$$

All of the components of **D** are zero, since there is no deformation here. As a result, $\dot{\gamma} = 0$ for this flow.

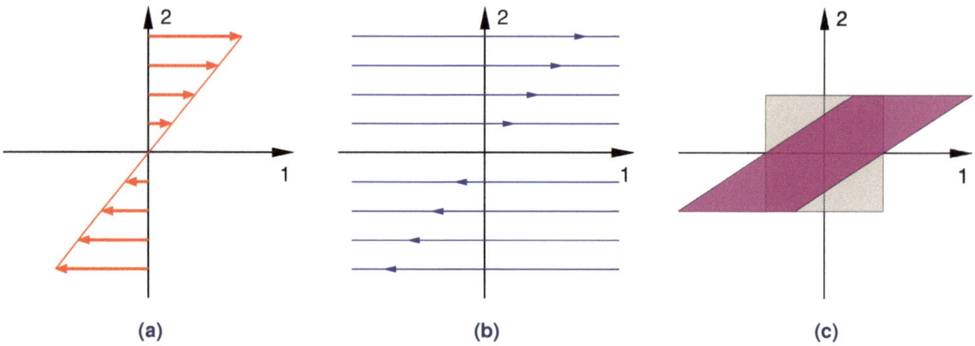

Figure 4.3 Simple shear flow for $\dot{\gamma}_S > 0$, showing (a) velocity vectors on the 2 axis, (b) streamlines and (c) an initial fluid element (gray) and its shape at a later time (color).

The magnitude of the vorticity tensor components are controlled by Ω, the angular velocity, and we see that $W_{12} = -W_{21}$ are non-zero for motion in the 1–2 plane, corresponding to rotation around the 3 axis. In a similar way, if we had rotation around the 1 axis then $W_{23} = -W_{32}$ would be non-zero, and $W_{31} = -W_{13}$ would be non-zero for rotation around the 2 axis. ■

Example 4.1.3 Simple shear flow

Find the velocity gradient, rate-of-deformation and vorticity tensors for simple shear flow. The velocity field is

$$v_1 = \dot{\gamma}_S x_2 \tag{4.29}$$

$$v_2 = 0 \tag{4.30}$$

$$v_3 = 0 \tag{4.31}$$

where $\dot{\gamma}_S$ is the shear rate. Also, find the eigenvalues and eigenvectors of **D** for this flow, and compare **D** in the principal axes to planar extension.

As Fig. 4.3 shows, the fluid moves only in the 1 direction, but with a velocity that is proportional to x_2. This produces a shearing motion, which changes the shape of fluid elements as shown in Fig. 4.3(c).

The velocity gradient tensor for simple shear is

$$\mathbf{L} = \begin{bmatrix} 0 & \dot{\gamma}_S & 0 \\ 0 & 0 & 0 \\ 0 & 0 & 0 \end{bmatrix} \tag{4.32}$$

so the rate-of-deformation and vorticity tensors are

$$\mathbf{D} = \begin{bmatrix} 0 & \dot{\gamma}_S/2 & 0 \\ \dot{\gamma}_S/2 & 0 & 0 \\ 0 & 0 & 0 \end{bmatrix} \quad \text{and} \quad \mathbf{W} = \begin{bmatrix} 0 & \dot{\gamma}_S/2 & 0 \\ -\dot{\gamma}_S/2 & 0 & 0 \\ 0 & 0 & 0 \end{bmatrix} \tag{4.33}$$

Now the off-diagonal components $D_{12} = D_{21}$ are non-zero, and in general we can interpret off-diagonal components of **D** as shearing deformations.

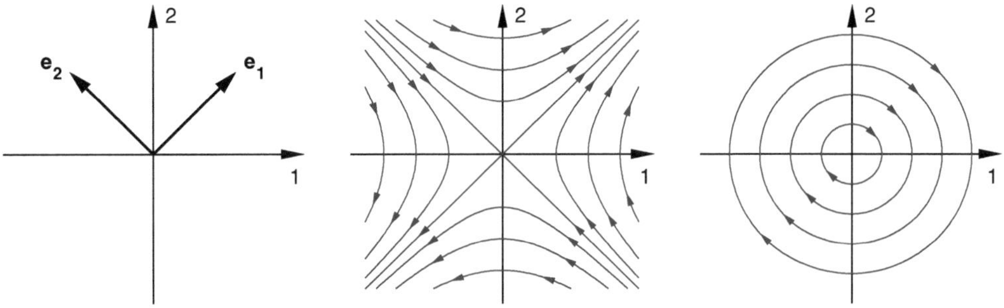

Figure 4.4 Analysis of the simple shear flow from Fig. 4.3. (a) The eigenvectors of **D**. (b) Streamlines for the motion associated with **D**, which is planar extension with stretching along the 45° axis and contraction along the -45° axis. (c) Streamlines for the motion associated with **W**, which is a clockwise rigid-body rotation. The flow in Fig. 4.3 is the sum of (b) and (c).

The scalar strain rate for this flow is

$$\dot{\gamma} = \dot{\gamma}_S \tag{4.34}$$

This is consistent with the usual definition of shear rate in a simple shearing flow: $\dot{\gamma} = \partial v_1/\partial x_2$. The factor of 2 in Eqn. (4.14) appears there to achieve this correspondence between the scalar magnitude of **D** and the shear rate in simple shear flow. In the future we will use $\dot{\gamma}$ rather than $\dot{\gamma}_S$ for the shear rate in simple shear.

The vorticity tensor **W** is also non-zero in simple shear. Comparing Eqns. (4.33) and (4.28), we see that this simple shear flow has a rigid-body rotation at an angular velocity of $-\dot{\gamma}_S/2$. This rotation is clockwise when $\dot{\gamma}_S > 0$.

An eigenvalue-eigenvector analysis of **D** from Eqn. (4.33) shows that two eigenvectors lie in the 1–2 plane at 45° and 135° to the 1 axis, and the corresponding eigenvalues are $\dot{\gamma}_S/2$ and $-\dot{\gamma}_S/2$, respectively. Figure 4.4(a) shows these eigenvectors. The third eigenvalue is zero and the corresponding eigenvector is parallel to the 3 axis.

In this principal coordinate system the rate-of-deformation tensor for simple shear flow is

$$\overline{\mathbf{D}} = \begin{bmatrix} \dot{\gamma}_S/2 & 0 & 0 \\ 0 & -\dot{\gamma}_S/2 & 0 \\ 0 & 0 & 0 \end{bmatrix} \tag{4.35}$$

This is identical to the rate-of-deformation tensor for planar extensional flow, Eqn. (4.22), if we choose $\dot{\varepsilon}_P = \dot{\gamma}_S/2$. That is, the deformation rate for simple shear is planar extension on the ±45° axes. Figure 4.4(b) shows the streamlines corresponding to this deformation by itself. However, simple shearing flow also has a non-zero vorticity, and Fig. 4.4(c) shows the streamlines for that motion. The full simple shearing flow in Fig. 4.3 is the sum of the motions in Figs. 4.4(b) and (c). ■

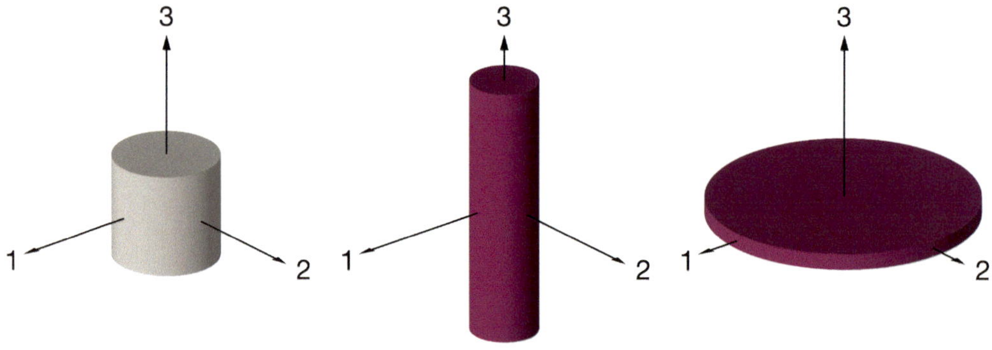

Figure 4.5 Deformation in three-dimensional extensional flows. (a) Initial shape. (b) Deformed shape in uniaxial extension. (c) Deformed shape in biaxial extension.

Example 4.1.4 Uniaxial extensional flow

Find **L**, **D** and **W** for uniaxial extensional flow, where the velocity field is

$$v_1 = -(\dot{\varepsilon}_U/2)x_1 \tag{4.36}$$

$$v_2 = -(\dot{\varepsilon}_U/2)x_2 \tag{4.37}$$

$$v_3 = \dot{\varepsilon}_U\, x_3 \tag{4.38}$$

For $\dot{\varepsilon}_U > 0$ the material stretches along the 3 axis and contracts equally in all directions in the 1–2 plane. Figure 4.5(a) shows a cylindrical initial material element and Fig. 4.5(b) shows the deformed shape at a later time.

The velocity gradient tensor for this flow is

$$\mathbf{L} = \begin{bmatrix} -\dot{\varepsilon}_U/2 & 0 & 0 \\ 0 & -\dot{\varepsilon}_U/2 & 0 \\ 0 & 0 & \dot{\varepsilon}_U \end{bmatrix} \tag{4.39}$$

and the rate-of-deformation and vorticity tensors are

$$\mathbf{D} = \begin{bmatrix} -\dot{\varepsilon}_U/2 & 0 & 0 \\ 0 & -\dot{\varepsilon}_U/2 & 0 \\ 0 & 0 & \dot{\varepsilon}_U \end{bmatrix} \quad \text{and} \quad \mathbf{W} = \begin{bmatrix} 0 & 0 & 0 \\ 0 & 0 & 0 \\ 0 & 0 & 0 \end{bmatrix} \tag{4.40}$$

Just as with planar extension, the diagonal elements of **D** describe the stretching rates along the coordinate axes: extension at a rate $\dot{\varepsilon}_U$ along the 3 axis and contraction at a rate $\dot{\varepsilon}_U/2$ along the 1 and 2 axes. The magnitude of each lateral contraction rate must be half the magnitude of the extension rate to maintain constant volume and keep $\operatorname{tr}\mathbf{D} = 0$.

The vorticity is zero in uniaxial extension, and the scalar strain rate is

$$\dot{\gamma} = \sqrt{3}\,\dot{\varepsilon}_U \tag{4.41}$$

■

Example 4.1.5 Biaxial extensional flow

Find **L**, **D** and **W** for biaxial extensional flow, where the velocity field is

$$v_1 = \dot{\varepsilon}_B\, x_1 \tag{4.42}$$

$$v_2 = \dot{\varepsilon}_B\, x_2 \tag{4.43}$$

$$v_3 = -2\dot{\varepsilon}_B\, x_3 \tag{4.44}$$

In biaxial extension the material is stretched at a rate $\dot{\varepsilon}_B$ along the 1 and 2 axes (and in all directions in the 1–2 plane), while it contracts along the 3 axis. This type of flow can occur in compression molding. Figure 4.5(c) shows the deformed shape of the initial material element in Fig. 4.5(a) for this flow.

The velocity gradient tensor is

$$\mathbf{L} = \begin{bmatrix} \dot{\varepsilon}_B & 0 & 0 \\ 0 & \dot{\varepsilon}_B & 0 \\ 0 & 0 & -2\dot{\varepsilon}_B \end{bmatrix} \tag{4.45}$$

and the rate-of-deformation and vorticity are

$$\mathbf{D} = \begin{bmatrix} \dot{\varepsilon}_B & 0 & 0 \\ 0 & \dot{\varepsilon}_B & 0 \\ 0 & 0 & -2\dot{\varepsilon}_B \end{bmatrix} \quad \text{and} \quad \mathbf{W} = \begin{bmatrix} 0 & 0 & 0 \\ 0 & 0 & 0 \\ 0 & 0 & 0 \end{bmatrix} \tag{4.46}$$

Again, the diagonal elements of **D** show the stretching rates along the coordinate axes, and $\text{tr}\,\mathbf{D} = 0$. This flow also has zero vorticity, and the scalar strain rate is

$$\dot{\gamma} = 2\sqrt{3}\,\dot{\varepsilon}_B \tag{4.47}$$

■

4.1.2 Finite Deformation Measures

The rate-of-deformation and vorticity tensors will be the primary tools we use to calculate fiber orientation. However, because **D** and **W** describe instantaneous *rates* of shape change and rotation, they only tell us about the shape and angles an infinitesimal time later. Occasionally it will be useful to calculate changes over a finite amount of time in a single step. To do this we will need to describe finite amounts of deformation.

Finite deformation measures describe the difference between two configurations of a body, and these two states can be separated by any amount of time. We will use time $t = 0$ at the initial state or *reference configuration*, when the material has not deformed. The position of any point in the material in the initial state, say point P in Fig. 4.6(a), is denoted by $\mathbf{x}'$ or x'_j. At some later time t the material is in the *deformed configuration*, and the position of any material point in this configuration is $\mathbf{x}$ or x_i.

4.1.2.1 Displacement Functions

We can specify the motion of the material by specifying the **x** coordinates as a function of $\mathbf{x}'$ and t. These functions are called *displacement functions*, and specifying them is equivalent to specifying the velocity distribution $\mathbf{v}(\mathbf{x}, t)$.

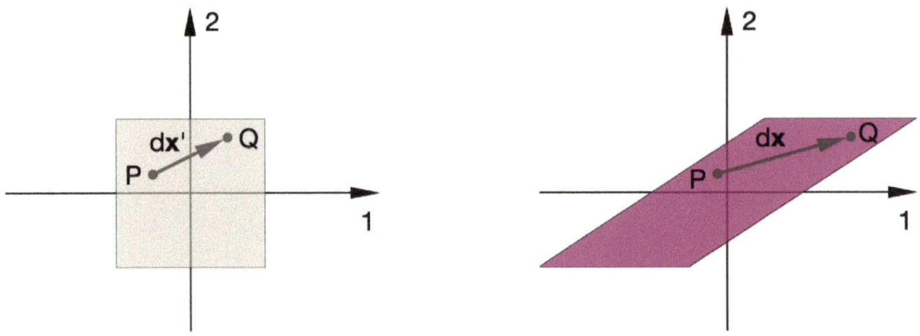

(a) Initial (b) Deformed

Figure 4.6 A body in an initial configuration at time = 0 and a deformed configuration at time = t. P and Q are material points that are connected by the material vector $d\mathbf{x}'$ in the initial state and $d\mathbf{x}$ in the deformed state.

The deformation in Fig. 4.6 is simple shearing, just as in Fig. 4.3. In this case it is easy to look at the velocity distributions in Eqns. (4.29–4.31) and write the displacement functions:

$$x_1 = x_1' + \dot{\gamma} t x_2' \tag{4.48}$$

$$x_2 = x_2' \tag{4.49}$$

$$x_3 = x_3' \tag{4.50}$$

Consistent with Fig. 4.3, any material point retains its original x_2 and x_3 coordinates, and its x_1 coordinate changes with time in proportion to x_2.

The displacement functions can be inverted to give the initial position $\mathbf{x}'$ as a function of the position $\mathbf{x}$ at time t. For our simple-shear example the inversion is straightforward:

$$x_1' = x_1' - \dot{\gamma} t x_2 \tag{4.51}$$

$$x_2' = x_2 \tag{4.52}$$

$$x_3' = x_3 \tag{4.53}$$

For any type of deformation, if the velocity field is known then the displacement functions can be found by integrating the differential equation

$$\dot{\mathbf{x}} = \mathbf{v}(\mathbf{x}, t) \qquad \text{with} \qquad \mathbf{x}(t = 0) = \mathbf{x}' \tag{4.54}$$

In this book a dot over a variable, like $\dot{\mathbf{x}}$, means the time derivative of $\mathbf{x}$ following a material point, and is called a *material derivative*. Equation (4.54) says that the velocity $\mathbf{v}$ of any material point P is the material derivative of $\mathbf{x}$ at that point.

4.1.2.2 Deformation Gradient Tensors

Once the displacement functions are known, we begin to analyze the deformation by finding the *deformation gradient tensor* $\mathbf{F}$. This contains the derivatives of $\mathbf{x}$ with respect to $\mathbf{x}'$:

$$\mathbf{F} = \frac{\partial \mathbf{x}}{\partial \mathbf{x}'} \qquad \text{or} \qquad F_{ij} = \frac{\partial x_i}{\partial x_j'} \tag{4.55}$$

We will also use the *inverse deformation gradient tensor* $\mathbf{F}^{-1}$, which gives the derivatives of $\mathbf{x}'$ with respect to $\mathbf{x}$:

$$\mathbf{F}^{-1} = \frac{\partial \mathbf{x}'}{\partial \mathbf{x}} \qquad \text{or} \qquad F_{ij}^{-1} = \frac{\partial x_i'}{\partial x_j} \tag{4.56}$$

The inverse deformation gradient can be computed from the inverted displacement functions (e.g., Eqns. (4.51–4.53)), or by taking the tensor inverse of $\mathbf{F}$, since $\mathbf{F}\cdot\mathbf{F}^{-1} = \mathbf{I}$.

For our simple-shear example, the deformation gradient and inverse deformation gradient tensors are

$$\mathbf{F} = \begin{bmatrix} 1 & \dot{\gamma} t & 0 \\ 0 & 1 & 0 \\ 0 & 0 & 1 \end{bmatrix} \qquad \text{and} \qquad \mathbf{F}^{-1} = \begin{bmatrix} 1 & -\dot{\gamma} t & 0 \\ 0 & 1 & 0 \\ 0 & 0 & 1 \end{bmatrix} \tag{4.57}$$

In this example $\mathbf{F}$ and $\mathbf{F}^{-1}$ do not depend on $\mathbf{x}$ or $\mathbf{x}'$, indicating that this is a uniform deformation. In general the deformation gradient and its inverse will depend on position.

When $t = 0$ and there is no deformation, both deformation gradient tensors equal the identity tensor:

$$\mathbf{F}(t=0) = \mathbf{I} \qquad \text{and} \qquad \mathbf{F}^{-1}(t=0) = \mathbf{I} \tag{4.58}$$

This holds for any deformation.

Now consider two neighboring material points P and Q, as shown in Fig. 4.6. In the initial configuration they are connected by a material vector $\mathrm{d}\mathbf{x}'$, and at time t that vector has transformed to $\mathrm{d}\mathbf{x}$. These two vectors are related by the deformation gradient:

$$\mathrm{d}\mathbf{x} = \mathbf{F}\cdot\mathrm{d}\mathbf{x}' \qquad \text{or} \qquad \mathrm{d}x_i = F_{ij}\,\mathrm{d}x_j' \tag{4.59}$$

Equation (4.59) is an important equation for fiber orientation, and we will refer to it as the *deformation of a material line*. Here is an example for simple shear flow.

Example 4.1.6 Material vectors in finite simple shear

For simple shear deformation, find the material vectors $\mathrm{d}\mathbf{x}$ when $\mathrm{d}\mathbf{x}' = [1,0,0]^T$ and $[0,1,0]^T$.

Figure 4.7 shows these two material vectors in both their initial and deformed configurations. Equation (4.57) gives the deformation gradient tensor for simple shear, and for simplicity we use $\gamma = \dot{\gamma} t$, which is the total shear strain. Now apply Eqn. (4.59) for the two cases.

The first $\mathrm{d}\mathbf{x}'$ vector points in the 1 direction, and we have

$$\mathrm{d}\mathbf{x} = \mathbf{F}\cdot\mathrm{d}\mathbf{x}' = \begin{bmatrix} 1 & \gamma & 0 \\ 0 & 1 & 0 \\ 0 & 0 & 1 \end{bmatrix} \begin{Bmatrix} 1 \\ 0 \\ 0 \end{Bmatrix} = \begin{Bmatrix} 1 \\ 0 \\ 0 \end{Bmatrix} \tag{4.60}$$

This material vector does not change either its length or its orientation.

The second $\mathrm{d}\mathbf{x}'$ vector points in the 2 direction, and we have

$$\mathrm{d}\mathbf{x} = \mathbf{F}\cdot\mathrm{d}\mathbf{x}' = \begin{bmatrix} 1 & \gamma & 0 \\ 0 & 1 & 0 \\ 0 & 0 & 1 \end{bmatrix} \begin{Bmatrix} 0 \\ 1 \\ 0 \end{Bmatrix} = \begin{Bmatrix} \gamma \\ 1 \\ 0 \end{Bmatrix} \tag{4.61}$$

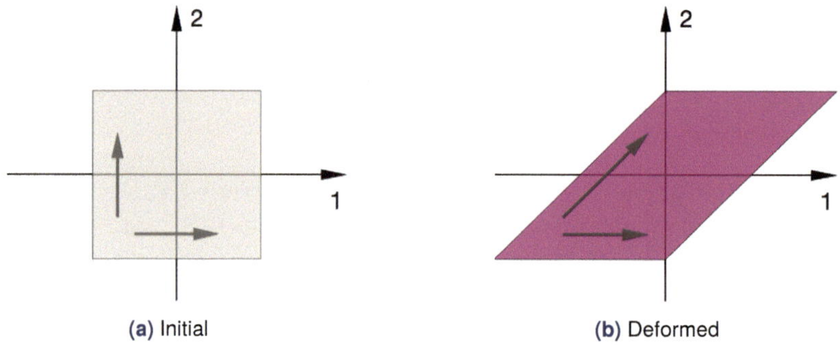

Figure 4.7 Initial and deformed configurations of a body in simple shear for $\gamma = 1$, and the two material vectors from Example 4.1.6.

This material vector has changed orientation, and its length has increased from unity to $\sqrt{1+\gamma^2}$. Deformation can cause a material vector to change length, as well as orientation.

Note that the deformation is uniform in this example (i.e., $\mathbf{F}$ does not change with position) so the initial positions of the vectors do not affect the results. ■

The inverse deformation gradient $\mathbf{F}^{-1}$ allows us to determine the original vector $\mathrm{d}\mathbf{x}'$ from the deformed vector $\mathrm{d}\mathbf{x}$:

$$\mathrm{d}\mathbf{x}' = \mathbf{F}^{-1} \cdot \mathrm{d}\mathbf{x} \qquad \text{or} \qquad \mathrm{d}x'_i = F^{-1}_{ij}\,\mathrm{d}x_j \tag{4.62}$$

If a material has constant density, then the determinants of both deformation gradient tensors equal unity:

$$\det \mathbf{F} = 1 \qquad\qquad \det \mathbf{F}^{-1} = 1 \tag{4.63}$$

The deformation gradient tensor can also be found using the velocity gradient tensor, since

$$\dot{\mathbf{F}} = \mathbf{L} \cdot \mathbf{F} \quad \text{or} \quad \dot{F}_{ij} = L_{ik} F_{kj} \qquad \text{with} \qquad \mathbf{F} = \mathbf{I} \quad \text{or} \quad F_{ij} = \delta_{ij} \quad \text{at} \quad t = 0 \tag{4.64}$$

Again, the dot in $\dot{\mathbf{F}}$ indicates a time derivative following a material point.

4.1.2.3 Stretching of a Material Line

Suppose that we know $\mathrm{d}\mathbf{x}$ and want to find how much the material in that direction had been stretched from its original length. Let $\mathrm{d}\ell$ be the length of $\mathrm{d}\mathbf{x}$, and let $\mathrm{d}\ell'$ be the length of $\mathrm{d}\mathbf{x}'$. These lengths are given by

$$(\mathrm{d}\ell')^2 = \mathrm{d}\mathbf{x}' \cdot \mathrm{d}\mathbf{x}' \qquad \text{and} \qquad (\mathrm{d}\ell)^2 = \mathrm{d}\mathbf{x} \cdot \mathrm{d}\mathbf{x} \tag{4.65}$$

We want to find the ratio of these two lengths, $\mathrm{d}\ell' / \mathrm{d}\ell$, which is called the *stretch ratio.* A stretch ratio greater than unity means that the material vector $\mathrm{d}\mathbf{x}$ is longer than $\mathrm{d}\mathbf{x}'$, and a ratio less than unity means that $\mathrm{d}\mathbf{x}$ is shorter than $\mathrm{d}\mathbf{x}'$.

Divide the second part of Eqn. (4.65) by the first part and use Eqn. (4.62) in the denominator.

$$\left(\frac{\mathrm{d}\ell}{\mathrm{d}\ell'}\right)^2 = \frac{\mathrm{d}\mathbf{x} \cdot \mathrm{d}\mathbf{x}}{\mathrm{d}\mathbf{x} \cdot \left(\mathbf{F}^{-T} \cdot \mathbf{F}^{-1}\right) \cdot \mathrm{d}\mathbf{x}} \tag{4.66}$$

$\mathbf{F}^{-T}$ is the transpose of $\mathbf{F}^{-1}$ and we have used $\left(\mathbf{F}^{-1}\cdot\mathbf{p}\right)^T = \mathbf{p}\cdot\mathbf{F}^{-T}$.

We can specify the direction in the deformed configuration using a unit vector $\mathbf{p} = \mathrm{d}\mathbf{x}/\,\mathrm{d}\ell$. Divide both the numerator and denominator on the right-hand side of Eqn. (4.66) by $(\mathrm{d}\ell)^2$ and use $\mathbf{p}\cdot\mathbf{p} = 1$, since $\mathbf{p}$ is a unit vector. This gives

$$\left(\frac{\mathrm{d}\ell}{\mathrm{d}\ell'}\right)^2 = \frac{1}{\mathbf{p}\cdot\left(\mathbf{F}^{-T}\cdot\mathbf{F}^{-1}\right)\cdot\mathbf{p}} \tag{4.67}$$

The quantity in parentheses in the denominator of Eqn. (4.67) is a tensor that allows us to compute the stretching or contraction of a material line along any direction in the deformed configuration. This tensor provides a useful measure of deformation, and it is called the *Piola tensor*[2]. We will give it the symbol $\mathbf{B}^{-1}$ and its definition is

$$\mathbf{B}^{-1} = \mathbf{F}^{-T}\cdot\mathbf{F}^{-1} \qquad \text{or} \qquad B^{-1}_{ij} = F^{-1}_{ki}F^{-1}_{kj} = \frac{\partial x'_k}{\partial x_i}\frac{\partial x'_k}{\partial x_j} \tag{4.68}$$

Using this in Eqn. (4.67) gives the final result for the stretch ratio of a material line that is oriented along the direction $\mathbf{p}$ in the deformed configuration:

$$\frac{\mathrm{d}\ell}{\mathrm{d}\ell'} = \frac{1}{\left(\mathbf{p}\cdot\mathbf{B}^{-1}\cdot\mathbf{p}\right)^{1/2}} \tag{4.69}$$

The Piola tensor will appear in the natural closure approximation and in the Jeffery distribution function. Just like the deformation gradient tensors, $\mathbf{B}^{-1} = \mathbf{I}$ when $t = 0$ and there is no deformation.

For completeness, a closely related deformation tensor is the Green tensor[3] $\mathbf{B}$, defined as

$$\mathbf{B} = \mathbf{F}\cdot\mathbf{F}^T \qquad \text{or} \qquad B_{ij} = F_{ik}F_{jk} = \frac{\partial x_i}{\partial x'_k}\frac{\partial x_j}{\partial x'_k} \tag{4.70}$$

As the notation suggests, $\mathbf{B}^{-1}$ is the inverse of $\mathbf{B}$.

If a material has constant density, then the determinants of the Piola and Green tensors also equal unity:

$$\det\mathbf{B} = 1 \qquad\qquad \det\mathbf{B}^{-1} = 1 \tag{4.71}$$

4.1.2.4 Basic Deformations

For simple shearing deformation, using the deformation gradient tensors from Eqn. (4.57) in Eqns. (4.70) and (4.68) gives the Green and Piola tensors for simple shear. Again using $\gamma = \dot{\gamma}t$, these are:

$$\mathbf{B} = \begin{bmatrix} 1+\gamma^2 & \gamma & 0 \\ \gamma & 1 & 0 \\ 0 & 0 & 1 \end{bmatrix} \qquad \text{and} \qquad \mathbf{B}^{-1} = \begin{bmatrix} 1 & \gamma & 0 \\ \gamma & 1+\gamma^2 & 0 \\ 0 & 0 & 1 \end{bmatrix} \tag{4.72}$$

Here are a few more examples of the finite deformation tensors.

---

2 This name for the Piola tensor is recommended by IUPAC [KSW+98]. Some texts call this the Almansi tensor.

3 IUPAC [KSW+98] also recommends this name for the Green tensor, though many texts call it the Finger tensor.

Example 4.1.7 Finite planar extension

Find $\mathbf{F}$, $\mathbf{F}^{-1}$, $\mathbf{B}$ and $\mathbf{B}^{-1}$ for planar extensional deformation. The displacement functions are

$$x_1 = x'_1\, e^{\dot{\varepsilon}_P t} \tag{4.73}$$

$$x_2 = x'_2\, e^{-\dot{\varepsilon}_P t} \tag{4.74}$$

$$x_3 = x'_3 \tag{4.75}$$

Note that the x_1 coordinate of a material point must increase exponentially with time to maintain a constant stretching rate $\dot{\varepsilon}_P$. Equations (4.73–4.75) can be obtained by substituting the velocity expressions for planar extension, Eqns. (4.18–4.20), into Eqn. (4.54) and integrating.

Using the definitions Eqns. (4.55) and (4.56), the deformation gradient tensors are

$$\mathbf{F} = \begin{bmatrix} e^{\dot{\varepsilon}_P t} & 0 & 0 \\ 0 & e^{-\dot{\varepsilon}_P t} & 0 \\ 0 & 0 & 1 \end{bmatrix} \quad \text{and} \quad \mathbf{F}^{-1} = \begin{bmatrix} e^{-\dot{\varepsilon}_P t} & 0 & 0 \\ 0 & e^{\dot{\varepsilon}_P t} & 0 \\ 0 & 0 & 1 \end{bmatrix} \tag{4.76}$$

Then, using Eqns. (4.70) and (4.68), the Green and Piola tensors are

$$\mathbf{B} = \begin{bmatrix} e^{2\dot{\varepsilon}_P t} & 0 & 0 \\ 0 & e^{-2\dot{\varepsilon}_P t} & 0 \\ 0 & 0 & 1 \end{bmatrix} \quad \text{and} \quad \mathbf{B}^{-1} = \begin{bmatrix} e^{-2\dot{\varepsilon}_P t} & 0 & 0 \\ 0 & e^{2\dot{\varepsilon}_P t} & 0 \\ 0 & 0 & 1 \end{bmatrix} \tag{4.77}$$

In this example we see that the deformation gradient tensor $\mathbf{F}$ reflects the stretching of the material along the 1 axis (which is exponential in time) and the corresponding contraction along the 2 axis. The diagonal elements of the $\mathbf{B}$ tensor are the squares of these stretch ratios, and the diagonal elements of $\mathbf{B}^{-1}$ are the reciprocals of the squared stretch ratios, which is consistent with Eqn. (4.69). ■

Example 4.1.8 Finite rigid-body rotation

Find $\mathbf{F}$, $\mathbf{F}^{-1}$, $\mathbf{B}$ and $\mathbf{B}^{-1}$ for rigid-body rotation. The displacement functions that correspond to the velocities in Eqns. (4.24–4.26) are

$$x_1 = x'_1 \cos(\Omega t) - x'_2 \sin(\Omega t) \tag{4.78}$$

$$x_2 = x'_1 \sin(\Omega t) + x'_2 \cos(\Omega t) \tag{4.79}$$

$$x_3 = x'_3 \tag{4.80}$$

Recall that Ω is the rotation rate about the 3 axis, so Ωt is the angle by which the material has rotated at time t.

Using Eqns. (4.55) and (4.56), the deformation gradient tensors are

$$\mathbf{F} = \begin{bmatrix} \cos(\Omega t) & -\sin(\Omega t) & 0 \\ \sin(\Omega t) & \cos(\Omega t) & 0 \\ 0 & 0 & 1 \end{bmatrix} \quad \text{and} \quad \mathbf{F}^{-1} = \begin{bmatrix} \cos(\Omega t) & \sin(\Omega t) & 0 \\ -\sin(\Omega t) & \cos(\Omega t) & 0 \\ 0 & 0 & 1 \end{bmatrix} \tag{4.81}$$

Using this $\mathbf{F}$ tensor in Eqn. (4.59) gives a vector $\mathbf{dx}$ that is the vector $\mathbf{dx}'$ rotated by an angle Ωt (counter-clockwise if $\Omega t > 0$). Similarly, $\mathbf{F}^{-1}$ rotates any $\mathbf{x}$ in the opposite direction by Ωt to recover the original $\mathbf{dx}'$. The deformation gradient tensor $\mathbf{F}$ and its inverse contain any rigid-body rotation that is part of the motion.

Now, using Eqns. (4.70) and (4.68), the Green and Piola tensors are

$$\mathbf{B} = \begin{bmatrix} 1 & 0 & 0 \\ 0 & 1 & 0 \\ 0 & 0 & 1 \end{bmatrix} \quad \text{and} \quad \mathbf{B}^{-1} = \begin{bmatrix} 1 & 0 & 0 \\ 0 & 1 & 0 \\ 0 & 0 & 1 \end{bmatrix} \tag{4.82}$$

Both $\mathbf{B}$ and $\mathbf{B}^{-1}$ equal the identity tensor $\mathbf{I}$, indicating that there is no change of shape and no stretching along any direction. This is consistent with the idea of a *rigid*-body rotation. ■

These examples show that the deformation gradient tensors $\mathbf{F}$ and $\mathbf{F}^{-1}$ contain information about both stretching and rotation, but that the Piola tensor $\mathbf{B}^{-1}$ and the Green tensor $\mathbf{B}$ contain information only about stretching. More precisely, the Piola and Green tensors reveal stretching along any material direction in the deformed configuration, regardless of the amount of rigid-body rotation between the reference and deformed configurations.

■ 4.2 Jeffery's Equation

In 1922, English mathematician George Barker Jeffery published a paper concerning the motion of ellipsoidal particles immersed in a viscous fluid [Jef22]. Jeffery was interested in exact solutions to physical problems, and he developed an exact solution for the flow of a Newtonian fluid around a small solid particle having an ellipsoidal shape. His solution applies when the particle is small enough and the fluid viscosity large enough that inertial effects may be neglected. By requiring the particle to experience zero net force and zero net torque, Jeffery derived expressions for the translation and rotation of the particle. This result is the foundation for all modeling of flow-induced fiber orientation, and it reveals the main features of fiber orientation in any flow field.

4.2.1 Rate Form

An ellipsoid is a three-dimensional shape whose surface is given by

$$\frac{x_1^2}{a^2} + \frac{x_2^2}{b^2} + \frac{x_3^2}{c^2} = 1 \tag{4.83}$$

where a, b and c are the semi-axis lengths of the ellipsoid. In this expression the principal axes of the ellipsoid are aligned with the coordinate axes, but the shape can be rotated to any orientation.

While Jeffery's solution includes general ellipsoids, $a \neq b \neq c$, we will focus on *spheroids*, or ellipsoids of revolution, where $b = c$. These shapes are axisymmetric about the axis corresponding to a, and we can describe their orientation with a unit vector $\mathbf{p}$. Figures 4.8(a) and (b) show two examples. We will be primarily concerned with slender, rod-like shapes where $a > b$ (prolate spheroids), but Jeffery's equation also applies to disk-like shapes where $a < b$ (oblate spheroids).

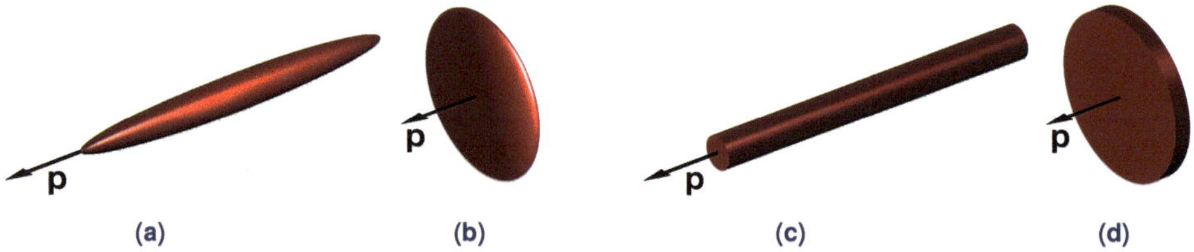

Figure 4.8 Particles that follow Jeffery's equation: (a) prolate spheroid, $a > b$; (b) oblate spheroid, $a < b$; (c) cylinder; (d) disk.

Jeffery also assumed that the fluid velocity in the vicinity of the particle varies linearly with position, so the velocity gradient **L** seen by the particle is uniform.

The first result of Jeffery is that the centroid of the particle moves with the unperturbed fluid velocity at that point. If we let $\mathbf{x}_c$ denote the centroid position, then

$$\dot{\mathbf{x}}_c = \mathbf{v}(\mathbf{x}_c) \tag{4.84}$$

Here the dot indicates a time derivative following the particle. Thus, the first insight from Jeffery's analysis is:

In Jeffery's solution, particle centroids translate with the fluid.

Jeffery also solved for the rotation of the particle, and his solution gives the rate of change of **p** in what we now call Jeffery's equation:

$$\dot{\mathbf{p}} = \mathbf{W}\cdot\mathbf{p} + \xi\left(\mathbf{D}\cdot\mathbf{p} - \mathbf{D}:\mathbf{ppp}\right) \quad \text{or} \quad \dot{p}_i = W_{ij}p_j + \xi\left(D_{ij}p_j - D_{jk}p_k p_j p_i\right) \tag{4.85}$$

Here **W** and **D** are the vorticity and rate-of-deformation tensors from Section 4.1 and ξ is a constant called the *particle shape factor*. It is given by

$$\xi = \frac{(a/b)^2 - 1}{(a/b)^2 + 1} \tag{4.86}$$

The particle shape factor depends on the ratio of the axis lengths a/b, but not on the absolute size of the particle. We will examine the effects of particle shape in Section 4.2.4.

The first term on the right-hand side of Eqn. (4.85) represents the effect of rigid-body rotation on fiber orientation. If the fluid motion has a non-zero rotation rate, then this term adds that to the rotation rate of the fiber.

The second term on the right-hand side of Eqn. (4.85) represents the effect of fluid deformation on orientation. If the particle is fiber-shaped ($a > b$) then $\xi > 1$ and this term pulls the fiber toward the direction of maximum stretching rate. The $\mathbf{D}:\mathbf{ppp}$ term may look unfamiliar, but recall that **pp** is a tensor, so $\mathbf{D}:\mathbf{pp} = D_{jk}p_k p_j$ is a double-dot product of two tensors, which produces a scalar. This scalar multiplies the final vector **p**, so $(\mathbf{D}:\mathbf{pp})\mathbf{p}$ is a vector that is parallel to **p**. This term subtracts off the portion of $\mathbf{D}\cdot\mathbf{p}$ that is parallel to **p**, ensuring that $\dot{\mathbf{p}}$ is always perpendicular to **p** and that **p** does not change length.

The form of Jeffery's equation that gives the time derivative of the orientation vector $\mathbf{p}$ is the one we will use for most purposes. If, instead, one wants the angular velocity vector $\boldsymbol{\omega}$ of the fiber, that is related to $\dot{\mathbf{p}}$ by

$$\boldsymbol{\omega} = \mathbf{p} \times \dot{\mathbf{p}} \tag{4.87}$$

4.2.2 Example Solutions and the Two Rules of Fiber Orientation

To build some understanding of Jeffery's equation, we next examine some analytical solutions to Eqn. (4.85). To keep the mathematics simple, we will consider planar orientation in the 1–2 plane ($p_3 = 0$) and planar flows ($v_3 = 0$ and neither v_1 nor v_2 vary in the 3 direction). For planar orientation, the direction of the fiber is completely described by the angle ϕ.

These solutions will illustrate two general rules of flow-induced fiber orientation, which will also hold for three-dimensional flows and suspensions with multiple, interacting fibers.

Using the assumptions of planar orientation and planar flow in Eqn. (4.85), one can derive an equation for the time derivative of ϕ in any planar flow (see Appendix B.1):

$$\dot{\phi} = -W_{12} - \xi D_{11} \sin(2\phi) + \xi D_{12} \cos(2\phi) \tag{4.88}$$

In this section we will use $\xi = 1$, which is the limit for a very slender rod-like particle, $a/b \rightarrow \infty$. The corresponding solutions for very thin disk-shaped particles would use $\xi = -1$.

4.2.2.1 Planar Orientation in Simple Shear

First, consider simple shear flow in the 1–2 plane, as in Example 4.1.3. For this flow $W_{12} = D_{12} = \dot{\gamma}/2$ and $D_{11} = 0$, so Eqn. (4.88) simplifies to

$$\dot{\phi} = -\dot{\gamma} \sin^2 \phi \tag{4.89}$$

We can see that $\phi = 0$ and $\phi = 180°$ are steady-state orientations, since $\dot{\phi} = 0$ there. The particle rotates most rapidly at $\phi = 90°$, and more slowly as it approaches $\phi = 0$.

This equation is readily integrated to give $\phi(t)$. Using the initial condition that $\phi = \phi_0$ at $t = 0$, we find

$$\cot \phi = \cot \phi_0 + \dot{\gamma} t \tag{4.90}$$

This solution is graphed as the red line in Fig. 4.9 for $\phi_0 = 170°$, and Fig. 4.10 shows the orientations at several times for a range of initial conditions. Regardless of the initial orientation, all fibers eventually approach a steady-state orientation in the flow direction, though for a fiber with ϕ_0 close to 180° this could take a long time.

These results illustrate the first rule of fiber orientation:

Rule 1: Shearing flows align fibers in the **flow** direction.

A subtlety of simple shear flow for $\xi = 1$ is that, while $\phi = 0$ is a steady-state solution, its stability is one-sided. If a fiber at $\phi = 0$ is perturbed to a positive value of ϕ, it returns asymptotically to $\phi = 0$. However, if a fiber is perturbed to a negative value of ϕ, it will rotate clockwise away from $\phi = 0$ and settle asymptotically at $\phi = 180°$.

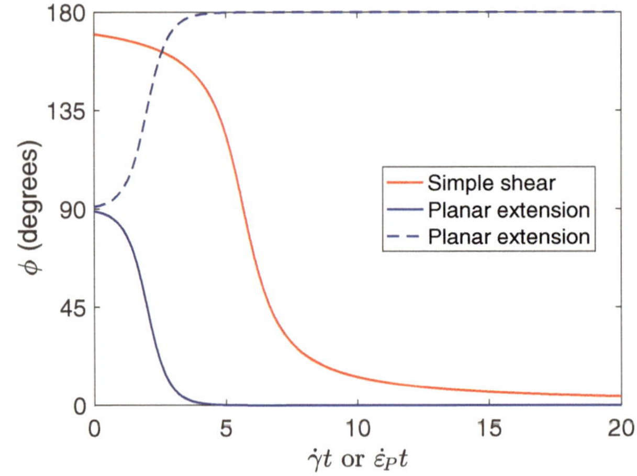

Figure 4.9 Solutions of Jeffery's equation for a particle with $\xi = 1$ in simple shear flow and planar extension. Initial orientations are 170° for simple shear and 89° (solid line) or 91° (dashed line) for planar extension.

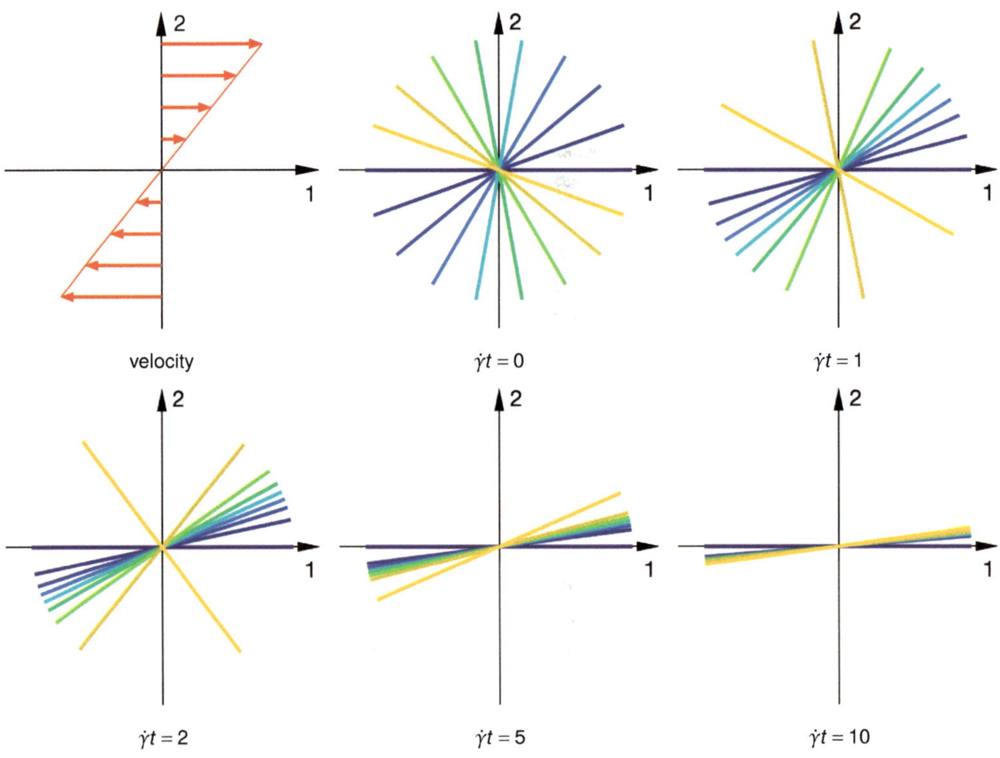

Figure 4.10 Orientation history of Jeffery fibers with $\xi = 1$ in simple shear flow with a shear rate of $\dot{\gamma}$. Each color represents a different initial orientation and all fibers lie in the 1–2 plane.

4.2.2.2 Planar Orientation in Planar Extensional Flow

Next, consider planar extensional flow in the 1–2 plane, as in Example 4.1.1. In this flow $W_{12} = D_{12} = 0$ and $D_{11} = \dot{\varepsilon}_P$, the extension rate. Now Eqn. (4.88) simplifies to

$$\dot{\phi} = -\dot{\varepsilon}_P \sin(2\phi) \tag{4.91}$$

Again this is readily integrated to give ϕ as a function of time. Using ϕ_0 as the initial condition, the solution is

$$\cot\phi = \cot\phi_0 \; e^{2\dot{\varepsilon}_P t} \tag{4.92}$$

The orientations of particles with different initial conditions are shown in Fig. 4.11 at selected times, and the solutions for two different initial orientations are plotted in blue in Fig. 4.9.

Just as with simple shear flow, $\phi = 0$ and $\phi = 180°$ are steady-state solutions, and all fibers asymptotically approach these orientations. However, unlike simple shear flow, these steady states are fully stable. A fiber at $\phi = 0$ or $\phi = 180°$ can be perturbed in either direction and will return asymptotically to its stable orientation.

From Eqn. (4.91) we can see that $\phi = 90°$ and $\phi = 270°$ are also steady-state solutions (i.e., $\dot{\phi} = 0$ there), but these steady states are *unstable*. A small perturbation in ϕ to an angle smaller than

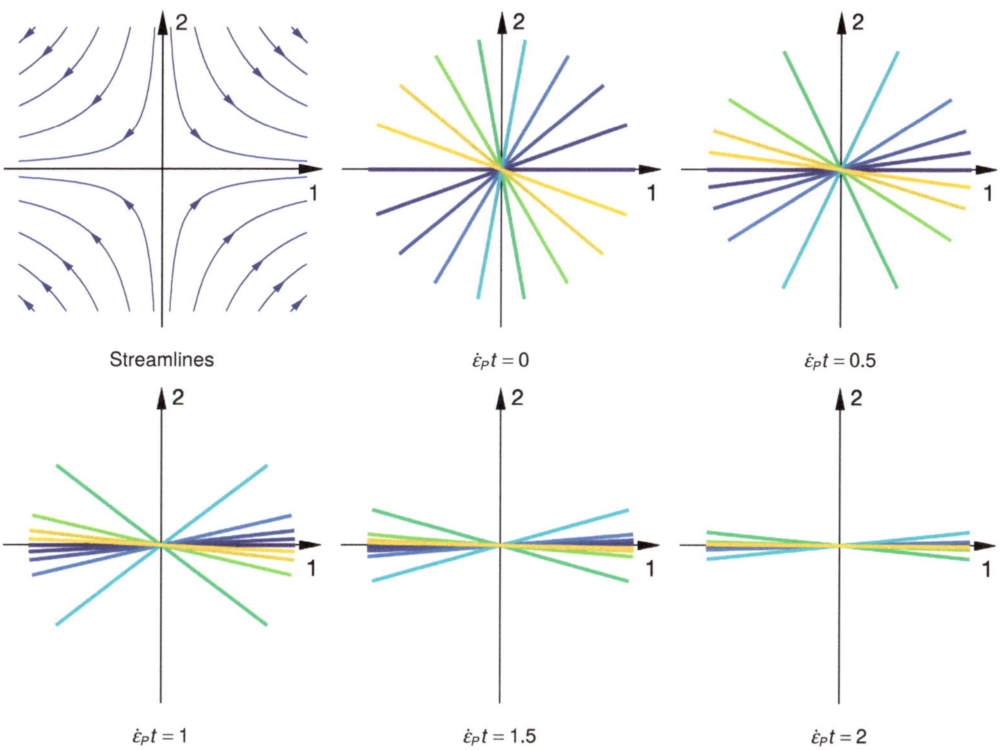

Figure 4.11 Orientation history of Jeffery fibers with $\xi = 1$ in planar extensional flow with elongation rate $\dot{\varepsilon}_P$. Each color represents a different initial orientation and all fibers lie in the 1–2 plane.

90° causes the fiber to move toward $\phi = 0$, and a small perturbation to an angle larger than 90° causes the fiber to move toward $\phi = 180°$. Figure 4.9 illustrates this by showing solutions for initial angles of 89° and 91°. This small difference in initial condition produces a large change in the trajectory of the orientation. However, in both cases the particle goes to one of the stable steady-state orientations.

This example illustrates the second rule of flow-induced fiber orientation:

Rule 2: Stretching flows align fibers in the **stretching** direction.

It is important to remember that the direction of maximum stretching is not necessarily the direction of flow. We will see an important example of this in Section 4.3.

4.2.3 Numerical Solutions

While there are additional analytical solutions to Jeffery's equation for uniform flows (e.g., [TM51]), for computer-aided engineering we are usually interested in flows where the velocity gradient seen by a particle varies with time. For those cases, Jeffery's equation is solved numerically.

Numerical solution methods work with vectors or tensors by solving for their scalar components. In that context, Eqn. (4.85) for a single fiber can be viewed as a set of three ordinary differential equations for the three components of $\mathbf{p}$. Written in matrix form, these are

$$\begin{aligned}\frac{\mathrm{d}}{\mathrm{d}t}\begin{Bmatrix} p_1 \\ p_2 \\ p_3 \end{Bmatrix} &= \begin{bmatrix} W_{11} & W_{12} & W_{13} \\ W_{21} & W_{22} & W_{23} \\ W_{31} & W_{32} & W_{33} \end{bmatrix}\begin{Bmatrix} p_1 \\ p_2 \\ p_3 \end{Bmatrix} \\ &+ \xi\left(\begin{bmatrix} D_{11} & D_{12} & D_{13} \\ D_{21} & D_{22} & D_{23} \\ D_{31} & D_{32} & D_{33} \end{bmatrix}\begin{Bmatrix} p_1 \\ p_2 \\ p_3 \end{Bmatrix} - \left(\sum_{i=1}^{3}\sum_{j=1}^{3} D_{ij}p_i p_j\right)\begin{Bmatrix} p_1 \\ p_2 \\ p_3 \end{Bmatrix}\right)\end{aligned} \tag{4.93}$$

On the second line, the double-sum term in parentheses is $\mathbf{D}:\mathbf{pp}$, which is a scalar. Multiplying the vector $\mathbf{p}$ by this quantity gives $\mathbf{D}:\mathbf{ppp}$.

These equations can be integrated numerically using many different schemes. For a single fiber, a fourth-order Runge–Kutta scheme is both efficient and accurate, and is available in MATLAB as the function `ode45`. This function uses a fifth-order Runge–Kutta scheme for error estimation, and will adaptively reduce the time step as needed to meet any desired error tolerance.

To implement this in MATLAB, one first writes a function that calculates the time derivatives of p_1, p_2 and p_3 using Eqn. (4.93). Here we have named that function `pDot` and it is saved in the file `pDot.m`. This function is also available in the accompanying software. Because MATLAB operates directly with matrices, the code is very compact.

```
function dpdt = pDot(p, L, xi)
%   dpdt = pDot(p, L, xi) gives the time rate of change of the orientation
%   vector p (3x1) according to Jeffery's equation, for a velocity
%   gradient tensor L (3x3) and a scalar particle shape factor xi.
```

```
W = (L-L')/2;                    % Vorticity tensor
D = (L+L')/2;                    % Rate-of-deformation tensor
dpdt = W*p + xi*(D*p - (p'*D*p)*p);  % Jeffery's equation

return
end
```

Next, a short script is written to set up the necessary input information and call `ode45` to compute the solution. For example:

```
%% Solve Jeffery's equation for a single fiber
% Input:
pzero = [0, 1, 0.5]';          % Initial condition; should be 3x1
pzero = pzero/norm(pzero);     % Ensures that pzero has unit length

L = zeros(3);                  % Velocity gradient tensor
L(1,2) = 1;                    % Simple shear flow

xi = 1.0;                      % Particle shape factor

tspan = 0:0.05:20;             % Find the solution at these times

% Solution tolerances for ode45
options = odeset('RelTol', 1e-6, 'AbsTol', 1e-9);

%% Primary calculation
[t, p] = ode45(@(t,p) pDot(p, L, xi), tspan, pzero, options);

% Recover (theta, phi) angles
theta = acos(p(:,3));
phi   = atan(p(:,2)./p(:,1));

%% Simple plot of results, with theta and phi in degrees
plot(t, theta*180/pi, 'r-', t, phi*180/pi, 'b-')
xlabel('time')
ylabel('angle␣(degrees)')
legend('\theta', '\phi')
```

Here we have used the `odeset` command to tighten the default error tolerances for `ode45` and get a highly accurate solution. The solution is returned in the matrices `t` and `p`, where `p(i,j)` is the `j` component of **p** at time `t(i)`. If the angles (θ, ϕ) are desired, they can be computed from `p` as shown.

With this tool we can solve Jeffery's equation for three-dimensional orientation with any value of ξ and any type of flow.

4.2.4 Particle Shape Factors

Let us return to the particle shape factor ξ, which was given in Eqn. (4.86). Recall that ξ depends only on the ratio a/b, where a and b are the semi-axis lengths of the spheroidal particle. Figure 4.12 shows how ξ varies with a/b:

- For **rod-like particles** like Fig. 4.8(a), $a/b > 1$ and $0 < \xi < 1$. As the particle becomes long and slender, $a/b \to \infty$ and $\xi \to 1$.

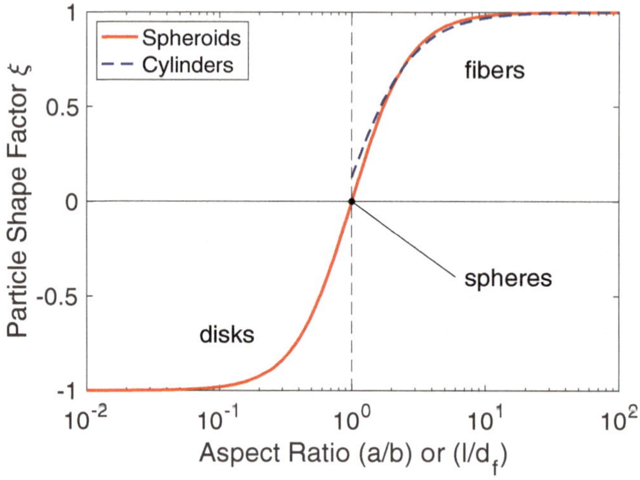

Figure 4.12 Particle shape factor ξ versus particle aspect ratio. The red line is $\xi(a/b)$ for spheroidal particles, Eqn. (4.86), and the blue line is $\xi(\ell/d_f)$ for cylindrical fibers using Eqns. (4.94) and (4.95).

- For **disk-like particles** like Fig. 4.8(b), $a/b < 1$ and $-1 < \xi < 0$. As the particle becomes thin compared to its large diameter, $a/b \rightarrow 0$ and $\xi \rightarrow -1$.
- A **sphere** is a spheroid with $a/b = 1$, so its motion is also modeled by Jeffery's equation. For a sphere, $\xi = 0$.

Jeffery solved for ellipsoidal particles, but what about cylindrical particles like fibers? Symmetry arguments show that Jeffery's equation (4.85) should apply to *any* particle whose shape is axisymmetric [Bre62], provided the value of ξ is chosen correctly. This means that both cylindrical fibers and circular disks (Figs. 4.8(c) and (d)) can be modeled using Jeffery's equation. Experiments on single particles show that a rigid, cylindrical fiber follows Jeffery's motion as accurately as one can measure [TM51]. For a single fiber in a Newtonian fluid, Jeffery's equation is exact.

We still need to find ξ for cylindrical fibers. Let ℓ be the length of the cylinder and d_f its diameter, and let r_e be the effective value of a/b for the cylinder. r_e is called the *equivalent ellipsoidal axis ratio.* Since ξ depends only on a/b, we expect r_e to be a function of the fiber aspect ratio ℓ/d_f. While there are some theoretical and numerical calculations for this [Cox71, ZESAJMS11], a useful result is a correlation of experimental data over a wide range of fiber aspect ratios [HP75]:

$$r_e = 1.14(\ell/d_f)^{0.844} \tag{4.94}$$

The shape factor is then found from using r_e in Eqn. (4.86),

$$\xi = \frac{r_e^2 - 1}{r_e^2 + 1} \tag{4.95}$$

This correlation is shown as the dashed blue line in Fig. 4.12. Once the fiber aspect ratio exceeds about 10, ξ is very close to unity. As the figure shows, using $r_e \simeq \ell/d_f$ is a useful rough approximation for the behavior of cylindrical fibers.

4.2.5 Jeffery Orbits

We used $\xi = 1$ in Section 4.2.2, but for any real fiber ξ is always less than unity. If a fiber is reasonably slender, this difference is small: if $\ell/d_f \geq 10$ then $\xi \geq 0.97$. While a value of ξ that is slightly less than unity has very little effect in extensional flow, it has a large effect in simple shear flow.

Building on our planar-orientation analysis from Section 4.2.2, if a fiber has finite aspect ratio so that $0 < \xi < 1$, then in simple shear flow Eqn. (4.88) becomes

$$\dot{\phi} = -\frac{\dot{\gamma}}{2}\left(1 - \xi\cos(2\phi)\right) \tag{4.96}$$

Using some trigonometric identities this can be re-written as

$$\dot{\phi} = -\frac{\dot{\gamma}}{2}\left[(1+\xi)\sin^2\phi + (1-\xi)\cos^2\phi\right] \tag{4.97}$$

Now we see that the $\sin^2\phi$ term from Eqn. (4.89) is slightly altered when ξ is close to unity, but there is a new term proportional to $\cos^2\phi$ that was not present for $\xi = 1$. With this term the fiber continues to rotate at $\phi = 0$, though more slowly than at other angles because $\cos^2\phi$ is multiplied by the small coefficient $(1-\xi)$. For $-1 < \xi < 1$ there is no steady-state value of ϕ. Instead, the fiber executes a periodic motion known as a *Jeffery orbit*.

The solution to Eqn. (4.97) for an initial condition of $\phi = 0$ is

$$\cot\phi = r_e \cot\left(\frac{-\dot{\gamma}t}{r_e + r_e^{-1}}\right) \tag{4.98}$$

This solution also applies when the fiber does not lie in the 1–2 plane. From a full 3-D solution [TM51], the out-of-plane angle is

$$\tan\theta = \frac{\tan\theta_0}{\sqrt{r_e^2\sin^2\phi + \cos^2\phi}} \tag{4.99}$$

with $\theta = \theta_0$ when $\phi = 0$. This solution is periodic and, from Eqn. (4.98), the period is

$$T = \frac{2\pi}{\dot{\gamma}}\left(r_e + r_e^{-1}\right) \tag{4.100}$$

Figure 4.13 shows θ and ϕ vs. time for two values of the equivalent ellipsoidal axis ratio r_e. The motion is periodic, and the period increases as the particle becomes more slender and the axis ratio increases. The fibers rotate rapidly when ϕ is near 90°, but slowly when ϕ is near 0. For large values of r_e the fiber rotates very slowly near $\phi = 0$, then flips over rapidly to $\phi = 180°$.

While a finite-length fiber has a periodic Jeffery orbit in simple shear, it spends most of its time aligned in the flow direction, and our first rule of fiber orientation still applies.

Figure 4.14 shows Jeffery orbits for several different values of θ_0, plotted on the unit sphere. The fibers all have $r_e = 15$, and the magenta orbit matches Fig. 4.13(a). The orbits are closed, so the initial orientation of a fiber and its value of r_e completely define its Jeffery orbit. In the figure, the points on each orbit are equally spaced in time, and it is apparent that all three fibers spend most of their time oriented close to the flow (x_1) direction. Notice also that, while the magenta and blue orbits have a significant out-of-plane orientation when $\phi = 90°$, the Jeffery orbits bring these fibers close the $\theta = 90°$ as ϕ approaches zero.

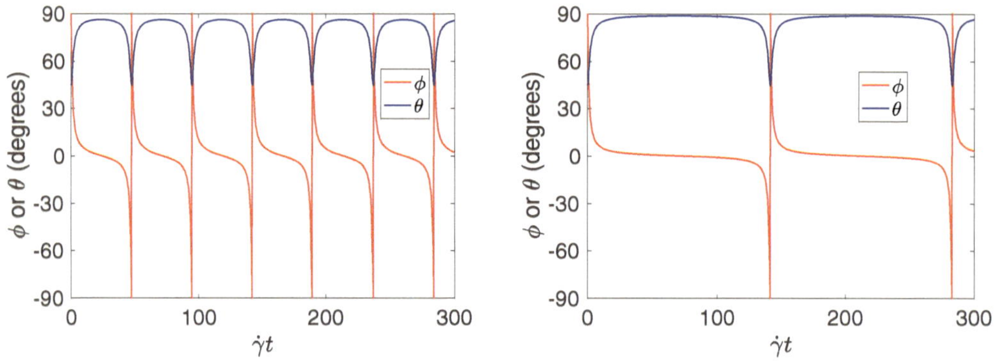

(a) $r_e = 15$, $\xi = 0.9910$ (b) $r_e = 45$, $\xi = 0.9995$

Figure 4.13 Orientation angles vs. time in simple shear flow for fibers with different aspect ratios. Both fibers are executing a Jeffery orbit. The shear rate is $\dot{\gamma}$ and initial conditions are $\phi_0 = 90°$ and $\theta_0 = 45°$.

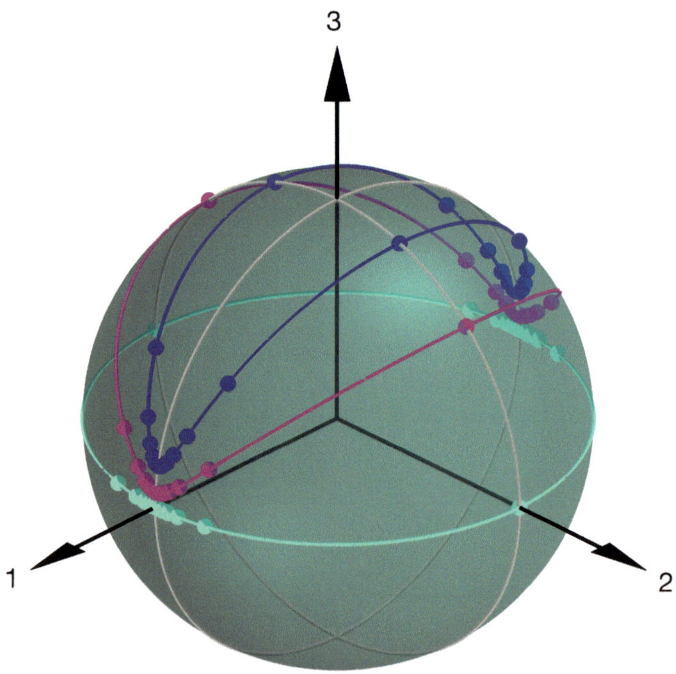

Figure 4.14 Jeffery orbits in 1–2 simple shear flow for a fiber with $r_e = 15$, $\xi = 0.991$. The three orbits have $\theta = 90°$ (cyan), $\theta = 45°$ (magenta) and $\theta = 20°$ (blue) at $\phi = 90°$. Points on each orbit are equally spaced in time.

4.2.6 Strain-Time Scaling and Reversibility

In analytical solutions of Jeffery's equation for simple shear flow like Eqns. (4.90) and (4.98), the fiber orientation at any time depends only on the initial orientation and on the product $\dot{\gamma}t$. The change in orientation for a Jeffery fiber depends on the accumulated strain $\dot{\gamma}t$, but the rate of straining by itself does not affect the results.

The same statement applies for any type of deformation. For instance, in the Jeffery solution for planar elongation, Eqn. (4.92), the orientation depends on $\dot{\varepsilon}_P t$. This occurs because every term on the right-hand side of Jeffery's equation (4.85) is linear in the velocity gradient **L**, from which **D** and **W** are derived.

The time dependence of a Jeffery solution can always be replaced by a strain dependence, so we will call this property *strain-time scaling*. With this scaling, we can plot transient solutions to Jeffery's equation with $\dot{\gamma}t$ or $\dot{\varepsilon}_P t$ as the independent variable, and the plot will be valid for any deformation rate.

This fact that $\dot{\mathbf{p}}$ is linear in **L** also means that the motion of a Jeffery fiber is *reversible.* If the flow is reversed, then the fiber retraces it previous positions and orientations, and eventually returns to its initial state.

4.2.7 Deformation Form

So far we have written Jeffery's equation in a rate form, which gives the orientation by integrating over time, starting from an initial condition. As an alternative, we can write Jeffery's equation in a finite-deformation form. This allows the orientation at any time t to be computed from the initial orientation in a single step.

To see the relationship between deformation and fiber orientation, consider the initial material element shown in Fig. 4.15(a). A material vector $\mathrm{d}\mathbf{x}'$ connects material points P and Q. When the material is deformed to the state in Fig. 4.15(b), the material vector is transformed according to Eqn. (4.59), which we repeat here:

$$\mathrm{d}\mathbf{x} = \mathbf{F} \cdot \mathrm{d}\mathbf{x}' \tag{4.101}$$

Recall that **F** is the deformation gradient tensor, defined in Eqn. (4.55). If we know the tensor **F** that connects the initial configuration with the deformed configuration at time t, then we can find $\mathrm{d}\mathbf{x}(t)$.

Now let us construct a vector **p** that matches the direction of $\mathrm{d}\mathbf{x}$ at all times t but maintains unit length. In the initial configuration this vector must be

$$\mathbf{p}' = \frac{\mathrm{d}\mathbf{x}'}{|\mathrm{d}\mathbf{x}'|} \tag{4.102}$$

That is, $\mathbf{p}'$ points in the same direction as $\mathrm{d}\mathbf{x}'$, but has unit length.

The vector **p** can match the direction of $\mathrm{d}\mathbf{x}$ at all later times t if $\mathbf{p} = \mathbf{F} \cdot \mathbf{p}'$, but that would allow **p** to change length. To maintain **p** at unit length, we divide this product by its magnitude.

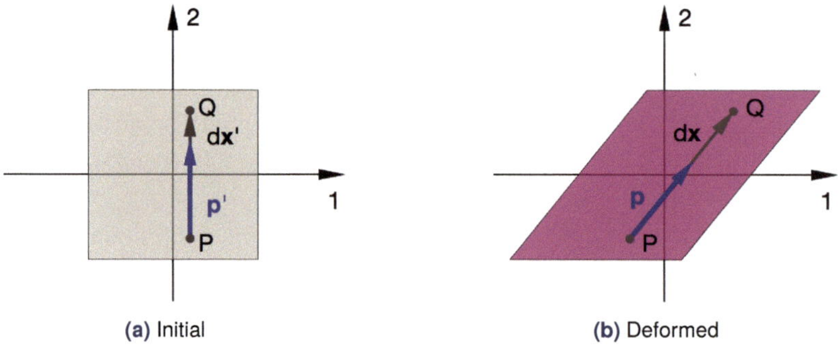

Figure 4.15 Initial and deformed states of a material vector $d\mathbf{x}$ and an orientation vector $\mathbf{p}$ for a Jeffery particle with $\xi = 1$. If $\mathbf{p}$ is initially parallel to $d\mathbf{x}$ it remains parallel at all times but does not change length.

This gives

$$\mathbf{p} = \frac{\mathbf{F}\cdot\mathbf{p}'}{|\mathbf{F}\cdot\mathbf{p}'|} \tag{4.103}$$

Now we have an expression for $\mathbf{p}$ that follows the orientation of a material vector for any finite deformation but does not change length.

How is this related to Jeffery's equation? By taking the time derivative of Eqn. (4.103) following the particle[4], we obtain an expression for $\dot{\mathbf{p}}$:

$$\begin{aligned}\dot{\mathbf{p}} &= \mathbf{L}\cdot\mathbf{p} - \mathbf{L}:\mathbf{ppp} \\ &= \mathbf{W}\cdot\mathbf{p} + \mathbf{D}\cdot\mathbf{p} - \mathbf{D}:\mathbf{ppp}\end{aligned} \tag{4.104}$$

In the second line we have used Eqn. (4.4) to replace $\mathbf{L}$ with $\mathbf{W}$ and $\mathbf{D}$. Also, $\mathbf{W}:\mathbf{pp} = 0$ because $\mathbf{W}$ is anti-symmetric. Equation (4.104) is exactly Jeffery's equation (4.85) for $\xi = 1$. Thus, Eqn. (4.103) gives the reorientation of a Jeffery particle in the limit of $r_e \to \infty$. This provides a useful insight:

> The orientation of a Jeffery particle in the limit of infinite aspect ratio matches the orientation of a material line, except that the particle does not change length.

We can also extend this finite-deformation analysis to particles with $\xi < 1$. Comparing Eqn. (4.104) to Eqn. (4.85), we see that a Jeffery particle reorients in response to a velocity gradient $(\mathbf{W}+\mathbf{D})$ like a material line would reorient in response to a velocity gradient $(\mathbf{W}+\xi\mathbf{D})$. With this in mind, instead of calculating $\mathbf{F}$ using $\dot{\mathbf{F}} = \mathbf{L}\cdot\mathbf{F}$, we can calculate a modified deformation gradient tensor $\hat{\mathbf{F}}$ using

$$\dot{\hat{\mathbf{F}}} = (\mathbf{W}+\xi\mathbf{D})\cdot\hat{\mathbf{F}} \qquad \text{with} \qquad \hat{\mathbf{F}} = \mathbf{I} \quad \text{at} \quad t = 0 \tag{4.105}$$

[4] The derivative of the left-hand side of (4.103) is $\dot{\mathbf{p}}$. On the right-hand side the time derivative of $\mathbf{p}'$ is zero, since $\mathbf{p}'$ does not change with t, and the time derivative of $\mathbf{F}$ is $\mathbf{L}\cdot\mathbf{F}$ from Eqn. (4.64). Evaluating the derivative at $t = 0$, where $\mathbf{p}' = \mathbf{p}$ and $\mathbf{F} = \mathbf{I}$, gives Eqn. (4.104).

One then uses the modified deformation gradient $\hat{\mathbf{F}}$ in Eqn. (4.103), so that

$$\mathbf{p} = \frac{\hat{\mathbf{F}} \cdot \mathbf{p}'}{|\hat{\mathbf{F}} \cdot \mathbf{p}'|} \tag{4.106}$$

Equations (4.105) and (4.106) give exactly the same results as the rate form of Jeffery's equation, Eqn. (4.85), including Jeffery orbits in simple shear flow.

Of course one still has to integrate Eqn. (4.105) to find $\hat{\mathbf{F}}$, so this approach does not eliminate the need to solve differential equations. However, this result provides the basis for the Jeffery distribution function and the natural closure (Sections 2.4.1.2, 2.4.2 and 5.4.2.3), which are important tools for modeling fiber orientation.

4.3 Orientation Patterns in Injection and Compression Molding

4.3.1 Two Rules that Explain Everything

Jeffery's equation explains *all* the fiber orientation patterns that typically occur in injection and compression molded parts, at least qualitatively. By "qualitatively" we mean that Jeffery's equation provides the principal direction of the orientation. Quantitative orientation predictions require some extensions to Jeffery's equation, which are discussed in Chapt. 5.

In fact, we do not even need to solve Jeffery's equation again to understand orientation patterns. The two rules of orientation provided by Jeffery's equation are sufficient. From Section 4.2.2, these are:

Rule 1: Shearing flows align fibers in the flow direction.

Rule 2: Stretching flows align fibers in the stretching direction.

Most injection molded parts are typically thin in one direction compared to the other dimensions of the part. As we saw in Chapt. 3, this type of molding has a core/shell structure: a core layer at the center of the part is sandwiched between two shell layers. Some moldings may also have thin skin layers at the surface, outside the shell layers, though skin layers are not always present.

In this section we look at flow fields typical of injection molding and show how the two rules explain the shell, core and skin-layer orientations. We also discuss compression molding, where the squeezing motion of the mold is also important.

4.3.2 Shell Layers in Injection Molding

In an injection molded part the shell layers are always oriented in the flow direction.

To see why this is, consider a long, thin strip that is filled from one end. Let the length of the strip be L, the width W, and the thickness $2h$. Use the coordinate system shown in Fig. 4.16, where the 1 axis is the flow direction, the 2 axis is the crossflow direction and the 3 axis is the thickness direction. This will be our standard choice of coordinate directions for analyzing molded parts. We also choose $x_3 = 0$ on the midplane of the part, so that $x_3/h = \pm 1$ on the upper and lower mold walls.

The strip is "thin" for flow-analysis purposes if $h \ll L$ and $h \ll W$. If this geometry is being filled with a Newtonian fluid of constant viscosity, the velocity field is

$$v_1 = \frac{3\overline{v}}{2}\left(1 - \frac{x_3^2}{h^2}\right) \quad \text{and} \quad v_2 = v_3 = 0 \tag{4.107}$$

where $\overline{v}$ is the average velocity of the fluid in the 1 direction (i.e., the filling speed). Figure 4.17 shows this velocity profile, which is parabolic. This velocity solution is worked out in many fluid mechanics texts (e.g., [DT01]) and we will not show the details here.

For a Jeffery analysis we need the components of the velocity gradient tensor, $L_{ij} = \partial v_i / \partial x_j$. In this case the only non-zero component of $\mathbf{L}$ is

$$L_{13} = \frac{\partial v_1}{\partial x_3} = -\frac{3\overline{v} x_3}{h^2} \tag{4.108}$$

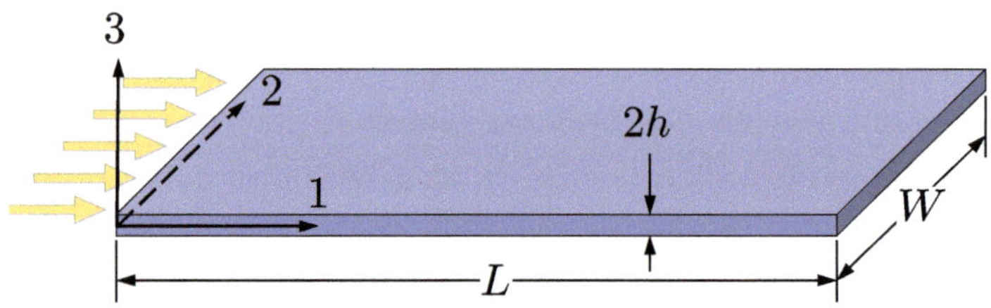

Figure 4.16 Geometry and coordinate system for an end-gated strip.

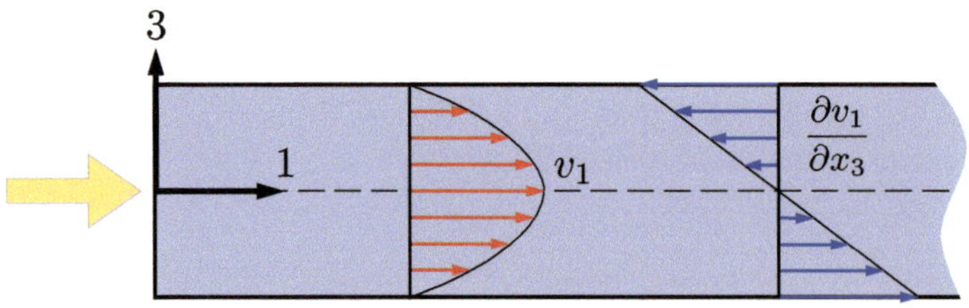

Figure 4.17 Velocity and velocity gradient in an end-gated strip. Newtonian fluid with constant viscosity.

This is a simple shear flow, and the shear rate $\partial v_1/\partial x_3$ varies linearly with x_3. Figure 4.17 also shows the profile of the shear rate.

The first rule of fiber orientation is that shearing flows align fibers in the shearing direction, so we expect that in this flow the fibers will align in the 1 direction. This is exactly what happens in the shell layer. However, on the midplane, $x_3 = 0$, the velocity gradient is zero. Thus, at the midplane, Jeffery's equation tells us that $\dot{\mathbf{p}} = 0$, and a fiber with its centroid at $x_3 = 0$ translates along the midplane without changing orientation. In practice, this means that fibers on or near the midplane keep the orientation they had when they entered the mold cavity, an orientation that depends on the details of the gate geometry. These fibers form the core layer. In contrast, closer to the mold walls there is significant shearing, and the fibers align in the flow direction. Strong shearing near the mold walls forms the shell layer. The shearing direction is the flow direction, so shell layers are always aligned with the local flow direction during mold filling.

Of course molten polymers are non-Newtonian, and injection molding flows are highly non-isothermal, with some polymer solidifying on the mold walls during filling. This does alter the shapes of the velocity and shear-rate profiles. Even so, the flow in a strip-shaped mold is still dominated by simple shear deformation, so shell layers aligned in the flow direction still form. The relative thickness of the shell layers compared to the core will depend on the details of the shear-rate profile, and we will see in Section 5.7 that the distance from the gate also affects core thickness. However, flow near any wall with no slip is always simple shear flow, so shell layers aligned in the flow direction are present in almost all injection molded parts.

4.3.3 Core Orientation in Injection Molding

The core layer in an injection molded part may be oriented along the flow, across the flow, or it may have a random-in-plane orientation. The direction of the core orientation depends on whether the flow field is converging, diverging or parallel, when viewed along the 3 axis.

A convenient geometry for understanding this is the center-gated disk. This mold cavity is a circular disk with outer radius R and thickness $2h$, and it is injected through a gate at the center of the disk. Figure 4.18 shows an example of this geometry that uses a large sprue gate.

We assume that the flow field is radially symmetric, and that the disk is thin compared to its radius, $2h \ll R$. We use cylindrical coordinates (r, θ, z) and place $z = 0$ on the midplane of the disk.

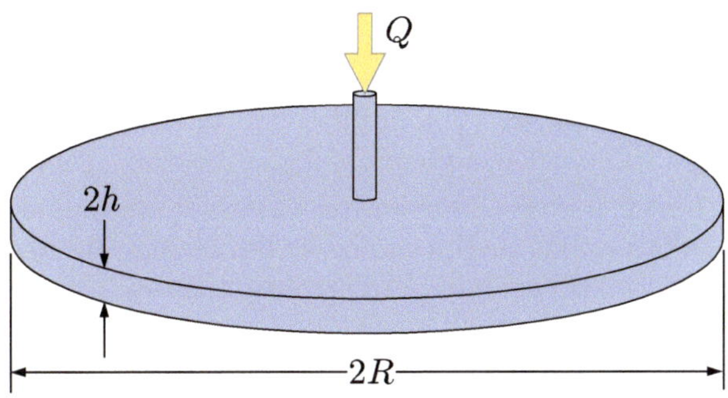

Figure 4.18 Geometry for a center-gated disk.

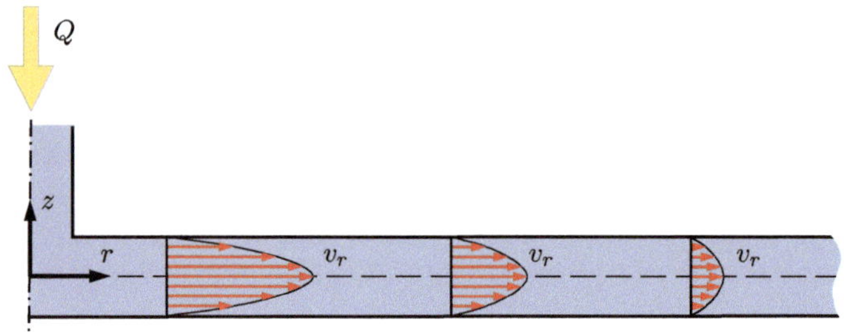

Figure 4.19 Velocity distribution in a center-gated disk. Newtonian fluid with constant viscosity.

Again using a Newtonian fluid of constant viscosity, the velocity field is (e.g., [DT01])

$$v_r = \frac{1}{r}\frac{3Q}{8\pi h}\left(1 - \frac{z^2}{h^2}\right) \qquad \text{and} \qquad v_\theta = v_z = 0 \tag{4.109}$$

Here Q is the volume flow rate of material entering the gate. This solution applies away from the gate, say for $r \geq 5h$.

The velocity distribution, shown in Fig. 4.19, has a parabolic shape in the z direction at any radius r, just like the strip flow, so there is a shearing component to this flow. The magnitude of the radial velocity v_r decreases as r increases, since the same volume flow rate Q is driving the motion at all values of r.

To see what Jeffery's equation says about fiber orientation in this flow, we need the components of **D** and **W**. Equations for these components in a cylindrical coordinate system are given in Appendix A.6. Using those expressions we find

$$\mathbf{D} = \frac{1}{r^2}\frac{3Q}{8\pi h}\begin{bmatrix} -\left(1 - z^2/h^2\right) & 0 & -rz/h^2 \\ 0 & \left(1 - z^2/h^2\right) & 0 \\ -rz/h^2 & 0 & 0 \end{bmatrix} \tag{4.110}$$

$$\mathbf{W} = \frac{1}{r^2}\frac{3Q}{8\pi h}\begin{bmatrix} 0 & 0 & -rz/h^2 \\ 0 & 0 & 0 \\ rz/h^2 & 0 & 0 \end{bmatrix} \tag{4.111}$$

The r–z shearing flow from the parabolic velocity profile is present, since $D_{rz} = W_{rz}$. Also, just as in the strip flow, the shearing rate is zero on the midplane ($z = 0$) and largest at the mold walls.

In addition, D_{rr} and $D_{\theta\theta}$ are non-zero, so the center-gated disk flow also has planar extensional deformation in the r–θ plane. $D_{\theta\theta}$ is positive and D_{rr} is negative, so the fluid is stretching in the θ direction, across the flow, and contracting in the r direction, along the flow.

This crossflow stretching motion occurs in any flow where the streamlines are diverging. Figure 4.20(a) shows how this works. The streamlines in this flow are diverging, and two edges of the gray fluid element lie on streamlines. As this element moves in the flow its edges remain on their respective streamlines, so the fluid must stretch in the circumferential (crossflow) direction. If the motion takes place in the plane of the figure, the fluid element must contract in the radial (flow) direction to maintain constant volume. The result is that, after some flow, the fluid element has the magenta shape shown in the figure. Thus, we can say that:

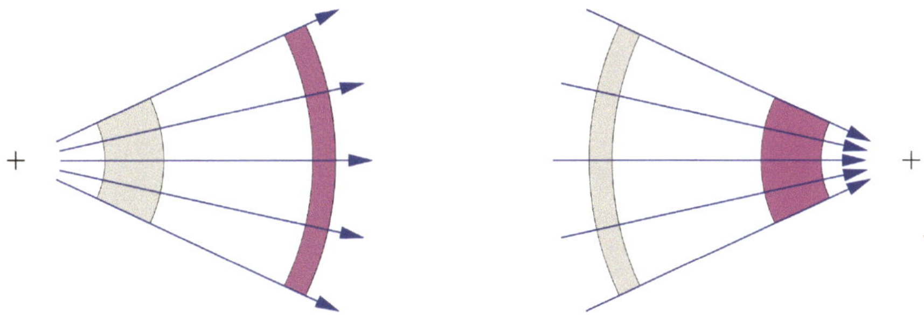

(a) Diverging flow (b) Converging flow

Figure 4.20 Material deformation in diverging and converging planar flows. Blue lines are streamlines. A fluid element with the gray initial shape deforms into the magenta shape as it flows.

Diverging flows stretch the fluid in the **crossflow** direction.

The opposite happens when the streamlines are converging. As shown in Fig. 4.20(b), a fluid element in a converging flow contracts in the crossflow direction to stay on its streamlines, and stretches in the flow direction to maintain constant volume. Thus:

Converging flows stretch the fluid in the **flow** direction.

Returning to the center-gated disk, on the midplane, $z = 0$, the shearing rate is zero and the stretching rate is maximum, so fibers on the midplane only see the planar extensional part of the deformation. The deformation of a material element on the midplane is exactly like Fig. 4.20(a), so the fluid is stretched in the crossflow direction. Using our second rule, this creates a core layer in which fibers are primarily aligned across the flow, in the θ direction.

At the mold walls, $z = \pm h$, the shearing rate is maximum and the stretching rate goes to zero. The flow here is simple shear flow, so fibers close to the mold walls will align in the flow direction. This creates the usual flow-aligned shell layers.

At intermediate heights, $0 < z < h$, both the stretching and shearing motions are present, and the two types of deformation compete to determine the main direction of orientation. The ratio of the shear rate to the stretching rate, $|D_{rz}/D_{\theta\theta}|$, determines which flow wins this competition. Closer to the midplane, the stretching flow dominates. Closer to the walls, the shearing flow dominates. As we saw in Section 3.1.3, the transition from crossflow orientation in the core to flow-aligned fibers in the shell occurs gradually across the thickness of the part.

If a diverging flow can create a core with fibers oriented across the flow direction, a converging flow will align the core fibers in the flow direction. A common example of this is an injection molded tensile-test specimen, often called a dogbone. If the specimen is gated at one end, as in Fig. 4.21, then the flow entering the gauge section is converging. This aligns the core fibers in the flow direction within the gauge section. Fibers in the shell layer are already aligned in this direction, so the gauge section will have fibers aligned in the direction of the tensile test across its entire thickness.

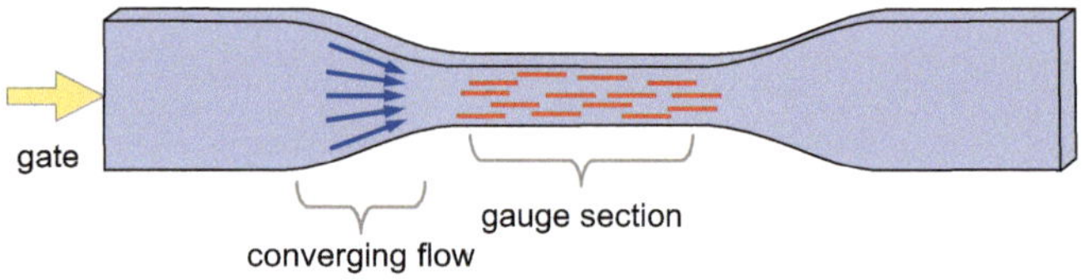

Figure 4.21 Injection molded tensile specimen, showing how converging flow creates a flow-aligned core in the gauge section.

In a molded dogbone, aligned fibers in the gauge section maximize the elastic modulus and tensile strength in the testing direction, but minimize the modulus and strength perpendicular to the test direction. Since very few molded parts will have fibers strongly aligned across the entire part thickness *and* be loaded primarily in that direction, the mechanical properties determined from a molded dogbone specimen are *not* representative of the properties likely to occur in most parts molded from that same material. A much better way to measure properties of fiber reinforced injection molding materials is to mold larger plates and cut out tensile test specimens in both the flow and crossflow directions [ISO17].

If the streamlines viewed along the 3 axis are parallel (i.e., neither converging nor diverging) then there is neither shearing nor stretching on the midplane. The strip shape from Section 4.3.2 is a good example of this. With parallel streamlines, the core orientation is carried along by the flow but does not change. This can produce a random-in-plane core orientation or, if the upstream flow is strongly converging or diverging, the core can be flow-aligned or crossflow-aligned, respectively.

4.3.4 Skin Layers and the Fountain Flow

Some injection moldings also have a skin layer near the outer surface, where the fibers are less flow-aligned than in the shell layer. To understand skin formation, we first need to understand the fountain flow.

In a thin injection molding, the velocity is nearly parallel to the mold surfaces over most of the flow. However, this is not true at the flow front, where the polymer/fiber mixture is displacing air from the mold. When viewed from the side, the flow front will take on a characteristic shape, which is nearly semi-circular for a fluid without fibers. This shape will advance at the gapwise-average fluid velocity $\overline{v}$. However, as we saw with the strip in Section 4.3.2, behind the flow front there is a profile of velocity across the thickness, and the fluid near the midplane moves faster than $\overline{v}$. This situation is sketched in Fig. 4.22(a). Because of this speed difference, fluid near the midplane catches up to the moving flow front. As this material gets close to the front it must decelerate to a speed of $\overline{v}$. The streamlines then spread in the gapwise direction, and material from the midplane rolls onto the mold walls. This flow near the front is called the *fountain flow*.

A common way to portray the fountain flow is to draw streamlines in a coordinate frame that moves with velocity $\overline{v}$, as in Fig. 4.22(b). In this frame the flow front appears to be stationary, the mold walls move backward with a velocity $-\overline{v}$, and the velocity field in the fountain region is steady. As a result, the streamlines in this flow are also particle path lines. (The streamlines in Fig. 4.22(a) are not particle path lines, because the flow field is unsteady in that reference

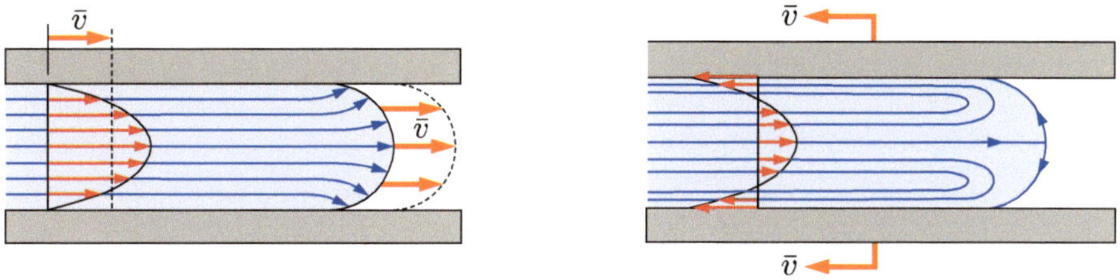

(a) Fixed frame (b) Moving frame

Figure 4.22 Side view of the fountain flow in (a) a coordinate frame fixed in the mold and (b) a coordinate frame that moves with the flow front.

frame.) In this view the motion of material from the midplane to the walls is more apparent. In fact, material deposited directly on the wall comes from the midplane, and material that ends up a bit inside the upper wall comes from just above the midplane. While we will make good use of this picture of the fountain flow, remember that the fluid velocity in the fixed frame is aways in the direction of filling. Relative to the mold, fluid never moves backwards toward the gate.

Fibers are carried along with the fluid, and we need to know something about deformation in the fountain flow to understand how it influences fiber orientation. The full details of fountain-flow deformation are complex [CBM87], but in general any material passing through the fountain flow is flipped over 180°, stretched and then sheared. Figure 4.23(a) gives an idea of this deformation.

By the time material reaches the flow front in an injection mold, the fibers all lie close to the 1–2 plane. If fibers near the core are oriented in the flow direction, the fountain flow will do little to change their orientation. However, if core fibers approaching the flow front have random-in-plane or crossflow orientation, then after passing through the fountain flow they will be partially aligned in the flow direction. These fibers will, however, be less flow-aligned than fibers in the shell. Fibers from the core that have passed through the fountain region and are now close to the mold wall form the skin layers.

> Skin layers form when random-in-plane or crossflow core fibers pass through the fountain flow at the flow front.

However, some molded parts do not have a skin layer, even though they do have random-in-plane or crossflow core orientation. How can this be? The answer depends on what happens to the material *after* it has passed through the fountain flow, and on the formation of frozen layers. For thermoplastic injection molding, the mold walls are much colder than the injected polymer melt. The polymer begins to solidify as soon as it contacts the wall, and frozen layers start growing behind the flow front. If the frozen layers grow fast enough relative to the filling, then the orientation created by the fountain flow near the mold wall will be frozen in before it experiences much additional shearing. Then the final part will have a skin layer that is less oriented than the shell. This deformation is sketched in Fig. 4.23(a). However, if the filling is fast compared to the growth of frozen layers, as in Fig. 4.23(b), then core fibers that pass through the fountain flow subsequently experience a great deal of shear flow, since shearing

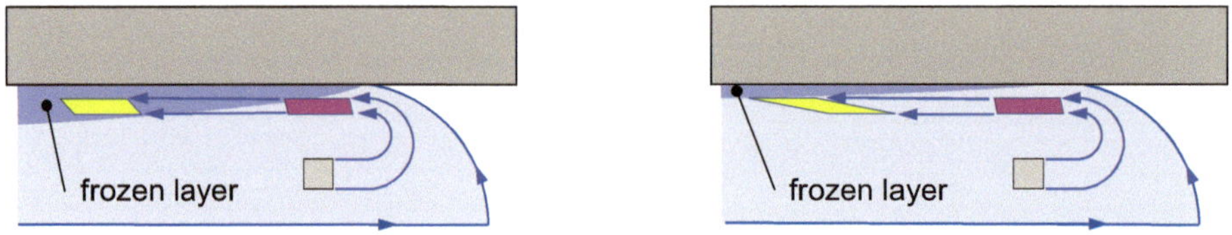

(a) Slow filling (b) Fast filling

Figure 4.23 Schematic of material deformation in the fountain flow, for slow and fast filling, as seen in the moving frame. In both diagrams the initial gray material deforms into the magenta shape due to the fountain flow, then into the yellow shape due to shear flow near the wall.

is strong near the wall. These fibers can achieve the same degree of flow-alignment as the shell layer. In that case, the skin layer is indistinguishable from the shell, and the molded part appears not to have a skin.

This explanation of skin formation is relevant when the fiber length is smaller than the cavity thickness. This is usually true of conventional short fiber injection molding compounds (SFTs). When the fibers are longer than the cavity thickness, fibers cannot execute the flipping motion of the fountain flow and the flow pattern at the flow front is much different. For example, with sheet molding compound, where the fibers are much longer than the part thickness, there is no fountain flow and no skin layer of the type described here.

In principle, CAE software can predict the presence or absence of skin layers, as least for SFT materials. However, software may not predict a skin layer, even when one forms, if the mesh is too coarse or the flow field in the fountain region is not modeled accurately. A special numerical treatment that can capture skin orientation in thin-cavity flows is described in [BT92a, BT92b], and similar methods are used in some CAE codes.

4.3.5 The Mystery of Thick Plaques

In early research on fiber orientation, thicker parts were observed to have much thicker core layers than expected. This occurred in the ISO plaque geometry [ISO17], Fig. 4.24(a), where the combination of a thick flood gate and a thin film gate was intended to create unidirectional flow during mold filling.

As Fig. 4.24(b) shows, the 2 mm thick plaque had the expected shell layers, a core occupying approximately 25% of the part thickness, and random-in-plane orientation at the midplane. In contrast, the 6 mm thick plaque had a much thicker core, with strong crossflow orientation on the midplane. However, calculations based on unidirectional flow said that the orientation in both plaques should look like the 2 mm results. Was some special mechanism operating in the thick plaques and affecting the orientation?

This mystery was resolved when the experimenters made short-shot moldings, to see the actual mold filling pattern. The 2 mm thick plaques, Fig. 4.25(a), filled as expected, with the flow fronts becoming straight soon after the polymer entered the film gate. However, the mold design was such that the thickness of the film gate increased as the cavity thickness increased. As a result, the 6 mm thick plaques, Fig. 4.25(b), initially filled in a radial pattern, with semicircular flow fronts, and only developed a straight flow front near the end of filling.

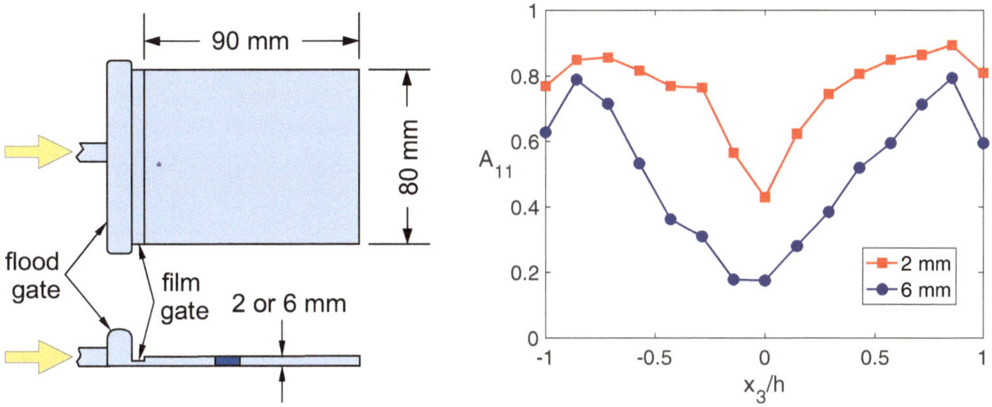

(a) Geometry **(b)** Measured orientation

Figure 4.24 Geometry of ISO plaque and measured orientation for two different plaque thicknesses. PBT with 30% by weight glass fiber. Orientation is measured between 30 and 40 mm from the gate (dark shaded region) along the plaque centerline. Data from [OWN03].

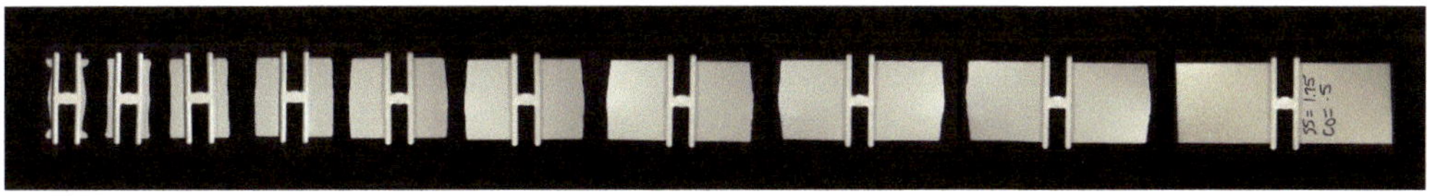

(a) 2 mm thick

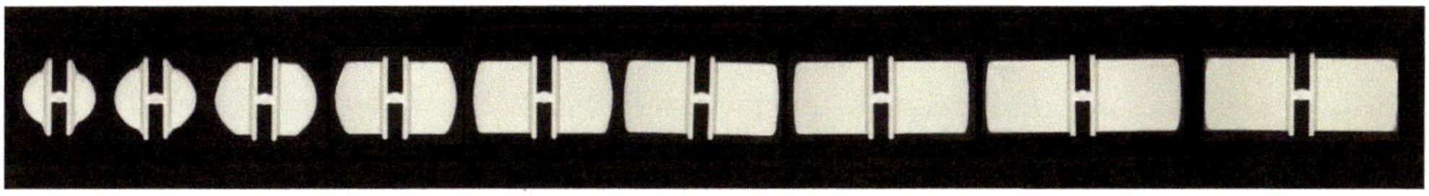

(b) 6 mm thick

Figure 4.25 Short shots of ISO plaques that are (a) 2 mm thick and (b) 6 mm thick. The mold has two identical cavities and is injected through a sprue at the center. From [Wan07].

When viewed from the top, the actual filling patterns in the two plaques were as shown in Fig. 4.26: the 2 mm thick plaque filled like a strip, while the flow near the gate of the 6 mm plaque had diverging streamlines like a center-gated disk. This generated a core with a strong crossflow orientation just inside the gate in the 6 mm plaques, and this core orientation persisted along the flow length.

When the mold filling simulation for the 6 mm thick parts was adjusted to look like Fig. 4.26(b), the fiber orientation predictions matched the experiments [Wan07]. Thus, once the flow patterns in both mold cavities were understood, the two rules of fiber orientation explained the fiber orientation data, and the mystery was solved.

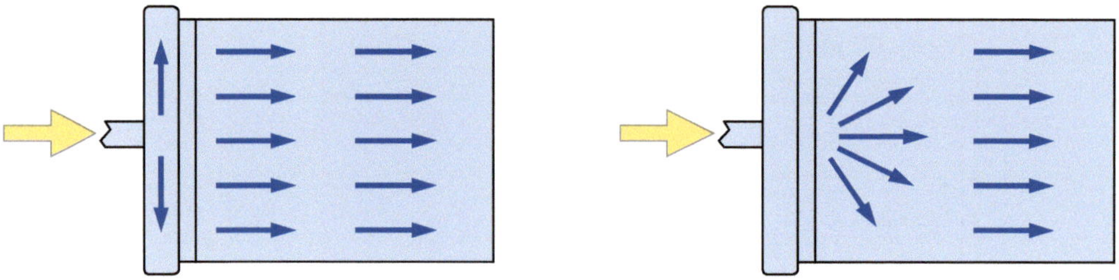

(a) 2 mm thick (b) 6 mm thick

Figure 4.26 Top-view flow patterns in thin and thick ISO plaques [Wan07].

4.3.6 Compression Molding

So far we have considered injection molding, where the flow is driven by pressure from the injection unit. In contrast, flow during compression molding is generated by squeezing the mold halves together. While the velocity of the mold is small compared to the in-plane velocity, the squeezing motion changes the way we interpret converging and diverging flows.

A side view of the geometry we will analyze is shown in Fig. 4.27. The instantaneous cavity thickness is $2h(t)$, the upper and lower mold surfaces are parallel to the 1–2 plane and the mold closes along the 3 direction. Normally the lower mold is stationary and the upper mold moves downward to shape the part. However, we will take advantage of symmetry by choosing a coordinate system whose origin is halfway between the upper and lower mold surfaces. In this coordinate system the lower mold surface moves upward with velocity $-\dot{h}$ and the upper mold surface moves downward at the same speed. When the mold is closing, $\mathrm{d}h/\mathrm{d}t = \dot{h} < 0$, so the numerical value of $(-\dot{h})$ is positive. Lateral flow of the charge is eventually limited by shear edges, where the upper mold fits closely into the lower mold.

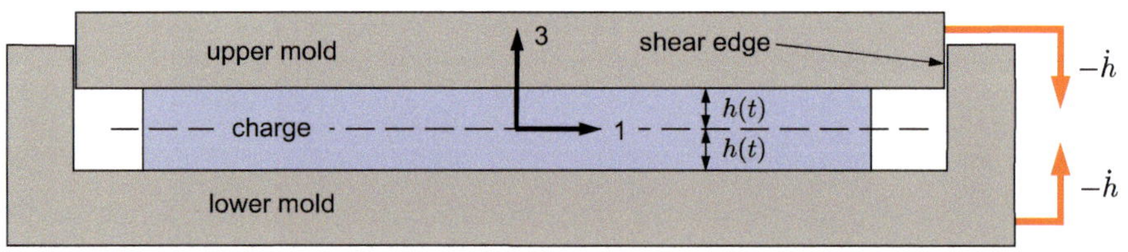

Figure 4.27 Geometry and coordinate system for compression molding flows, side view.

4.3.6.1 Lubricated Squeeze Flow

We first consider lubricated squeeze flow, where the fiber/polymer mixture is assumed to slip over the mold surface with no friction. This is an idealization, but it is a good approximation in thin parts made from sheet molding compound (SMC) [BC85]. This may occur in part because the molding compound has a resin-rich surface, and in part because the resin viscosity is low in a thin layer at the mold surface, where the cold SMC is heated by the hot mold.

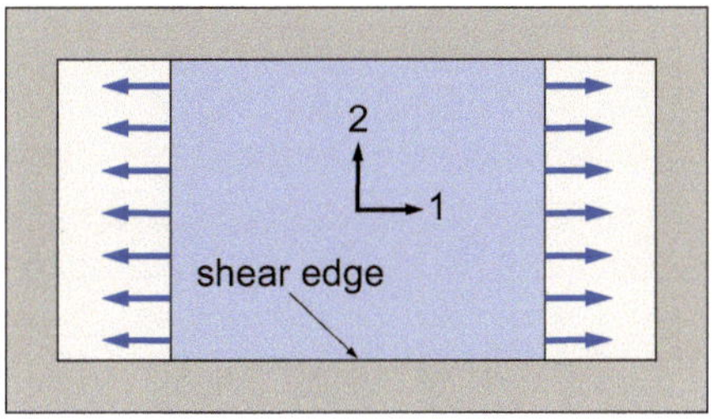

(a) Planar squeeze flow

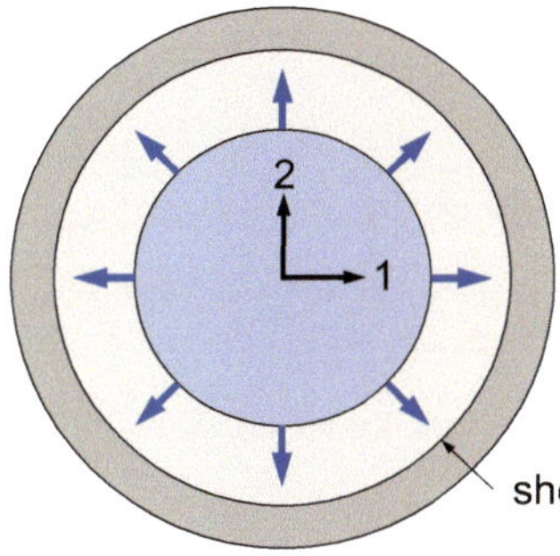

(b) Radial squeeze flow

Figure 4.28 Flow cases for compression molding, top view.

If the mold is rectangular and the initial width of the charge is equal to the mold width, then we have the **planar squeeze flow** shown in Fig. 4.28(a). The charge is squeezed in the 3 direction, flows outward in the 1 direction, and there is no motion in the 2 direction. Thus, the flow is planar in the 1–3 plane. The velocity distribution in this case is

$$v_1 = -\frac{\dot{h}}{h}x_1 \qquad v_2 = 0 \qquad v_3 = \frac{\dot{h}}{h}x_3 \tag{4.112}$$

The 1-direction velocity distribution is sketched in Fig. 4.29. We are assuming that the fluid slips at the walls with zero shear stress, so $\partial v_1/\partial x_3 = 0$.

The velocity gradient tensor **L** is independent of position in this flow, the vorticity tensor **W** is zero, and the rate-of-deformation tensor is

$$\mathbf{D} = -\frac{\dot{h}}{h}\begin{bmatrix} 1 & 0 & 0 \\ 0 & 0 & 0 \\ 0 & 0 & -1 \end{bmatrix} \tag{4.113}$$

Thus, the entire charge undergoes a uniform *planar extension* as in Example 4.1.1. In this case the material stretches in the 1 direction (because $-\dot{h}/h > 0$) and contracts in the 3 direction.

Looking at the top view of this flow, Fig. 4.28(a), one might be tempted to say that the streamlines are parallel so there is no stretching in the 1–2 plane. This would miss the fact that the fluid is contracting in the 3 direction due to the squeezing motion of the mold, and is therefore stretching in the 1 direction. Our rules about stretching directions for converging and diverging streamlines are still correct, but we must account for the squeezing motion when applying those rules. In this flow the streamlines in the 1–3 plane look exactly like Fig. 4.1(b), which correspond to stretching in the 1 direction and contraction in the 3 direction.

Since fibers in SMC are all oriented close to the plane of the part, there is very little 3-direction orientation even in the initial charge. Still, this planar squeezing action draws fibers toward the 1 direction because of stretching in that direction.

Another important compression molding flow is a disk-shaped charge that is not constrained by shear edges, as shown in Fig. 4.28b. This is a **radial squeeze flow** when viewed from above. Using cylindrical coordinates, the lubricated squeeze flow solution for this geometry is

$$v_r = -\frac{\dot{h}}{2h}r \qquad v_\theta = 0 \qquad v_z = \frac{\dot{h}}{h}z \tag{4.114}$$

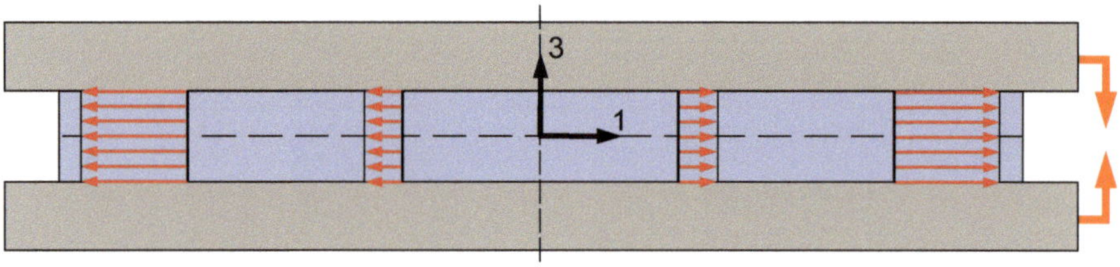

Figure 4.29 Horizontal velocity distribution $v_1(x_1)$ for lubricated squeeze flow.

The side view for $v_r(r,z)$ again looks like Fig. 4.29, though the ratio of v_r to $-\dot{h}/h$ is half as large as for planar squeeze flow.

Applying Eqns. (A.71–A.79) for **D** and **W** in cylindrical coordinates, we find that the vorticity is zero in this flow and the rate-of-deformation tensor is

$$\mathbf{D} = -\frac{\dot{h}}{h}\begin{bmatrix} 1/2 & 0 & 0 \\ 0 & 1/2 & 0 \\ 0 & 0 & -1 \end{bmatrix} \tag{4.115}$$

Again, **D** does not depend on position, but now the flow is *biaxial extension*, stretching equally in the r and θ directions (or, equivalently, in the 1 and 2 directions) and contracting in the z direction.

If the initial charge has any 3-direction orientation, this flow rotate the fibers into the 1–2 plane. Since the material is stretched equally in the 1 and 2 directions, there will be little change in the in-plane orientation. If the initial charge is randomly oriented in the 1–2 plane it will remain that way during this flow.

4.3.6.2 Non-Lubricated Squeeze Flow

Compression molding flows become more complex when the material does not slip at the mold surface. This is often referred to as *non-lubricated squeeze flow*, or NLSF.

As with our injection molding examples, we examine solutions for a Newtonian fluid with constant viscosity. We also assume that the gap height h is small compared to the lateral dimensions of the charge. With these assumptions the velocity solution for **non-lubricated planar squeeze flow** is

$$v_1 = -\frac{3}{2}\frac{\dot{h}}{h}x_1\left(1-\frac{x_3^2}{h^2}\right) \qquad v_2 = 0 \qquad v_3 = \frac{3}{2}\frac{\dot{h}}{h}\left(x_3 - \frac{x_3^3}{3h^2}\right) \tag{4.116}$$

The in-plane component v_1 has the same parabolic profile as we saw in the injection molded strip, Eqn. (4.107), and the velocity also increases with x_1 just as in the lubricated squeeze flow. This velocity distribution is sketched in Fig. 4.30. The gapwise velocity x_3 is more complicated, though it still has $v_3 = 0$ at $x_3 = 0$ and $v_3 = \dot{h}$ at $x_3 = h$.

Working out the velocity gradient tensor for this flow, we find

$$\mathbf{L} = -\frac{3}{2}\frac{\dot{h}}{h}\begin{bmatrix} \left(1-x_3^2/h^2\right) & 0 & -2x_1x_3/h^2 \\ 0 & 0 & 0 \\ 0 & 0 & -\left(1-x_3^2/h^2\right) \end{bmatrix} \tag{4.117}$$

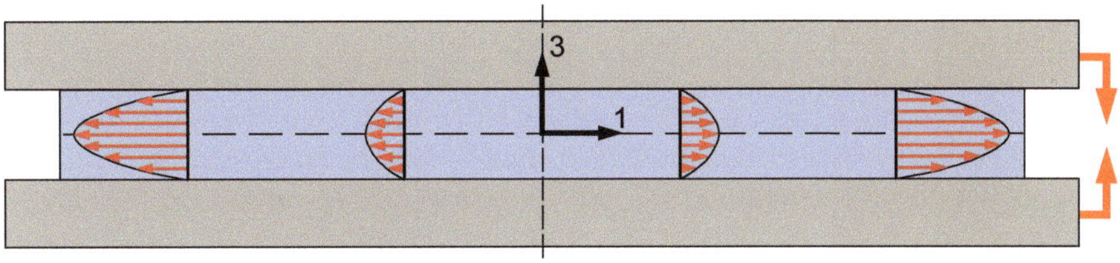

Figure 4.30 Horizontal velocity distribution $v_1(x_1,x_3)$ for non-lubricated squeeze flow.

This is a combination of planar extensional flow (L_{11} and L_{33}) and simple shear flow (L_{13}). The elongation rate is independent of x_1 but varies across the thickness in a parabolic fashion, just like the gapwise velocity. The shear rate is zero at the midplane ($x_3 = 0$) and maximum at the mold walls, varying linearly with x_3 just as in the injection molded strip. In addition, the shear rate increases with x_1, since the velocity also increases in that direction.

The ratio between the elongation rate and the shear rate varies over the charge in a complex fashion. However, both types of deformation tend to align fibers in the 1 direction, which is both the stretching direction for the extensional flow and the flow direction for the shear flow. Thus, we would expect this flow to orient fibers in the flow direction.

We can also have a **non-lubricated radial squeeze flow**. Now the velocity distribution is

$$v_r = -\frac{\dot{h}}{h}\frac{3r}{4}\left(1-\frac{z^2}{h^2}\right) \qquad v_\theta = 0 \qquad v_z = \frac{3}{2}\frac{\dot{h}}{h}\left(z-\frac{z^3}{3h^2}\right) \tag{4.118}$$

Again the gapwise profile is parabolic and the magnitude of velocity increases with r, so the general picture is like Fig. 4.30. Contrast this with the radial flow in an injection molded disk in Fig. 4.19. In the injection molding case the flow rate Q is the same at every radius, so the velocity decreases with r. In compression molding the flow rate increases with r, because the squeezing motion operates over the entire area inside that radius.

Following through on the details, the rate-of-deformation and vorticity tensors for this flow are

$$\mathbf{D} = -\frac{3}{4}\frac{\dot{h}}{h}\begin{bmatrix} \left(1-z^2/h^2\right) & 0 & -2rz/h^2 \\ 0 & \left(1-z^2/h^2\right) & 0 \\ -2rz/h^2 & 0 & -2\left(1-z^2/h^2\right) \end{bmatrix} \tag{4.119}$$

$$\mathbf{W} = -\frac{3}{4}\frac{\dot{h}}{h}\begin{bmatrix} 0 & 0 & -2rz/h^2 \\ 0 & 0 & 0 \\ 2rz/h^2 & 0 & 0 \end{bmatrix} \tag{4.120}$$

This is a combination of biaxial extension and simple shear. The extensional flow stretches the material equally in the r and θ (or 1 and 2) directions and contracts the material in the z (or 3) direction. The shearing motion is an r–z (or 1–3) shear flow. Biaxial extension dominates close to the midplane ($z = 0$) and near the centerline (r not much larger than h), while shearing dominates close to the mold walls ($z = h$) and at radii $r \gg h$. If the molding is thin, with overall radius $R \gg h$, shearing deformation will dominate and tend to align the fibers radially.

4.4 Chapter Summary

Single, isolated fibers translate with the fluid velocity at their centroid, and their rotation is accurately described by Jeffery's equation. We primarily use the rate form of Jeffery's equation (4.85), which gives $\dot{\mathbf{p}}$ as a function of the current orientation $\mathbf{p}$ and the rate-of-deformation and vorticity tensors $\mathbf{D}$ and $\mathbf{W}$. Jeffery's equation can also be written in a deformation form, Eqn. (4.103), where the final orientation $\mathbf{p}$ is a function of the initial orientation $\mathbf{p}'$ and the deformation gradient tensor $\mathbf{F}$.

The particle shape factor ξ in Jeffery's equation depends on the fiber aspect ratio. For $\xi = 1$, the limit as $\ell / d_f \to \infty$, a fiber rotates like a material line but does not change length. Fibers with a finite aspect ratio ($0 < \xi < 1$) have a periodic motion in simple shear called a Jeffery orbit, but spend most of their time oriented in the flow direction. Disk-shaped particles can be modeled using $\xi < 1$, with $\xi = -1$ being the limit for a thin disk.

Jeffery's equation provides two rules of fiber orientation: shearing flows align fibers in the flow direction, and stretching flows align fibers in the stretching direction. Using these rules, we can explain the typical orientation pattern observed in injection molded parts: two shell layers of flow-oriented fibers surrounding a core layer. The shell layers are aligned in the flow direction, while the core orientation may be in the flow direction, the crossflow direction, or random in plane, depending on whether the flow is converging, diverging, or parallel. Compression molding flows follow the same rules, though one must account for the squeezing deformation from mold closing.

5 Flow Orientation of Groups of Fibers

While Jeffery's equation accurately describes the motion of a single fiber, no practical composite has just one fiber. Instead, real composites have large numbers of fibers, and it is impractical to track each fiber individually. Instead, computer-aided engineering tools predict the second-order orientation tensor at each point in the part. This chapter develops the models used to do that.

The first step is to derive an equation for the orientation tensor when each fibers follows Jeffery's equation. This introduces the closure problem, where we need to approximate the fourth-order orientation tensor in terms of the second-order orientation tensor. We also show how to set up a solution for the full orientation distribution function, initially for Jeffery's equation. Distribution-function solutions require far more computing that tensor solutions, so they are not used in engineering software. However, they are useful in forming closure approximations. The main options for closure approximations are also discussed here.

The combination of the Jeffery equation for **A** and a good closure approximation is a powerful tool, but its predictions do not match the degree of fiber alignment observed in practice. Each fiber in a typical discontinuous fiber composite interacts with many neighboring fibers, which alters its orientation behavior. These interactions cannot be modeled exactly, so empirical modifications to Jeffery's equation are used to represent their effects. These modifications include rotary diffusion models, which adjust the steady-state orientation for a given flow, and slow kinetics models, which alter the rate at which the orientation moves toward this steady state. All of these models are used in commercial software. This chapter focuses on understanding the options for modeling orientation, knowing what each option can and cannot do, and choosing suitable parameters for a given material.

5.1 Jeffery's Equation for the Orientation Tensor

Suppose we wanted to know the orientation tensor versus time $\mathbf{A}(t)$ for a group of fibers where each fiber follows Jeffery's equation. All of the fibers have the same particle shape factor ξ and all experience the same velocity gradient $\mathbf{L}$, but each fiber has its own initial orientation $\mathbf{p}(t = 0)$. We can use the initial $\mathbf{p}$ values to calculate the initial second-orientation tensor $\mathbf{A}(t = 0)$. In principle, we could integrate Jeffery's equation for each fiber and use those $\mathbf{p}$ vectors to find $\mathbf{A}(t)$ at any later time. However, for a large number of fibers this would require a great deal of calculation. It would be much more efficient if we could find $\mathbf{A}(t)$ without tracking each individual fiber.

5.1.1 Derivation

What we want is an equation for $\dot{\mathbf{A}}$, the time derivative of $\mathbf{A}$ following a material point. To derive this, it is helpful to think of the orientation average $\langle \mathbf{pp} \rangle$ as an arithmetic average over a group of discrete $\mathbf{p}$ vectors, as in Eqn. (2.19). However, now each $\mathbf{p}$ vector can change over time as the flow reorients the fibers[1].

The derivation starts by taking the material derivative of the definition of the second-order orientation tensor, $\mathbf{A} = \langle \mathbf{pp} \rangle$:

$$\dot{\mathbf{A}} = \langle \dot{\overline{\mathbf{pp}}} \rangle \tag{5.1}$$

Using the product rule this is

$$\dot{\mathbf{A}} = \langle \dot{\mathbf{p}}\mathbf{p} \rangle + \langle \mathbf{p}\dot{\mathbf{p}} \rangle \tag{5.2}$$

Jeffery's result tells us that fibers translate with the fluid motion, so the derivative following a fiber is identical to the material derivative. That lets us substitute $\dot{\mathbf{p}}$ from Jeffery's equation, Eqn. (4.85), to obtain

$$\dot{\mathbf{A}} = \big\langle \mathbf{W}\cdot\mathbf{pp} + \xi\mathbf{D}\cdot\mathbf{pp} - \xi(\mathbf{D}:\mathbf{ppp})\mathbf{p} \big\rangle + \big\langle \mathbf{p}\mathbf{W}\cdot\mathbf{p} + \xi\mathbf{p}\mathbf{D}\cdot\mathbf{p} - \xi\mathbf{p}(\mathbf{D}:\mathbf{ppp}) \big\rangle \tag{5.3}$$

We can use the identity $\mathbf{W}\cdot\mathbf{p} = \mathbf{p}\cdot\mathbf{W}^T$ to do rearrangements like $\mathbf{p}\mathbf{W}\cdot\mathbf{p} = \mathbf{pp}\cdot\mathbf{W}^T$. Also, recall that $(\mathbf{D}:\mathbf{pp})$ is a scalar, so it can appear in any order in a product. This gives

$$\dot{\mathbf{A}} = \big\langle \mathbf{W}\cdot\mathbf{pp} + \xi\mathbf{D}\cdot\mathbf{pp} - \xi(\mathbf{D}:\mathbf{pp})\mathbf{pp} \big\rangle + \big\langle \mathbf{pp}\cdot\mathbf{W}^T + \xi\mathbf{pp}\cdot\mathbf{D}^T - \xi(\mathbf{D}:\mathbf{pp})\,\mathbf{pp} \big\rangle \tag{5.4}$$

Neither $\mathbf{D}$ nor $\mathbf{W}$ depend on $\mathbf{p}$, so they can come outside the brackets of the orientation average. We also rearrange the right-hand side to group similar terms.

$$\dot{\mathbf{A}} = \mathbf{W}\cdot\langle \mathbf{pp} \rangle + \langle \mathbf{pp} \rangle\cdot\mathbf{W}^T + \xi\left(\mathbf{D}\cdot\langle \mathbf{pp} \rangle + \langle \mathbf{pp} \rangle\cdot\mathbf{D}^T - 2\mathbf{D}:\langle \mathbf{pppp} \rangle\right) \tag{5.5}$$

Now the orientation averages only contain $\langle \mathbf{pp} \rangle$, which is the second-order tensor $\mathbf{A}$, and $\langle \mathbf{pppp} \rangle$, which is the fourth-order orientation tensor $\mathbb{A}$. Substitute these in, and also recall that the rate-of-deformation tensor is symmetric, $\mathbf{D}^T = \mathbf{D}$, and the vorticity tensor is anti-symmetric, $\mathbf{W}^T = -\mathbf{W}$. This gives the final result:

$$\dot{\mathbf{A}} = \mathbf{W}\cdot\mathbf{A} - \mathbf{A}\cdot\mathbf{W} + \xi(\mathbf{D}\cdot\mathbf{A} + \mathbf{A}\cdot\mathbf{D} - 2\mathbf{D}:\mathbb{A}) \qquad \text{Jeffery fibers} \tag{5.6}$$

This is an exact transformation of Jeffery's equation, and all the Jeffery terms are here. If the flow has rigid-body rotation, the terms proportional to $\mathbf{W}$ will rotate the $\mathbf{A}$ tensor. The terms proportional to $\mathbf{D}$ will draw the fibers toward the direction of maximum stretch rate. And the term containing $\mathbf{D}:\mathbb{A}$ will keep each $\mathbf{p}$ vector at unit length.

We can integrate Eqn. (5.6), starting from an initial orientation tensor $\mathbf{A}(t=0)$, to find $\mathbf{A}(t)$ at any later time. The result will be the same as if we had started with the complete set of

[1] A more formal approach to this derivation is to use the integral definition of the orientation average, Eqn. (2.33), treat $\mathbf{p}$ as an independent variable, and allow $\psi(\mathbf{p})$ to vary in time as the fibers change orientation. The mathematics of that derivation are more detailed (see Appendix B.2), but the result is the same.

initial **p** vectors, integrated each one using Jeffery's equation, and computed $\mathbf{A}(t)$ from the final orientation vectors. Equation (5.6) allows us to model dozens, thousands or millions of fibers using a single orientation tensor. This greatly reduces computational effort, and has made the prediction of fiber orientation a practical possibility for computer-aided engineering.

We still have to ensure that the predictions match experiments, which will require adding some new terms to Eqn. (5.6). Later sections in this chapter will cover those adjustments to the model.

We also need to deal with the presence of the fourth-order tensor in Eqn. (5.6), since we do not actually know $\mathbb{A}$. This term is treated using a *closure approximation*, which approximates $\mathbb{A}$ at any time using the current value of **A**. There are several closure approximations one might use. The simplest is called the *quadratic closure*, and it is

$$\mathbb{A}^{Q}_{ijkl} = A_{ij} A_{kl} \tag{5.7}$$

With a closure approximation, Eqn. (5.6) can be solved for any flow. Other, more accurate closures are used in commercial software, and we will discuss the best closure options in Section 5.4. For the moment, the quadratic closure allows us to explore solutions to the tensor form of Jeffery's equation.

5.1.2 Numerical Solution for a Material Point

The numerical solution of Eqn. (5.6) for a single material point proceeds much like the numerical solution of Jeffery's equation discussed in Section 4.2.3. We view Eqn. (5.6) as a set of ordinary differential equations for the components of **A**, and use an integration scheme like the fourth-order Runge–Kutta method (`ode45` in MATLAB).

Here we will use the symmetry of **A** and solve for six components. This works well in MATLAB, which is most efficient when operating on entire matrices. It is also possible to use the trace property of **A** to eliminate another component, leaving only five variables. This is common in commercial software, where the gain in efficiency is worth the additional programming complexity.

As in Section 4.2.3, we write a function to calculate the time derivative, $\dot{\mathbf{A}}$. Here is a function that does that for Jeffery's equation using the quadratic closure.

```
function Aderiv = AdotJeffQuad(Av, L, xi)
%   Aderiv = AdotJeffQuad(Av,L,xi) returns the derivatives Aderiv of the
%   components of the second-order orientation tensor when all fibers
%   follow Jeffery's equation and the quadratic closure is used.  Av
%   contains the orientation tensor components in a column vector, using
%   contracted notation.  L is the velocity gradient tensor (3x3) and xi is
%   the particle shape factor. Aderiv is a 6x1 column vector, also in
%   contracted notation.

A  = vec2tens(Av);              % A in 3x3 matrix form.
A4 = Av*Av';                    % Quadratic closure for the 4th-order
                                % orientation tensor, (6x6) contracted form
R4 = diag([1 1 1 2 2 2]);       % Used for contracted-notation products
W  = (L-L')/2;                  % Vorticity tensor
D  = (L+L')/2;                  % Rate-of-deformation tensor
```

```
% Main equation
Aderiv = tens2vec(W*A - A*W + xi*(D*A + A*D)) - xi*2*A4*R4*tens2vec(D);

return
end
```

This code uses the functions tens2vec and vec2tens from the accompanying software. As shown in Appendix A.5, these functions convert symmetric second-order tensors between the 3×3 matrix form typically used for tensors and the 6×1 column-vector form used in contracted notation. Appendix A.4 explains why the R4 matrix is needed when computing $\mathbb{A} : \mathbf{D}$ in contracted notation.

A short script is used to set up the input variables, call the ode45 function to integrate the differential equation and plot the results.

```
%% Example solution of the orientation tensor equations

% Problem specification
L = zeros(3,3); L(1,3) = 1;     % Velocity gradient, 1--3 simple shear flow
xi = 1;                         % Particle shape factor
tspan = 0:0.20:20;              % Solution times
Avzero = tens2vec(eye(3)/3);    % Initial condition: isotropic orientation

% ODE integration options
options = odeset('RelTol', 1e-6, 'AbsTol', 1e-9);

% Integrate the ODE
[t, Av] = ode45(@(t,Av) AdotJeffQuad(Av,L,xi), tspan, Avzero, options);

% Simple plot of results
figure
plot(t,Av(:,1),'r.-', t,Av(:,3),'b.-', t,Av(:,5),'g.-')
xlabel('$\dot{\gamma}_t$', 'Interpreter', 'LaTeX')
ylabel('A_{ij}')
legend('A_{11}', 'A_{33}', 'A_{13}', 'Location', 'East')
```

Figure 5.1(a) shows this solution. The flow is simple shear, with flow in the 1 direction and shearing in the 3 direction, $v_1 = \dot{\gamma} x_3$, $v_2 = v_3 = 0$. The initial condition is isotropic and all fibers have $\xi = 1$.

A_{11} rises with time and approaches a steady-state value of unity, while A_{33} falls with time and approaches zero at steady state. In this particular example, $A_{22} = A_{33}$. These results are consistent with our first rule of fiber orientation: shear flows orient fibers in the flow direction.

The A_{13} component starts at zero and initially rises, suggesting that one of the principal axes of orientation is between the 1 and 3 direction. At long times A_{13} approaches zero, and it eventually goes to zero. The A_{23} and A_{13} components are zero at all times, so they are not shown.

Once the tensor components have been found, additional information can be calculated at any time step. Figure 5.1(b) shows the largest eigenvalue λ_1 and the direction of its corresponding eigenvector $\mathbf{e}_1$. The eigenvalue λ_1 rises more rapidly than A_{11} and reaches a steady-state value of unity. This tells us that the fibers quickly align around the $\mathbf{e}_1$ axis, and are almost perfectly aligned at steady state. χ_1 is the angle between $\mathbf{e}_1$ and the 1 axis. This eigenvector lies in the 1–3 plane, and first appears at 45° to the 1 axis, along the principal stretch direction of $\mathbf{D}$. It

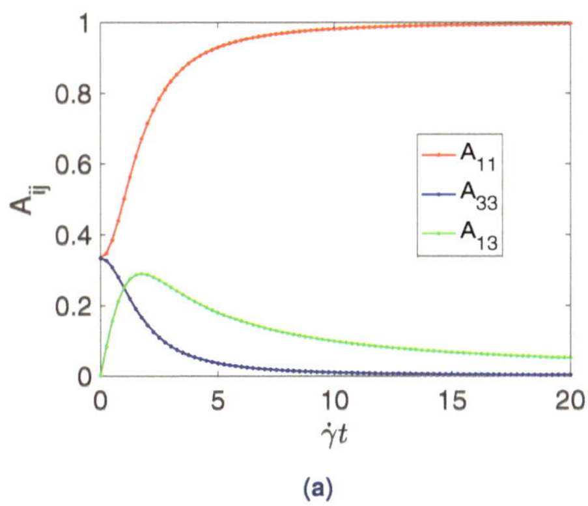

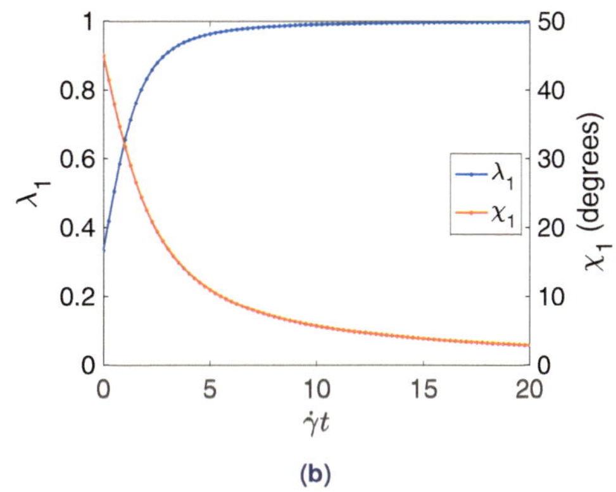

Figure 5.1 Numerical solution to the tensor form of Jeffery's equation, Eqn. (5.6), for simple shear flow using the quadratic closure. Fibers have $\xi = 1$ and the initial orientation is isotropic. (a) Tensor components. In this example $A_{22} = A_{33}$, and $A_{13} = A_{23} = 0$. (b) Largest eigenvalue λ_1, and the angle χ_1 between the corresponding eigenvector and the 1 axis.

rotates toward the 1 axis, but settles toward zero more gradually than λ_1 approaches unity. This is consistent with the behavior of A_{13}. Remember that these results are for Jeffery fibers with $\xi = 1$, using the quadratic closure.

While this is a 3-D solution, the general features are consistent with the planar solution for simple shear in Section 4.2.2.1.

5.1.3 Spatial Fields of Orientation

The solution technique in the preceding section is suitable for a single material point, but fiber orientation in a molded part varies with position and, during filling, with time. In CAE software we want to solve for this spatial field of orientation, $\mathbf{A}(\mathbf{x}, t)$.

It is possible to build up a picture of the spatial field of orientation by repeating the single-point solution for many different points. However, a better approach is to solve for the entire spatial field using the finite element method or some other numerical scheme. This requires a change of viewpoint, from following $\mathbf{A}$ at a moving material point to solving for $\mathbf{A}$ at fixed points in space. In a finite element solution these fixed points are the nodes in the finite element mesh.

The tensor form of Jeffery's equation (5.6) is easily transformed from a material-point expression to spatial-field expression by making the following substitution on the left-hand side:

$$\dot{\mathbf{A}} \equiv \frac{\partial \mathbf{A}}{\partial t} + \mathbf{v} \cdot \nabla \mathbf{A} \qquad \text{or} \qquad \dot{A}_{ij} \equiv \frac{\partial A_{ij}}{\partial t} + v_k \frac{\partial A_{ij}}{\partial x_k} \tag{5.8}$$

The right-hand side of this equation expresses the material derivative $\dot{\mathbf{A}}$ in terms of the spatial field $\mathbf{A}(\mathbf{x}, t)$. Using this on the left-hand side of Eqn. (5.6), or any of the other orientation equations in this chapter, gives a partial differential equation that can be solved for $\mathbf{A}(\mathbf{x}, t)$. Results for this type of solution are shown in Figs. 1.5(a), 6.9 and 6.11. The material derivative is also written as $\mathrm{D}\mathbf{A}/\mathrm{D}t$, and we will use that form where the dot notation is not suitable.

This chapter focuses on the right-hand sides of equations like (5.6), and we will use single-point solutions to explore the behavior of each model. The left-hand sides will be written as $\dot{\mathbf{A}}$, knowing that, with Eqn. (5.8), we can use any such equation in a spatial field solution.

5.2 Solving for the Distribution Function

The tensor form of Jeffery's equation is useful, but it requires a closure approximation for the fourth-order tensor and this may introduce some inaccuracies. An alternative approach that does *not* require a closure is to solve directly for the distribution function $\psi(\mathbf{p}, t)$. These solutions generally require too much computation to be useful for mold filling simulations[2]. However, solutions for a single material point can easily be computed, and they provide a valuable check on tensor-based results and closure approximations. We will see in Section 5.4 that data from distribution function solutions can also be used to produce very accurate closures.

5.2.1 Planar Orientation

To set up a solution for $\psi(\mathbf{p})$ we need an additional equation that expresses conservation of the probability distribution function as the fibers change orientation. This conservation equation is easiest to understand in planar orientation, so we first derive it for that case.

Suppose that we have a planar distribution function $\psi_\phi(\phi)$ at time t, and all the fibers are moving with some known rotation rate $\dot{\phi}(\phi)$. Consider a small increment of angle between ϕ and $\phi + \mathrm{d}\phi$, as shown in Fig. 5.2. At time t the probability of a fiber lying in this angle range is $\psi_\phi(\phi)\,\mathrm{d}\phi$.

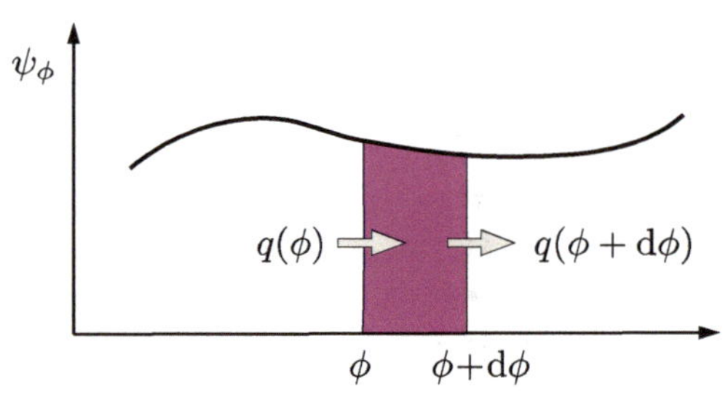

Figure 5.2 A planar distribution function ψ_ϕ and the flux of probability into and out of a small angle increment between ϕ and $\phi + \mathrm{d}\phi$.

If the fibers are changing orientation with time then this probability will change with time, either because fibers from other orientations move into this angle range, or because fibers within this angle range move out. Think of the angle increment between ϕ and $\phi + \mathrm{d}\phi$ as a control volume, and let $q(\phi)$ represent the flux of probability at any angle ϕ. We can then write

2 Though exceptions are starting to appear: see [FMAA20].

a conservation equation for the probability as

$$\frac{\mathrm{D}}{\mathrm{D}t}\left(\psi_\phi \,\mathrm{d}\phi\right) = q(\phi) - q(\phi + \mathrm{d}\phi) \tag{5.9}$$

The material derivative $\mathrm{D}/\mathrm{D}t$ appears on the left-hand side because fibers translate with the fluid, so ψ_ϕ does as well. For a single-point solution, $\mathrm{D}/\mathrm{D}t$ functions like $\partial/\partial t$.

Now divide both sides by $\mathrm{d}\phi$ and take the limit as $\mathrm{d}\phi$ goes to zero. This gives the general result for planar orientation:

$$\frac{\mathrm{D}\psi_\phi}{\mathrm{D}t} = -\frac{\partial q}{\partial \phi} \tag{5.10}$$

This is called the *continuity equation*, and it expresses the conservation of probability as fibers change orientation. Equation (5.10) applies to all fiber suspensions with planar orientation, whatever the flux q or fiber motion $\dot{\phi}$.

For fibers that follow Jeffery's equation, the probability flux $q(\phi)$ is the product of the density, ψ_ϕ, of fibers with orientation ϕ and the speed at which the fibers are changing orientation, $\dot{\phi}$. Thus, the flux due to Jeffery motion is

$$q^J = \psi_\phi \dot{\phi} \tag{5.11}$$

Using this in the general continuity equation (5.10) gives

$$\frac{\mathrm{D}\psi_\phi}{\mathrm{D}t} = -\frac{\partial}{\partial \phi}\left(\psi_\phi \dot{\phi}\right) \tag{5.12}$$

We can substitute Eqn. (4.88) for $\dot{\phi}$ and obtain an explicit expression,

$$\frac{\mathrm{D}\psi_\phi}{\mathrm{D}t} = -\frac{\partial}{\partial \phi}\Big(\psi_\phi \left[-W_{12} - \xi D_{11}\sin(2\phi) + \xi D_{12}\cos(2\phi)\right]\Big) \tag{5.13}$$

This can be solved numerically for a single material point, for example using a finite difference method, to give solutions for $\psi_\phi(\phi, t)$ in any planar flow field. A function that does this, `solvePsi2D`, is available in the accompanying software.

Figure 5.3(a) shows an example solution for fibers with $\xi = 1$ in planar extensional flow. Compare these results to Fig. 4.11. The initial condition is random in plane, so ψ_ϕ is initially constant. As the flow proceeds, the fibers increasingly orient toward the stretching direction, and ψ_ϕ develops a peak at $\phi = 0$. At the same time, the number of fibers near $\phi = 90°$ and $-90°$ decreases. This particular example uses Jeffery's equation for $\dot{\phi}$ and starts from a random-in-plane orientation, so the distributions in Fig. 5.3 match the planar Jeffery distribution of Eqn. (5.67) exactly, within the accuracy of the numerical solution scheme. However, the general approach also applies to any orientation model.

In Fig. 5.3(b), tensor components are computed from the distribution function results at each time step, and these are compared to solutions of the tensor equation (5.6) using the quadratic closure approximation. Both results show the correct trends: fibers gradually align in the 1 direction, which is the direction of greatest positive stretching, so A_{11} increases toward unity and A_{22} decreases toward zero. A_{12} is zero in this example.

The tensor calculation using the quadratic closure goes in the proper direction and reaches the correct steady state, but A_{11} rises too quickly and A_{22} falls too quickly compared to the direct solution for ψ_ϕ. This quantitative error is introduced by the closure approximation, and illustrates the need for more accurate closures. Closure approximations can also cause steady-state errors.

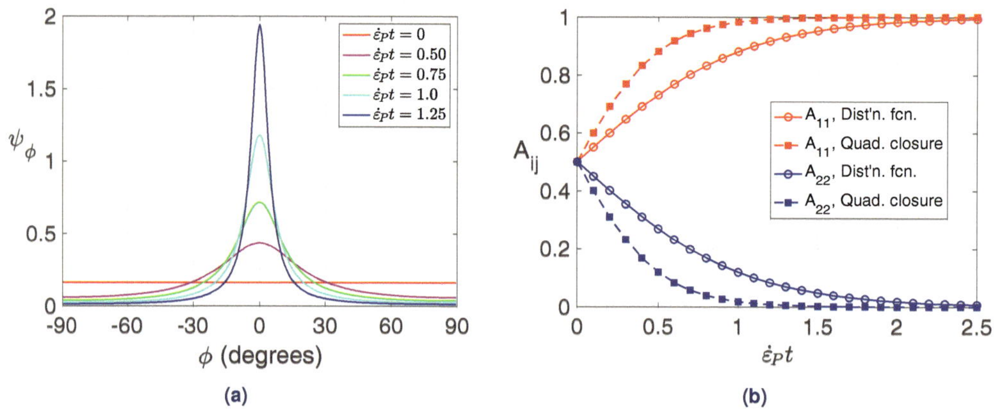

Figure 5.3 (a) Direct solutions for the planar orientation distribution function $\psi_\phi(\phi,t)$ in planar extensional flow with elongation rate $\dot\varepsilon_P$. Jeffery fibers with $\xi = 1$. Initial condition is random in plane. (b) Comparison of the direct solution to a tensor-based solution using the quadratic closure.

5.2.2 3-D Orientation

The 3-D version of the continuity equation (5.12) is derived by similar arguments, and the general form is

$$\frac{D\psi}{Dt} = -\nabla_s \cdot \mathbf{q} \tag{5.14}$$

Here $\mathbf{q}$ is the probability flux, which is now a vector tangent to the surface of the unit sphere. The flux due to Jeffery motion is

$$\mathbf{q}^J = \psi\,\dot{\mathbf{p}} \tag{5.15}$$

If we extract $\dot\theta$ and $\dot\phi$ from the Jeffery solution, $\mathbf{q}^J$ in spherical coordinates is

$$\mathbf{q}^J = \psi\,\dot\theta\,\hat{\boldsymbol{\delta}}_\theta + \psi\,\sin\theta\,\dot\phi\,\hat{\boldsymbol{\delta}}_\phi \tag{5.16}$$

The ∇_s symbol in Eqn. (5.14) is the gradient operator on the surface of the unit sphere. To better understand ∇_s, recall that the gradient operator ∇ in 3-D space is, in Cartesian coordinates,

$$\nabla = \hat{\boldsymbol{\delta}}_1\frac{\partial}{\partial x_1} + \hat{\boldsymbol{\delta}}_2\frac{\partial}{\partial x_2} + \hat{\boldsymbol{\delta}}_3\frac{\partial}{\partial x_3} \tag{5.17}$$

while in spherical coordinates the gradient operator is

$$\nabla = \hat{\boldsymbol{\delta}}_r\frac{\partial}{\partial r} + \hat{\boldsymbol{\delta}}_\theta\frac{1}{r}\frac{\partial}{\partial\theta} + \hat{\boldsymbol{\delta}}_\phi\frac{1}{r\sin\theta}\frac{\partial}{\partial\phi} \tag{5.18}$$

For any function f that varies with position $\mathbf{x}$, the gradient of the function ∇f is a vector pointing in the direction in which f increases most rapidly, with magnitude equal to the rate of increase in that direction.

Now consider a function like $\psi(\mathbf{p})$ that is defined only on the surface of the unit sphere. On this surface $r = 1$ and, since the function is confined to the surface, $\partial\psi/\partial r$ has no meaning. To find the gradient of such a function we need an operator like Eqn. (5.18), but with $r = 1$ and the $\partial/\partial r$ term omitted. This is the surface gradient operator,

$$\nabla_s = \hat{\boldsymbol{\delta}}_\theta \frac{\partial}{\partial \theta} + \hat{\boldsymbol{\delta}}_\phi \frac{1}{\sin\theta} \frac{\partial}{\partial \phi} \tag{5.19}$$

If this operator is applied to a scalar function such as $\psi(\mathbf{p})$, the resulting gradient $\nabla_s \psi$ will have vector components in the θ and ϕ directions, which are everywhere tangent to the surface, but there will be no r component. In some literature, ∇_s is written as $\partial/\partial\mathbf{p}$.

In principle we can substitute the Jeffery flux from Eqn. (5.16) into the continuity equation (5.14) and solve for the distribution function $\psi(\mathbf{p}, t)$ for any flow field. The details of setting up such an equation analytically are explained in [Bay91].

In practice it is common to solve for $\psi(\theta, \phi, t)$ using a finite difference grid in θ–ϕ coordinates, and it is easier to write a continuity expression for a finite-sized control volume in that grid, and develop the finite difference equations from that.

Figure 5.4(a) shows a finite difference grid on the unit sphere. The θ–ϕ space is divided into control volumes or cells, each with dimensions $\Delta\theta$ by $\Delta\phi$. A node is placed at the center of each cell, and the value of ψ is tracked at each node. Let ψ_P be the nodal value of ψ at one such cell, and let $A_P = \sin\theta_P \,\mathrm{d}\theta\, \mathrm{d}\phi$ be the surface area of the cell. Label the boundaries of the cell as n, s, e and w, as shown in Fig. 5.4(b)[3]. Let q_n, q_s, q_e and q_w be the fluxes of probability at these cell boundaries. Also, let $\dot{\theta}_n$, $\dot{\theta}_s$, $\sin\theta_e\, \dot{\phi}_e$ and $\sin\theta_w\, \dot{\phi}_w$ be the fiber velocities at the boundaries, and ψ_n, ψ_s, ψ_e and ψ_w be the distribution function values at the boundaries. q_n and q_s are θ-direction fluxes, while q_e and q_w are ϕ-direction fluxes. Using Eqn. (5.16), these are

$$q_n = \psi_n\, \dot{\theta}_n \tag{5.20}$$

$$q_s = \psi_s\, \dot{\theta}_s \tag{5.21}$$

$$q_e = \psi_e\, \sin\theta_e\, \dot{\phi}_e \tag{5.22}$$

$$q_w = \psi_w\, \sin\theta_w\, \dot{\phi}_w \tag{5.23}$$

Taking account of the length of each boundary edge, the conservation equation for the cell is

$$A_P \frac{\mathrm{D}\psi_P}{\mathrm{D}t} = q_n \sin\theta_n\, \Delta\phi - q_s \sin\theta_s\, \Delta\phi + q_w\, \Delta\theta - q_e\, \Delta\theta \tag{5.24}$$

This is a discrete version of the continuity equation for 3-D orientation, Eqn. (5.14). Inserting finite difference approximations for ψ_n, ψ_s, ψ_e and ψ_w in terms of the nodal values in neighboring cells gives a set of equations that can be solved to find ψ_P at every cell at the next time step. Further details of the equation setup and solutions procedures are available in [Bay91, FHH$^+$08]. `solvePsi3D`, a MATLAB function that implements Eqn. (5.24), is available in the accompanying software[4]. We will build on this equation and see some solutions in Section 5.3.2.

While some efficiency can be gained by only meshing half of the sphere and using the periodicity property of ψ, a typical grid for a single-point solution of $\psi(\mathbf{p})$ might have 1,600 to

---

3 Here, n, s, e, w were chosen to correspond to the compass directions, as if Fig. 5.4(a) were a globe. However, [Bay91] and [FHH$^+$08] define these directions differently.

4 *https://github.com/charlestucker3/Fiber-Orientation-Tools*

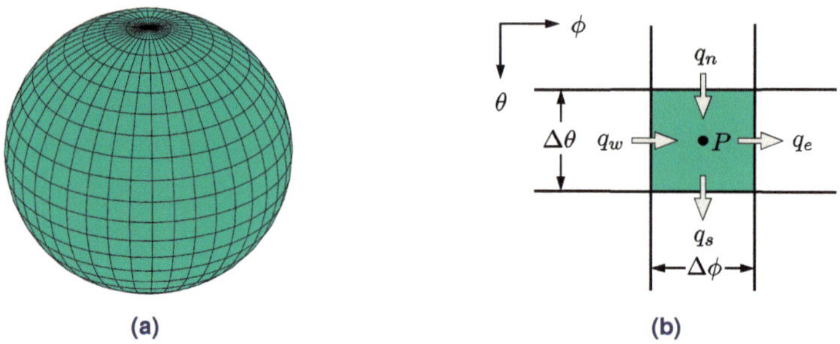

Figure 5.4 (a) A finite difference grid on the unit sphere, as used for numerical solutions of $\psi(\mathbf{p})$ for 3-D orientation. (b) A cell from that grid in θ–ϕ space, showing the fluxes of probability at the north, south, east and west cell boundaries.

8,100 nodes covering the unit sphere. Solving a single-point problem of this size can take several seconds, which is not a long calculation. However, the 3-D mesh for simulating a molded part can easily have one million nodes in physical space. A full distribution function calculation for molded part could easily require solving 10^9 or even 10^{10} orientation variables at each time step. In contrast, the tensor approach needs just five variables at each spatial node. Tensor models can easily be a thousand times more efficient than distribution function models, which is why tensor models are used in computer-aided engineering. Distribution function solutions for a single material point are useful primarily for exploring orientation models and forming closure approximations.

5.3 Isotropic Rotary Diffusion: The Folgar–Tucker Model

While Jeffery's equation gives excellent information about the *qualitative* nature of fiber orientation for any flow field, it cannot predict orientation with useful *quantitative* accuracy. To remedy this, different modifications have been made to Jeffery's equation.

The first feature to modify is steady-state orientation, especially in simple shear flow. We saw in Section 4.3.2 that the shell layer in an injection molded part has very large shear strains, so fiber orientation in the shell is close to the steady-state orientation in simple shear flow. Measurements typically show A_{11} in the range of 0.7 to 0.9 in the shell, where flow is in the 1 direction. However, as we saw in Fig. 5.1, Jeffery's equation goes to $A_{11} = 1$ for large strains in simple shear flow. This is a signicant difference. To remedy this issue, Folgar and Tucker [FT84] proposed adding *rotary diffusion* to the Jeffery model.

The inspiration for this model is Brownian motion. When microscopic spherical particles are suspended in water, individual particles can be seen to have a random translational motion even when the suspension is at rest. This movement, called Brownian motion, is a consequence of collisions between molecules of the suspending fluid and the particles. The collisions occur at random, and if a particle is small enough it will translate slightly before experiencing a collision on the opposite side. If one region of the suspension has a higher concentration of particles than a neighboring region, there will be a net movement of particles from the high-concentration region to the low-concentration region. The net motion follows the mathematics of spatial diffusion.

If the tiny particles are rod-like in shape, collisions with molecules of the fluid also change their orientation. This is called *rotary Brownian motion.* If the suspension is deforming, the rotary Brownian motion is superimposed on the Jeffery motion. Each individual particle has a somewhat erratic movement; however, the net effect of Brownian motion is for particles to move away from the areas of high orientation and toward the areas of low orientation. This is true rotary diffusion.

In discontinuous fiber composites, the fibers are too large for true rotary diffusion to be significant. However, in a composite material, fibers interact with one another, either by direct mechanical contact or by altering the local flow field in the polymer matrix. Modeling all of these interactions is a formidable task. Rather than do that, Folgar and Tucker suggested that the effect of fiber-fiber interactions was to perturb the orientations of the interacting fibers. If these perturbations occur equally in all directions for all orientations, their net effect should be like rotary diffusion.

5.3.1 Planar Orientation Distribution Function

The mathematics of rotary diffusion is most easily understood with planar orientation. The continuity equation (5.10) still applies, and the Jeffery contribution to the orientation flux is given by Eqn. (5.11).

The random motions of diffusion create a flux whose magnitude is proportional to the gradient of concentration and whose direction is opposite to the gradient. For rotary diffusion in planar orientation this diffusive contribution is

$$q^d = -D_r \frac{\partial \psi_\phi}{\partial \phi} \tag{5.25}$$

The constant of proportionality D_r is called the *rotary diffusivity.*

If D_r were constant then rotary diffusion would operate at a constant rate, and the fiber orientation would change even when the suspension is not deforming. This is not what happens in practice. Instead, when the fluid motion stops the fiber motion stops. To capture this effect, Folgar and Tucker proposed that the rotary diffusivity D_r should be proportional to the scalar strain rate,

$$D_r = C_I \dot{\gamma} \tag{5.26}$$

Now $D_r = 0$ when $\dot{\gamma} = 0$, so when the suspension is not deforming and the Jeffery motion stops, the diffusive motion stops too. This assumption also makes the steady-state orientation independent of the strain rate. Since rotary diffusion is a model for fiber-fiber interactions, the constant of proportionality C_I is called the *interaction coefficient.*

To create a full model for planar orientation, we assume that the total flux is the sum of Jeffery and rotary diffusion terms,

$$q = q^J + q^d \tag{5.27}$$

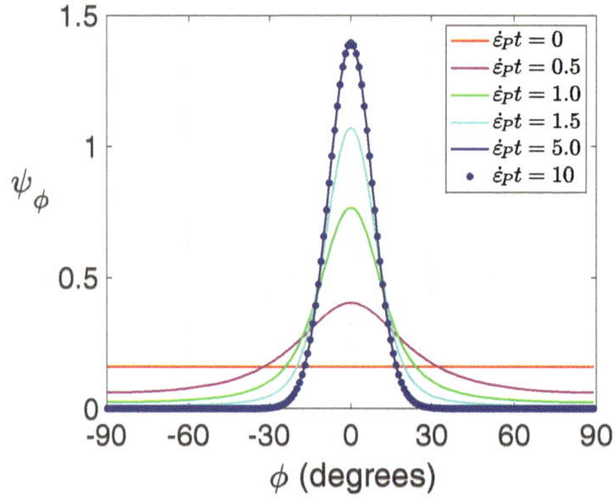

(a) Planar extension

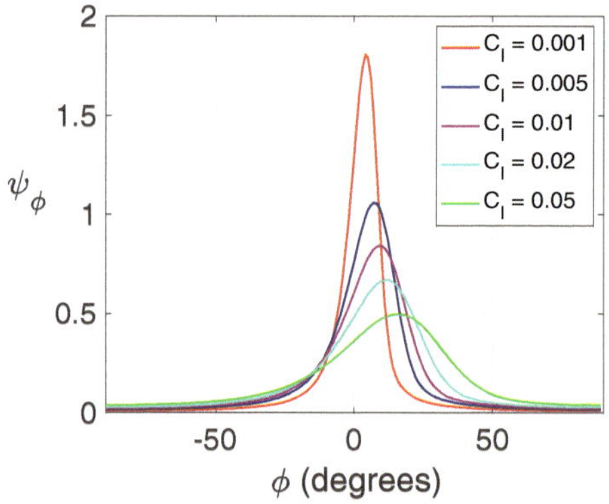

(b) Steady simple shear

Figure 5.5 Numerical solutions of the orientation distribution function ψ_ϕ for the Folgar–Tucker model in planar orientation. (a) Transient solution in planar extensional flow, $C_I = 0.02$. A steady state has been reached by $\dot{\varepsilon}_P t = 5$. (b) Steady-state solutions in simple shear flow for different values of C_I. $\xi = 1$ in both cases.

Then we use the Jeffery flux (5.11) and the diffusive flux (5.25), with Eqn. (5.26) for the diffusivity, in the continuity equation for planar orientation, Eqn. (5.10). The result is

$$\frac{D\psi_\phi}{Dt} = -\frac{\partial}{\partial \phi}\left(\psi_\phi \dot{\phi}\right) + C_I \dot{\gamma} \frac{\partial^2 \psi_\phi}{\partial \phi^2} \tag{5.28}$$

Substituting the Jeffery expression for $\dot{\phi}$, Eqn. (4.88), gives an equation that can be solved for $\psi_\phi(\phi, t)$. Finite difference methods are again useful for this purpose.

Figure 5.5(a) shows a transient solution in planar extensional flow for the Folgar–Tucker model, using `solvePsi2D` from the accompanying software. Like the Jeffery solution in Fig. 5.3(a), an orientation peak builds around $\phi = 0$ and grows sharper with time. However, unlike the Jeffery model, ψ_ϕ reaches a steady state that is *not* fully aligned. The solution at $\dot{\varepsilon}_P t = 10$ is identical to the solution at $\dot{\varepsilon}_P t = 5$. In contrast, the steady-state solution to the Jeffery model is for all fibers to reach $\phi = 0$.

The degree of alignment at steady state can be tuned by adjusting the value of the interaction coefficient C_I. Figure 5.5(b) shows steady-state solutions in simple shear flow for a range of interaction coefficients. As C_I increases, the peak in orientation becomes broader and the fibers are less aligned at steady state. Also, in simple shear flow the peak in ψ_ϕ shifts to larger values of ϕ as C_I increases.

As Fig. 5.5 shows, even a small value of the interaction coefficient can have a significant effect on the steady-state orientation. C_I values of 0.01 or less are typically used to match the behavior of real composites. When adjusting the value of C_I to match data, it is helpful to think of C_I on a logarithmic scale, and make initial adjustments by factors of five or ten.

5.3.2 3-D Orientation Distribution Function

The rotary diffusion model can easily be extended to 3-D orientation. The 3-D orientation version of Eqn. (5.25) gives the diffusive part of the orientation flux as

$$\mathbf{q}^d = -C_I \dot{\gamma}\ \nabla_s \psi \tag{5.29}$$

We use this together with the Jeffery flux (5.15) in the 3-D continuity equation (5.14) to get an equation that can be solved for $\psi(\mathbf{p}, t)$:

$$\begin{aligned} \frac{\mathrm{D}\psi}{\mathrm{D}t} &= -\nabla_s \cdot (\psi \dot{\mathbf{p}} - C_I \dot{\gamma}\ \nabla_s \psi) \\ &= -\nabla_s \cdot (\psi \dot{\mathbf{p}}) + C_I \dot{\gamma}\ \nabla_s^2 \psi \end{aligned} \tag{5.30}$$

To form a numerical solution, it is helpful to use the expression for ∇_s in spherical coordinates from Eqn (5.19) to obtain the θ and ϕ components of the diffusive flux. These components are

$$q_\theta^d = -C_I \dot{\gamma}\ \frac{\partial \psi}{\partial \theta} \tag{5.31}$$

$$q_\phi^d = -C_I \dot{\gamma}\ \frac{1}{\sin\theta} \frac{\partial \psi}{\partial \phi} \tag{5.32}$$

Referring to Fig. 5.4(b), one adds q_θ^d terms to the north and south fluxes in Eqns. (5.20) and (5.21), and q_ϕ^d terms to the east and west fluxes in Eqns. (5.22) and (5.23). These total fluxes are used in the continuity equation (5.24), and the numerical solution proceeds as before.

Figure 5.6 shows an example solution of the distribution function using `solvePsi3D` from the accompanying software. The interaction coefficient is $C_I = 0.02$, a larger value than is typically needed in practice. The flow is 1–3 simple shear, $v_1 = \dot{\gamma} x_3$ and $v_2 = v_3 = 0$, and the initial condition is isotropic. At $\dot{\gamma} t = 1$ a region of preferred orientation is just starting to develop between the 1 and the 3 axes, near the direction of maximum stretching. At $\dot{\gamma} t = 2$ the orientation is clearly concentrated in this region, which has shifted toward the 1 axis. From $\dot{\gamma} t = 2$ to $\dot{\gamma} t = 15$ the change in ψ is modest, with stronger alignment and additional movement of the alignment region toward the flow direction (the 1 axis). The orientation at $\dot{\gamma} t = 15$ is nearly at steady state.

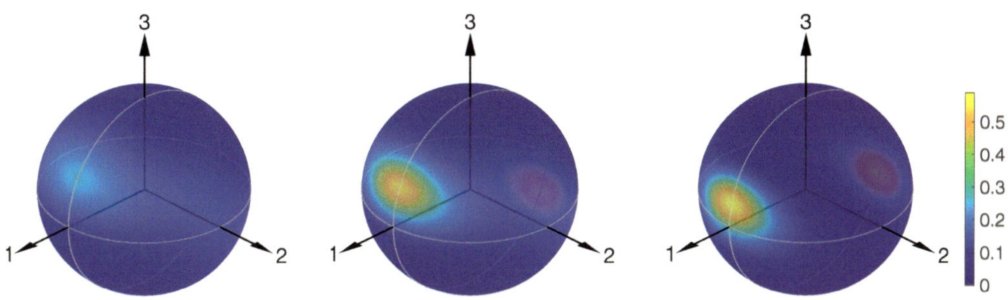

(a) $\dot{\gamma} t = 1$ (b) $\dot{\gamma} t = 2$ (c) $\dot{\gamma} t = 15$

Figure 5.6 Orientation distribution functions at different times for the Folgar–Tucker model in simple shear flow, $v_1 = \dot{\gamma} x_3$. $C_I = 0.02$, $\xi = 1$ and the initial condition is isotropic. The color scale is the same for all images.

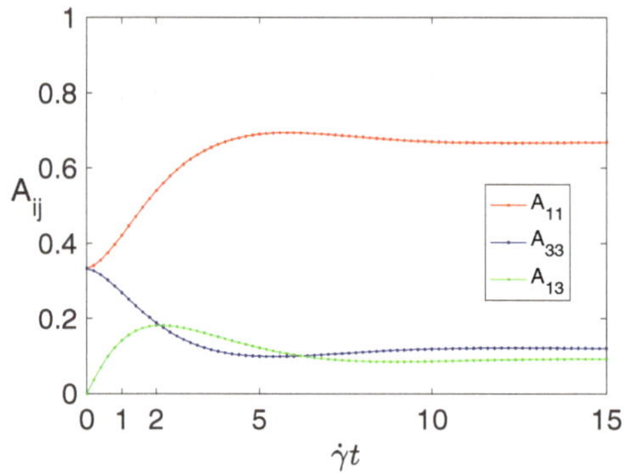

Figure 5.7 Tensor components vs. time for the simple shear flow example in Fig. 5.6. Distribution function solution, $\xi = 1$ and $C_I = 0.02$.

This is confirmed in Fig. 5.7, which shows the tensor components vs. time for this calculation. The final value of A_{11} is 0.668, very different from the near-perfect alignment produced by Jeffery's equation alone. The rotary diffusion term has given us the type of result we were looking for.

5.3.3 Tensor Equation

While distribution function calculations show many details of the model, for computer-aided engineering purposes we would prefer an orientation-tensor model. Once we have Eqn. (5.30), we can derive an equation of change for the orientation tensor for the 3-D Folgar–Tucker model. The derivation is sketched in Appendix B.2, and the result is

$$\dot{\mathbf{A}} = \mathbf{W}\cdot\mathbf{A} - \mathbf{A}\cdot\mathbf{W} + \xi(\mathbf{D}\cdot\mathbf{A} + \mathbf{A}\cdot\mathbf{D} - 2\mathbf{D}:\mathbb{A}) + 2C_I\dot{\gamma}(\mathbf{I} - 3\mathbf{A}) \qquad (5.33)$$

This equation was the first practical tool for modeling fiber orientation in commercial injection molding codes, and it continues to be useful. All other tensor-based models for fiber orientation are refinements of this one.

Comparing Eqn. (5.33) to the Jeffery form of Eqn. (5.6) shows that the Jeffery terms are still present, and that a new term has been added to represent the rotary diffusion. If only the diffusion term were present, the rate of change of the orientation tensor would be

$$\dot{\mathbf{A}}^d = 6C_I\dot{\gamma}\left(\frac{\mathbf{I}}{3} - \mathbf{A}\right) \qquad \text{diffusion only} \qquad (5.34)$$

This equation reaches a steady state ($\dot{\mathbf{A}} = 0$) when $\mathbf{A} = \mathbf{I}/3$ and the orientation is isotropic. The effect of rotary diffusion is to pull the orientation toward isotropy, or randomness. The full Folgar–Tucker model (5.33) reaches a steady state for a given flow when the randomizing effects of diffusion exactly balance the aligning effects of Jeffery's equation. The smaller the value of C_I, the more aligned the fibers must be for this to occur.

Section 4.2.6 discussed the fact that Jeffery's equation obeys a strain-time scaling, where the orientation at any time depends on the accumulated strain, e.g., $\dot{\gamma}t$ for simple shear flow, but the strain rate by itself does not affect the results. The Folgar–Tucker model has this same property, because all the terms on the right-hand side of Eqn. (5.33) are linear in either **D**, **W** or $\dot{\gamma}$. However, while Jeffery's equation was reversible, the rotary diffusion term makes the Folgar–Tucker model *irreversible*. If the flow is reversed, the orientation tensor **A** will not retrace its previous values. When a diffusion term is present, a flowing suspension gradually "forgets" its original orientation state.

The behavior of Eqn. (5.33) depends the closure approximation used for $\mathbb{A}$, so we examine closures next and then provide advice on choosing the value of C_I.

5.4 Closure Approximations

While Eqn. (5.33) is a powerful tool, we cannot use it without knowing the fourth-order tensor $\mathbb{A}$ at each time step. It is possible to derive a similar equation for $\dot{\mathbb{A}}$ using the approach of Section 5.1, but that equation contains the sixth-order orientation tensor $\mathscr{A}$, Eqn. (2.98). Similarly, the equation for the sixth-order tensor contains the eighth-order tensor, and so forth.

To resolve this dilemma we replace $\mathbb{A}$ in Eqn. (5.6) with an approximation. Such functions are called *closure approximations*, since they close the infinite set of equations for **A**, $\mathbb{A}$, $\mathscr{A}$, etc.

There is no unique way to form a closure approximation, and a variety of closures have been used to model fiber orientation. The choice of closure affects the accuracy of the orientation calculation, and also the values of orientation parameters like C_I. The quadratic and hybrid closures were the first forms used in fiber orientation modeling, and they are simple to program. More accurate forms were then developed by fitting the closure to detailed information about typical orientation distribution functions. The best fitted closures are quite accurate and are now standard in computer-aided engineering software.

All the closures discussed in this section are implemented in `closeA4` in the accompanying software.

In this section we study closure approximations in the context of fiber orientation modeling, but they also play a key role in the mechanical property predictions discussed in Chapt. 8. Fortunately, the best closures for flow-induced orientation also work well for mechanical properties.

5.4.1 Quadratic, Linear and Hybrid Closures

All closure approximations give the fourth-order orientation tensor $\mathbb{A}$ as a function of the second-order tensor **A**. The simplest such form is the *quadratic closure* [Doi81],

$$\mathbb{A}^{Q}_{ijkl} = A_{ij} A_{kl} \tag{5.35}$$

This form does not have all the symmetries of the exact fourth-order orientation tensor, Eqn. (2.67), but it does have the symmetries of an elasticity tensor, Eqn. (2.68), and that is adequate for a closure approximation [CT95]. The quadratic closure has the contraction property of the fourth-order tensor, $\mathbb{A}_{ijkk} = A_{ij}$, Eqn. (2.69).

The quadratic closure is simple, and we saw in Section 5.2.1 that it gives reasonable results when used in Eqn. (5.6). It is also exact when all the fibers are aligned in a single direction. However, the quadratic closure is not as accurate as other choices.

An alternative way to form a closure [Han62] is to use all the products of **A** and **I** that are linear in **A**. We can also use the unit tensor **I** (whose components are δ_{ij}) in forming closures, since it is isotropic and thus introduces no new directional information. The result is called the *linear closure*:

$$\mathbb{A}^{\mathrm{L}}_{ijkl} = -\frac{1}{35}\left(\delta_{ij}\delta_{kl} + \delta_{ik}\delta_{jl} + \delta_{il}\delta_{jk}\right) + \frac{1}{7}\left(A_{ij}\delta_{kl} + A_{ik}\delta_{jl} + A_{il}\delta_{jk} + A_{jl}\delta_{ik} + A_{jk}\delta_{il} + A_{kl}\delta_{ij}\right) \tag{5.36}$$

The form of each term ensures that the closure has the full symmetries of the fourth-order tensor, and the numerical coefficients ensure that $\mathbb{A}_{ijkk} = A_{ij}$. For planar orientation the form is the same, but the indices i, j, k and l take only the values 1 and 2 and the numerical factors are $-1/24$ and $1/6$.

The linear closure is exact for isotropic orientation and is accurate when the orientation state is only slightly perturbed from isotropy. It is equivalent to assuming that the distribution function is given exactly by a second-order Fourier or spherical harmonic reconstruction based on the second-order tensor (see Section 2.4). However, the linear closure is inaccurate if the fibers are even moderately aligned, and will produce non-physical solutions in some flows. For this reason, the linear closure is rarely used by itself.

One can, however, mix the linear closure with the quadratic closure to create a closure that is accurate in both the aligned and random limits. This is the idea behind the *hybrid closure* [AT87, AT90], which is

$$\mathbb{A}^{\mathrm{H}}_{ijkl} = (1-f)\mathbb{A}^{\mathrm{L}}_{ijkl} + f\mathbb{A}^{\mathrm{Q}}_{ijkl} \tag{5.37}$$

The scalar factor f should depend on some scalar invariant of **A**, be equal to unity for fully aligned fibers, and be zero for an isotropic orientation. For 3-D orientation states, the recommended form is [AT90]

$$f = 1 - 27\det\mathbf{A} \tag{5.38}$$

For all planar orientation states $\det\mathbf{A} = 0$, so the hybrid closure for planar orientation uses [AT87]

$$f = 2A_{ij}A_{ji} - 1 \tag{5.39}$$

5.4.2 Fitted Closures

An alternate approach to forming closure approximations, which has proven very successful, is to fit the closure to data based on a set of orientation distribution functions. There are two forms for the fitting functions, one based on eigenvalues and eigenvectors [CT95], and another based on the invariants of the second-order tensor [CK02]. We cover those general forms first, then discuss options for gathering the distribution function data.

5.4.2.1 Eigenvalue-Based Orthotropic Fitting

From the discussion in Section 2.3.4.2, we can think of the second-order orientation tensor **A** as being composed of three eigenvalues λ_1, λ_2 and λ_3, and three corresponding eigenvectors $\mathbf{e}_1$, $\mathbf{e}_2$ and $\mathbf{e}_3$. The eigenvalues provide information about the shape of the orientation state and the eigenvectors tell how that shape is oriented in space.

The eigenvalue-based approach to closures uses these ideas to simplify the fitting process. We start in the principal axis system of the second-order tensor (i.e., the coordinate axes corresponding to the eigenvectors). In this coordinate system the second-order tensor is

$$\bar{\mathbf{A}} = \begin{bmatrix} \lambda_1 & 0 & 0 \\ 0 & \lambda_2 & 0 \\ 0 & 0 & \lambda_3 \end{bmatrix} \tag{5.40}$$

Since the trace of **A** equals unity,

$$\lambda_1 + \lambda_2 + \lambda_3 = 1 \tag{5.41}$$

and only two of the eigenvalues are independent. We choose λ_1 and λ_2 as the independent eigenvalues, and recover λ_3 from Eqn. (5.41) when needed.

Each eigenvalue cannot be less than zero or greater than unity. As a result, the space of all possible values of (λ_1, λ_2) is the shaded area in Fig. 5.8(a). We further choose to number the eigenvalues and eigenvectors so that

$$\lambda_1 \geq \lambda_2 \geq \lambda_3 \tag{5.42}$$

Now the space of possible values for (λ_1, λ_2) is the magenta triangle in Fig. 5.8(a). We will call this the *UBT triangle*.

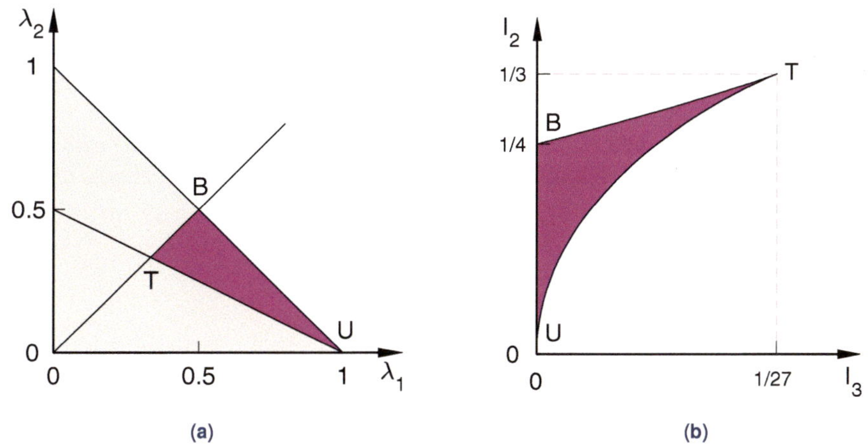

Figure 5.8 The space of all possible orientation states (in color) in terms of (a) eigenvalues and (b) invariants of the second-order orientation tensor. U, B and T represent uniaxial, biaxial and triaxial orientation states, respectively.

The vertices of this triangle represent important orientation states. The U vertex, where $\lambda_1 = 1$ and $\lambda_2 = \lambda_3 = 0$, represents *uniaxial* orientation, where all fibers are aligned in a single direction—in this case along the direction of $\mathbf{e}_1$. The B vertex, where $\lambda_1 = \lambda_2 = 1/2$ and $\lambda_3 = 0$, represents *biaxial* orientation, where fibers all lie in the $\mathbf{e}_1$–$\mathbf{e}_2$ plane and are equally divided between the two axes. Random-in-plane orientation falls on this point. The T vertex, where $\lambda_1 = \lambda_2 = \lambda_3 = 1/3$, represents *triaxial* orientation, where the fibers are evenly divided between the three coordinate directions. The isotropic (3-D random) orientation state falls on this point.

The edges of the UBT triangle also have meaning. On the UB edge we have $\lambda_3 = 0$, so this edge contains all *planar* orientation states, from random-in-plane at B to fully aligned at U. The TU and TB edges contain all the orientation states that are *transversely isotropic*, which means that the orientation has no preferred direction in a plane that is normal (transverse) to a single axis. On the TU edge, $\lambda_1 > \lambda_2 = \lambda_3$, so the fibers are preferentially aligned toward the $\mathbf{e}_1$ axis and have no preferred direction in the $\mathbf{e}_2$–$\mathbf{e}_3$ plane. On the TB edge, $\lambda_1 = \lambda_2 > \lambda_3$, so the fibers are aligned toward the $\mathbf{e}_1$–$\mathbf{e}_2$ plane with no preferred direction in that plane, and $\mathbf{e}_3$ is the symmetry axis.

The independent variables in any fitting process are λ_1 and λ_2, and the fitting functions must span the UBT triangle. What are the dependent variables for the fit?

The fourth-order tensor $\mathbb{A}_{ijkl}$ has 14 independent components, but the number of dependent variables can be greatly reduced by recognizing that any fourth-order tensor coming from a closure approximation must be *orthotropic*. In solid mechanics (see Section 8.1.4), an orthotropic material is one whose properties have three orthogonal planes of symmetry, which are defined by three mutually perpendicular symmetry axes. For example, a laminated composite plate with a 0°/90°/symmetric layup is orthotropic, with symmetry axes corresponding to the two fiber directions and the vector normal to the plate.

A closure approximation, in which $\mathbb{A}$ is formed from $\mathbf{A}$, can only draw its directional information from the eigenvectors of $\mathbf{A}$. If any other directional information were introduced, the closure approximation would change if it were calculated using different coordinate axes. The directional information of $\mathbf{A}$ is contained in its eigenvectors, which are orthogonal. As a result, $\mathbf{A}$ is orthotropic, so $\mathbb{A}$ must be orthotropic as well. This guarantees that many of the components of the fourth-order tensor are zero, when that tensor is expressed in the principal axes of the second-order tensor. Let $\bar{\mathbb{A}}$ be the fourth-order tensor in the principal axes of $\mathbf{A}$. Using contracted notation, this is

$$[\bar{\mathbb{A}}] = \begin{bmatrix} \bar{\mathbb{A}}_{11} & \bar{\mathbb{A}}_{12} & \bar{\mathbb{A}}_{13} & 0 & 0 & 0 \\ \bar{\mathbb{A}}_{21} & \bar{\mathbb{A}}_{22} & \bar{\mathbb{A}}_{23} & 0 & 0 & 0 \\ \bar{\mathbb{A}}_{31} & \bar{\mathbb{A}}_{32} & \bar{\mathbb{A}}_{33} & 0 & 0 & 0 \\ 0 & 0 & 0 & \bar{\mathbb{A}}_{44} & 0 & 0 \\ 0 & 0 & 0 & 0 & \bar{\mathbb{A}}_{55} & 0 \\ 0 & 0 & 0 & 0 & 0 & \bar{\mathbb{A}}_{66} \end{bmatrix} \tag{5.43}$$

In addition, the complete symmetry of the fourth-order tensor, Eqn. (2.67), tells us that

$$\bar{\mathbb{A}}_{12} = \bar{\mathbb{A}}_{21} = \bar{\mathbb{A}}_{66} \tag{5.44}$$

$$\bar{\mathbb{A}}_{13} = \bar{\mathbb{A}}_{31} = \bar{\mathbb{A}}_{55} \tag{5.45}$$

$$\bar{\mathbb{A}}_{23} = \bar{\mathbb{A}}_{32} = \bar{\mathbb{A}}_{44} \tag{5.46}$$

That leaves six independent components. We can also apply the contraction property of the fourth-order tensor, Eqn. (2.69), to get

$$\bar{\mathbb{A}}_{11} + \bar{\mathbb{A}}_{12} + \bar{\mathbb{A}}_{13} = \lambda_1 \quad (5.47)$$

$$\bar{\mathbb{A}}_{21} + \bar{\mathbb{A}}_{22} + \bar{\mathbb{A}}_{23} = \lambda_2 \quad (5.48)$$

$$\bar{\mathbb{A}}_{31} + \bar{\mathbb{A}}_{32} + \bar{\mathbb{A}}_{33} = \lambda_3 \quad (5.49)$$

Now, of the twelve non-zero components of $\bar{\mathbb{A}}$ in Eqn. (5.43), only three are independent. This allows us to define a closure approximation by fitting just three components of $\mathbb{A}$. Traditionally this is done as

$$\bar{\mathbb{A}}_{11} = f_1(\lambda_1, \lambda_2) \qquad \bar{\mathbb{A}}_{22} = f_2(\lambda_1, \lambda_2) \qquad \bar{\mathbb{A}}_{33} = f_3(\lambda_1, \lambda_2) \quad (5.50)$$

Any closure computed this way is called an EBOF closure, for *eigenvalue-based orthotropic fitted.*

As a simple example, assume that the B vertex of the UBT triangle has random-in-plane orientation and the T vertex has isotropic (3-D random) orientation, then interpolate linearly over the triangle. (There is only one possible orientation state for the U vertex: fully aligned.) The result is called the ORS closure (for *orthotropic smooth*) [CT95]:

$$\bar{\mathbb{A}}_{11} = -0.15 + 1.15\,\lambda_1 - 0.10\,\lambda_2 \quad (5.51)$$

$$\bar{\mathbb{A}}_{22} = -0.15 + 0.15\,\lambda_1 + 0.90\,\lambda_2 \quad (5.52)$$

$$\bar{\mathbb{A}}_{33} = 0.60 - 0.60\,\lambda_1 - 0.60\,\lambda_2 \quad (5.53)$$

Once the fitting functions f_1, f_2 and f_3 have been determined, the calculation of $\mathbb{A}$ proceeds as follows:

- Compute the eigenvalues and eigenvectors of **A**, numbering them so that $\lambda_1 \geq \lambda_2 \geq \lambda_3$.
- Compute $\bar{\mathbb{A}}_{11}$, $\bar{\mathbb{A}}_{22}$ and $\bar{\mathbb{A}}_{33}$ using the fitting functions.
- Compute the remaining non-zero components of $\bar{\mathbb{A}}$ using Eqns. (5.44–5.49), and assemble the full $[\bar{\mathbb{A}}]$ tensor using Eqn. (5.43).
- Construct the rotation matrix **Q** between the principal axes and the laboratory axes using the eigenvectors of **A**. (The eigenvectors are the columns of **Q**; see Eqn. (2.55).)
- Transform $\bar{\mathbb{A}}$ to the laboratory coordinates using the fourth-order tensor transformation rule. In tensor notation this is

$$\mathbb{A}_{ijkl} = Q_{im} Q_{jn} Q_{ko} Q_{lp} \bar{\mathbb{A}}_{mnop} \quad (5.54)$$

  Appendix A.4.5 shows how to compute this coordinate transformation in contracted notation. The result is the fourth-order tensor $\mathbb{A}$ corresponding to the second-order tensor **A**.

5.4.2.2 Invariant-Based Orthotropic Fitting

An alternate approach that does not require the computation of eigenvalues and eigenvectors is to use the invariants of the second-order tensor as the independent variables.

Recall from Section 4.1.1.1 that any symmetric second-order tensor has three scalar invariants. For closures we use the versions in Eqns. (4.15–4.17). These invariants of **A**, and their relationship to the eigenvalues, are

$$I_1 = A_{ii} = \lambda_1 + \lambda_2 + \lambda_3 \tag{5.55}$$

$$I_2 = \frac{1}{2}\left[(A_{ii})^2 - \left(A_{ij}A_{ji}\right)\right] = \lambda_1\lambda_2 + \lambda_2\lambda_3 + \lambda_3\lambda_1 \tag{5.56}$$

$$I_3 = \det \mathbf{A} = \lambda_1\lambda_2\lambda_3 \tag{5.57}$$

$I_1 = 1$ for all orientation states, so I_2 and I_3 are the independent variables for fitted closures. Figure 5.8(b) shows the space of all possible orientation states in terms of these invariants.

The most general expression for $\mathbb{A}$ in terms of **A** can be written as [VCD94, CK02]

$$\begin{aligned}\mathbb{A}_{ijkl} = {} & \beta_1\, S(\delta_{ij}\delta_{kl}) + \beta_2\, S(\delta_{ij}A_{kl}) + \beta_3\, S(A_{ij}A_{kl}) + \beta_4\, S(\delta_{ij}A_{km}A_{ml}) \\ & + \beta_5\, S(A_{ij}A_{km}A_{ml}) + \beta_6\, S(A_{im}A_{mj}A_{kn}A_{nl})\end{aligned} \tag{5.58}$$

Here the $S(.)$ operator gives the fully symmetric part of its argument. For any fourth-order tensor $\mathbb{T}$, this is

$$\begin{aligned}S(\mathbb{T}_{ijkl}) = \frac{1}{24}\Big[\; & \mathbb{T}_{ijkl} + \mathbb{T}_{ijlk} + \mathbb{T}_{ikjl} + \mathbb{T}_{iklj} + \mathbb{T}_{iljk} + \mathbb{T}_{ilkj} \\ & \mathbb{T}_{jikl} + \mathbb{T}_{jilk} + \mathbb{T}_{jkil} + \mathbb{T}_{jkli} + \mathbb{T}_{jlik} + \mathbb{T}_{jlki} \\ & \mathbb{T}_{kijl} + \mathbb{T}_{kilj} + \mathbb{T}_{kjil} + \mathbb{T}_{kjli} + \mathbb{T}_{klij} + \mathbb{T}_{klji} \\ & \mathbb{T}_{lijk} + \mathbb{T}_{likj} + \mathbb{T}_{ljik} + \mathbb{T}_{ljki} + \mathbb{T}_{lkij} + \mathbb{T}_{lkji}\;\Big]\end{aligned} \tag{5.59}$$

Using $S(.)$ in Eqn. (5.58) guarantees that the fourth-order tensor will be fully symmetric. Also, since Eqn. (5.58) uses only the second-order tensor on the right-hand side, the resulting fourth-order tensor will be orthotropic.

The six β_is in Eqn. (5.58) are scalar functions of (I_2, I_3), and fitting these functions creates a particular closure. Using the contraction property of the fourth-order tensor, Eqn. (2.69), only three of these functions are independent. Thus, the invariant-based fitting approach requires the fitting of three scalar functions of two independent variables, just like the eigenvalue-based approach.

Chung and Kwon [CK02] explored various options, and recommend that β_3, β_4 and β_6 be the primary fitting variables. They then give equations for β_1, β_2 and β_5 in terms of those three variables plus I_2 and I_3, equations we will not repeat here.

For the invariant-based form, a closure calculation involves finding I_2 and I_3 from the second-order tensor components, using the fitting functions to find β_3, β_4 and β_5, filling in the other β_is, then using those values in Eqn. (5.58) to produce $\mathbb{A}$. All calculations are carried out in the laboratory coordinates, so no eigenvectors or coordinate transformations are required.

Any closure computed this way is called an IBOF closure, for *invariant-based orthotropic fitted.*

5.4.2.3 The Jeffery Distribution and the Natural Closure

We still need some data to which we can fit the closure functions. One option is to use the distribution function that arises for an initially isotropic set of fibers that all follow Jeffery's equation with the same shape factor ξ. This is the *Jeffery distribution*, and it has an analytical expression that we saw in Section 2.4. The resulting closure approximation is called the *natural closure* [VD93, VCD94].

The derivation of the Jeffery distribution starts with the deformation form of Jeffery's equation. From the analysis in Section 4.2.7 we know that when a fiber with $\xi = 1$ and initial orientation $\mathbf{p}'$ is subjected to a deformation gradient $\mathbf{F}$, it changes orientation like a material line but does not change length:

$$\mathbf{p} = \frac{\mathbf{F} \cdot \mathbf{p}'}{|\mathbf{F} \cdot \mathbf{p}'|} \tag{5.60}$$

We also saw that a fiber with $\xi \neq 1$ can be represented by the same equation if we modify the deformation gradient tensor $\mathbf{F}$ as in Eqns. (4.105) and (4.106). We will proceed with the $\xi = 1$ form, but the results can be applied for any value of ξ.

Now consider an initial orientation vector $\mathbf{p}'$ and a few nearby orientation vectors that define an incremental area $\mathbf{dp}'$ on the surface of the unit sphere, as shown in Fig. 5.9(a). When the material is deformed, each vector will reorient according to Eqn. (5.60), defining a new area $\mathbf{dp}$ on the unit sphere.

For Jeffery motion, any fibers oriented within $\mathbf{dp}'$ in the initial state will be also oriented within $\mathbf{dp}$ in the deformed state. Let $\psi'(\mathbf{p}')$ be the initial orientation distribution function and $\psi(\mathbf{p})$ the distribution in the deformed state. The probability of a fiber being oriented within $\mathbf{dp}$ must equal the probability of a fiber being oriented within $\mathbf{dp}'$, so the two distribution functions are connected by

$$\psi'(\mathbf{p}')\, \mathbf{dp}' = \psi(\mathbf{p})\, \mathbf{dp} \tag{5.61}$$

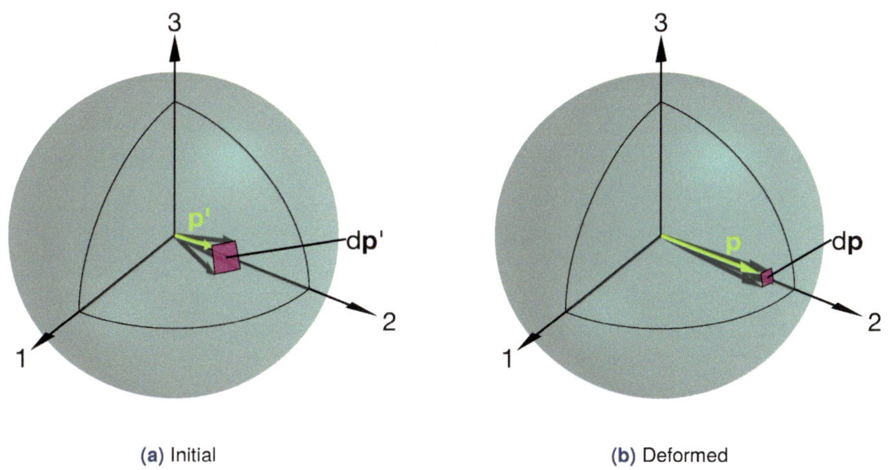

Figure 5.9 (a) An orientation vector $\mathbf{p}'$ (in yellow) and three nearby vectors (gray) defining an area $\mathbf{dp}'$ on the unit sphere in the initial configuration. (b) The corresponding vector $\mathbf{p}$, neighboring vectors and area $\mathbf{dp}$ in the deformed configuration. In this example the deformation is uniaxial extension with stretching along the 2 axis.

We also see from Fig. 5.9 that the area $\mathrm{d}\mathbf{p}$ may be different from $\mathrm{d}\mathbf{p}'$. A detailed analysis [SL96] shows that these two areas are related by the amount a material line along the direction $\mathbf{p}$ was stretched by the deformation $\mathbf{F}$. Following Section 4.1.2.3, stretching of a material line is described by the stretch ratio $(\mathrm{d}\ell/\mathrm{d}\ell')$, which we know depends on $\mathbf{p}$. The relationship for the areas $\mathrm{d}\mathbf{p}$ and $\mathrm{d}\mathbf{p}'$ is

$$\mathrm{d}\mathbf{p} = \frac{\mathrm{d}\mathbf{p}'}{(\mathrm{d}\ell/\mathrm{d}\ell')^3} \tag{5.62}$$

The more the material stretches along the $\mathbf{p}$ direction, the smaller $\mathrm{d}\mathbf{p}$ is relative to $\mathrm{d}\mathbf{p}'$. Using Eqn. (4.69) to replace $(\mathrm{d}\ell/\mathrm{d}\ell')$ makes the dependence on $\mathbf{p}$ clear:

$$\mathrm{d}\mathbf{p} = \mathrm{d}\mathbf{p}'\left(\mathbf{p}\cdot\mathbf{B}^{-1}\cdot\mathbf{p}\right)^{3/2} \tag{5.63}$$

The deformation enters through the Piola tensor $\mathbf{B}^{-1}$ from Eqn. (4.68).

Combining Eqns. (5.61) and (5.63) gives $\psi(\mathbf{p})$ as

$$\psi(\mathbf{p}) = \frac{\psi'(\mathbf{p}')}{\left(\mathbf{p}\cdot\mathbf{B}^{-1}\cdot\mathbf{p}\right)^{3/2}} \tag{5.64}$$

Thus, the greater the stretching along the direction $\mathbf{p}$, the larger $\psi(\mathbf{p})$ is compared to $\psi'(\mathbf{p}')$.

To create the Jeffery distribution we assume that the orientation distribution was isotropic (3-D random) at some time in the past, and choose that as the initial configuration. This gives $\psi'(\mathbf{p}') = 1/4\pi$, so the Jeffery distribution is

$$\psi(\mathbf{p}) = \frac{1}{4\pi\left(\mathbf{p}\cdot\mathbf{B}^{-1}\cdot\mathbf{p}\right)^{3/2}} \quad . \tag{5.65}$$

This produces the type of orientation distribution functions shown in Fig. 2.5, and the choice of the tensor $\mathbf{B}^{-1}$ controls the shape of the distribution function. The only restriction is that $\det\mathbf{B}^{-1} = 1$, since we are interested in deformations at constant volume.

This distribution function was first developed by Dinh and Armstrong [DA84] for fibers with $\xi = 1$. Lipscomb et al. [LDHB88] showed that it also applied for any ξ, and other authors have explored it in detail [SL96, MSHJS11].

For any given $\mathbf{B}^{-1}$ we can use the Jeffery distribution to calculate the second-order and fourth-order orientation tensors. The principal axes of $\mathbf{A}$ must match the principal axes of $\mathbf{B}^{-1}$, and the fourth-order tensor is orthotropic around these axes. There are exact formulas relating $\mathbf{A}$ and $\mathbb{A}$ to $\mathbf{B}^{-1}$ [MSHJS11], but these involve elliptic integrals and there is no analytical formula for $\mathbb{A}$ (or even $\mathbf{B}^{-1}$) in terms of $\mathbf{A}$.

Instead, to generate a fitted closure, one chooses a set of $\mathbf{B}^{-1}$ values, and calculates the second-order and fourth-order tensors for each case. A well-chosen set of $\mathbf{B}^{-1}$ values will uniformly populate the UBT triangle (or the corresponding space in terms of (I_2, I_3)) from Fig. 5.8. The functions f_1, f_2, f_3 (or β_3, β_4, β_6) are then fitted to these points, typically using a high-order polynomial. Following the EBOF procedure from Section 5.4.2.1 or the IBOF procedure from Section 5.4.2.2 then generates the closure.

The closure based on the Jeffery distribution is called the *natural closure*, and was developed by Verleye and Dupret [VD93]. Their original 3-D closure [VCD94] used an IBOF form. An EBOF version was developed by Wetzel [Wet98], and details are reported by VerWeyst [Ver98]. That version is labelled as the ORE closure, and it is used for all natural-closure calculations in this book. Its coefficients are available in `closeA4` in the accompanying software. The IBOF and EBOF versions of the natural closure are identical, except for small numerical differences in the fitting functions.

For **planar orientation**, Verleye and Dupret developed a separate version of the natural closure [VD93]. In this case the initial orientation state is random-in-plane, $\psi'_\phi(\phi') = 1/2\pi$, and deformation only occurs in the plane of the orientation. Note that planar orientation states occupy the UB edge of the UBT triangle (see Fig. 5.8), so it is only necessary to analyze orientation states along that line.

If we restrict attention to the principal axes of the orientation (which are also the principal axes of the deformation) then the deformation gradient tensor has a single scalar parameter F and, to maintain $\det \mathbf{F} = 1$, must have the form

$$\mathbf{F} = \begin{bmatrix} F & 0 \\ 0 & 1/F \end{bmatrix} \tag{5.66}$$

The assumptions of the Jeffery distribution then lead to the orientation distribution function,

$$\psi_\phi(\phi) = \frac{F^2}{2\pi\left[(1-F^4)\cos^2(\phi) + F^4\right]} \tag{5.67}$$

To rotate this from the principal-axis system to the laboratory axes, one can replace ϕ on the right-hand side with $(\phi - \hat{\phi})$, where $\hat{\phi}$ describes the orientation of the eigenvector $\mathbf{e}_1$. This recovers Eqn. (2.102).

For planar orientation, Verleye and Dupret [VD93] developed an analytical relationship between the second-order and fourth-order tensors. In the principal-axis system (and thus useful for an EBOF closure) this is

$$\bar{\mathbb{A}}_{11} = \frac{1}{2}\lambda_1 (1 + \lambda_1) \tag{5.68}$$

$$\bar{\mathbb{A}}_{22} = \frac{1}{2} (1 - \lambda_1)(2 - \lambda_1) \tag{5.69}$$

$$\bar{\mathbb{A}}_{12} = \frac{1}{2}\lambda_1 (1 - \lambda_1) \tag{5.70}$$

The IBOF version of the planar natural closure is

$$\mathbb{A}_{ijkl} = \frac{1}{4}(1 - II_A)\, S(\delta_{ij}\delta_{kl}) + S(A_{ij}A_{kl}) \tag{5.71}$$

This version uses the second invariant as in Eqn. (4.12), $II_A = A_{mn}A_{nm}$. Also, the indices i, j, k, l only take on the values 1 and 2 since, for planar orientation, any tensor component is zero if one or more of its indices equals 3.

5.4.2.4 Closures Using Numerical Distribution Function Solutions

Another source of fitting data is numerical solutions of the distribution function. A flow is selected, and at each time step one uses $\psi(\mathbf{p}, t)$ to calculate the second-order tensor $\mathbf{A}$ and the corresponding fourth-order tensor $\mathbb{A}$. The usual approach is to solve the history of orientation for several different flows, using an isotropic initial condition. The set of flows should provide points in all parts of the UBT triangle. For example, uniaxial extensional flow will generate points along the TU edge, while biaxial extension will generate points along the TB edge. Simple shear flow creates a path into the interior of the triangle. Additional paths are generated by superimposing simple shear flow with some type of extensional flow, for example 1–2 simple shear with uniaxial extension in the 3 direction, and by varying the ratio of shear rate to stretch rate [CT95, CK01, CK02]. Figure 5.10 shows an example. Some additional flows would be needed to have a good data set for closure fitting.

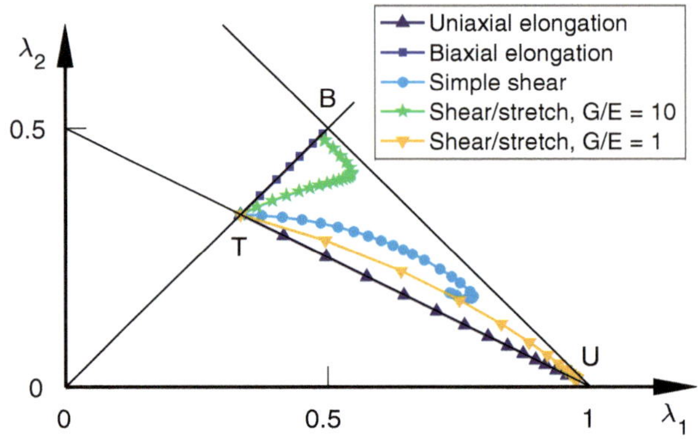

Figure 5.10 Paths generated over the UBT triangle by distribution function solutions for different flow fields. All paths begin at the T vertex. Folgar–Tucker model with $C_I = 0.01$ and $\xi = 1$. The shearing/stretching flows have $v_1 = -Ex_1 + Gx_2$, $v_2 = -Ex_2$ and $v_3 = Ex_3$.

Once the points are generated on the UBT triangle, an eigenvalue-based or invariant-based closure can be fitted. Figure 5.11 shows the fitted functions for $\bar{\mathbb{A}}_{11}$, $\bar{\mathbb{A}}_{22}$ and $\bar{\mathbb{A}}_{33}$ for the natural closure. While the curvature in these surfaces appears to be subtle on this scale, these functions are fourth-order polynomials in λ_1 and λ_2. The corresponding functions for other fitted closures look very similar at this scale.

The first closures fitted to distribution function data were EBOF versions created by Cintra and Tucker [CT95]. Their ORF (*orthotropic fitted*) and ORL (*orthotropic low-C_I*) closures are still used occasionally. Chung and Kwon [CK01] provided an improved fit called ORW that has good accuracy, though the EBOF fit to the natural closure (ORE) should be slightly better. The IBOF approach was developed in detail by Chung and Kwon [CK02]. The closure presented in their paper has been widely used, and is usually called "the IBOF closure."

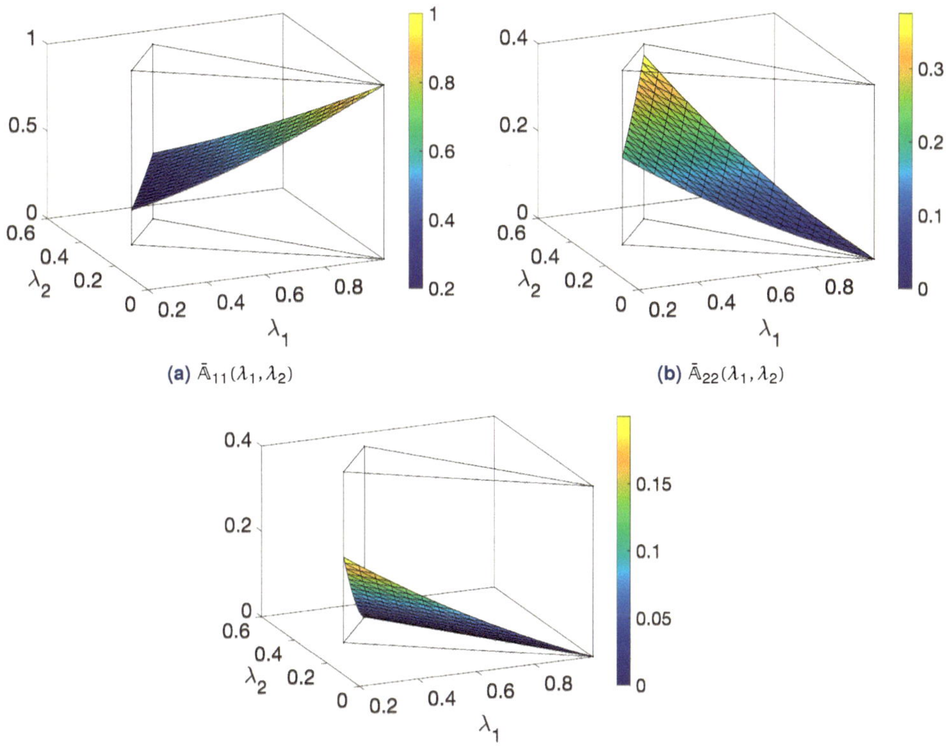

(a) $\bar{\mathbb{A}}_{11}(\lambda_1, \lambda_2)$

(b) $\bar{\mathbb{A}}_{22}(\lambda_1, \lambda_2)$

(c) $\bar{\mathbb{A}}_{33}(\lambda_1, \lambda_2)$

Figure 5.11 Fitted functions for the principal fourth-order components over the UBT triangle for the natural closure.

5.4.3 Comparison of Closure Approximations

The standard way to test a closure approximation is to do a single-point calculation using Eqn. (5.33) with the closure, and compare the results to the full distribution-function solution. The accompanying software provides functions `Adot2` and `closeA4` that implement the Folgar–Tucker model and different closure approximations, as well as `solvePsi3D` for the distribution function. Here we show some tests using simple shear flow. While a thorough closure test would include a variety of flows, simple shear provides a demanding test and is the most important flow for fiber orientation.

5.4.3.1 Equal Interaction Coefficients

Figure 5.12 shows a test of the natural, IBOF and hybrid closures for the Folgar–Tucker model with $C_I = 0.005$ and $\xi = 1$, using an isotropic initial condition. All three closures follow the general trend of the distribution function calculations, with A_{11} rising and A_{33} falling to their steady-state values. A_{13} rises quickly to a peak, then falls to a lower steady-state value. This behavior is consistent with the principal direction of orientation building initially between the 1 and 3 axes and then settling to a steady-state direction close to the 1 axis, as in Fig. 5.6.

All three closures have significant steady-state errors, with the hybrid closure having much larger errors than the natural or IBOF closures. The natural and IBOF closures follow the initial transient rise of A_{11} more closely than the hybrid closure. They also exhibit a slight overshoot in A_{11} before settling to steady state, though the overshoot in A_{11} is much more pronounced in the distribution function results.

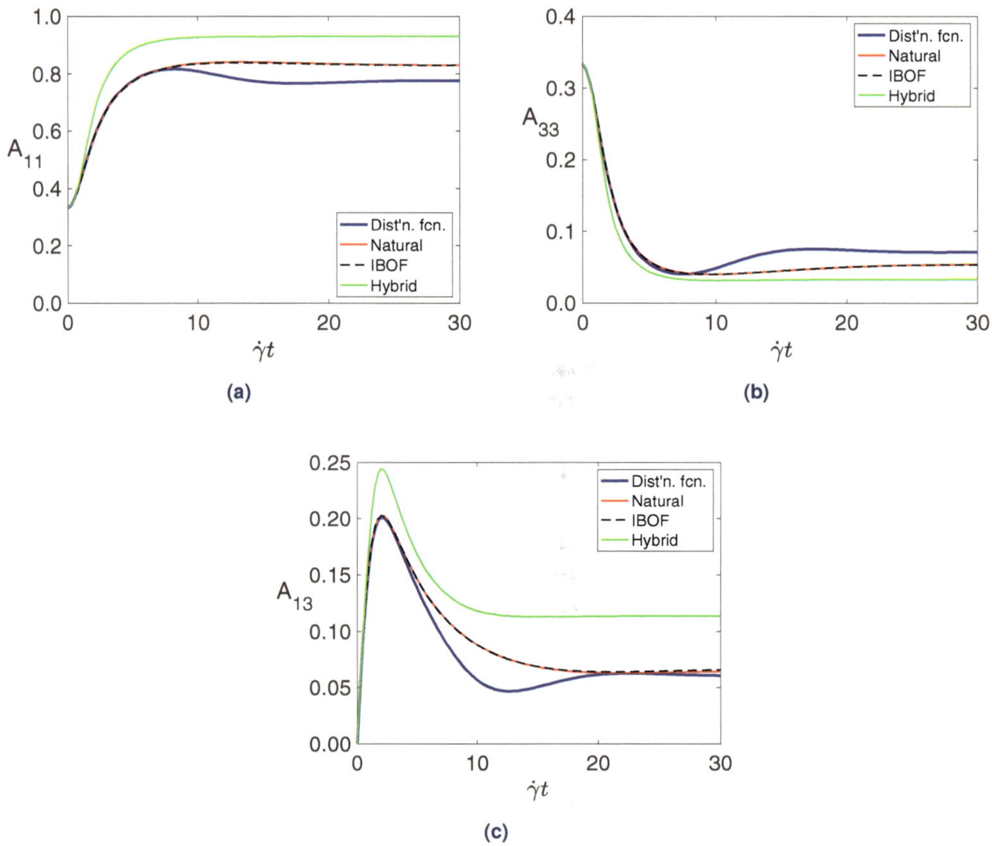

Figure 5.12 Transient behavior of different closure approximations in 1–3 simple shear flow compared to distribution function results. All calculations use the Folgar–Tucker model with $C_I = 0.005$ and $\xi = 1$.

5.4.3.2 Tuned Interaction Coefficients

It is tempting to say that the most accurate closure approximation is the one that matches the distribution function result using the same value of C_I. However, we do not know if rotary diffusion truly represents the dynamics of concentrated fiber suspensions, so there is no guarantee that the distribution function results match the behavior of a real suspension. Since our goal is to predict the behavior of real composites, and since there is no theory to predict the value of the interaction coefficient, it is standard practice to adjust the value of C_I to match the model to some initial data for the composite. Using this C_I value, one should then get useful predictions for the many orientation states that might arise in a molded part.

Given that the value of the interaction coefficient is adjusted when matching experimental data, it makes sense to adjust it for each closure approximation when testing closures. A standard practice is to adjust C_I to match the steady-state value of A_{11} in simple shear flow. Figure 5.13 shows a test of the same three closures as in Fig. 5.12, but with the C_I value for each calculation adjusted to achieve $A_{11} = 0.80$ in steady simple shear.

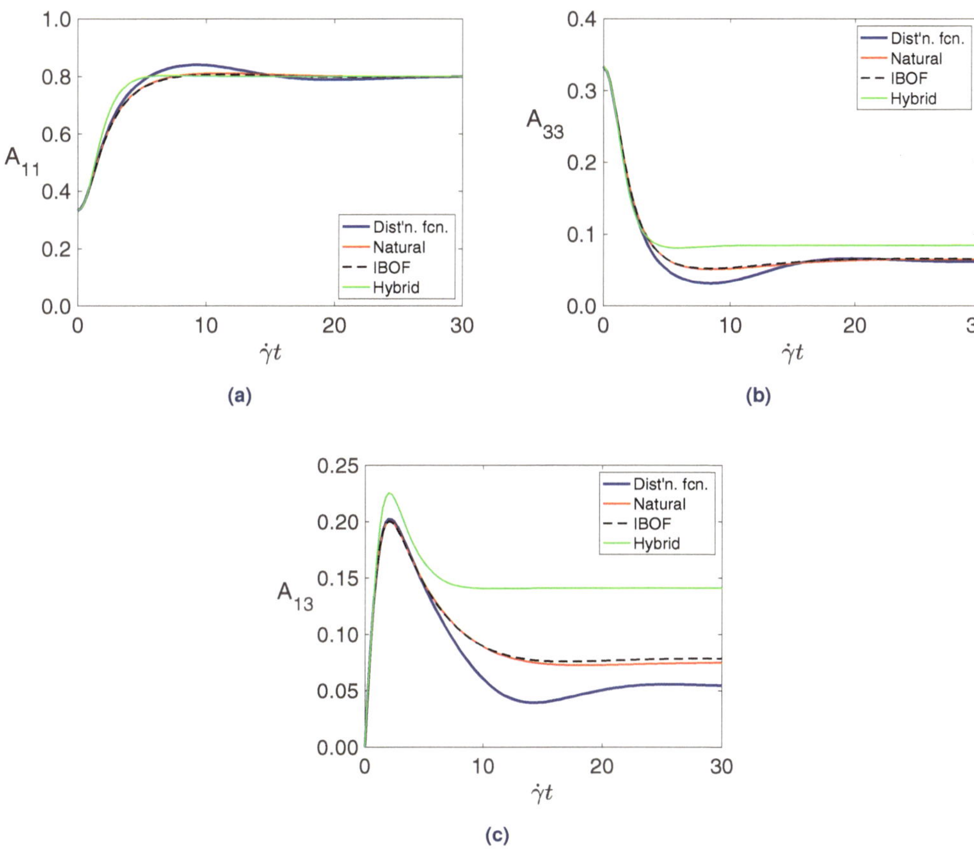

Figure 5.13 Comparison of closure approximations and distribution function results in 1–3 simple shear flow when the C_I value for each closure is adjusted to achieve $A_{11} = 0.80$ at steady state. The interaction coefficient values are given in Table 5.1.

Table 5.1 Interaction coefficients used in Fig. 5.13 to achieve $A_{11} = 0.80$ in steady simple shear flow

Model	C_I
Distribution function	3.41×10^{-3}
Natural closure	7.37×10^{-3}
IBOF closure	7.68×10^{-3}
Hybrid closure	2.57×10^{-2}

Now the errors in A_{11} are similar for all three closures, with the hybrid closure reaching steady state too soon. The errors in A_{33} are also small, especially for the natural and IBOF closures, which match the steady-state value of A_{33} quite well. Errors in A_{13} are a bit larger, but note the expanded vertical scale on this figure.

The C_I value for each set of curves is shown in Table 5.1. In this case the natural and IBOF closures need similar values of the interaction coefficient, and both values are about twice as large as the distribution function value. The hybrid closure needs a C_I value that is almost an order of magnitude larger than the distribution function value to match the steady-state value of A_{11}. This emphasizes an important point about the interaction coefficient:

> The value of the interaction coefficient C_I depends on both the composite being modeled and the closure approximation being used.

When giving the C_I value used to model a composite, one should always specify the closure approximation.

A complete picture of the relationship between C_I, the closure approximation and the steady-state orientation in simple shear flow appears in Fig. 5.14. As $C_I \to 0$ all closures (and the distribution function results) approach a flow-aligned orientation, $A_{11} = 1$ and $A_{33} = A_{13} = 0$. This is the Jeffery result for $\xi = 1$. As C_I becomes very large, rotary diffusion dominates the fiber motion and the orientation approaches the isotropic state, $A_{11} = A_{33} = 1/3$ and $A_{13} = 0$.

Between these extremes the choice of closure approximation has a large influence on the C_I value. This effect is largest in the range of $A_{11} = 0.70$ to 0.90, which is the typical range for discontinuous fiber composites.

If one has experimental data for A_{11} in simple shear flow for a particular composite, Fig. 5.14(a) can be used (together with a closure choice) to choose a value of C_I. Since there is only one parameter to adjust in the Folgar–Tucker model, there is no way to adjust the steady-state values of A_{33} or A_{13}. One must simply accept the values associated with the chosen interaction coefficient.

5.4.3.3 Particle Shape Factor

While we are testing closures, let us look at the effect of the particle shape factor ξ. Recall from Section 4.2.4 that, for cylindrical fibers, ξ depends on the aspect ratio ℓ/d_f. Also, from Section 4.2.5, a fiber with a finite aspect ratio (and thus $\xi < 1$) executes a Jeffery orbit in simple shear flow, rotating periodically rather than settling monotonically to a steady state.

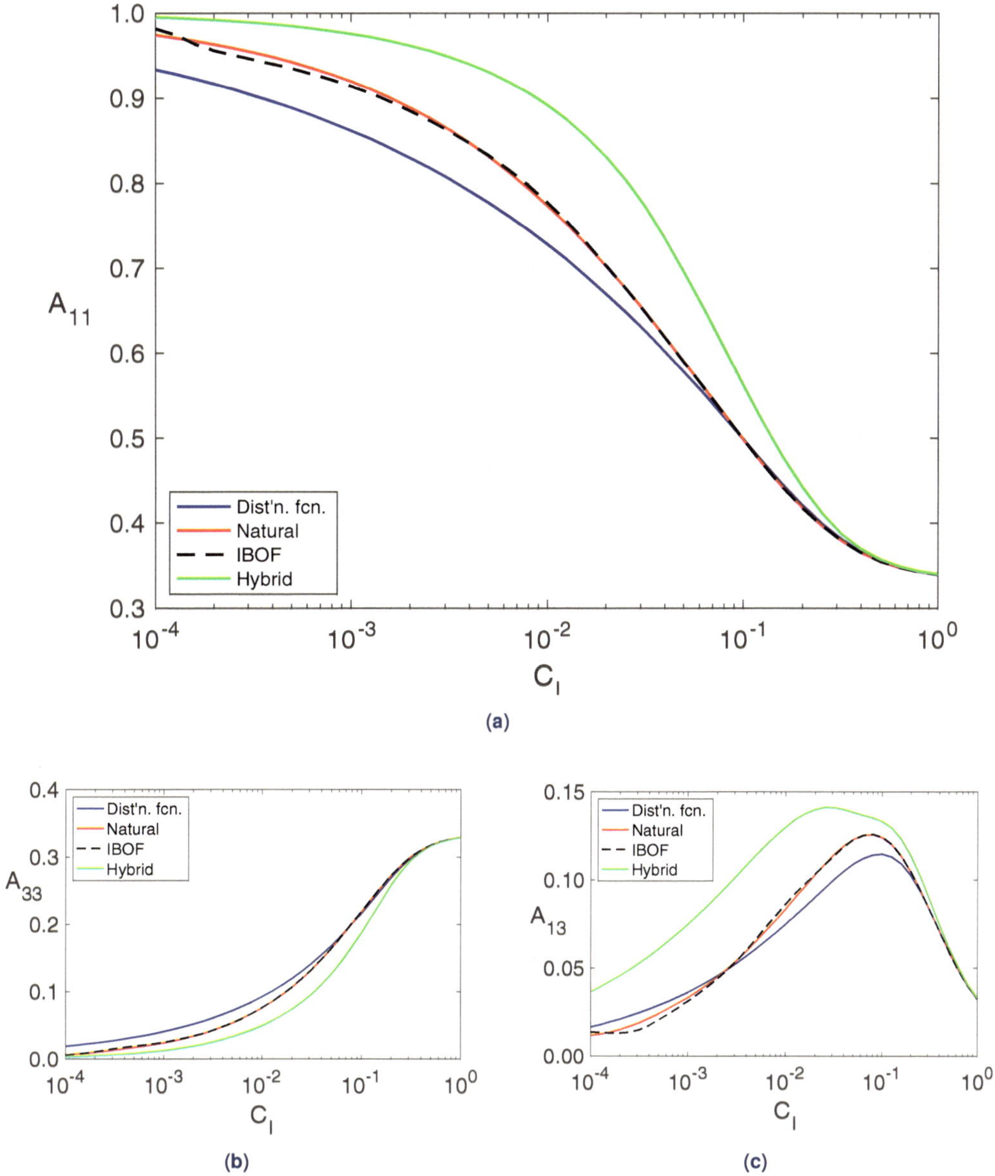

Figure 5.14 Steady-state tensor components in 1–3 simple shear flow vs. the interaction coefficient C_I for the Folgar–Tucker model with $\xi = 1$. Different closure approximations are compared to the distribution function results.

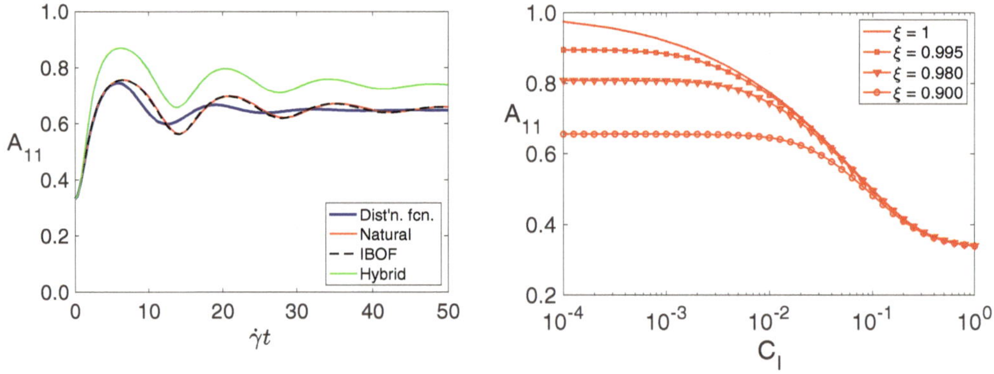

(a) Transient, $\xi = 0.90$, $C_I = 0.005$ (b) Steady state, natural closure

Figure 5.15 Behavior of the Folgar–Tucker model in 1–3 simple shear flow for $\xi < 1$. The values of $\xi = (1, 0.995, 0.980, 0.900)$ correspond approximately to $\ell/d_f = (\infty, 30, 15, 5)$.

In Fig. 5.15(a) we see A_{11} vs. time in simple shear flow for a fiber with $\xi = 0.90$. This value of ξ corresponds approximately to $\ell/d_f = 5$, which is a low aspect ratio. We have also used the same C_I value for the distribution function calculations and all closures. The results have some of the features of Fig. 5.12(a): orientation rises quickly and settles to a steady state, and the natural and IBOF closures match the distribution function results better than the hybrid closure. However, now A_{11} shows significant oscillations.

These oscillations result from the periodic motion of the fibers as they execute their Jeffery orbits. A_{11} grows initially because the fibers perpendicular to the flow rotate quickly, while the fibers aligned with the flow rotate slowly. The fast-rotating fibers catch up to the slow-rotating fibers, and alignment builds in the flow direction. However, eventually the fibers that were initially oriented in the flow direction move away from that axis, and A_{11} decreases. If there were no rotary diffusion ($C_I = 0$), every fiber would return to its original orientation at the end of each Jeffery period, the orientation at the end of each period would match the initial orientation, and the oscillations in A_{11} would be un-damped. Rotary diffusion tends to randomize fiber orientation, so setting C_I to a value greater than zero damps the oscillations, and the orientation eventually settles to a steady state. The larger the value of C_I, the greater the damping.

The effects of ξ and C_I on steady-state orientation in simple shear flow are shown in Fig. 5.15(b). Here we use the natural closure, but results for the IBOF closure are nearly identical. $\xi = 0.995$ and 0.980 correspond to $\ell/d_f = 30$ and 15 respectively, which are representative of the average fiber length in short fiber injection molded composites.

For larger values of C_I the rotary diffusion dominates the steady-state orientation, and ξ has very little effect. For each ξ there is a value of C_I below which ξ dominates the steady-state orientation and C_I has no effect; the smaller the value of ξ, the smaller the steady-state A_{11} in this range. What this figure does not show is that oscillations like those in Fig. 5.15(a) persist for very long times when the interaction coefficient is small, say $C_I < 10^{-3}$.

From a modeling standpoint, the oscillations that accompany values of $\xi < 1$ are a considerable disadvantage. In a full-field solution they will produce spatial oscillations in orientation, and the wavelength of these oscillations will go to zero as one approaches any mold wall where a no-slip boundary condition is imposed. This, in turn, can prevent some numerical solutions methods from converging. Moreover, using $\xi < 1$ does not provide any particular advantage with respect to steady-state orientation, at least in simple shear flow, since the A_{11} values in Fig. 5.15(b) can easily be obtained with $\xi = 1$ by adjusting C_I. Finally, injection molded composites typically have a wide range of fiber lengths. This will dampen any Jeffery oscillations in **A**, since fibers of different lengths have different periods, and longer fibers may restrain neighboring fibers from executing their Jeffery orbits. For these reasons, it is standard practice to set $\xi = 1$ and to adjust C_I (or other parameters to be discussed next) to match the orientation model to the real composite.

5.5 Choosing the Value of the Interaction Coefficient

For the Folgar–Tucker model, a good strategy is to choose C_I to match the flow-direction alignment A_{11} at steady state in simple shear flow. For injection molded composites, the necessary experimental data can be obtained from the shell regions of a simple molding such as an end-gated strip. A plate-like molding with uniform thickness is ideal. The best measurement location is far downstream of the gate, but not too close to the edge of the molding, where the fountain flow or edge effects can change the orientation.

At the desired location, one measures orientation across the thickness of the part as described in Chapt. 3. Figure 5.16 shows an example data set of this type. From this data, ignore the core (near the midplane of the part) and focus on the shell layers. If the part has skin layers at the surfaces, ignore those too. The orientation in the shell layers should be nearly independent of position, though there will be noise in the data. Select the shell points, and compute the average of A_{11} from those points. This gives an experimental value of A_{11} in steady simple shear flow.

Using that value and the same closure approximation as you will use in simulating this material, go to Fig. 5.14(a) and read off the value of C_I. If more precision is desired, the accompanying software provides a function `fitCI` that returns C_I, A_{33} and A_{13} for a given A_{11} and closure.

For the data shown in Fig. 5.16, we identify the points in the shaded areas as the shell. This sample does not have skin layers. Averaging the values for points in the two shell regions gives $A_{11} = 0.826$ in steady simple shear flow. This value is shown as the dashed line in the figure. Then, for the natural closure, Fig. 5.14(a) gives $C_I = 0.0053$.

For this C_I value and the natural closure, the steady-state orientation in 1–3 simple shear flow is

$$\mathbf{A} = \begin{bmatrix} 0.826 & 0.000 & 0.066 \\ 0.000 & 0.119 & 0.000 \\ 0.066 & 0.000 & 0.055 \end{bmatrix} \tag{5.72}$$

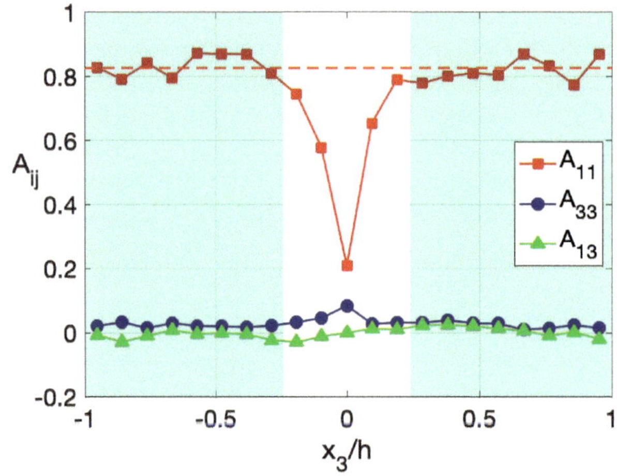

Figure 5.16 Measured orientation for an injection molded plate of polycarbonate with 30% by weight of short glass fibers. Data from [VTFO99]. The shaded areas are the shell regions and the dashed line is the average value of A_{11} in the shell.

Note that the Folgar–Tucker model does not provide any ability to adjust A_{33} or A_{13}. For the data in Fig. 5.16 the average values in the shell layer are $A_{33} = 0.023$ and $A_{13} = 0.008$ (accounting for the fact that A_{13} should be negative for $x_3 > 0$ and positive for $x_3 < 0$ in this part). The values in Eqn. (5.72) are a reasonable match, so the Folgar–Tucker model is useful for this material.

The corresponding distribution function (recovered using the Jeffery distribution) is shown in Fig. 5.17(a). The relatively small value of the interaction coefficient causes the flow-direction peak in $\psi(\mathbf{p})$ to spread slightly, as required to match the data for A_{11}, but the peak does not shift much toward the 3 axis, which is the shearing direction.

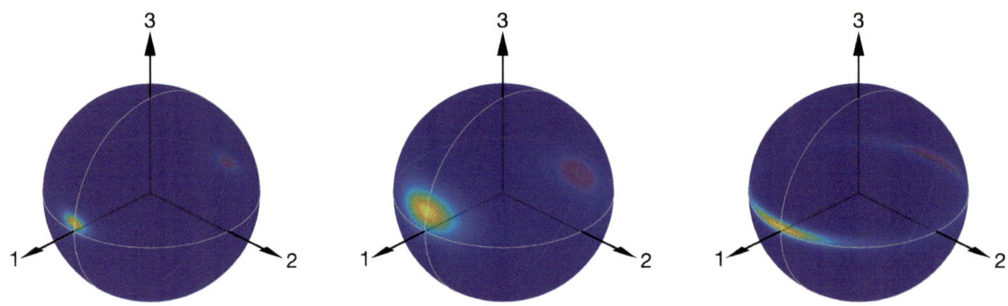

(a) SFT **(b)** LFT, Folgar–Tucker fit **(c)** LFT, actual

Figure 5.17 Orientation distribution functions in 1–3 simple shear flow, $v_1 = \dot{\gamma} x_3$ corresponding to (a) the shell layer of a short fiber thermoplastic, Fig. 5.16 and Eqn. (5.72); (b) Folgar–Tucker fit to $A_{11} = 0.653$, Eqn. (5.74); (c) the actual average shell-layer orientation in Fig. 5.18, Eqn. (5.73).

5.6 Anisotropic Rotary Diffusion

5.6.1 Motivation and Concept

While the Folgar–Tucker model tends to work well for injection molded short fiber/thermoplastics (SFTs), it is not always accurate for long fiber/thermoplastic materials (LFTs). LFT materials have average fiber lengths greater than 1 mm in molded parts, and the longest fibers can be 10 to 12 mm long.

LFTs tend to have less flow-direction alignment in the shell layer than short fiber materials, but their out-of-plane orientation is similar to SFTs. Figure 5.18 shows an example. This data is for an LFT with 40% by weight of long glass fibers in polypropylene [Phe09].

Following the procedure of Section 5.5, we identify the shaded areas in Fig. 5.18(a) as the shell layers. This part has significant skin layers, so the two data points on each side closest to the part surfaces are assigned to the skin. The choice of the boundary between shell and core is guided by the fact that LFT materials tend to have thicker cores and thinner shells that SFT materials. We can easily compute averages for A_{11} and A_{33} for the shell data points.

The data for A_{13} in Fig. 5.18(b) is harder to interpret. As the figure shows, in a strip geometry like this ISO plaque, any simulation will predict $A_{13} < 0$ for points above the midplane ($x_3 > 0$), and $A_{13} > 0$ for points below the midplane ($x_3 < 0$). However, almost all points in this data set have the opposite sign, a situation that is not unusual. For this case we ignore the A_{13} data and, for now, assign an average value for the shell that is similar in magnitude to A_{33}. We will see later in this section that, when choosing model parameters, it is common to match A_{11} and A_{33}, but allow A_{13} to be determined by the model.

With this treatment, the experimental shell orientation for the region where $A_{13} > 0$ is

$$\mathbf{A} = \begin{bmatrix} 0.653 & 0.000 & 0.030 \\ 0.000 & 0.327 & 0.000 \\ 0.030 & 0.000 & 0.020 \end{bmatrix} \tag{5.73}$$

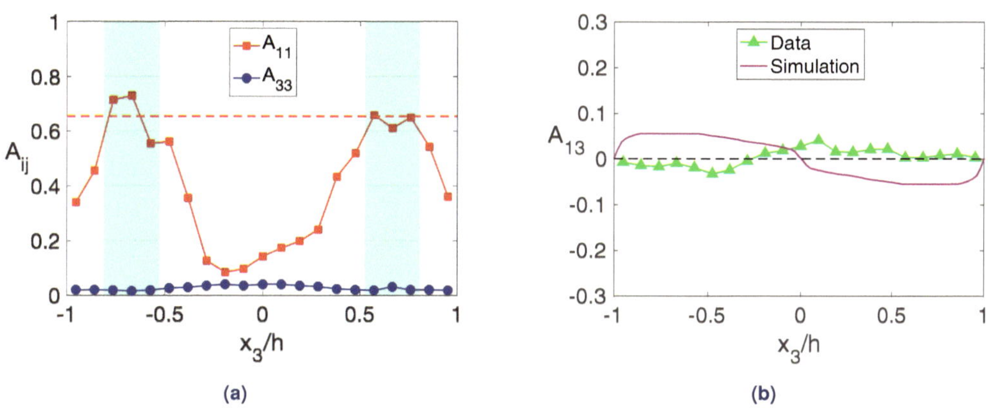

Figure 5.18 Measured orientation for an injection molded ISO plaque of polypropylene with 40% by weight of long glass fibers. The shaded areas in (a) are selected as the shell regions and the dashed line is the average value of A_{11} in the shell. The simulation line in (b) is typical of what a simulation would predict for A_{13}. Data from [Phe09], sample PNNLAS3I, location B.

This is a much different orientation state than the one we saw for short fibers in Eqn. (5.72). There is less flow-direction alignment, but the out-of-plane alignment, as described by A_{33}, is still very small.

If we follow the procedure from Section 5.5 and fit the Folgar–Tucker model to this data, matching the experimental value for A_{11} in steady simple shear and using the natural closure, we find $C_I = 0.0303$. The full shell-layer orientation tensor predicted by the Folgar–Tucker model is then

$$\mathbf{A} = \begin{bmatrix} 0.653 & 0.000 & 0.113 \\ 0.000 & 0.217 & 0.000 \\ 0.113 & 0.000 & 0.130 \end{bmatrix} \tag{5.74}$$

While Folgar–Tucker can easily match this low value of A_{11}, compared to Eqn. (5.73) it substantially over-predicts A_{33}. Also, since $A_{11} + A_{22} + A_{33} = 1$, it under-predicts A_{22}. The value of A_{13} is also large compared to the data.

Some insight into the problem is provided by the distribution functions for these two orientation tensors. The Folgar–Tucker solution is shown in Fig. 5.17(b). In order to achieve $A_{11} = 0.653$, the model spreads the orientation peak in both the 2 and 3 directions. This occurs because the rotary diffusion in the Folgar–Tucker model is *isotropic*, so it operates equally in all directions tangent to the surface of the unit sphere. The model also moves the orientation peak substantially toward the 3 axis.

A distribution function that matches the experimental data in Eqn. (5.73) is shown in Fig. 5.17(c). For a model to reproduce this type of orientation, it must spread the orientation peak strongly in the 1–2 plane, but not spread the orientation as much in the 3 direction. That is, we need the rotary diffusion terms to be direction-dependent, or *anisotropic*.

A family of models that meets this need has been developed. They are called ARD models, for *anisotropic rotary diffusion*. ARD models replace the scalar interaction coefficient C_I with a rotary diffusion tensor **C**. This tensor in turn depends on other direction-dependent quantities such as **A** and **D**. Several different ways of forming **C** have been proposed, and we will cover them in Section 5.6.3. But before examining how **C** is formed, we first need to incorporate a tensor diffusivity into the equation for $\dot{\mathbf{A}}$.

5.6.2 General Equations

5.6.2.1 Tensor Equation

The general expression for $\dot{\mathbf{A}}$ for a model with anisotropic rotary diffusion was developed by Phelps and Tucker [PT09, Phe09], and is

$$\begin{aligned}\dot{\mathbf{A}} &= \mathbf{W}\cdot\mathbf{A} - \mathbf{A}\cdot\mathbf{W} + \xi(\mathbf{D}\cdot\mathbf{A} + \mathbf{A}\cdot\mathbf{D} - 2\mathbb{A}:\mathbf{D}) \\ &\quad + \dot{\gamma}\left[2\mathbf{C} - 2(\operatorname{tr}\mathbf{C})\mathbf{A} - 5(\mathbf{C}\cdot\mathbf{A} + \mathbf{A}\cdot\mathbf{C}) + 10\mathbb{A}:\mathbf{C}\right]\end{aligned} \tag{5.75}$$

The usual Jeffery terms appear on the first line, but the rotary diffusion term on the second line is complicated, and it is difficult to see how **C** will affect the orientation.

This equation is easier to understand if we introduce the following auxiliary variables suggested by Favaloro [Fav18, FT19]:

$$\hat{C}_I = \text{tr}\mathbf{C}/3 \qquad \hat{\mathbf{A}} = \mathbf{C}/\text{tr}\mathbf{C} \qquad \hat{\mathbf{D}} = \xi\mathbf{D} - 5\dot{\gamma}\mathbf{C} \tag{5.76}$$

When these are substituted into Eqn. (5.75), we obtain

$$\dot{\mathbf{A}} = \mathbf{W}\cdot\mathbf{A} - \mathbf{A}\cdot\mathbf{W} + \hat{\mathbf{D}}\cdot\mathbf{A} + \mathbf{A}\cdot\hat{\mathbf{D}} - 2\mathbb{A}:\hat{\mathbf{D}} + 6\hat{C}_I\dot{\gamma}(\hat{\mathbf{A}} - \mathbf{A}) \tag{5.77}$$

This looks exactly like the Folgar–Tucker model, Eqn. (5.33), with $\xi = 1$, except that $\hat{\mathbf{D}}$ is substituted for $\mathbf{D}$, $\hat{C}_I$ is substituted for C_I, and $\hat{\mathbf{A}}$ appears in place of $\mathbf{I}/3$. We can then see that $\hat{\mathbf{D}}$ is a modified rate-of-deformation tensor, $\hat{C}_I$ is the effective interaction coefficient, and the last term instantaneously draws the orientation tensor toward $\hat{\mathbf{A}}$, rather than always drawing it directly toward isotropic orientation, $\mathbf{I}/3$. Because of this, we will refer to $\hat{\mathbf{A}}$ as the *target orientation tensor*.

With Eqns. (5.76) and (5.77), we can implement any ARD model, once we have chosen an expression for the rotary diffusion tensor $\mathbf{C}$. These equations, and all of the options for $\mathbf{C}$ in Section 5.6.3, are implemented in the `Adot2` function in the accompanying software.

5.6.2.2 Distribution Function Equation

Before discussing choices for $\mathbf{C}$, let us look at how anisotropic rotary diffusion appears in distribution function calculations.

Recall from Sections 5.2.2 and 5.3.2 that $\psi(\mathbf{p}, t)$ changes in response to the flux $\mathbf{q}(\mathbf{p})$ in orientation space. The Jeffery motion of the particle provides a flux of $\mathbf{q}^J = \psi\,\dot{\mathbf{p}}$, and for the Folgar–Tucker model there is a diffusive component of the flux given by Eqn. (5.29), which we repeat here:

$$\mathbf{q}^d = -C_I\dot{\gamma}\,\nabla_s\psi \tag{5.78}$$

Adding these two flux components together, $\mathbf{q} = \mathbf{q}^J + \mathbf{q}^d$, and using the result in the continuity equation (5.14) gives an equation that can be solved to find $\psi(\mathbf{p}, t)$ for any flow field.

We can see that Eqn. (5.78) gives isotropic rotary diffusion, because the surface gradient ∇_s of ψ is multiplied by scalar quantities. This ensures that the surface vector $\mathbf{q}^d$ is parallel to $\nabla_s\psi$ and, because of the minus sign, points in the opposite direction.

For anisotropic rotary diffusion we want a similar equation, but with a tensor multiplying $\nabla_s\psi$. We cannot use the rotary diffusion tensor $\mathbf{C}$ for this purpose, because $\mathbf{C}$ is a space tensor that lives in (x_1, x_2, x_3) physical space, while $\nabla_s\psi$ and $\mathbf{q}^d$ are surface vectors that live in the orientation space of the unit sphere. Instead, we need a way to produce a *surface* rotary diffusion tensor, call this $\mathbf{D}^r$, at each point $\mathbf{p}$ in orientation space. While $\mathbf{C}$ is 3×3 and has nine components, $\mathbf{D}^r$ is 2×2 and has four components. If we use spherical coordinates, then these components are $D^r_{\theta\theta}$, $D^r_{\theta\phi}$, $D^r_{\phi\theta}$ and $D^r_{\phi\phi}$.

The tool that transforms a space tensor into a surface tensor on a unit sphere is the mixed tensor [Ari62],

$$\begin{aligned}\mathbf{T} = &\cos\theta\,\cos\phi\;\hat{\boldsymbol{\delta}}_1\hat{\boldsymbol{\delta}}_\theta + \cos\theta\sin\phi\;\hat{\boldsymbol{\delta}}_2\hat{\boldsymbol{\delta}}_\theta - \sin\theta\;\hat{\boldsymbol{\delta}}_3\hat{\boldsymbol{\delta}}_\theta \\ &-\sin\phi\;\hat{\boldsymbol{\delta}}_1\hat{\boldsymbol{\delta}}_\phi + \cos\phi\;\hat{\boldsymbol{\delta}}_2\hat{\boldsymbol{\delta}}_\phi + 0\;\hat{\boldsymbol{\delta}}_3\hat{\boldsymbol{\delta}}_\phi\end{aligned} \tag{5.79}$$

The components of $\mathbf{T}$ are a 2×3 matrix whose rows correspond to the spherical coordinate angles (θ, ϕ) and whose columns correspond to the spatial coordinates (x_1, x_2, x_3). Each point (θ, ϕ) on the unit sphere has its own $\mathbf{T}$ tensor.

Now, for any point on the unit sphere, the local surface rotary diffusion tensor is

$$\mathbf{D}^r = \mathbf{T}^T \cdot \mathbf{C} \cdot \mathbf{T} \tag{5.80}$$

and the flux due to anisotropic rotary diffusion is

$$\mathbf{q}^d = -\dot{\gamma}\, \mathbf{D}^r \cdot \nabla_s \psi \tag{5.81}$$

Written out in component form, this is

$$q_\theta^d = -\dot{\gamma}\, D_{\theta\theta}^r \frac{\partial \psi}{\partial \theta} - \dot{\gamma}\, D_{\theta\phi}^r \frac{1}{\sin\theta} \frac{\partial \psi}{\partial \phi} \tag{5.82}$$

$$q_\phi^d = -\dot{\gamma}\, D_{\phi\theta}^r \frac{\partial \psi}{\partial \theta} - \dot{\gamma}\, D_{\phi\phi}^r \frac{1}{\sin\theta} \frac{\partial \psi}{\partial \phi} \tag{5.83}$$

Using these expressions in place of Eqns. (5.31) and (5.32) gives an equation that can be solved for $\psi(\mathbf{p}, t)$ for any ARD model. The `solvePsiARD` function in the accompanying software does this.

In a diffusion model the flux is always from higher concentrations (higher values of ψ) to lower concentrations. This will be true provided $\mathbf{D}^r$ is symmetric and its eigenvalues are non-negative. This, in turn, requires that $\mathbf{C}$ be symmetric and have non-negative eigenvalues. Any valid model for $\mathbf{C}$ must meet this requirement.

One last point is that the diffusive flux $\mathbf{q}^d$ must equal zero when the probability gradient $\nabla_s \psi$ is zero, for both isotropic and anisotropic rotary diffusion. Equations (5.81–5.83) obviously have this property. What is less obvious, but can be proved, is that the rotary diffusion term in the tensor form Eqn. (5.75) will give $\dot{\mathbf{A}}^d = 0$ when the orientation is isotropic, $\mathbf{A} = \mathbf{I}/3$. This is true for any rotary diffusion tensor $\mathbf{C}$ that is symmetric.

5.6.3 Choices for the Rotary Diffusion Tensor

We now need to develop an expression for the anisotropic rotary diffusion tensor $\mathbf{C}$. This tensor must inherit its anisotropy from other direction-dependent information in the problem. The usual assumption [PT09] is that $\mathbf{C}$ may depend on the orientation state through $\mathbf{A}$ and on the rate-of-deformation tensor $\mathbf{D}$. In fact, since overall rate of diffusion is already proportional to $\dot{\gamma}$, we allow $\mathbf{C}$ to depend on the dimensionless tensor $\mathbf{D}^*$, where

$$\mathbf{D}^* \equiv \frac{\mathbf{D}}{\dot{\gamma}} \tag{5.84}$$

Thus, we assume that $\mathbf{C} = f(\mathbf{A}, \mathbf{D}^*)$.

5.6.3.1 Five-Constant Model

In the initial development of ARD models, Phelps and Tucker [PT09] used the form

$$\mathbf{C} = b_1 \mathbf{I} + b_2 \mathbf{A} + b_3 \mathbf{A}^2 + b_4 \mathbf{D}^* + b_5 (\mathbf{D}^*)^2 \tag{5.85}$$

Here the five b_is are scalar constants. This form can match experimental data well, however, some combinations of the b_i values make Eqn. (5.75) unstable and lead to non-physical solutions. Phelps and Tucker matched a steady-state orientation of $A_{11} = 0.65$, $A_{33} = 0.01$, $A_{13} = 0.03$ in a long glass fiber/PP composite using $(b_1, b_2, b_3, b_4, b_5) = (1.924 \times 10^{-4}, 5.839 \times 10^{-3}, 4.0 \times 10^{-2}, 1.168 \times 10^{-5}, 0)$. Those values have been used by others for similar materials. However, fitting the five-constant model to new data is time-consuming. When a set of parameters must be fitted to new data, one of the ARD models below is recommended.

5.6.3.2 Folgar–Tucker Model

The Folgar–Tucker model from Section 5.3 is recovered from Eqn. (5.85) by keeping only the first term and setting $b_1 = C_I$:

$$\mathbf{C}^{\mathrm{FT}} = C_I\,\mathbf{I} \tag{5.86}$$

Of course this model has isotropic rotary diffusion. We include it here for comparison to other ARD models.

5.6.3.3 iARD Model

Tseng and co-workers were the first to develop an alternative to Eqn. (5.85) that was stable for a wide range of model parameters. Their *improved ARD* (iARD) model is [TCH16]

$$\mathbf{C}^{\mathrm{iARD}} = C_I^{\mathrm{iARD}}\left[\mathbf{I} - 4C_M(\mathbf{D}^*)^2\right] \tag{5.87}$$

This model has two parameters: the interaction coefficient and a new parameter C_M that controls anisotropy. The iARD model can be viewed as a special form of Eqn. (5.85) that uses only the b_1 and b_5 terms.

The value of C_M is typically chosen to be between 0 (which gives isotropic rotary diffusion) and 1 (which gives maximum anisotropy). The iARD model is well-behaved in simple shear flow for $0 \le C_M \le 1$. However, in some extensional flows it produces non-physical results if $C_M > 3/4$ [Fav18]. This problem is most severe in biaxial extensional flow, which occurs frequently in compression molding. Thus, while $C_M \simeq 1$ is commonly used in injection molding analyses, it is wise to keep $C_M \le 3/4$, especially for modeling compression molding and general 3-D flows.

5.6.3.4 MRD and pARD Models

Rather than simplifying Eqn. (5.85), some models form $\mathbf{C}$ in the eigensystem of the orientation tensor $\mathbf{A}$. This approach is similar to the EBOF approach for fitting closure approximations, though the math is simpler because $\mathbf{C}$ is a second-order tensor.

Recall from Section 2.3.4 that we can write the orientation tensor in terms of its eigenvalues and eigenvectors,

$$\mathbf{A} = \lambda_1\mathbf{e}_1\mathbf{e}_1 + \lambda_2\mathbf{e}_2\mathbf{e}_2 + \lambda_3\mathbf{e}_3\mathbf{e}_3 \tag{5.88}$$

This form is valid whether the eigenvectors $\mathbf{e}_1$, $\mathbf{e}_2$, $\mathbf{e}_3$ are aligned with the coordinate axes or not.

If we choose a model where $\mathbf{C}$ depends on $\mathbf{A}$, then the eigenvectors of $\mathbf{C}$ must be identical to the eigenvectors of $\mathbf{A}$. Mathematically, we say that the $\mathbf{C}$ tensor is *coaxial* with $\mathbf{A}$. Now we only need to choose the eigenvalues of $\mathbf{C}$. We retain the convention that the eigenvalues of $\mathbf{A}$ are numbered so that $\lambda_1 \geq \lambda_2 \geq \lambda_3$. This allows us to construct $\mathbf{C}$ so that, for example, the largest eigenvalue of $\mathbf{C}$ corresponds to the same eigenvector as the largest eigenvalue of $\mathbf{A}$.

Two ARD models have been developed using this idea. The first is the *Moldflow Rotational Diffusion* model, or MRD [BYR+17]:

$$\mathbf{C}^{\text{MRD}} = C_I^{\text{MRD}}(D_1\mathbf{e}_1\mathbf{e}_1 + D_2\mathbf{e}_2\mathbf{e}_2 + D_3\mathbf{e}_3\mathbf{e}_3) \tag{5.89}$$

where D_1, D_2, $D_3 \geq 0$. While the MRD model has four parameters, we can choose $D_1 = 1$ with no loss of generality, so MRD is effectively a three-parameter model. Default values are $D_1 = 1$, $D_2 = 0.50$ and $D_3 = 0.30$ [Aut19], and it is standard to choose $D_3 \leq D_2 \leq 1$. The MRD model reduces to isotropic rotary diffusion (the Folgar–Tucker model) when $D_1 = D_2 = D_3$, and it gives maximum anisotropy when $D_3 = 0$.

The second model based in the principal axes of the orientation tensor is the *principal ARD* model [TCH18b], knows as pARD. This model has two parameters, an interaction coefficient and a parameter Ω that falls in the range $0.5 \leq \Omega \leq 1$. The pARD rotary diffusion tensor is

$$\mathbf{C}^{\text{pARD}} = C_I^{\text{pARD}}\left[\mathbf{e}_1\mathbf{e}_1 + \Omega\mathbf{e}_2\mathbf{e}_2 + (1-\Omega)\mathbf{e}_3\mathbf{e}_3\right] \tag{5.90}$$

This model gives maximum anisotropy at $\Omega = 1$, and minimum anisotropy when $\Omega = 0.5$. The pARD model cannot be reduced to isotropic rotary diffusion by adjusting Ω, though as Fig. 5.19 shows, it comes close.

The pARD model has slightly less flexibility than the MRD model, since the $\mathbf{e}_2\mathbf{e}_2$ and $\mathbf{e}_3\mathbf{e}_3$ terms cannot be adjusted independently. However, it still produces useful results.

5.6.3.5 Wang or Two-Constant Model

Wang [Wan18] found that keeping only the b_1 and b_3 terms in Eqn. (5.85) provided useful solutions that were stable for a wide range of parameter values. Since this is a modification of the original Phelps–Tucker ARD model, [Fav18] called this the WPT model. It gives the rotary diffusion tensor as

$$\mathbf{C}^{\text{Wang}} = b_1\mathbf{I} + b_3\mathbf{A}^2 \tag{5.91}$$

This is a two-parameter model. It can easily be implemented in any code that uses Eqn. (5.85) by setting $b_2 = b_4 = b_5 = 0$.

Note that the $\mathbf{C}$ tensor in the Wang model is coaxial with $\mathbf{A}$, just like the MRD and pARD models. However, in the Wang model the eigenvalues of $\mathbf{C}$ change as the orientation state changes.

To make comparisons with the other models more convenient, we rewrite this model as [Fav18]

$$\mathbf{C}^{\text{Wang}} = C_I^{\text{Wang}}\left[(1-w)\mathbf{I} + w\mathbf{A}^2\right] \tag{5.92}$$

where the model parameters are now $C_I^{\text{Wang}} = b_1 + b_3$ and $w = b_3/(b_1 + b_3)$. Choosing $w = 0$ gives isotropic rotary diffusion, and $w = 1$ gives maximum anisotropy.

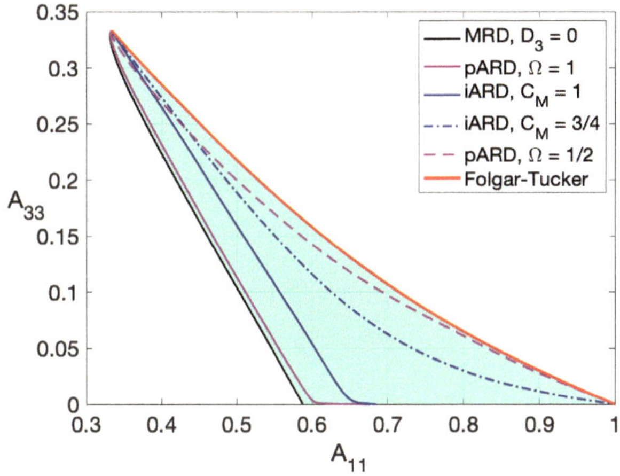

Figure 5.19 Combinations of (A_{11}, A_{33}) in 1–3 steady simple shear flow that can be accessed by ARD models and $\xi = 1$. The red Folgar–Tucker line represents isotropic rotary diffusion, and other solid lines give the minimum values of A_{33} for any given A_{11}. The blue dash-dot line is the minimum A_{33} for iARD to avoid non-physical solutions. The magenta dashed line is the maximum A_{33} for pARD. The iARD and pARD curves use the IBOF closure, while Folgar–Tucker and MRD curves use the natural closure. Minimum A_{33} for the Wang model, not shown, is close to the MRD line.

5.6.4 Choosing ARD Model Parameters

One approach to choosing ARD model parameters is to match desired values of A_{11} and A_{33} in steady simple shear flow, and we discuss that in detail here. The process is similar to setting C_I for the Folgar–Tucker model, but has a few additional considerations.

Some researchers have used optimization methods together with mold filling simulations to obtain the best match to a wider range of orientation data (e.g., [WRB20]). Any such match is equivalent to fitting *some* pair of (A_{11}, A_{33}) values in steady simple shear, in which case the discussion below is still useful.

To set ARD parameters to match steady simple shear data:

1. **Choose the desired values of A_{11} and A_{33}.**

   These are determined from experimental data as in Section 5.6.1.

2. **Ensure that the desired values are accessible to the model.**

   Not every combination of (A_{11}, A_{33}) can be achieved by a given ARD model. If the desired values are not accessible, the parameter-fitting calculation will fail or the calculation will produce non-physical values of the model parameters. Non-physical model parameters are those that produce negative diffusivity in some directions. This, in turn, can make fiber orientation calculations unstable, sometimes wildly so.

   The accessible combinations of (A_{11}, A_{33}) for the most popular ARD models are shown in Fig. 5.19. The solid red line shows the combinations available with the Folgar–Tucker model, and the iARD, MRD and Wang models all can reach these values. The pARD model does not reduce to isotropic diffusion, and the closest it can come is with $\Omega = 1/2$. However, the difference between this line and the Folgar–Tucker line is modest.

The shaded area below the Folgar–Tucker line contains steady-state orientations that ARD models can match. The entire shaded region is available with the MRD model, and the pARD model covers most of this region as well. The iARD model has a smaller accessible region, even if we allow $C_M \leq 1$. If the iARD model is restricted to $C_M \leq 3/4$, which avoids non-physical results in any flow, its accessible region is much smaller. The Wang model is not shown in this figure, but its lower limit is similar to the MRD curve.

If the desired combination of (A_{11}, A_{33}) is not accessible, revise the values to reach the accessible region. It is unlikely that the true orientation state of a real composite would fall outside of the shaded region in Fig. 5.19, but experimental data could. If this happens, consider increasing both A_{11} and A_{33}. Remember that increasing either value will decrease A_{22} at steady state.

The example from the start of this section had $(A_{11}, A_{33}) = (0.653, 0.020)$. This is readily accessible for MRD and pARD (and Wang), and is also accessible region for iARD if we allow $C_M > 3/4$. Thus, all four models can be fitted to this data.

3. **Adjust two model parameters.**

Now, find the model parameters that will give the desired (A_{11}, A_{33}) combination at steady state in 1–3 simple shear flow. For the two-parameter models this also determines A_{13} at steady state and, as we will see in a moment, the recommended treatment for the three-parameter MRD model is to fix D_2 at a reasonable value and treat MRD as a two-parameter model as well.

Figure 5.20 provides a starting point for choosing ARD model parameters, and in many cases the values from one of these charts may be sufficient. These graphs show the two model parameters as a function of the target steady-state value of A_{11}, when the target $A_{33} = 0.02$. As expected, the interaction coefficient increases as the desired A_{11} decreases, just as in the Folgar–Tucker model. The anisotropy also increases as A_{11} decreases (C_M increasing, Ω increasing, D_3 decreasing, w increasing), to keep the out-of-plane diffusion small as the in-plane diffusion increases. Notice also that the iARD model cannot go below about $A_{11} = 0.63$ with this value of A_{33}. If we restrict $C_M \geq 0.75$ (the blue dash-dot line in Fig. 5.20(a)) then iARD cannot go below $A_{11} = 0.85$. For other desired values of A_{33} the parameters maps provided by Favaloro and Tucker [FT19] may be helpful.

If more precise parameters are needed, they can be found numerically. The MATLAB function `fitARD` in the accompanying software[5] does this for any of the ARD models discussed here.

Why should the MRD model, which has three parameters, be treated like a two-parameter model? One might expect that the three parameters could be selected to match desired values of A_{11}, A_{33} and A_{13}. However, once A_{11} and A_{33} have been matched, the value of A_{13} at steady state is insensitive to the model parameters. Figure 5.21(a) illustrates the issue. C_I^{MRD} and D_3 can be adjusted to give the desired A_{11} and A_{33} values for *any* legal value of D_2, and the resulting A_{13} value is the same in all cases. Thus, a good strategy for choosing MRD parameters is to fix D_2 and treat MRD as a two-parameter model, adjusting C_I^{MRD} and D_3 to fit target values of A_{11} and A_{13}. `fitARD` uses this approach.

D_2 does affect other aspects of the model behavior. Figure 5.21(b) shows the transient solution in simple shear flow for a few different values of D_2. In each case, the other parameters

5 *http://github.com/charlestucker3/Fiber-Orientation-Tools*

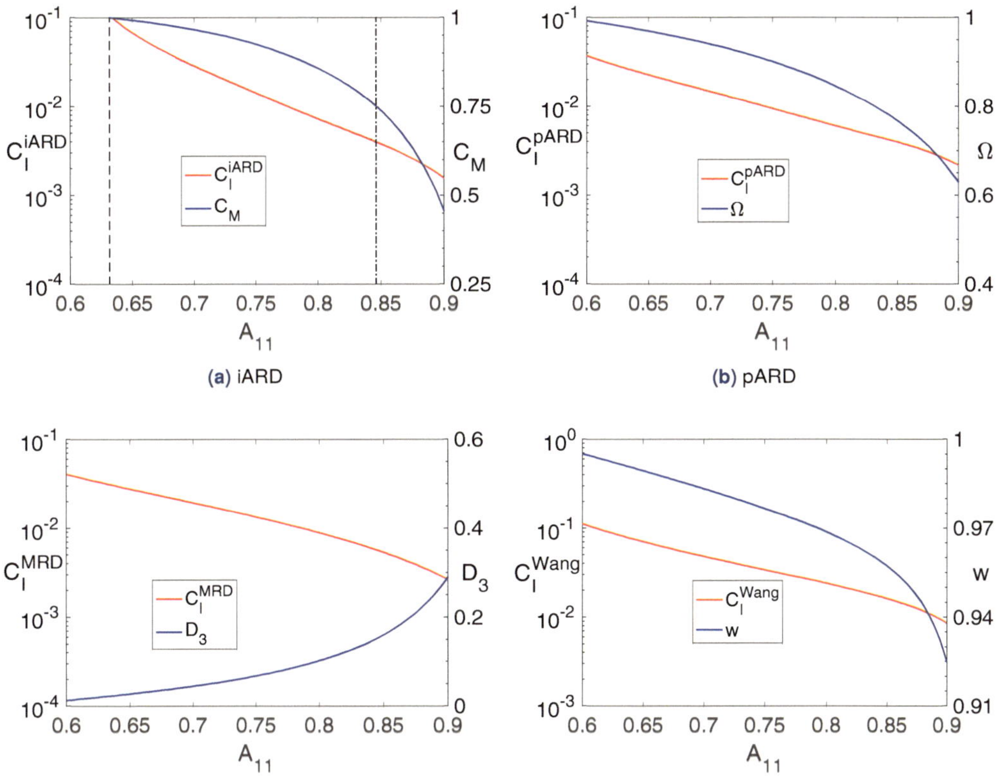

(a) iARD (b) pARD

(c) MRD (d) Wang

Figure 5.20 ARD model parameters to achieve various values of A_{11} in 1–3 steady simple shear with $A_{33} = 0.020$. iARD and pARD plots use the IBOF closure, while MRD and Wang plots use the natural closure. The MRD model uses $D_2 = 0.50$.

are chosen to match to the same steady-state orientation. Here we see that D_2 affects the amount overshoot in A_{11}. Experiments suggest that this overshoot is small in real materials, so it is best to keep D_2 toward the higher end of its range. The Autodesk Moldflow Insight default value of $D_2 = 0.5$ is a good choice.

4. **Ensure that the parameters are valid.**

   It is always wise to check the parameter values and ensure that they fall in the valid range for the model. This is particularly important when a numerical procedure like `fitARD` has been used to fit the parameters. Invalid parameters can give unstable, unpredictable and even non-physical results when used to calculate fiber orientation. Table 5.2 summarizes the valid ranges for the ARD models discussed here. Remember that the original range of C_M for the iARD model can produce non-physical results in some flows, while the recommended range ensures physical results in all flow but restricts the accessible range of the model. If invalid parameters are found, revise the desired values of A_{11} and A_{33} and try again.

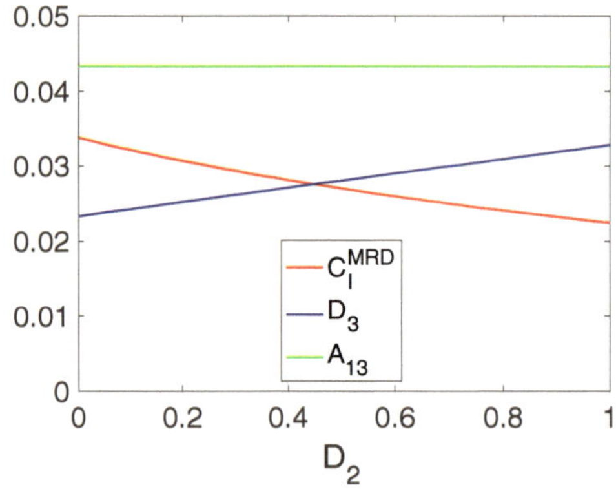

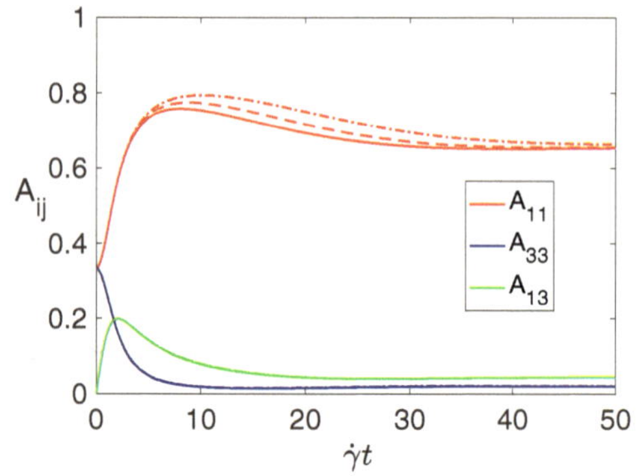

Figure 5.21 (a) Parameters obtained by fitting the MRD model to $(A_{11}, A_{33}) = (0.653, 0.020)$ for different values of D_2. The resulting value of A_{13} is the same in all cases. (b) Transient solutions in simple shear flow using MRD for three of the parameter sets from (a). Solid lines: $D_2 = 1$, dashed lines $D_2 = 0.5$, dash-dot lines $D_2 = 0.2$.

Table 5.2 Valid parameter ranges for ARD models

Model	Parameter 1	Parameter 2
iARD (original)	$C_I^{\text{iARD}} \geq 0$	$0 \leq C_M \leq 1$
iARD (recommended)	$C_I^{\text{iARD}} \geq 0$	$0 \leq C_M \leq 0.75$
pARD	$C_I^{\text{pARD}} \geq 0$	$0.5 \leq \Omega \leq 1$
MRD	$C_I^{\text{MRD}} \geq 0$	$0 \leq D_3 \leq D_2 \leq 1$
Wang	$b_1 \geq 0$	$b_3 \geq 0$
Wang (alt. params.)	$C_I^{\text{Wang}} \geq 0$	$0 \leq w \leq 1$

Table 5.3 ARD model parameters to fit $(A_{11}, A_{33}) = (0.653, 0.020)$ in steady simple shear

Model	Closure	C_I	Parameter 2	A_{13}
iARD	IBOF	0.0627	$C_M = 0.9908$	0.0415
pARD	IBOF	0.0218	$\Omega = 0.9673$	0.0415
MRD, $D_2 = 0.5$	Natural	0.0270	$D_3 = 0.0281$	0.0433
Wang	Natural	0.0682	$w = 0.9890$	0.0433

Final results of this process for our example from the start of this section are shown in Table 5.3. The interaction coefficients are different from one model to another, which is not surprising since C_I has a different interpretation in each model. The second-parameter values are close to maximum anisotropy for each model, showing that the desired steady-state orientation requires strong anisotropy in the rotary diffusion.

Interestingly, the values of A_{13} appear to depend only on the closure approximation and the desired (A_{11}, A_{33}) values, and not on the ARD model. This occurs consistently in numerical calculations, though the author is not aware of any algebraic proof for this.

5.7 Transient Orientation and Slow Kinetics Models

5.7.1 Motivation

Up to this point, modifications of Jeffery's equation have focused on improving the match to *steady* orientation states, especially in simple shear flow. We now turn to the issue of representing transient orientation states, and matching experimental data for the time (or strain) required to reach the steady-state orientation from some initial state.

When fiber orientation models were first incorporated into mold filling simulations, researchers discovered that even in simple parts like an end-gated strip, the simulations predicted a core that was much thinner than the actual core.

Figure 5.22 shows an example. The part is an ISO plaque (Fig. 4.24). It has the usual core/shell structure, and $A_{11} > 0.8$ in the shell is consistent with the short glass fibers in this material. One would expect the Folgar–Tucker model to do a good job predicting the orientation. However, when the Folgar–Tucker model is used in a mold filling simulation, the resulting predictions (the dashed lines in Fig. 5.22) have a core that is far too thin compared to the measured values. This particular part has thick skin layers as well, which are also not accurately predicted.

This kind of error in the orientation prediction leads to inaccurate mechanical property predictions, as shown in Table 5.4. This table compares measured elastic moduli values with predictions of two types. Prediction A uses the experimental orientation data, while prediction B uses predicted orientation data, similar to the dashed lines in Fig. 5.22. E_{11} is the elastic modulus in the flow direction and E_{22} is the modulus in the crossflow direction. The moldings are end-gated strips that are identical in thickness and cross-section, but have different lengths. (Models for predicting mechanical properties are discussed in Chapt. 8.)

Prediction A, using measured fiber orientation, captures both the anisotropy of the parts and the differences between the two flow lengths. This indicates that the micromechanics model,

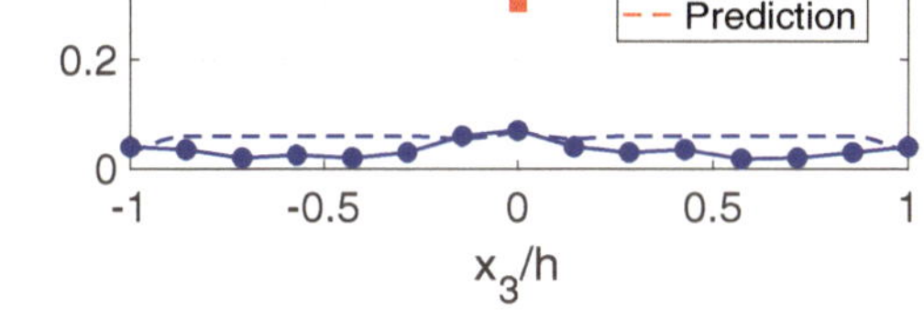

Figure 5.22 Measured orientation at the center of a 3 mm thick ISO plaque, compared to predictions (dashed lines) using the Folgar–Tucker model. The predicted core is much thinner than the actual core. PBT with 30% by weight glass fibers. Data and predictions from [Wan07].

Table 5.4 Measured and predicted elastic moduli for injection molded end-gated strips of different lengths L. Prediction A uses the experimental fiber orientation. Prediction B uses fiber orientation predicted from a mold filling simulation and the Folgar–Tucker model. All strips are molded from PBT reinforced with 30% by weight of glass fibers. Both strips are 76 mm wide and 3 mm thick. Data from [OT03].

Length (mm)	Property	Experiment (GPa)	Prediction A (GPa)	Prediction B (GPa)
125	E_{11}	7.7	7.5	8.6
125	E_{22}	5.4	4.8	4.3
279	E_{11}	8.8	9.1	8.6
279	E_{22}	4.6	4.4	4.3

which converts fiber orientation to elastic modulus, is reasonably accurate. However, for the 125 mm flow length, prediction B over-predicts E_{11} and under-predicts E_{22}. This indicates that the calculated orientation has too much fiber alignment in the flow direction and not enough alignment in the crossflow direction. This is consistent with the narrow predicted core in Fig. 5.22. Interestingly, for the longer flow path, the calculated orientation becomes more accurate.

This suggests that the calculated orientation is getting to steady state too soon. The calculated orientation is nearly at steady state across most of the part thickness for both flow lengths (hence the identical modulus predictions), and this is representative of the actual orientation in the longer part, but in the shorter part the actual orientation is farther from steady state.

Figure 5.23 gives a more detailed view of the issue. Both figures show A_{11} vs. position for a side view of a strip mold (as in Fig. 4.16) being filled with a Newtonian fluid with constant viscosity. The Folgar–Tucker model is used, and the inlet condition on the left edge has $A_{11} = 0.4$, which is typical of the orientation just inside the gate of an ISO plaque. As discussed in

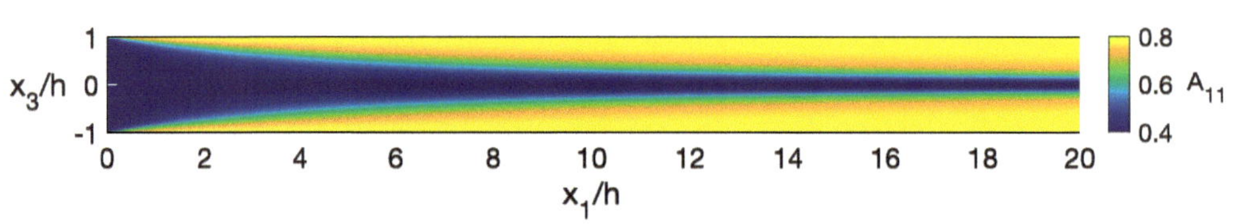

(a) Standard model

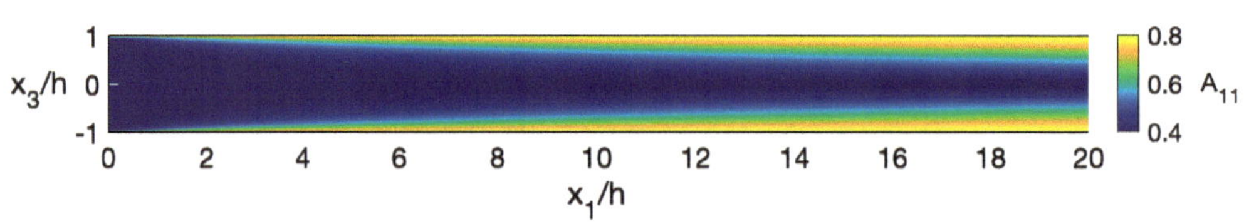

(b) RSC model, $\kappa = 0.2$

Figure 5.23 Contours of A_{11} for flow entering a strip mold, for a Newtonian fluid with constant viscosity. The inlet at $x_1 = 0$ has $A_{11} = 0.40$, $A_{33} = 0.05$ and $A_{13} = 0$. Flow is from left to right. Both calculations use the Folgar–Tucker model with $C_I = 0.01$ and the natural closure. The fountain flow is not modeled.

Section 4.3.2, material near the mold walls experiences a higher shear rate and lower velocity, so the orientation develops rapidly near the wall. Material near the midplane sees a lower shear rate and a higher velocity, so it travels much farther down the flow path before the fibers become oriented in the flow direction.

If all fibers follow the Folgar–Tucker model, the orientation develops as in Fig. 5.23(a). The shell layers, shown in yellow here, grow quickly, and by $x_1/h = 20$ (just 30 mm down the flow path in a 3 mm thick part) the core, shown in blue, is quite thin. This is the type of orientation pattern represented by the dashed lines in Fig. 5.22 and prediction B in Table 5.4.

In contrast, the experimental data in Fig. 5.22 and Table 5.4 suggest that the real situation is more like Fig. 5.23(b). The shell layers develop as before, increasing in thickness down the flow path, but the rate of increase is much smaller than the Folgar–Tucker model predicts. Material near the midplane also gradually becomes more oriented, but at a lower rate than the standard model suggests. As a result, the core occupies more of the part thickness for a longer time. If we extended the picture of Fig. 5.23(b) to larger values of x_1/h, we would eventually see a profile of A_{11} vs. x_3 much like the right side of Fig. 5.23(a).

While steady-state orientation in simple shear flow can be matched to experimental data by adjusting the interaction coefficient or the parameters of an ARD model, the transient behavior inherent in Jeffery's equation is too fast, and needs to be slowed down. We will refer to this phenomenon as *slow orientation kinetics*. Rheological experiments [SAC04, SCG$^+$04, WOT08, EBWVG09b, EVGBW10] provide additional evidence for slow kinetics, and we will see some of that data in Chapt. 6.

The physical mechanism behind slow orientation kinetics is currently unknown. There is speculation [Huy01] that under deformation a dense suspension of fibers forms a mesostructure, with some thin regions of reduced viscosity, perhaps due to low fiber concentration or shorter fiber lengths, and larger regions with higher viscosity that contain most of the fibers. If the low-viscosity layers are oriented in the planes of shear, then they could absorb most of the apparent macroscopic deformation, and the fiber-rich regions would deform much less. This would cause most of the fibers to change orientation more slowly than Jeffery or Folgar–Tucker would predict. While this explanation may be appealing, it is only speculation. The true source of slow kinetics is a mystery that awaits a compelling explanation backed by experimental evidence.

While we do not know where slow orientation kinetics comes from, it is nevertheless widely observed in injection molded composites. How can we adjust the model to behave like the real materials?

5.7.2 SRF/Slip Model

The simplest way to introduce slow kinetics into an orientation model is to multiply the entire right-hand side of the Folgar–Tucker model (5.33) by a scalar constant κ, where $0 \leq \kappa \leq 1$:

$$\dot{\mathbf{A}} = \kappa \left\{ \mathbf{W}\cdot\mathbf{A} - \mathbf{A}\cdot\mathbf{W} + \xi(\mathbf{D}\cdot\mathbf{A} + \mathbf{A}\cdot\mathbf{D} - 2\mathbf{D} : \mathbb{A}) + 2C_I\dot{\gamma}(\mathbf{I} - 3\mathbf{A}) \right\} \tag{5.93}$$

This gives good results in simple parts like injection molded strips [Huy01], and it is often used to match rheological data [SAC04, SCG$^+$04]. Huynh [Huy01] called this the *strain reduction factor* (SRF) model, where $1/\kappa$ is the strain reduction factor. It is also referred to as the *slip model*, and κ is called the slip factor.

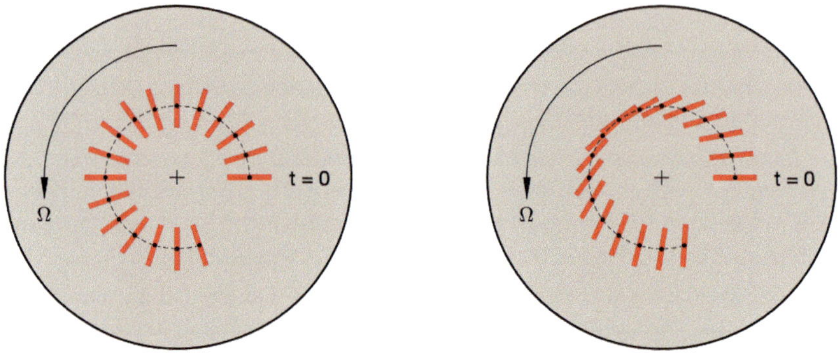

(a) Standard model (b) Strain reduction/slip model

Figure 5.24 The motion of a fiber in a solid matrix rotating at an angular velocity Ω, as predicted by (a) the standard Jeffery/Folgar–Tucker model and (b) the SRF/slip model with $\kappa = 0.3$.

The SRF/slip model is still used widely in comparing orientation models to rheological experiments, where the flow is simple shear flow. However, this model should *not* be used in mold filling simulations, because it is not *objective.*

A fiber orientation model, or any constitutive model in continuum mechanics, is objective if the behavior of the model is independent of the coordinate frame in which the model is written. The idea that constitutive equations should be objective is also called the *principle of coordinate frame indifference.*

To illustrate the problem with the strain reduction/slip model, consider a single fiber in a solid matrix. Place the matrix and fiber on a turntable and rotate the turntable at an angular velocity Ω. The centroid of the fiber will move in a circular path, and the fiber will rotate at an angular velocity Ω, producing the motion shown in Fig. 5.24(a).

What do our orientation models say about this case? The matrix is undergoing a rigid-body rotation like Example 4.1.2, so it has the **W** tensor in Eqn. (4.28), $\mathbf{D} = 0$ and $\dot{\gamma} = 0$. With these results, the Folgar–Tucker equation reduces to

$$\dot{\mathbf{A}} = \mathbf{W} \cdot \mathbf{A} - \mathbf{A} \cdot \mathbf{W} \tag{5.94}$$

The right-hand side rotates the orientation tensor at an angular velocity of Ω, and integrating this equation gives the fiber motion of Figure 5.24(a). This is just what we expect.

If we now apply the strain reduction/slip model to this case, Eqn. (5.93) reduces to

$$\dot{\mathbf{A}} = \kappa \left\{ \mathbf{W} \cdot \mathbf{A} - \mathbf{A} \cdot \mathbf{W} \right\} \tag{5.95}$$

This equation predicts that the orientation will rotate at a rate $\kappa\Omega$, a motion that is pictured in Fig. 5.24(b). But the fiber is locked into a solid matrix, which is rotating at a speed Ω. The model must be wrong!

While the strain reduction/slip model gives useful results for simple shear flow, it can easily produce non-physical results in other flows where $\mathbf{W} \neq 0$. Thus, it cannot be used in a mold filling simulation. We need an objective model instead.

5.7.3 Objective Models: RSC and RPR

Forming an objective equation that mimics the behavior of Eqn. (5.93) takes some work. If we leave the $(\mathbf{W}\cdot\mathbf{A}-\mathbf{A}\cdot\mathbf{W})$ term from Folgar–Tucker unchanged and multiply the remaining terms by κ, the model will be objective. However, this new equation is just the original Folgar–Tucker equation with ξ replaced by $\kappa\xi$ and C_I replaced by κC_I. We saw in Section 5.4.3.3 that using $\xi < 1$ will give oscillatory results in simple shear, so this is not a useful approach.

A model that is both well-behaved and objective was developed by Wang, O'Gara and Tucker [WOT08] and is called the RSC model, for *reduced strain closure*. The reason for this name will become clear in a moment.

The development of the RSC model starts by decomposing the second-order tensor **A** into its eigenvalues and eigenvectors, as we have done before. For this section we will *not* use the convention that repeated indices imply summation, and instead write out all sums explicitly. Now the orientation tensor expressed in terms of its eigenvalues and eigenvectors (5.88) is

$$\mathbf{A} = \sum_{i=1}^{3} \lambda_i \mathbf{e}_i \mathbf{e}_i \tag{5.96}$$

Since we are interested in $\dot{\mathbf{A}}$ we take the time derivative of both sides of this equation. Using the product rule on the right-hand side gives

$$\dot{\mathbf{A}} = \sum_{i=1}^{3} \left(\dot{\lambda}_i \mathbf{e}_i \mathbf{e}_i + \lambda_i \dot{\mathbf{e}}_i \mathbf{e}_i + \lambda_i \mathbf{e}_i \dot{\mathbf{e}}_i \right) \tag{5.97}$$

This expression shows that **A** can change over time, either because the eigenvalues are changing magnitude ($\dot{\lambda}_i$) or the eigenvectors are changing direction ($\dot{\mathbf{e}}_i$).

The idea of the RSC model is to slow down the rate of change of the eigenvalues λ_i but not change the equations for the rotation of the eigenvectors. Again using κ as the slow-down factor, this means that the time derivative of **A** in the RSC model will be

$$\dot{\mathbf{A}}^{\mathrm{RSC}} = \sum_{i=1}^{3} \left(\kappa \dot{\lambda}_i \mathbf{e}_i \mathbf{e}_i + \lambda_i \dot{\mathbf{e}}_i \mathbf{e}_i + \lambda_i \mathbf{e}_i \dot{\mathbf{e}}_i \right) \tag{5.98}$$

Using Eqn. (5.97), we can write this as

$$\dot{\mathbf{A}}^{\mathrm{RSC}} = \dot{\mathbf{A}} + (1-\kappa) \sum_{i=1}^{3} \dot{\lambda}_i \mathbf{e}_i \mathbf{e}_i \tag{5.99}$$

Note that $\dot{\mathbf{A}}$ on the right-hand side of this equation is the standard expression from the Folgar–Tucker model.

To develop this idea into a practical orientation model we need expressions for the eigenvalue derivatives $\dot{\lambda}_i$. To find $\dot{\lambda}_j$, substitute the Folgar–Tucker expression for $\dot{\mathbf{A}}$ into the left-hand side of Eqn. (5.97) and take $\mathbf{e}_j \cdot (\,) \cdot \mathbf{e}_j$ of both sides. The eigenvectors are mutually orthogonal, so $\mathbf{e}_i \cdot \mathbf{e}_j = \delta_{ij}$, and because each eigenvector has unit length it is orthogonal to its derivative, $\mathbf{e}_i \cdot \dot{\mathbf{e}}_i = 0$. Using this, one obtains

$$\dot{\lambda}_i = 2\xi(\lambda_i \mathbf{D} : \mathbf{e}_i \mathbf{e}_i - \mathbf{e}_i \mathbf{e}_i : \mathbb{A} : \mathbf{D}) + 2C_I \dot{\gamma}(1 - 3\lambda_i) \tag{5.100}$$

Equations for the eigenvectors can be derived by a similar process [WOT08]. For later reference, the equation for $\dot{\mathbf{e}}_1$ is

$$\dot{\mathbf{e}}_1 = \mathbf{W}\cdot\mathbf{e}_1 + \xi\left(\frac{\lambda_1+\lambda_2-2\bar{\mathbb{A}}_{1212}}{\lambda_1-\lambda_2}\right)\mathbf{D}:\mathbf{e}_1\mathbf{e}_2\mathbf{e}_2 + \xi\left(\frac{\lambda_1+\lambda_3-2\bar{\mathbb{A}}_{1313}}{\lambda_1-\lambda_3}\right)\mathbf{D}:\mathbf{e}_1\mathbf{e}_3\mathbf{e}_3 \tag{5.101}$$

In this equation, $\bar{\mathbb{A}}_{ijkl}$ are the components of the fourth-order tensor in the principal axes of $\mathbf{A}$. Expressions for $\dot{\mathbf{e}}_2$ and $\dot{\mathbf{e}}_3$ can be written by permuting the indices.

Using Eqn. (5.100) in (5.99), one can derive an explicit expression for $\dot{\mathbf{A}}^{\mathrm{RSC}}$. This leads to two new fourth-order tensors that are formed from the eigenvalues and eigenvectors of $\mathbf{A}$:

$$\mathbb{L} = \sum_{i=1}^{3}\lambda_i\mathbf{e}_i\mathbf{e}_i\mathbf{e}_i\mathbf{e}_i \quad \text{and} \quad \mathbb{M} = \sum_{i=1}^{3}\mathbf{e}_i\mathbf{e}_i\mathbf{e}_i\mathbf{e}_i \tag{5.102}$$

The final expression for the RSC version of the Folgar–Tucker model is then

$$\dot{\mathbf{A}}^{\mathrm{RSC}} = \mathbf{W}\cdot\mathbf{A}-\mathbf{A}\cdot\mathbf{W}+\xi\left\{\mathbf{D}\cdot\mathbf{A}+\mathbf{A}\cdot\mathbf{D}-2[\mathbb{A}+(1-\kappa)(\mathbb{L}-\mathbb{M}:\mathbb{A})]:\mathbf{D}\right\}+2\kappa C_I\dot{\gamma}(\mathbf{I}-3\mathbf{A}) \tag{5.103}$$

This looks like the original Folgar–Tucker expression, Eqn. (5.33), except that the rotary diffusion term is multiplied by κ and the fourth-order tensor $\mathbb{A}$ is replaced by the expression in square brackets. Modification of the fourth-order term can be implemented as part of the closure approximation, which is the source of the name *reduced strain closure*. We will examine the properties of this model in a moment.

An equivalent approach for slow orientation kinetics was developed by Tseng et al. [TCH13], and is called the *retarding principal rate* or RPR model. The concept is identical to the RSC model: reduce the growth rate of the eigenvalues without changing the time derivatives of the eigenvectors. However, the calculation path is different. In the RPR approach one first calculates $\dot{\mathbf{A}}$ for the standard model, then modifies that tensor according to Eqn. (5.99). This is done numerically at each time step, rather than algebraically as in the RSC model.

Before looking at the details of the RPR calculation, recall from Eqn. (2.55) that we can use the eigenvectors of $\mathbf{A}$ to define a coordinate transformation matrix $\mathbf{Q}$, such that the orientation tensor in its principal axis system, $\bar{\mathbf{A}}$, is

$$\mathbf{A} = \mathbf{Q}\cdot\bar{\mathbf{A}}\cdot\mathbf{Q}^T \tag{5.104}$$

Using the fact that $\mathbf{Q}$ is orthogonal (Eqn. (2.56)), we can transform in the opposite direction, from the laboratory axes to the principal axes of the orientation tensor, using

$$\bar{\mathbf{A}} = \mathbf{Q}^T\cdot\mathbf{A}\cdot\mathbf{Q} \tag{5.105}$$

To start the RPR modification, first calculate $\dot{\mathbf{A}}$ using the standard model. Also, find the eigenvectors of $\mathbf{A}$ and calculate $\mathbf{Q}$. Transform $\dot{\mathbf{A}}$ to the principal axes of $\mathbf{A}$:

$$\bar{\dot{\mathbf{A}}} = \mathbf{Q}^T\cdot\dot{\mathbf{A}}\cdot\mathbf{Q} \tag{5.106}$$

In this coordinate system the diagonal entries of $\bar{\dot{\mathbf{A}}}$ are the derivatives of the eigenvalues. That is, $\bar{\dot{A}}_{11} = \dot{\lambda}_1$, $\bar{\dot{A}}_{22} = \dot{\lambda}_2$ and $\bar{\dot{A}}_{33} = \dot{\lambda}_3$. This can be shown from Eqn. (5.97) and the fact that each $\dot{\mathbf{e}}_i$

is normal to its corresponding $\mathbf{e}_i$. The off-diagonal components of $\dot{\bar{\mathbf{A}}}$ come from the derivatives of the eigenvectors.

To slow down the growth of the eigenvalues (i.e., to *retard the principal rate*), multiply each diagonal entry of $\dot{\bar{\mathbf{A}}}$ by κ, while leaving the off-diagonal entries unchanged. This is equivalent to applying Eqn. (5.99), and produces the RPR version of $\dot{\bar{\mathbf{A}}}$:

$$\dot{\bar{\mathbf{A}}}^{\mathrm{RPR}} = \begin{bmatrix} \kappa\dot{\bar{A}}_{11} & \dot{\bar{A}}_{12} & \dot{\bar{A}}_{13} \\ \dot{\bar{A}}_{21} & \kappa\dot{\bar{A}}_{22} & \dot{\bar{A}}_{23} \\ \dot{\bar{A}}_{31} & \dot{\bar{A}}_{32} & \kappa\dot{\bar{A}}_{33} \end{bmatrix} \tag{5.107}$$

Finally, transform $\dot{\bar{\mathbf{A}}}^{\mathrm{RPR}}$ back to laboratory coordinates,

$$\dot{\mathbf{A}}^{\mathrm{RPR}} = \mathbf{Q}\cdot\dot{\bar{\mathbf{A}}}^{\mathrm{RPR}}\cdot\mathbf{Q}^T \tag{5.108}$$

This gives the RPR version of the time derivative of **A**.

The RPR and RSC models are different ways to calculate the same $\dot{\mathbf{A}}$, so results using these two models are identical.

The original papers on the RPR model [TCH13] use a factor α, equal to $(1-\kappa)$ in our notation. Thus, $\alpha = 0$ or $\kappa = 1$ gives standard orientation kinetics, while $\alpha = 0.9$ or $\kappa = 0.1$ gives orientation kinetics that are ten times slower than standard.

5.7.4 Choosing the RSC or RPR Parameter

The transient behavior of the RSC/RPR and the strain reduction/slip models are compared to the standard model in Fig. 5.25(a). Both models have a slower rise from the initial to the final state, as expected. The slip model exactly equals the standard response with the time axis stretched by a factor of $1/\kappa$, so it has the same overshoot in A_{11} and reaches the same steady state. The RSC model has the same general shape, but in this case it has no overshoot in A_{11} and a lower peak value of A_{13}.

It appears in Fig. 5.25(a) that the RSC model is going toward the same steady state as the standard model (with the same C_I value and the same closure). In fact, as Appendix B.3 shows:

> The RSC and RPR models have the same steady-state orientation as the standard model for the same **D**, **W**, rotary diffusion parameters and closure approximation.

As a result, one can adjust the RSC parameter κ (or the RPR parameter α) to match the transient behavior of the material without having to re-adjust C_I or any ARD parameters.

How does one actually choose κ? At this time the best option is to mold a simple part, like an end-gated strip or ISO plaque, measure the orientation profile across the part thickness, adjust the rotary diffusion parameters to match the shell orientation, then adjust κ to match the core width. Smaller values of κ will widen the core, and values closer to unity will narrow the core.

If possible, use experimental orientation data just inside the gate as an initial condition in the simulation, and compare predictions to measurements at one or more downstream points. Values of κ determined this way are often in the range of 0.10 to 0.03.

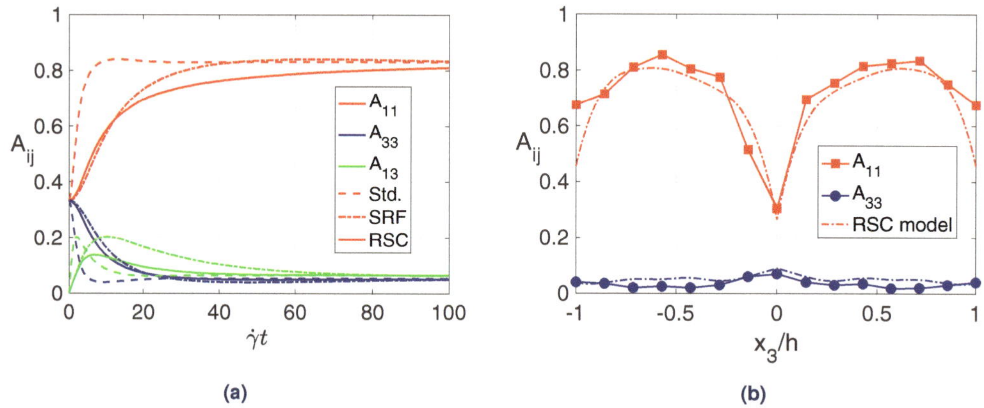

Figure 5.25 (a) Transient behavior in simple shear flow, Folgar–Tucker model with $C_I = 0.005$ and the natural closure. Standard model compared to SRF/slip and RSC models with $\kappa = 0.2$. (b) Experimental data from Fig. 5.22 (center of a 3 mm thick ISO plaque) compared to predictions (dash-dot lines) using the RSC model with $\kappa = 0.05$. PBT with 30% by weight glass fibers. Data and predictions from [Wan07].

In principle, one can also determine κ from a rheological measurement in transient shear flow. However, κ values determined this way typically fall between 0.20 and 0.50, which is larger than the values that match injection molded parts. This might be because the shear rates in transient rheological measurements are several orders of magnitude lower than in injection molding.

Figure 5.25(b) shows the experimental data from Fig. 5.22 compared to a prediction using the RSC model with $\kappa = 0.05$ [Wan07]. Now the core is predicted much more accurately. This particular mold filling simulation had a detailed treatment of the fountain flow, so using the RSC model improved the skin-layer predictions as well.

5.7.5 Combining Slow Kinetics with Anisotropic Rotary Diffusion

So far we have examined slow orientation kinetics models in the context of Folgar–Tucker type isotropic rotary diffusion. With a small extension, the RSC and RPR models can also be combined with any anisotropic rotary diffusion model.

The extension of RSC to ARD models was first developed by Phelps and Tucker [Phe09, PT09]. Their equations are more easily understood if we use the change of variables suggested by Favaloro, Eqn. (5.76).

Using the general ARD equation (5.77) in the eigenvalue form for $\dot{\mathbf{A}}$ (5.97), the rate of change of the eigenvalues for any ARD model is

$$\dot{\lambda}_i^{\text{ARD}} = 2\left(\lambda_i \hat{\mathbf{D}} : \mathbf{e}_i\mathbf{e}_i - \mathbf{e}_i\mathbf{e}_i : \mathbb{A} : \hat{\mathbf{D}}\right) + 6\hat{C}_I\dot{\gamma}\left(\hat{\mathbf{A}} : \mathbf{e}_i\mathbf{e}_i - \lambda_i\right) \tag{5.109}$$

Following through with the RSC analysis as in the previous section gives the time derivative of the second-order orientation tensor as

$$\dot{\mathbf{A}}^{\text{ARD, RSC}} = \mathbf{W}\cdot\mathbf{A} - \mathbf{A}\cdot\mathbf{W} + \hat{\mathbf{D}}\cdot\mathbf{A} + \mathbf{A}\cdot\hat{\mathbf{D}} - 2\left[\mathbb{A} + (1-\kappa)(\mathbb{L} - \mathbb{M}:\mathbb{A})\right]:\hat{\mathbf{D}} + 6\hat{C}_I\dot{\gamma}\left[\hat{\mathbf{A}} - (1-\kappa)\mathbb{M}:\hat{\mathbf{A}} - \kappa\mathbf{A}\right] \tag{5.110}$$

This is the most general equation for the time derivative of **A** in this book. It can be used to implement any ARD and RSC/RPR model, in any combination, by choosing κ and **C** appropriately. Isotropic rotary diffusion is recovered by setting $\mathbf{C} = C_I\mathbf{I}$, and standard (Jeffery) kinetics is recovered by setting $\kappa = 1$. This equation is implemented in the function `Adot2` in the accompanying software.

As a special case, if we choose an ARD model in which the rotary diffusion tensor **C** has the same eigenvectors as **A**, then $\mathbb{M}:\hat{\mathbf{A}} = \hat{\mathbf{A}}$ in Eqn. (5.110) and we obtain

$$\dot{\mathbf{A}}^{\text{ARD, RSC}} = \mathbf{W}\cdot\mathbf{A} - \mathbf{A}\cdot\mathbf{W} + \hat{\mathbf{D}}\cdot\mathbf{A} + \mathbf{A}\cdot\hat{\mathbf{D}} - 2\left[\mathbb{A} + (1-\kappa)(\mathbb{L} - \mathbb{M}:\mathbb{A})\right]:\hat{\mathbf{D}} + 6\kappa\hat{C}_I\dot{\gamma}\left(\hat{\mathbf{A}} - \mathbf{A}\right) \tag{5.111}$$

This looks like the RSC model for the Folgar–Tucker equation, Eqn. (5.103), with $\xi \to 1$, $\mathbf{D} \to \hat{\mathbf{D}}$, the rotary diffusion term multiplied by κ, and a target orientation tensor of $\hat{\mathbf{A}}$ in place of $\mathbf{I}/3$. This special form applies to the pARD, MRD and Wang versions of anisotropic rotary diffusion. However, the more general form of Eqn. (5.110) must be used for the iARD model and the five-constant form of Eqn. (5.85).

When the RPR form is used to treat slow kinetics, implementing an ARD model is straightforward. The standard-form ARD model (e.g., Eqn. (5.77)) is used to get the standard $\dot{\mathbf{A}}$, and the rest of the RPR calculation proceeds as before.

Adding the RSC or RPR treatment does not change the steady orientation states of an ARD model, as discussed in Section 5.7.4. Thus, one still chooses ARD parameters to get the desired steady-state orientation in simple shear flow, then chooses κ to match the experimental core thickness.

5.7.6 Flow-Type Dependence

5.7.6.1 Rationale

The RSC and RPR models initially used κ as a material constant. However, subsequent experiments showed that the value of the κ depends on whether the flow is shearing or extension. That is, κ depends on the *flow type*.

Figure 5.26 shows an example, using data from Lambert and Baird [LB16]. A composite with 10% by weight of long glass fibers in PP was subjected to simple shear flow using a sliding-plate rheometer, and to planar extension using lubricated squeeze flow (see Section 4.3.6.1). The samples were prepared by aligning pellets by hand in a compression mold, to create a repeatable initial orientation. Experiments were stopped at different amounts of deformation and the samples frozen, to obtain orientation data at different strain levels.

The figure compares this data to model calculations using the Folgar–Tucker/RSC model with $C_I = 0.010$ and the natural closure. In Fig. 5.26(a) we see that using $\kappa = 1$ (standard kinetics) provides a poor fit to the simple shear data, while reducing κ to 0.014 does a good job of fitting the data. However, the opposite is true for the extensional flow data in Fig. 5.26(b). A value of $\kappa = 1$ fits the extensional data well, while $\kappa = 0.14$ does not match the extensional data at all.

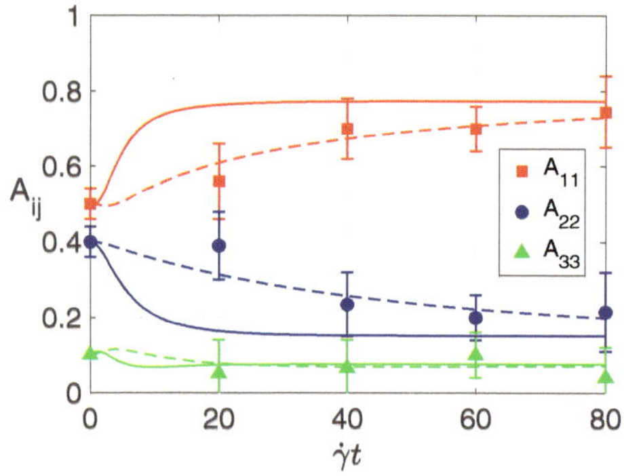

(a) Simple shear

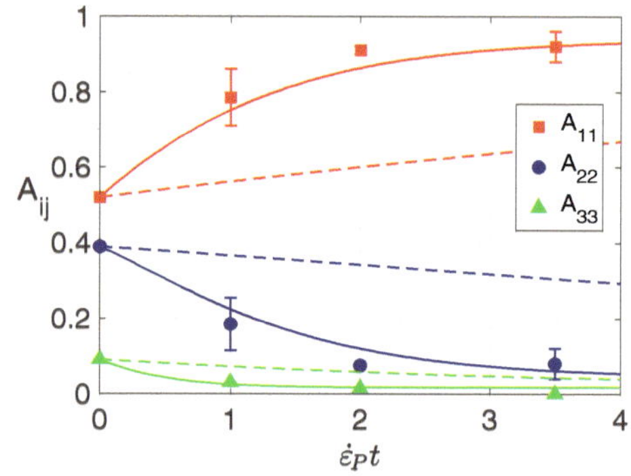

(b) Planar extension

Figure 5.26 Transient orientation for the same glass fiber/PP material in (a) simple shear flow and (b) planar extensional flow. Points are experimental data from [LB16]. Lines are the Folgar–Tucker/RSC model with $C_I = 0.010$ and the natural closure. Solid lines: $\kappa = 1$; dashed lines: $\kappa = 0.14$.

Many processing flows combine shear and extension. For example, the center-gated disk flow discussed in Section 4.3.3 and the non-lubricated squeeze flow in Section 4.3.6.2 combine shear flow near the mold walls with extensional flow near the midplane. For these geometries, the RSC or RPR model with a single value of κ will not be able to predict both the shell and core orientations [LWB17, CWB19b, KLC$^+$20b].

The physical reason for this flow-type dependence is not known, just as the reason for slow orientation kinetics in shear flow is not understood. We do know that in an extensional flow a material line (and thus a Jeffery fiber) is always rotating toward the direction of maximum stretching, so stretching along that material direction is persistent over time. In contrast, in simple shear a material line rotates quickly past the direction of maximum stretching and settles in a direction of zero stretching.

5.7.6.2 Models

It appears that we need two values of κ, one for extensional flow and one for shear flow. In fact, there are an infinite number of flows that are intermediate between simple shear and pure extension, so interpolating between the two values makes sense. This idea was first proposed by Chen, Wapperom and Baird [CWB19a].

To build such a model, we need a scalar parameter that describes whether the flow is shearing, extensional or somewhere in between. And, just like the RSC model, we would like the flow parameter to be objective.

Chen et al. [CWB19a] used an objective flow parameter based on [Ast79],

$$\beta = \frac{\mathbf{D}:\mathbf{D} + \overline{\mathbf{W}}:\overline{\mathbf{W}}}{\mathbf{D}:\mathbf{D} - \overline{\mathbf{W}}:\overline{\mathbf{W}}} \tag{5.112}$$

In this expression, $\overline{\mathbf{W}} = \mathbf{W} - \mathbf{\Omega}$, where $\mathbf{\Omega}$ gives the rate of rotation of the eigenvectors of $\mathbf{D}$. That is, if $\mathbf{d}_i$ represents the ith eigenvector of $\mathbf{D}$, then the rate of change of $\mathbf{d}_i$ is

$$\dot{\mathbf{d}}_i = \mathbf{\Omega} \cdot \mathbf{d}_i \tag{5.113}$$

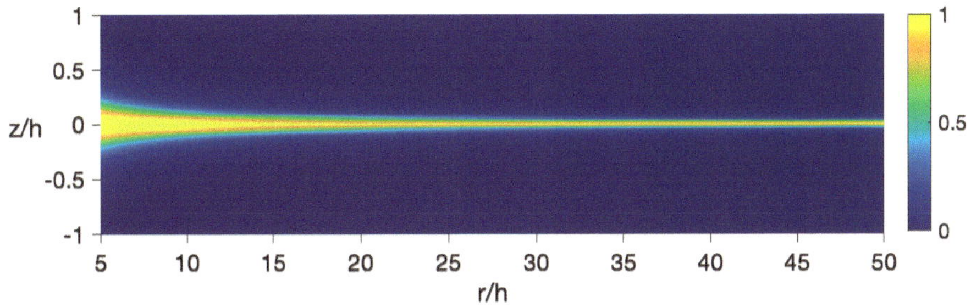

Figure 5.27 Flow parameter β for a center-gated disk, using the Newtonian velocity solution, Eqn. (4.109).

As we saw with the RSC model, the vorticity tensor $\mathbf{W}$ will change if we re-phrase the problem in a rotating coordinate system. However, $\overline{\mathbf{W}}$ is the vorticity referenced to a coordinate system aligned with the eigenvectors of $\mathbf{D}$, rather than to the laboratory axes. Thus, $\overline{\mathbf{W}}$ is objective, and so is β. Computing $\mathbf{\Omega}$ for an arbitrary flow requires some effort, and we will not go into the details here. $\mathbf{\Omega}$ is zero for a simple shear flow or a planar extensional flow.

Important values of β are

$$\beta = \begin{cases} -1 & \text{rigid-body rotation} \\ 0 & \text{simple shear flow} \\ 1 & \text{extensional flow} \end{cases} \tag{5.114}$$

We saw in Section 4.1.1.2 that simple shear flow ($\beta = 0$) is the superposition of rigid-body rotation and planar extension, where the rotation rate exactly matches the stretching rate. Flows with $0 < \beta \leq 1$ have more extension than rotation, while flows with $-1 \leq \beta < 0$ have more rotation than extension.

Figure 5.27 shows how β varies with position in a center-gated disk, using the Newtonian velocity field from Eqn. (4.109). As expected, the flow is simple shear ($\beta \to 0$) near the mold walls, and extensional ($\beta \to 1$) at the midplane. The region of strong extensional flow is relatively thin, and grows thinner as the radius increases.

The orientation model of Chen et al. has two values of the kinetic parameter: κ_e for extensional flow and κ_s for shear flow. The κ value at any point in the flow is interpolated between these values using β:

$$\kappa = \beta\,\kappa_e + (1-\beta)\,\kappa_s \tag{5.115}$$

The rest of the orientation model is unchanged.

Because Chen et al. studied non-lubricated squeeze flow, $0 \leq \beta \leq 1$ everywhere in their example, so the case of $\beta < 0$ was not discussed. However, one could easily choose to use $\kappa = \kappa_s$ for flows that are more rotational than simple shear, making $\beta = 0$ for those flows.

The Autodesk Moldflow Insight software takes a similar approach, but uses a different flow parameter [Lar85, MH09]:

$$\psi^{\text{AMI}} = 1 - \frac{\operatorname{tr}\left(\mathring{\mathbf{D}}^2\right)}{2\operatorname{tr}\left(\mathbf{D}^2\right)^2} \tag{5.116}$$

Here $\overset{\circ}{\mathbf{D}}$ is the *co-rotating* or *Jaumann derivative* of $\mathbf{D}$, defined as

$$\overset{\circ}{\mathbf{D}} = \dot{\mathbf{D}} - \mathbf{W} \cdot \mathbf{D} + \mathbf{D} \cdot \mathbf{W} \tag{5.117}$$

This derivative is the rate of change of a tensor ($\mathbf{D}$ in this case) seen by an observer who both translates and rotates with the fluid. The ψ^{AMI} parameter equals 0 for simple shear flow and 1 for elongational flow, though it goes to $-\infty$ for rigid-body rotation. In computations, any value of $\psi^{\text{AMI}} < 0$ is treated as 0, and κ is interpolated linearly between a simple shear value and an extensional flow value,

$$\kappa = \psi^{\text{AMI}} \, \kappa_e + \left(1 - \psi^{\text{AMI}}\right) \kappa_s \tag{5.118}$$

A slightly different flow-type model has been proposed by Kugler et al. [KDS+20]. They use a flow parameter proposed by Cheng and Manas-Zloczower [CMZ90],

$$M_z = \frac{|\mathbf{D}|}{|\mathbf{D}| + |\mathbf{W}|} \tag{5.119}$$

where $|\mathbf{D}| \equiv \sqrt{\mathbf{D} : \mathbf{D}}$ and $|\mathbf{W}| \equiv \sqrt{\mathbf{W} : \mathbf{W}}$ are scalar measures of the rate of deformation and vorticity, respectively. This parameter varies between zero and unity, with the following important values:

$$M_z = \begin{cases} 0 & \text{rigid-body rotation} \\ 0.5 & \text{simple shear flow} \\ 1 & \text{extensional flow} \end{cases} \tag{5.120}$$

Because the M_z parameter uses the vorticity tensor $\mathbf{W}$ it is not objective, so it has the same limitations as the SRF model. However, this does not appear to cause major errors in the cases studied.

The model of Kugler et al. not only uses two different values of κ for shear and extension, it also uses different values of C_I, or different ARD parameters. This gives some additional flexibility in fitting steady-state data in both shear and extension, provided that such data is available.

In their approach, the extensional parameters are used to compute a time derivative $\dot{\mathbf{A}}_e$, and the shear parameters are used to compute $\dot{\mathbf{A}}_s$. The final time derivative of the orientation tensor at any point and time is then interpolated as

$$\dot{\mathbf{A}} = 2(M_z - 0.5)\, \dot{\mathbf{A}}_e + 2(1 - M_z)\, \dot{\mathbf{A}}_s \tag{5.121}$$

Kugler et al. used the pARD/RSC model, but their scheme can be applied to any orientation model. They used the shear model, $\dot{\mathbf{A}} = \dot{\mathbf{A}}_s$, whenever Eqn. (5.119) gave $M_z < 0.5$.

5.7.6.3 Example

Figure 5.28 shows how using a flow-type dependence affects orientation predictions. Here the velocity field is the Newtonian solution for a center-gated disk from Eqn. (4.109), and the orientation at the inlet is representative of experimental data [CWB19b] for this geometry. The simulation space corresponds to Fig. 5.27, and the orientation profile in Fig. 5.28 is at the right-hand edge of Fig. 5.27.

Using a constant κ value typical of extensional flows, $\kappa_e = 1$, gives a core with strong crossflow alignment, but the core may be too thick. A constant κ typical of shearing flows, $\kappa_s = 0.20$,

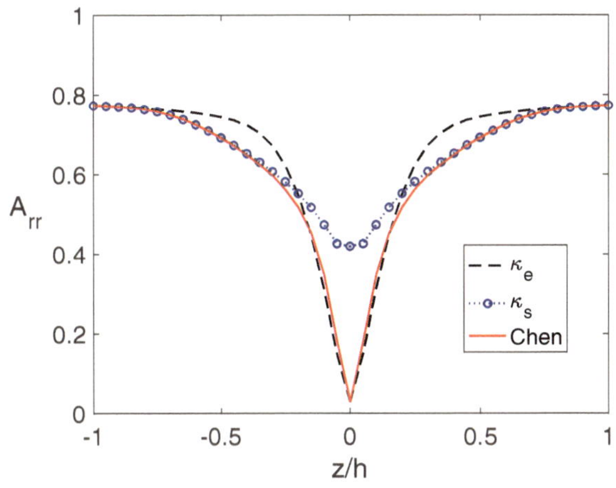

Figure 5.28 Predicted flow-direction orientation A_{rr} at $r/h = 50$ for a center-gated disk, using the Newtonian velocity solution, Eqn. (4.109). Folgar–Tucker/RSC model with $C_I = 0.010$ and the natural closure. The three curves have (a) constant $\kappa_e = 1$, (b) constant $\kappa_s = 0.20$ and (c) the flow-dependent model of Chen et al., Eqn. (5.115). The inlet at $r/h = 5$ has $A_{rr} = 0.65$, $A_{\theta\theta} = 0.30$ and $A_{zz} = 0.05$.

gives a more gradual transition from core to shell, but prevents the crossflow core orientation from developing at this radius. The model of Chen et al., which combines κ_e and κ_s using Eqn. (5.115), allows the transverse core to develop while keeping the core/shell transition gradual.

Models with flow-type dependence have improved the agreement between prediction and experiment in a center-gated disk [CWB19b] and in commercial parts [KDS$^+$20]. These flow-type models are available in most commercial software. Recommended practice is to use the flow-type option when it is available. The shear parameter κ_s is chosen as in Section 5.7.4, and the extensional parameter κ_e is either set to unity, or fitted to transient orientation measurements in extensional flow if such data is available.

5.8 Chapter Summary

Computer-aided engineering software tracks the distribution of orientations at each point using the second-order orientation tensor **A**. An equation for the rate of change of **A** can be developed from Jeffery's equation, but this equation also contains the fourth-order orientation tensor $\mathbb{A}$. To obtain useful solutions, a closure approximation is used to approximate $\mathbb{A}$ in terms of **A**.

Accurate closures have been developed by fitting distribution function data for a representative set of orientation states. The eigenvalue-based version of the natural closure (ORE) and the invariant-based fit of Chung and Kwon (IBOF) are among the best available closures, and are used in commercial software.

A tensor model based on Jeffery's equation alone predicts too much fiber alignment for large deformations. To obtain realistic steady-state orientations we add rotary diffusion terms to the model. The Folgar–Tucker model provides isotropic rotary diffusion controlled by a single parameter, the interaction coefficient C_I. For injection molded parts this coefficient is typically adjusted to match the flow-direction alignment in the shell layer. If isotropic diffusion predicts unacceptable values for out-of-plane orientation, any one of several anisotropic rotary diffusion (ARD) models can be employed. ARD models introduce one or two additional parameters.

Practical composites also transition from their initial orientation to a steady state more slowly than Jeffery-type models predict, especially in shear flow. This phenomenon is referred to as slow orientation kinetics. The RSC or RPR models can be used to match slow kinetics, introducing a "slow-down" parameter κ, where $0 < \kappa \leq 1$. The value of κ can be chosen by matching the core thickness in a molded plaque. The RSC or RPR models are combined with either the Folgar–Tucker or ARD models to obtain appropriate steady orientation states.

While slow orientation kinetics is a significant effect in simple shear flows, with κ values of 0.2 to 0.1 often observed, it does *not* occur in extensional flows. This can cause models with a constant κ to mispredict either the core orientation or the shell orientation of a molding. Models where κ varies with flow type (shear vs. extension) have been developed and show promise for improving prediction accuracy.

6 Suspension Rheology and Flow-Orientation Coupling

To predict fiber orientation in a manufacturing process, one must know the fluid velocity as a function of position and time. Solving for that velocity requires information about the flow properties, or rheology, of the fiber/polymer suspension. This chapter covers fiber suspension rheology and its relationship to velocity solutions.

We first review basic aspects of constitutive equations, which give the stress in a fluid as a function of its deformation. Most CAE software packages use a generalized Newtonian fluid model, fitted to data for the steady shear viscosity of the suspension. Because stress does not depend on fiber orientation in these models, the flow field can be calculated without knowing the orientation. This velocity solution is then used to find the fiber orientation. In such cases the flow and orientation are *decoupled*.

If the fibers are taken into account explicitly, then stress does depend on orientation. This occurs because fibers resist stretching along their axes more than they resist other types of deformation. We examine a rheological model of this type. Now the velocity solution depends on orientation, and the orientation still depends on the velocity. Hence, flow and orientation are *coupled*, and the two problems must be solved simultaneously. We show examples of coupled flow, where the presence of fibers alters the flow field. These effects can be predicted using an orientation-dependent rheological model. We also show situations where a decoupled solution is accurate, and give general criteria for using the decoupled approach.

A recent rheological model has a viscosity that depends on orientation, but is isotropic. We examine this model, which is intermediate between the generalized Newtonian fluid model and the full orientation-dependent model for fiber suspensions.

Finally, we look at transient rheological experiments. These experiments demonstrate the connection between fiber orientation and suspension stress, and provide information about fiber orientation models and their parameters.

6.1 Basics of Constitutive Equations

The rheology of a fluid or a fiber suspension is expressed using a *constitutive equation*. This gives the stress in the fluid as a function of deformation rate and any other variables of interest. Before examining fiber suspension behavior, we review a few basics of constitutive equations.

6.1.1 Total Stress and Extra Stress

Most readers are familiar with the stress tensor $\boldsymbol{\sigma}$, which describes forces transmitted through a material by physical contact. For a formal definition, let the vector $\mathbf{t}$ be the force per unit area acting on a surface that is normal to the unit vector $\mathbf{n}$. $\mathbf{t}$ is called the *surface traction.* The Euler–Cauchy stress principle states that there exists a second-order tensor $\boldsymbol{\sigma}$ such that

$$\mathbf{t} = \mathbf{n} \cdot \boldsymbol{\sigma} \qquad \text{or} \qquad t_j = n_i \sigma_{ij} \tag{6.1}$$

That is, $\boldsymbol{\sigma}$ provides a complete description of the contact forces at any point by giving the traction $\mathbf{t}$ for any orientation $\mathbf{n}$ of the surface. $\boldsymbol{\sigma}$ is called the *total stress tensor* or the Cauchy stress. With this definition, the component σ_{ij} is the force in the j direction acting on an area normal to the i direction. However, for most fluids the total stress is symmetric,

$$\boldsymbol{\sigma}^T = \boldsymbol{\sigma} \qquad \text{or} \qquad \sigma_{ij} = \sigma_{ji} \tag{6.2}$$

so σ_{ij} is also the force in the i direction acting on an area normal to the j direction.

If a material experiences only hydrostatic stress corresponding to a pressure p, then the total stress tensor is

$$\boldsymbol{\sigma} = -p\mathbf{I} = \begin{bmatrix} -p & 0 & 0 \\ 0 & -p & 0 \\ 0 & 0 & -p \end{bmatrix} \tag{6.3}$$

The minus sign appears because positive pressure corresponds to negative (compressive) normal stress.

A fluid with constant density does not change volume in response to a hydrostatic stress, so it is helpful to separate the total stress into two parts:

$$\boldsymbol{\sigma} = -p\,\mathbf{I} + \boldsymbol{\tau} \qquad \text{or} \qquad \sigma_{ij} = -p\delta_{ij} + \tau_{ij} \tag{6.4}$$

The new tensor $\boldsymbol{\tau}$ is called the *extra stress*, and we expect it to be related to the deformation of the fluid. $\boldsymbol{\tau}$ and $\boldsymbol{\sigma}$ differ only in their diagonal components; their off-diagonal components are equal. Thus, $\boldsymbol{\tau}$ is also symmetric.

6.1.2 Newtonian fluid

As an example of a constitutive equation, a Newtonian fluid with constant density has an extra stress given by

$$\boldsymbol{\tau} = 2\mu\mathbf{D} \tag{6.5}$$

where μ is the Newtonian viscosity. The viscosity may depend on temperature, but for a Newtonian fluid μ does not depend on the rate of deformation. As a result, for a Newtonian fluid the extra stress varies linearly with the rate of deformation.

A Newtonian fluid is also isotropic. Because the fluid itself has no preferred directions, any directionality in $\boldsymbol{\tau}$ comes exclusively from the directionality of the deformation rate $\mathbf{D}$.

6.1.3 Generalized Newtonian Fluid

Most polymer melts have a viscosity that depends on shear rate, which is one type of non-Newtonian behavior. The simplest constitutive equation that captures this is the *generalized Newtonian fluid* (GNF). GNF models have the form

$$\boldsymbol{\tau} = 2\eta \mathbf{D} \qquad \text{with} \qquad \eta = \eta(\dot{\gamma}) \tag{6.6}$$

For a generalized Newtonian fluid, the viscosity η depends on $\mathbf{D}$ through the scalar strain rate $\dot{\gamma}$ (Eqn. (4.14)). As we saw in Section 4.1.1.1, $\dot{\gamma}$ is an invariant of $\mathbf{D}$, so any GNF model will be objective.

Many GNF models exist, with different functions for $\eta(\dot{\gamma})$ [Mac94, BAH87]. One simple option is the *power-law fluid,*

$$\eta = m\dot{\gamma}^{n-1} \tag{6.7}$$

where m and n are constants. n is called the *power-law index,* and $n = 1$ recovers the Newtonian fluid. Most polymers have $n < 1$, so their viscosity decreases as the shear rate increases. Such materials are said to be *shear thinning.*

Most injection molding simulations use a GNF form like

$$\eta = \frac{\eta_0}{\left[1 + \left(\eta_0 \dot{\gamma}/\tau^*\right)^a\right]^{(1-n)/a}} \tag{6.8}$$

This has four material constants: η_0, n, τ^* and a. Choosing $a = 2$ gives the *Carreau model,* $a = 1 - n$ is the *Cross model,* and using a as an adjustable parameter is the *Carreau–Yasuda model.* All versions give a constant viscosity η_0 at low shear rates, a power-law behavior with index n at high shear rates, and a smooth transition in between. The smaller the value of a, the more gradual the transition.

For process modeling one must also capture the temperature dependence of viscosity. For the Cross/Carreau model one typically makes η_0 a function of temperature, often using the WLF equation, while n, a and τ^* are held constant [HC92].

6.1.4 Shear vs. Extensional Viscosity

The most common rheological measurements use simple shear flow. We discussed the kinematics of this flow in Example 4.1.3. Here we will use a slightly different coordinate system in which $v_1 = \dot{\gamma} x_3$ and $v_2 = v_3 = 0$.

For this flow, the shear viscosity η is defined using the steady-state stress as

$$\eta \equiv \frac{\tau_{13}}{\dot{\gamma}} \tag{6.9}$$

Some fluids also develop normal stresses in shear flow. These are described by two *normal stress differences.* In our current coordinate system these are

$$N_1 \equiv \tau_{11} - \tau_{33} \tag{6.10}$$

$$N_2 \equiv \tau_{33} - \tau_{22} \tag{6.11}$$

Individual normal components of the extra stress like τ_{11} or τ_{22} cannot be measured experimentally, because the partition between hydrostatic stress and extra stress in Eqn. (6.4) is arbitrary[1]. However, when one takes the *difference* between two normal stress components, the hydrostatic contributions cancel, so normal stress differences like N_1 and N_2 are well-defined and can be measured.

A Newtonian fluid or any generalized Newtonian fluid has $N_1 = N_2 = 0$ in simple shear flow. Molten polymers do have non-zero normal stresses in simple shear flow, and normal stresses are often regarded as a consequence of viscoelasticity. However, normal stresses can also arise from anisotropy, as we will see in a moment.

It is also possible to measure the *extensional viscosity* of a fluid. The fluid is subjected to a uniaxial extensional flow as in Example 4.1.4: $\dot{\varepsilon}_U$ is the elongation rate in the 3 direction and the material contracts at a rate $-\dot{\varepsilon}_U/2$ in the 1 and 2 directions. The extensional viscosity $\overline{\eta}$ is then defined as

$$\overline{\eta} \equiv \frac{\tau_{33} - \frac{1}{2}(\tau_{11} + \tau_{22})}{\dot{\varepsilon}_U} \tag{6.12}$$

The ratio of the extensional viscosity to the shear viscosity is know as the *Trouton ratio* R_T:

$$R_T \equiv \frac{\overline{\eta}}{\eta} \tag{6.13}$$

Using the Newtonian constitutive equation (6.5) in uniaxial extensional flow gives $\overline{\eta} = 3\mu$, so the Trouton ratio for a Newtonian fluid is always equal to 3. The same is true for a generalized Newtonian fluid, when the shear viscosity is evaluated at the scalar strain rate $\dot{\gamma}$ corresponding to the extensional flow (see Eqn. (4.41)). Polymer melts typically have Trouton ratios much greater than 3, so a high extensional viscosity is often interpreted as a sign of viscoelasticity. However, Trouton ratios greater than 3 can also arise in fiber suspensions due to anisotropy.

6.2 Steady Shear Viscosity of Fiber Suspensions

6.2.1 Measurements and Typical Data

For fiber reinforced injection molding compounds, the most common rheological measurement is steady shear viscosity.

It is best to obtain data in the range of shear rate that the material will see in the manufacturing process. Instruments for this are a capillary rheometer and a slit die attached to the injection unit of an injection molding machine.

In a capillary rheometer a piston pushes the material from a barrel through a small-diameter cylindrical tube (the capillary). The entire apparatus is heated to control the temperature of

[1] A single diagonal component of the total stress $\boldsymbol{\sigma}$, say σ_{11}, can be measured, but the partition between $\boldsymbol{\tau}$ and the pressure p is arbitrary, so $\tau_{11} = \sigma_{11} + p$ cannot be determined by itself.

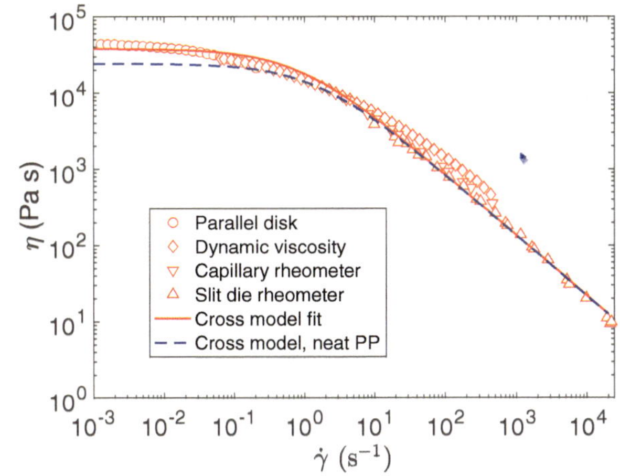

Figure 6.1 Steady shear viscosity of PP with 30% by weight of short glass fibers at 200 °C. The solid red line is a fit of the Cross model: $\eta_0 = 3.77 \times 10^4$ Pa s, $n = 0.2037$ and $\tau^* = 3.27 \times 10^4$ Pa. Dynamic viscosity data $\eta^*(\omega)$ using a parallel disk geometry was not included in the fit. The dashed blue line is a Cross model fitted to data for the same PP with no fibers (points not shown). Experimental data from [MCH+05].

the material. By analyzing the driving pressure as a function of flow rate, one can determine the shear viscosity as a function of shear rate, e.g., [Mac94].

Another good choice for fiber reinforced materials is to attach a slit die to the injection unit of an injection molding machine [MCH+05]. This is sometimes referred to as an on-line rheometer. Material is melted in the injection unit, then forced through the slit die and into the atmosphere. Pressure transducers in the die measure the pressure gradient as a function of flow rate, from which the viscosity can be calculated. Shear rates in the range of 10^1 to 10^4 s^{-1} can be achieved.

Figure 6.1 shows data obtained from capillary and slit die rheometers for a polypropylene (PP) with 30% by weight of short glass fibers. The material is shear thinning, with viscosity η decreasing as the shear rate $\dot{\gamma}$ increases. The figure also shows viscosity data at low shear rates measured using the torsional flow between parallel disks, as well as the dynamic viscosity η^* measured in small-amplitude oscillatory shear, also in a parallel disk geometry [MCH+05]. As the shear rate decreases, the viscosity levels off, approaching a plateau where the viscosity is constant. Measurements from the different instruments are also in agreement with one another.

A Cross model, Eqn. (6.8) fitted to the viscosity data (excluding the dynamic viscosity data) is shown as the solid red line in Fig. 6.1. The agreement is quite good, especially at the higher shear rates that are typical of injection molding. A full data set to support injection molding simulation would require similar measurements at multiple temperatures, so that the temperature-dependence of the viscosity could also be fitted. Practical data sets are not usually as extensive as this one.

The blue dashed line in Fig. 6.1 is a Cross model fitted to the viscosity of the same polypropylene with no fibers. While viscosity of the fiber suspension is higher than the PP viscosity at low shear rates, at the high shear rates the suspension viscosity is almost identical to the viscosity of the polymer without any fibers. This effect has been observed by several researchers (e.g., [MCH+05]) in suspensions where the suspending fluid is shear thinning. While fibers increase the shear viscosity of a Newtonian fluid, they also make the local shear rate of the fluid slightly higher than the applied shear rate. With a shear thinning fluid, this decreases the local fluid viscosity, which can offset the increase from the fibers.

6.2.2 Relationship to Fiber Orientation

In these experiments the measured viscosity is based on the shear stress at steady state, when the material has been subjected to a simple shear flow with large strain. Consequently, the fibers in these measurements should have the steady-state orientation for simple shear flow. The fibers near the walls of a slit die should be oriented much like the shell layer of an injection molded strip, and the same should be true for material near the walls of a capillary rheometer.

In simple shear flow the steady-state orientation will be primarily aligned in the flow direction, with a small amount of out-of-plane alignment. As we will see in Section 6.3, fibers parallel to a shear flow have little influence on the shear viscosity. Thus, a measurement of steady shear viscosity provides information about the orientation state where the fibers have the *least* influence on viscosity.

In contrast, the extensional viscosities of fiber suspensions are much higher than one would expect from the shear viscosity alone. Trouton ratios of 20 and 54 have been reported for PP with short glass fibers at low extension rates [FHPG+09, MCH+05]. An extensional viscosity measurement stretches the material in a consistent direction, so any fibers that are present will orient primarily in the stretching direction. In this orientation, the fibers contribute substantially to the stress, creating the high values of extensional viscosity.

As we saw in Section 6.1.4, a generalized Newtonian fluid model will always give a Trouton ratio of 3. To simultaneously represent both a low shear viscosity and a high extensional viscosity, we will need a different constitutive equation, one that is anisotropic.

■ 6.3 An Anisotropic Constitutive Equation for Fiber Suspensions

Here we develop a widely used constitutive equation for fiber suspensions that captures at least some of the anisotropy caused by the fibers. For this section we assume that viscous stresses are linear in the rate of deformation, which implies that the suspending fluid is Newtonian. Extensions to this model include shear-rate dependence [SV98, FHPG+09].

6.3.1 Fluid Motion Relative to the Fibers

A full derivation of a constitutive equation for fiber suspensions is beyond the scope of this book, but we can gain insight into the key physical phenomena by looking at the motion of the fluid relative to a fiber. Consider a slender, cylindrical fiber as shown in Fig. 6.2. Following

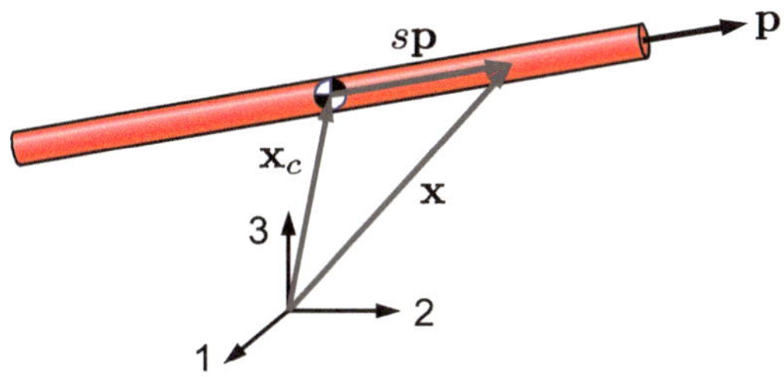

Figure 6.2 Nomenclature for analyzing fluid motion relative to a slender fiber.

Section 4.2, let $\mathbf{x}_c$ be the position of the centroid of the fiber. Any other point along the fiber axis has position $\mathbf{x}$, given by

$$\mathbf{x} = \mathbf{x}_c + s\mathbf{p} \tag{6.14}$$

where s measures distance along the fiber axis and $s = 0$ at the centroid. As the fiber moves, the velocity of the fiber at any point on its axis is the time derivative of this position,

$$\dot{\mathbf{x}} = \dot{\mathbf{x}}_c + s\dot{\mathbf{p}} \tag{6.15}$$

Note that s is constant for a point fixed on the fiber axis.

Turning to the fluid, the unperturbed fluid velocity at $\mathbf{x}$ is

$$\mathbf{v}(\mathbf{x}) = \mathbf{v}_0 + \mathbf{L}\cdot\mathbf{x} \tag{6.16}$$

In this expression $\mathbf{v}_0$ is the velocity at the origin of the coordinate system and $\mathbf{L}$ is the velocity gradient tensor. We saw in Eqn. (4.84) that the centroid velocity of a Jeffery fiber equals the fluid velocity at $\mathbf{x}_c$, so the fiber centroid velocity is

$$\dot{\mathbf{x}}_c = \mathbf{v}_0 + \mathbf{L}\cdot\mathbf{x}_c \tag{6.17}$$

We expect the viscous drag forces exerted by the fluid at any point on the fiber to depend on the fluid velocity relative to the fiber, $\mathbf{v}(\mathbf{x}) - \dot{\mathbf{x}}$. Using the above equations, this is

$$\begin{aligned}\mathbf{v}(\mathbf{x}) - \dot{\mathbf{x}} &= [\mathbf{v}_0 + \mathbf{L}\cdot\mathbf{x}] - \left[\dot{\mathbf{x}}_c + s\dot{\mathbf{p}}\right]\\ &= \left[\mathbf{v}_0 + \mathbf{L}\cdot(\mathbf{x}_c + s\mathbf{p})\right] - \left[\mathbf{v}_0 + \mathbf{L}\cdot\mathbf{x}_c + s\dot{\mathbf{p}}\right]\\ &= s\left[\mathbf{L}\cdot\mathbf{p} - \dot{\mathbf{p}}\right]\end{aligned} \tag{6.18}$$

For $\dot{\mathbf{p}}$ we substitute Jeffery's equation (4.85), using $\xi = 1$ for a slender fiber. The first two terms from that equation can be combined, since $\mathbf{L} = \mathbf{D} + \mathbf{W}$, so Jeffery's equation for $\xi = 1$ can be written as

$$\dot{\mathbf{p}} = \mathbf{L}\cdot\mathbf{p} - \mathbf{D}:\mathbf{ppp} \tag{6.19}$$

Using this in Eqn. (6.18) gives the desired result,

$$\mathbf{v}(\mathbf{x}) - \dot{\mathbf{x}} = s\,(\mathbf{D}:\mathbf{pp})\,\mathbf{p} \tag{6.20}$$

The quantity $(\mathbf{D}:\mathbf{pp})$ is a scalar that gives the stretching rate of the fluid along the $\mathbf{p}$ direction. The $\mathbf{p}$ outside the parentheses shows that the direction of this relative velocity is parallel to the fiber axis. Equation (6.20) shows that:

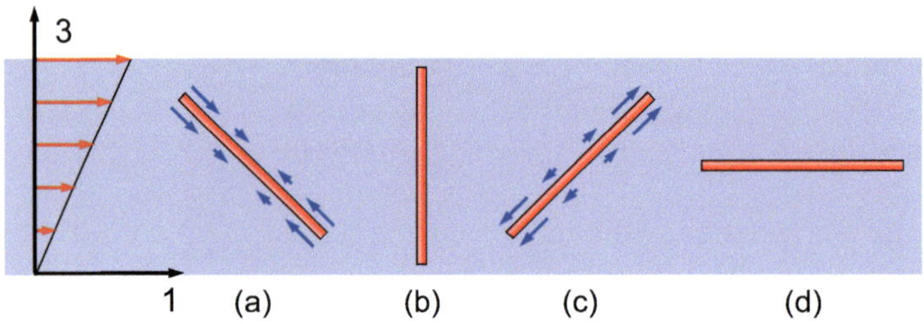

Figure 6.3 Velocity of the fluid relative to a fiber (blue arrows) according to Eqn. (6.20), for different fiber orientations in simple shear flow. Due to their orientation and rotational motion, fibers (b) and (d) have no relative velocity.

The velocity of the fluid relative to a fiber is primarily along the fiber axis, is proportional to the rate of stretching in the fiber direction, and is also proportional to the distance from the fiber centroid.

We have added "primarily" to this statement because fibers in a concentrated suspension do not follow Jeffery's equation exactly. Still, this is a significant result.

Figure 6.3 shows examples of this relative velocity for fibers in a simple shear flow. Fiber (a) is oriented at $-45°$ to the flow direction. For this orientation $\mathbf{D} : \mathbf{pp} < 0$, the fluid is contracting along the $\mathbf{p}$ direction, and this fiber will be in compression.

Fiber (b), at 90° to the flow, is rotating at a rate that gives no relative velocity, so this fiber develops no stress in simple shear.

Fiber (c) in Fig. 6.3 is oriented at +45° to the flow, and the relative motion of the fluid is from the center of the fiber toward the fiber ends. For this case $\mathbf{D} : \mathbf{pp} > 0$, so the fluid is stretching along the $\mathbf{p}$ direction. Consequently, this fiber will be in tension. The fluid stretching along this fiber's axis and the fluid contraction along the axis of fiber (a) are consistent with the stretching associated with $\mathbf{D}$ for a simple shear flow, Fig. 4.4(b).

For a simple shear flow, a fiber oriented in the flow direction like fiber (d) in Fig. 6.3 has $\mathbf{D} : \mathbf{pp} = 0$, so it has no relative velocity and develops no stress.

6.3.2 Stress with Fully Aligned Fibers

When this relative motion is used to develop a constitutive equation (e.g., [DA84]), the extra stress will have an isotropic contribution from the fluid, plus a contribution from the fibers that we expect to be anisotropic. If all of the fibers are aligned in a single direction $\mathbf{p}$ then the magnitude of the extra stress from the fibers will be proportional to ($\mathbf{D} : \mathbf{pp}$), since it is this quantity that creates tension or compression in the fibers. The tensorial "direction" of this stress will be $\mathbf{pp}$, which corresponds to a tensile or compressive stress in the direction of the fiber axis. Combining these two factors gives a stress contribution proportional to $\mathbf{D} : \mathbf{pppp}$. In a complete analysis the thickness of the fibers matters, and a term involving $\mathbf{D} \cdot \mathbf{pp}$ also appears. The result for aligned fibers is

$$\boldsymbol{\tau}^a(\mathbf{p}) = 2\eta_l \left[\mathbf{D} + N_p \mathbf{D} : \mathbf{pppp} + N_s \left(\mathbf{D} \cdot \mathbf{pp} + \mathbf{pp} \cdot \mathbf{D}\right)\right] \tag{6.21}$$

In this expression η_I is an isotropic viscosity and N_p is a dimensionless number describing the importance of stretching along the fiber axis relative to η_I. N_p is called the *particle number*. The N_s parameter is called the *shear number*, and is related to the difference in viscosity when the suspension is sheared parallel or perpendicular to the fibers. For typical fiber aspect ratios, $N_p \gg 1$ but $N_s \ll N_p$ (e.g., [Tuc91]). Thus, we expect the N_p term to dominate the anisotropic response, and the N_s term is usually neglected.

In Eqn. (6.21) the extra stress depends on both the rate of deformation and on the direction of the fibers **p**, so this constitutive equation is anisotropic. A few examples illustrate how the anisotropic N_p term operates in simple flows.

Example 6.3.1 Shearing parallel to aligned fibers

Find the extra stress $\boldsymbol{\tau}$ when a suspension with fibers aligned in the 1 direction is sheared parallel to that direction. Omit the N_s term. The velocity field is $v_1 = \dot{\gamma} x_3$, $v_2 = v_3 = 0$.

The flow is like Fig. 6.3, with all fibers oriented like fiber (d). The fiber orientation vector and **pp** tensor are

$$\mathbf{p} = \begin{Bmatrix} 1 \\ 0 \\ 0 \end{Bmatrix} \quad \text{and} \quad \mathbf{pp} = \begin{bmatrix} 1 & 0 & 0 \\ 0 & 0 & 0 \\ 0 & 0 & 0 \end{bmatrix} \tag{6.22}$$

and the rate-of-deformation tensor is

$$\mathbf{D} = \frac{\dot{\gamma}}{2} \begin{bmatrix} 0 & 0 & 1 \\ 0 & 0 & 0 \\ 1 & 0 & 0 \end{bmatrix} \tag{6.23}$$

Looking at Eqn. (6.21), N_p multiplies $\mathbf{D} : \mathbf{pppp}$. This is the product of $(\mathbf{D} : \mathbf{pp})$, which is a scalar, and **pp**, which is the tensor given in Eqn. (6.22). In index notation $(\mathbf{D} : \mathbf{pp}) = D_{ij} p_j p_i$, which is a sum of nine terms. However, for the **p** in Eqn. (6.22) the only non-zero term in this sum has $i = 1$ and $j = 1$. Thus, $\mathbf{D} : \mathbf{pp} = D_{11}$ for this fiber orientation.

For this simple shear flow $D_{11} = 0$, so the N_p term is multiplied by zero and Eqn. (6.21) reduces to

$$\boldsymbol{\tau} = 2\eta_I \mathbf{D} = \eta_I \dot{\gamma} \begin{bmatrix} 0 & 0 & 1 \\ 0 & 0 & 0 \\ 1 & 0 & 0 \end{bmatrix} \tag{6.24}$$

That is, $\tau_{13} = \tau_{31} = \eta_I \dot{\gamma}$ and all other τ_{ij} components are zero. This is the same stress as a Newtonian fluid with viscosity η_I. In this flow, fibers only affect the viscosity to the extent that they contribute to η_I. ■

Example 6.3.2 Planar stretching parallel to aligned fibers

Find the extra stress $\boldsymbol{\tau}$ when a suspension with fibers aligned in the 1 direction is stretched parallel to that direction in planar extension. Omit the N_s term. The velocity field is $v_1 = \dot{\varepsilon}_P x_1$, $v_2 = -\dot{\varepsilon}_P x_2$ and $v_3 = 0$.

p and **pp** are still given by Eqn. (6.22), and the rate-of-deformation tensor is

$$\mathbf{D} = \dot{\varepsilon}_P \begin{bmatrix} 1 & 0 & 0 \\ 0 & -1 & 0 \\ 0 & 0 & 0 \end{bmatrix} \tag{6.25}$$

In this flow D_{11} is not zero, so $\mathbf{D}:\mathbf{pp} = D_{11} = \dot{\varepsilon}_P$. Now the N_p term will contribute to the stress. This makes sense, because the fibers are aligned with the stretching direction and resist that deformation. Using this together with Eqns. (6.22)(b) and (6.25) in Eqn. (6.21), we find

$$\boldsymbol{\tau} = 2\eta_I \dot{\varepsilon}_P \left(\begin{bmatrix} 1 & 0 & 0 \\ 0 & -1 & 0 \\ 0 & 0 & 0 \end{bmatrix} + N_p \begin{bmatrix} 1 & 0 & 0 \\ 0 & 0 & 0 \\ 0 & 0 & 0 \end{bmatrix} \right) \tag{6.26}$$

It is not possible to measure a single diagonal component of $\boldsymbol{\tau}$ experimentally, but the difference between two diagonal components can be measured. In this example the relevant normal-stress difference is

$$\tau_{11} - \tau_{22} = \eta_I \dot{\varepsilon}_P (4 + 2N_p) \tag{6.27}$$

A Newtonian fluid would have $\tau_{11} - \tau_{22} = 4\mu\dot{\varepsilon}_P$ in this flow, so the fibers increase the viscosity of the suspension in this flow.

Note that reversing the flow, so that $\dot{\varepsilon}_P < 0$, corresponds to contraction along the fiber direction and places the fibers in compression. The stress difference $\tau_{11} - \tau_{22}$ then changes sign, but the N_p term is unchanged so the fibers still contribute to the stress. ■

These examples illustrate an important point:

In a fiber suspension, the main contribution of the fibers is to resist stretching or contraction along the fiber direction.

6.3.3 Stress with Distributed Orientation

While the aligned-fiber model is useful for understanding the way that fibers influence viscosity, a real suspension will have a distribution of fiber orientations. To treat this we average the aligned-fiber stress $\boldsymbol{\tau}^a(\mathbf{p})$ over all orientations, weighted by the orientation distribution function. That is, the overall stress with distributed orientation $\boldsymbol{\tau}$ is

$$\boldsymbol{\tau} = \langle \boldsymbol{\tau}^a(\mathbf{p}) \rangle \tag{6.28}$$

Performing this operation on Eqn. (6.21) and again omitting the N_s term[2] gives the result

$$\boldsymbol{\tau} = 2\eta_I \left[\mathbf{D} + N_p \mathbb{A} : \mathbf{D} \right] \tag{6.29}$$

In this model the stress depends on the current orientation state through the fourth-order tensor $\mathbb{A}$, which changes with time in most flows. If fiber orientation is modeled by calculating the second-order orientation tensor $\mathbf{A}$, then the fourth-order tensor in Eqn. (6.29) must be calculated using a closure approximation. As a result,

[2] Retaining the N_s term adds $2\eta_I N_s (\mathbf{D}\cdot\mathbf{A} + \mathbf{A}\cdot\mathbf{D})$ to the right-hand side of Eqn. (6.29).

The choice of closure approximation for $\mathbb{A}$ affects the stress in an orientation-dependent rheological model.

Here are two examples that illustrate the connection between the fourth-order orientation tensor and the stress in a fiber suspension.

Example 6.3.3 Transient stress in simple shear flow

Write an algebraic expression for the shear stress τ_{13} and the two normal stress differences N_1 and N_2 for a fiber suspension in simple shear flow, and identify the components of $\mathbb{A}$ that are important to these stresses. The velocity field is $v_1 = \dot{\gamma} x_3$, $v_2 = v_3 = 0$.

Equation (6.29) in index notation is

$$\tau_{ij} = 2\eta_I \left[D_{ij} + N_p \mathbb{A}_{ijkl} D_{lk} \right] \tag{6.30}$$

For any stress component τ_{ij}, the isotropic part will be proportional to the corresponding rate-of-deformation component D_{ij}, while the anisotropic component will be proportional to the sum $\mathbb{A}_{ijkl} D_{lk}$.

The rate-of-deformation tensor for this flow is given in Eqn. (6.23), and the only non-zero components of $\mathbf{D}$ are $D_{13} = D_{31}$. Using this, for any indices ij, in this flow we have

$$\mathbb{A}_{ijkl} D_{lk} = \mathbb{A}_{ij13} D_{31} + \mathbb{A}_{ij31} D_{13} = 2\mathbb{A}_{ij13} D_{13} = \mathbb{A}_{ij13} \dot{\gamma} \tag{6.31}$$

In this expression, symmetry of $\mathbb{A}$ and $\mathbf{D}$ was used to combine terms. Using this in Eqn. (6.30), we can write out the desired components of the extra stress:

$$\tau_{13} = \eta_I \dot{\gamma} \left[1 + 2N_p \mathbb{A}_{1313} \right] \tag{6.32}$$

$$N_1 = \tau_{11} - \tau_{33} = 2\eta_I \dot{\gamma} N_p \left(\mathbb{A}_{1113} - \mathbb{A}_{3313} \right) \tag{6.33}$$

$$N_2 = \tau_{33} - \tau_{22} = 2\eta_I \dot{\gamma} N_p \left(\mathbb{A}_{3313} - \mathbb{A}_{2213} \right) \tag{6.34}$$

Theses stresses will be time-dependent, because the orientation is time dependent. The fibers increase the shear stress τ_{13} only if there is some orientation in *both* the 1 and 3 directions, so that $\mathbb{A}_{1313}$ is non-zero. Not surprisingly, the fiber contribution to stress is sensitive to the orientation state.

Also, a fiber suspension will have non-zero normal stress differences N_1 and N_2 in simple shear flow. We often associate normal stresses in shear flow with viscoelasticity, and polymeric fluids have normal stress differences N_1 and N_2 that are proportional to $\dot{\gamma}^2$ at low shear rates. Viscoelastic normal stresses in slow flows are a second-order effect. However, fiber suspensions have normal stress differences in shear because they are *anisotropic*. For a fiber suspension N_1 and N_2 are proportional to the shear rate $\dot{\gamma}$, and are a first-order effect.

Figure 6.4 shows the transient orientation and stress calculated using these equations. The initial orientation state is isotropic, $\mathbf{A} = \mathbf{I}/3$, and the orientation evolves according to the Folgar–Tucker equation with $C_I = 0.01$ and $\kappa = 1$. The natural closure is used, and $N_p = 10$. The orientation is solved numerically, and the stresses at each time are calculated using Eqns. (6.32–6.34). The function `tauFiber` in the accompanying software performs this stress calculation.

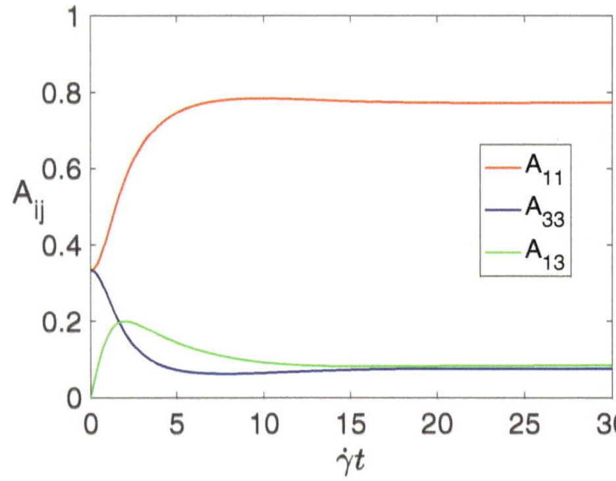

(a) Orientation

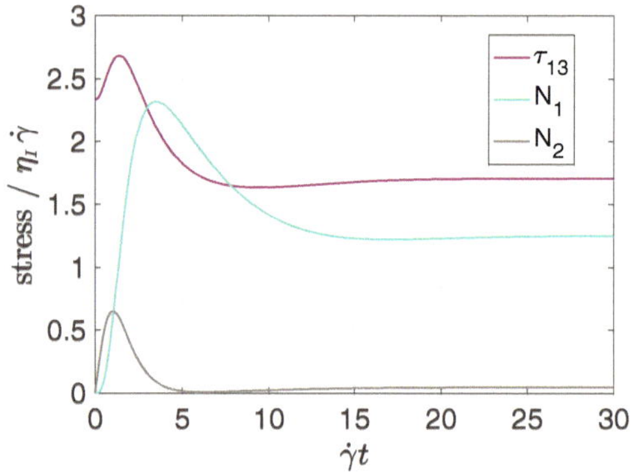

(b) Stress

Figure 6.4 Transient orientation and stress in simple shear flow for a fiber suspension with N_p = 10, using Eqns. (6.32–6.34). Folgar–Tucker model with C_I = 0.01, κ = 1 and the natural closure. Isotropic initial orientation.

The shear stress τ_{13} is initially more than twice as large as the stress in an isotropic fluid with viscosity η_I, and experiences a transient overshoot before settling to a steady-state value. The peak of the overshoot occurs near $\dot{\gamma}t = 1.5$, when A_{13} is nearing its peak value. At this instant the initially random fibers have started to cluster near 45° to the flow (as in Fig. 4.10 at $\dot{\gamma}t = 1$), where they make a maximum contribution to the shear stress. As the fibers rotate past this orientation and begin to lie down parallel to the flow, their contribution to the shear stress decreases. By $\dot{\gamma}t = 30$ the shear stress has nearly settled to its steady-state value. This is still 1.7 times larger than the isotropic value, because there is still some 3-direction fiber orientation as well as a great deal of 1-direction orientation.

The normal stress differences N_1 and N_2 are initially zero, because the orientation is isotropic and normal stresses in fiber suspensions come from anisotropy. As the fibers begin to align, N_1 grows rapidly, peaking near $\dot{\gamma}t = 3.5$ before settling to a steady state. N_2 has a similar behavior but is much smaller in magnitude. These stress transients are driven by the transient orientation.

■

Example 6.3.4 Transient stress in planar extensional flow

Write an algebraic expression for the stress difference $\tau_{11} - \tau_{22}$ for a fiber suspension in planar extensional flow, and identify the components of $\mathbb{A}$ that are important. The velocity field is $v_1 = \dot{\varepsilon}_P x_1$, $v_2 = -\dot{\varepsilon}_P x_2$ and $v_3 = 0$.

We again use Eqn. (6.30), but now the rate-of-deformation tensor is given by Eqn. (6.25). The non-zero components of $\mathbf{D}$ are $D_{11} = \dot{\varepsilon}_P$ and $D_{22} = -\dot{\varepsilon}_P$, so now the product $\mathbb{A} : \mathbf{D}$ reduces to

$$\mathbb{A}_{ijkl} D_{lk} = \mathbb{A}_{ij11} D_{11} + \mathbb{A}_{ij22} D_{22} = \dot{\varepsilon}_P (\mathbb{A}_{ij11} - \mathbb{A}_{ij22}) \tag{6.35}$$

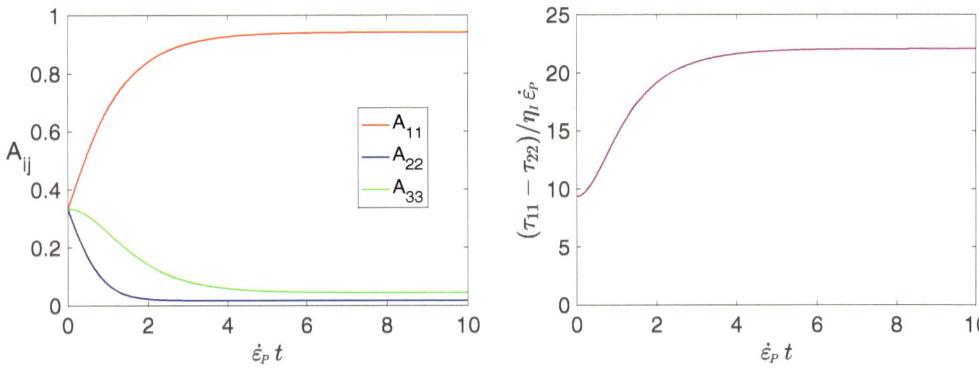

(a) Orientation (b) Stress

Figure 6.5 Transient orientation and extra stress in planar extensional flow for a fiber suspension with $N_p = 10$, using Eqns. (6.38). Folgar–Tucker model with $C_I = 0.01$, $\kappa = 1$ and the natural closure. Isotropic initial orientation.

Using this expression in Eqn. (6.30), the relevant components of the extra stress are

$$\tau_{11} = 2\eta_I\,\dot{\varepsilon}_P\left[\ \ 1 + N_p(\mathbb{A}_{1111} - \mathbb{A}_{1122})\right] \tag{6.36}$$

$$\tau_{22} = 2\eta_I\,\dot{\varepsilon}_P\left[-1 + N_p(\mathbb{A}_{2211} - \mathbb{A}_{2222})\right] \tag{6.37}$$

and the measurable stress difference is

$$\tau_{11} - \tau_{22} = \eta_I\,\dot{\varepsilon}_P\left[4 + 2N_p(\mathbb{A}_{1111} - 2\mathbb{A}_{1122} + \mathbb{A}_{2222})\right] \tag{6.38}$$

An example calculation of this stress is shown in Fig. 6.5. The orientation equation is again integrated numerically, and the results at each time step used to evaluate the extra stresses. Other parameters are the same as the previous example.

The stress difference is initially larger than the isotropic value of $\tau_{11} - \tau_{22} = 4\eta_I\,\dot{\varepsilon}_P$, because even with a random orientation some fibers are aligned close to the stretching direction (the 1 axis) and others are aligned close to the direction of contraction (the 2 axis). As the flow proceeds, more fibers shift toward the 1 axis, where they are better able to resist stretching, and the stress difference rises. The orientation in this example is three-dimensional ($A_{33} \neq 0$), but the planar fiber motion in Fig. 4.11 is still helpful in visualizing how fiber orientation changes over time.

■

6.3.4 Theoretical Predictions for N_p

It would be useful if the particle number N_p and the isotropic viscosity η_I could be predicted from information such as the fiber volume fraction, fiber aspect ratio and matrix viscosity. Such theories exist for dilute suspensions of ellipsoids in a Newtonian fluid [LDHB88, Pet99]. However, a fiber suspension is only dilute if the average distance between fiber centroids is large compared to the fiber length. This requires

$$\phi_f < \left(\frac{d_f}{\ell}\right)^2 \tag{6.39}$$

where ϕ_f is the fiber volume fraction. This is well below useful levels for composite materials. For example, if the fiber aspect ratio is $\ell/d_f = 10$, the fiber volume fraction would have to be well below 1% for the suspension to be dilute. A suspension with $\ell/d_f = 100$ would need $\phi_f < 0.01\%$ to be dilute.

One model for higher concentrations was developed by Dinh and Armstrong [DA84], and provides some insight into the possible values of N_p. This model applies for semi-concentrated suspensions, which means

$$\left(\frac{d_f}{\ell}\right)^2 < \phi_f < \left(\frac{d_f}{\ell}\right) \tag{6.40}$$

Practical composites are often more concentrated than the upper limit, but the trends from this model are still useful. It gives the particle number as

$$N_p = \frac{\phi_f(\ell/d_f)^2}{3\ln(2h_f/d_f)} \tag{6.41}$$

In this expression h_f is the distance between a fiber and its nearest neighbors, and is given by

$$\frac{2h_f}{d_f} = \begin{cases} \sqrt{\pi/\phi_f} & \text{for aligned fibers} \\ (\pi d_f)/(2\phi_f\ell) & \text{for random fibers} \end{cases} \tag{6.42}$$

The Dinh and Armstrong model treats the fiber as a slender body, so fibers oriented in the flow direction in simple shear do not affect the shear viscosity, and the model predicts that η_I is equal to the viscosity of the suspending fluid.

Figure 6.6 shows predictions for N_p from Eqn. (6.41) using aligned fiber spacing. N_p can be much greater than unity, especially when the fiber aspect ratio is large. This emphasizes the strong anisotropy of the viscosity of fiber suspensions.

In Fig. 6.6 the practical range for discontinuous fiber composites, say $0.1 \leq \phi_f \leq 0.5$ and $10 \leq \ell/d_f \leq 1000$, lies in the concentrated regime. The Dinh–Armstrong model is not strictly valid in

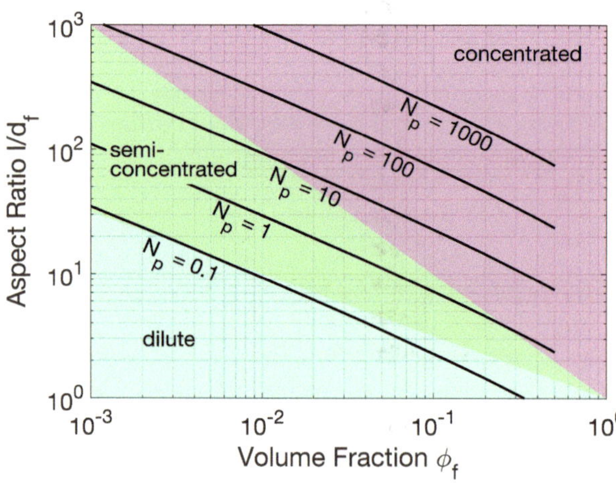

Figure 6.6 Predictions of the particle number N_p for different fiber volume fractions and aspect ratios, using the Dinh–Armstrong model with aligned-fiber spacing. Shaded areas show the concentration regimes for fiber suspensions using Eqn. (6.40).

this regime, and of course the suspending fluid in most practical composites is non-Newtonian as well. Concentrated fiber suspensions can also have a stress contribution from fiber-fiber contacts. The main value of models like Eqn. (6.41) is to provide qualitative understanding of the influence of volume fraction and aspect ratio on suspension rheology. Whenever possible, it is best to treat N_p and η_I as adjustable parameters and find their values by fitting rheological measurements.

At the extreme end of the concentration spectrum, there are composites where the fibers are highly aligned and the volume fraction approaches the packing limit of aligned cylinders. Models for the rheology of these materials [PHB⁺91, PCSŠ94] are available, and these models are actually somewhat easier to construct because the arrangement of the fibers relative to their neighbors is better understood. These hyperconcentrated suspensions are strongly anisotropic, as one would expect.

6.4 Solving Flow Problems

6.4.1 Governing Equations

Now that we have a constitutive equation for fiber suspensions, we are ready to formulate a full flow problem. The equations that follow will assume that the suspension has constant density.

Any fluid flow solution must satisfy conservation of mass at each point in the flow. This is expressed by a *mass balance* equation, also known as the *continuity equation.* If the density ρ is constant, the mass balance equation reduces to

$$\nabla \cdot \mathbf{v} = 0 \tag{6.43}$$

This is identical to requiring that $\mathrm{tr}\,\mathbf{D} = 0$. All of the basic flows in Section 4.1.1.2 and all of the examples in Section 4.3 satisfy this condition.

Newton's second law is expressed in flow problems as a balance of momentum. In terms of the pressure and extra stress, the general *momentum balance* equation (also called the *equation of motion*) is

$$\rho \frac{\mathrm{D}\mathbf{v}}{\mathrm{D}t} = -\nabla p + \nabla \cdot \boldsymbol{\tau} + \rho \mathbf{b} \tag{6.44}$$

Here $\mathrm{D}/\mathrm{D}t$ is the material derivative, which is expanded using the right-hand side of Eqn. (5.8), and $\mathbf{b}$ is the body force per unit mass. If gravity is the only body force then $\mathbf{b} = \mathbf{g}$, the acceleration of gravity ($9.82\,\mathrm{m/s^2}$, pointing down).

To this we add a stress constitutive equation, such as the Newtonian fluid (6.5) or the fiber suspension equation (6.29). We also add an equation for fiber orientation, such as the Folgar–Tucker equation (5.33) or any of the other orientation models in Chapt. 5.

Table 6.1 shows the number of scalar equations and the number of scalar variables (independent vector and tensor components) in the problem. Here we have used the symmetry of $\boldsymbol{\tau}$ to get the minimum number of independent extra-stress components (six), and both the symmetry and the trace property of $\mathbf{A}$ to get the minimum number of independent orientation tensor

Table 6.1 Equations and variables appearing in flow problems for fiber suspensions

Equation or Variable	Scalar Equations	Scalar Variables
Mass balance (6.43)	1	
Momentum balance (6.44)	3	
Constitutive equation (6.29)	6	
Orientation model, e.g. (5.33)	5	
Velocity $\mathbf{v}$		3
Pressure p		1
Extra stress $\boldsymbol{\tau}$		6
Orientation tensor $\mathbf{A}$		5
Total	**15**	**15**

components (five). Although there are many variables in the problem (fifteen!), we have the same number of equations so the problem can be solved.

In practice it is common to eliminate the extra-stress components algebraically by substituting the constitutive equation everywhere that $\boldsymbol{\tau}$ appears. That leaves nine variables and nine equations.

6.4.2 Boundary and Initial Conditions

To complete the setup for a flow problem we also need boundary and initial conditions. Some of the more common boundary conditions for velocity, pressure and stress are:

- Solid wall, no slip: $\mathbf{v} = \mathbf{v}_{\text{wall}}$, where $\mathbf{v}_{\text{wall}}$ is the wall velocity (which could be zero).
- Inflow boundary: Either $\mathbf{v}$ is specified or the surface traction $\mathbf{t}$ is specified. If $\mathbf{n}$ is the outwardly-directed unit normal vector on the boundary and an inlet pressure p_{in} is applied, $\mathbf{t} = -p_{\text{in}}\mathbf{n}$.
- Outflow boundary: Either $\mathbf{v}$ or $\mathbf{t}$ can be specified. However, $\mathbf{v}$ should not be specified at both the inflow and outflow boundaries.
- Moving flow front: Usually no traction, $\mathbf{t} = 0$. If surface tension is significant compared to the viscous stresses, a surface-tension boundary condition can be applied (see, e.g., [DT01]).

The terms on the left-hand side of the momentum balance equation (6.44) represent the inertia of the fluid. In most polymer processing problems, inertia is negligible compared to the viscous stresses and the pressure gradient, so the inertia terms can be neglected. This removes the time derivatives from the mass and momentum balance equations, so the velocity at any time can be calculated without reference to previous times, and no initial condition for velocity is needed. If inertia is not negligible then an initial condition on velocity, $\mathbf{v}(\mathbf{x}, t = 0)$, must also be specified.

The orientation tensor must be specified on any inflow boundary. However, no orientation boundary condition is required on walls, outflow boundaries or moving flow fronts. An initial condition for the orientation tensor is required for points in the domain that are filled at $t = 0$.

With suitable boundary and initial conditions, the flow problem is well-posed and can be solved.

6.4.3 Solution Schemes: Coupled and Decoupled

If one chooses a constitutive equation that does not depend on fiber orientation, such as a Newtonian or generalized Newtonian fluid, then the mass balance and momentum balance equations can be solved without knowing the fiber orientation. The resulting velocity can then be used in the orientation equation, which can be solved separately. In a model of this type, the flow and orientation have been *decoupled*. The flow always affects the orientation, but in a decoupled model the orientation does not affect the flow.

If one uses a constitutive equation where the stress depends on orientation, like Eqn. (6.29), then the flow and orientation equations must be solved simultaneously. In a model of this type, the flow and orientation are *coupled*.

Coupled problems must be solved numerically, and at the time of this writing there are no commercial software programs for solving coupled flow problems. Coupled flow problems are highly nonlinear, so it is common for each problem to need some manual adjustment of numerical parameters or solution strategies to obtain a converged solution. More information about the numerical solution of coupled problems appears in [Ver98, RM01, VT02, KGBF21].

6.4.4 Temperature Effects

So far we have discussed problems where the temperature and the material properties are constant. However, injection molding and other manufacturing processes are highly non-isothermal, with large variations in viscosity between the injection temperature and the mold wall temperature. For these cases the viscosity model will include a dependence on temperature, and an additional equation is needed to find the temperature as a function of position and time, $T(\mathbf{x}, t)$. The governing equation is an *energy balance* which, for a material with constant density and isotropic thermal conductivity k is

$$\rho c_p \left(\frac{\partial T}{\partial t} + \mathbf{v} \cdot \nabla T \right) = \nabla \cdot (k \nabla T) + \boldsymbol{\tau} : \mathbf{D} + \rho \dot{R} \tag{6.45}$$

Here c_p is the specific heat at constant pressure and $\dot{R}$ is the rate of internal heat generation per unit mass, for instance by a curing reaction in a thermosetting polymer.

The left-hand side of Eqn. (6.45) contains the unsteady term $\rho c_p \, \partial T / \partial t$, as well as an advection term $\rho c_p \mathbf{v} \cdot \nabla T$ that represents heat transported by fluid motion. The first term on the right-hand side represents heat conduction, and $\boldsymbol{\tau} : \mathbf{D}$ is the rate at which mechanical work is converted to heat by viscous dissipation.

While a non-isothermal solution using temperature-dependent viscosity is standard in computer-aided engineering software, this adds little to our discussion of fiber suspension flow, so we will focus on isothermal flows. The models discussed here can be adapted to non-isothermal flows by adding the energy balance equation to the problem and including the temperature dependence of the polymer matrix viscosity.

6.5 Example Solutions

6.5.1 Axisymmetric Contraction

A striking demonstration of flow-orientation coupling is the flow through an axisymmetric contraction. Lipscomb, Denn, Hur and Boger [LDHB88] pumped corn syrup through a large cylindrical tube that reduced to a smaller cylindrical tube (Fig. 6.7). The ratio of the tube diameters was 4.5 : 1. A small amount of reflective pigment particles was added to the mixture, and a sheet of laser light used to illuminate a plane passing through the symmetry axis of the tubes. A long-exposure photograph of that plane then showed the fluid streamlines as bright streaks.

Figure 6.8(a) shows the streamlines at the entrance to the contraction for a fluid with no fibers. The flow comes straight down the large tube, and the streamlines only begin to bend when they are close to the entrance to the small tube. There are small vortices in the corners of the large tube, with a slow recirculating flow.

In contrast, when even a tiny amount of long fibers are added to the fluid, the flow pattern is much different. In Fig. 6.8(b) the streamlines begin to angle toward the entrance of the small tube sooner, and the corner vortices are larger, with fluid in the vortices circulating more rapidly. The flow in the upstream tube is simple shear flow and the contraction flow has con-

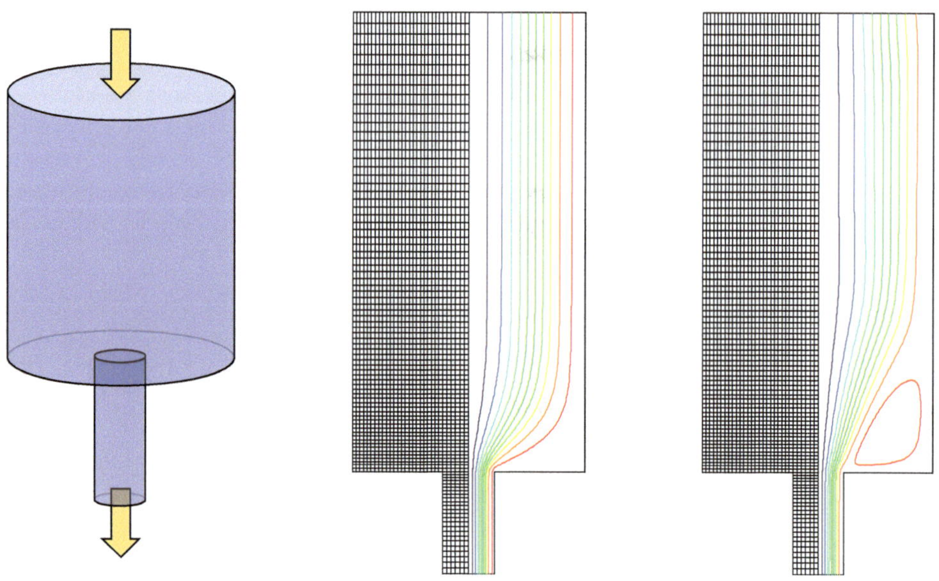

(a) Geometry (b) $N_p = 0$ (c) $N_p = 6$

Figure 6.7 Flow geometry, finite element meshes and computed streamlines for flow in an axisymmetric contraction (b) without fibers and (c) with fibers. Finite element calculations by VerWeyst [Ver98].

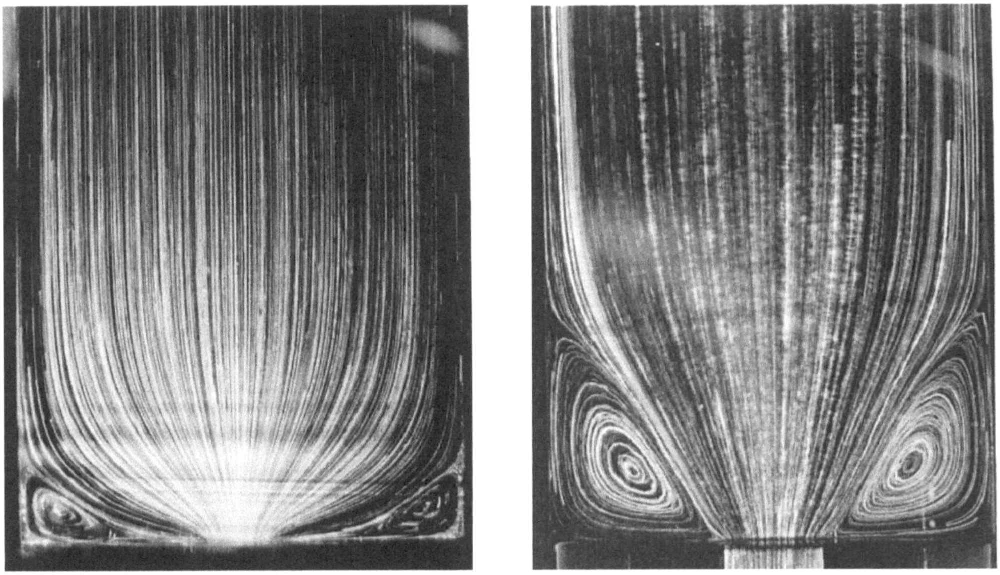

(a) $N_p = 0$ (b) $N_p = 6$

Figure 6.8 Experimental streamlines for flow in an axisymmetric contraction (a) without fibers and (b) with fibers. Flow is from top to bottom. Figures from [LDHB88], used by permission from Elsevier.

verging streamlines, so the fibers are nearly aligned with the streamlines. This suspension has only 0.110% by volume of glass fibers with an aspect ratio of $\ell / d_f = 276$. Even a modest amount of fibers can substantially alter both the flow field and the orientation pattern.

The anisotropic constitutive equation (6.29) does an excellent job of predicting the effect of fibers on this flow. Calculations using an orientation tensor model and Eqn. (6.29) are shown in Figs. 6.7 and 6.9 [Ver98, VT02]. The solution assumes axisymmetric flow and orientation. Finite element meshes are shown in Figs. 6.7(b) and (c) along with the steady-state streamlines. More details of the streamlines and results for the fiber orientation are shown in Fig. 6.9. Here the orientation is displayed as a vector in the direction of the eigenvector of **A** that has the largest eigenvalue, and the length and color of each line is determined by the eigenvalue λ_1.

These calculations correctly reproduce the experimental streamlines from Fig. 6.8, for both the Newtonian fluid and the fiber suspension. The large corner vortex and smaller entry angle of the streamlines for the fiber suspension are accurately modeled. Where the streamlines are converging, the main direction of orientation is in the flow direction, and the fibers are highly aligned as they enter the smaller tube. The original paper by Lipscomb et al. [LDHB88] shows finite element calculations that have similar agreement with the experiments, using a model where the fibers are always aligned with the flow.

In this problem the fibers primarily align with the local flow direction, both because the upstream flow is simple shear and because the flow near the contraction has converging streamlines. Then, compared to the Newtonian fluid, the fibers add a strong resistance to flow-direction stretching where the streamlines are converging, without changing the resistance for shearing parallel to the streamlines. The fiber suspension "prefers" a more gradual contraction

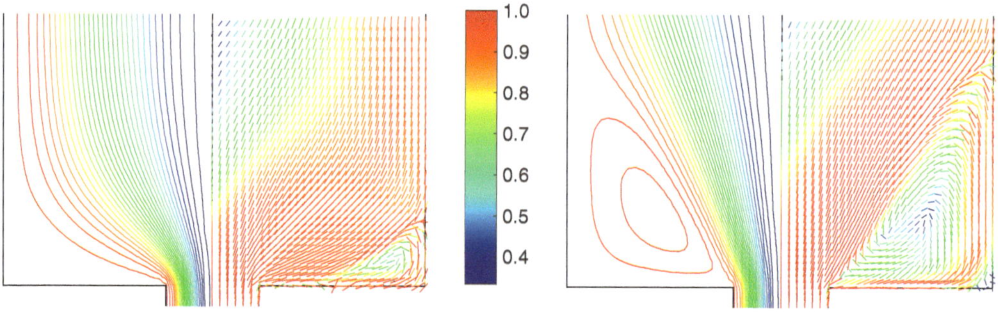

(a) $N_p = 0$ (b) $N_p = 6$

Figure 6.9 Streamlines and orientation vectors for flow in an axisymmetric contraction. Orientation vectors show the magnitude of the largest eigenvalue λ_1 (colorbar) and the direction of the associated eigenvector. Calculations by VerWeyst [Ver98].

of the streamlines, in order to spread out the flow-direction stretching at the expense of some additional shearing.

The contraction flow is one example where the presence of fibers creates a large, qualitative change in the flow pattern. It has become a standard validation problem for software that solves coupled flow and orientation.

6.5.2 Anisotropic Squeeze Flow

Another experiment where the presence of fibers has a clear visible effect on the flow is the squeeze flow of anisotropic fiber suspensions. Ericsson, Toll and Månson [ETM97] squeezed charges of a polypropylene/glass fiber composite between parallel platens. The fibers were 12 mm long and 12 µm in diameter, so they had $\ell/d_f = 1{,}000$. The initial sheets were made by a paper-making process that imparted a high degree of initial orientation to the fibers, such that $A_{11} \simeq 0.80$. The orientation state was nearly planar in the 1–2 plane.

The samples began as circles, with initial diameters from 35 to 75 mm, and were approximately 3.5 mm thick. The top view was initially like Fig. 4.28(b), and the side view was like Fig. 4.30.

If a circular disk of any isotropic material is squeezed in this way, it remains circular as it expands in radius and decreases in thickness. However, the samples in these experiments resist stretching in the 1 direction much more than in the 2 direction, so they expand rapidly in the 2 direction, and only slightly in the 1 direction. Calculated shape changes for $N_p = 100$ are shown in Fig. 6.10(a). These calculations assume a lubricated squeeze flow, where the material strains homogeneously, but the charge shapes with non-lubricated squeeze flow are similar.

Because the material is stretching primarily in the 2 direction (and contracting in the thickness direction), the fibers tend to rotate toward the 2 direction. This causes A_{22} to increase with time and A_{11} to decrease. Calculations for the orientation are shown in Fig. 6.10(b). In this figure, R_0 is the initial radius of the charge and R_2 is the distance from the center to the edge of the charge along the 2 axis.

While this change in fiber orientation is not dramatic, it does decrease the anisotropy of the material, making the flow less anisotropic in its later stages. Detailed analyses using the rheo-

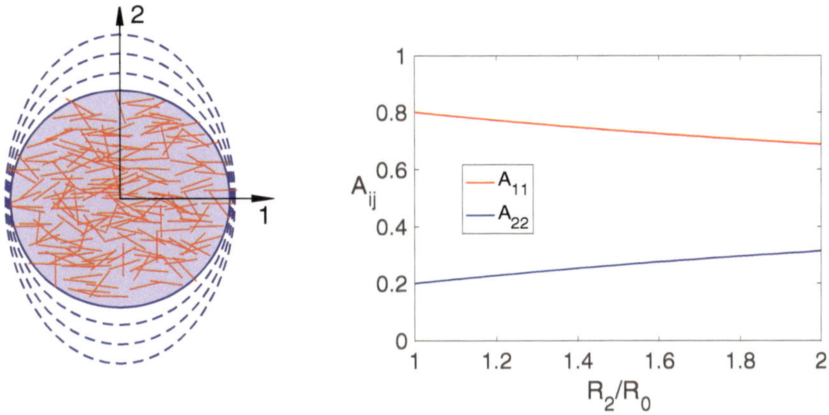

(a) (b)

Figure 6.10 Squeeze flow of an anisotropic fiber suspension with N_p = 100. Squeezing is along the 3 direction and the fibers are planar in the 1–2 plane. (a) Top view of initial charge (filled circle) and subsequent shapes (dashed lines). Red lines are representative of fibers in the initial charge. (b) Orientation in the charge as a function of deformation.

logical model of Eqn. (6.29) show excellent agreement with experimental charge shapes, provided that the change in fiber orientation over time is included in the calculation and the fourth-order orientation tensor is accurately represented [ETM97, SFP18].

6.6 Decoupled Flow as a Special Case

While the two preceding examples show that the presence of fibers can alter the flow, there are many cases where the influence of fibers is small, even in a coupled flow solution. If we can anticipate when this will occur, we will know when a decoupled model is likely to produce useful results.

We begin this section with an example, then discuss general criteria for decoupling.

6.6.1 Center-Gated Disk

Consider a center-gated disk, as shown in Fig. 4.18. This is a representative geometry for injection molding, and we know from Section 4.3.3 that both shearing and stretching motions are present in this flow. Equation (4.109) gave the velocity distribution for a Newtonian fluid with constant temperature (a decoupled solution). What happens when fibers are present and we have flow-orientation coupling?

VerWeyst and Tucker [Ver98, VT02] calculated coupled flow-orientation solutions for this geometry using Eqn. (6.29) as the constitutive equation. Thermal effects were neglected, and a steady-state solution was found for a disk with $R/h = 60$. The sprue diameter is equal to the thickness of the disk. Other parameters are $C_I = 0.001$, $\kappa = 1$, and the natural closure was used.

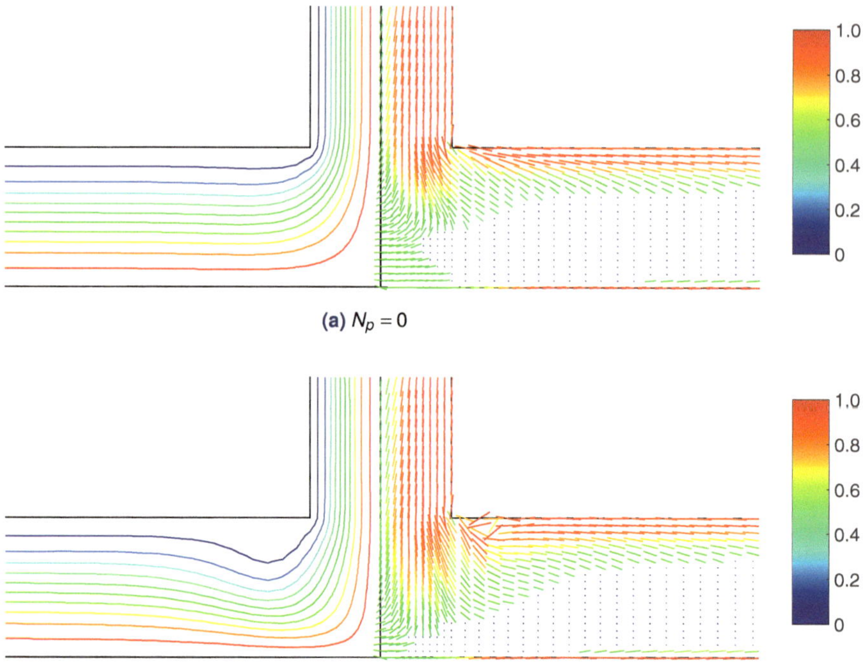

Figure 6.11 Streamlines (left) and fiber orientation vectors (right) near the center of a center-gated disk. Streamlines are equally spaced. Orientation vectors show the magnitude of the largest eigenvalue in the r–z plane (colorbars) and the direction of the associated eigenvector. Blue dots indicate θ-direction orientation. Folgar–Tucker model with $C_I = 0.001$, $\kappa = 1$, natural closure. Calculations by VerWeyst [Ver98].

The velocity and orientation fields are assumed to be axisymmetric. The coupled-flow solution used $N_p = 19.35$, which corresponds to $\ell / d_f = 20$ and $\phi_f = 0.2$ for the Dinh–Armstrong model. A decoupled solution ($N_p = 0$) was also calculated.

Figure 6.11 shows the streamlines and the fiber orientation near the center, where the sprue joins the disk. The central line in these figures is the symmetry axis and the radial axis is horizontal. This is only a small part of the solution domain. The sprue is long enough that the orientation in the sprue approaching the gate is fully developed, with fibers oriented strongly in the flow direction.

For the decoupled case, Fig. 6.11(a), the flow from the sprue enters the cavity and turns the corner smoothly. Near the upper wall of the disk the fibers near the sprue wall make a similar turn, and a shell-like layer develops immediately on the upper wall. Fibers at the midplane of the disk come from the mid-radius of the sprue, and quickly acquire a crossflow orientation. This is due to the diverging nature of the flow when viewed from the top. Fibers from the centerline of the sprue arrive at the lower wall of the disk and have a similar crossflow orientation. Moving outward from the centerline, we can see a shell layer of flow-aligned fibers starting to develop on the lower wall of the disk. The orientation in the disk is not symmetric about the midplane within the limits of this figure.

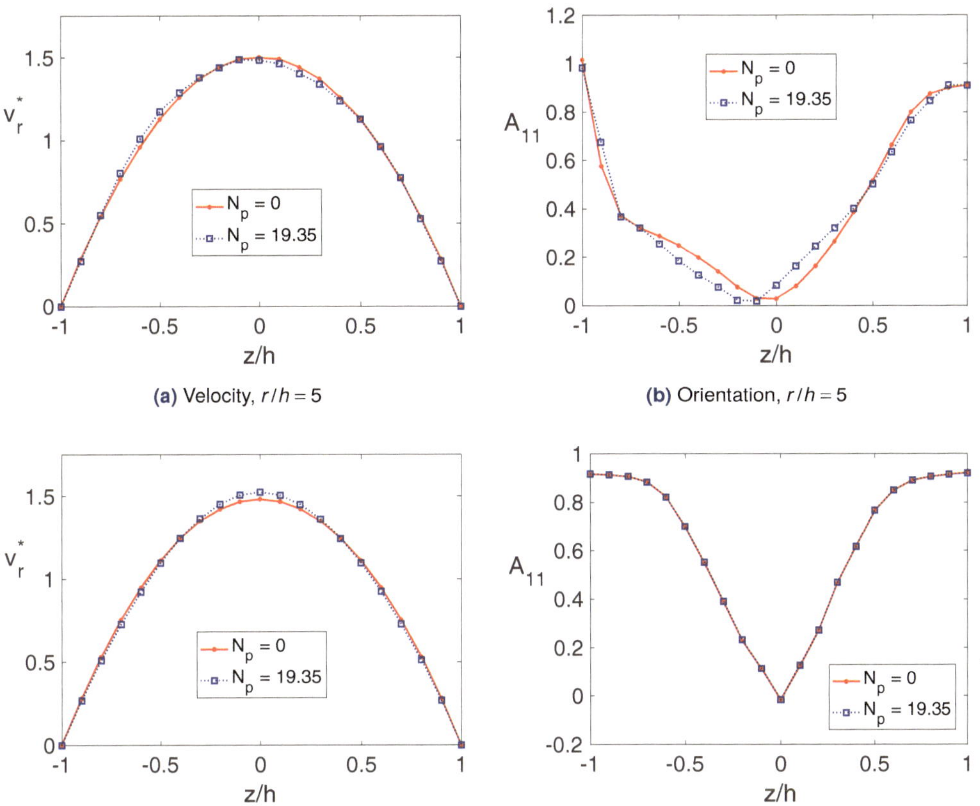

(a) Velocity, $r/h = 5$

(b) Orientation, $r/h = 5$

(c) Velocity, $r/h = 50$

(d) Orientation, $r/h = 50$

Figure 6.12 Profiles of radial velocity and flow-direction orientation at two positions in a center-gated disk for decoupled and coupled flow and orientation. Radial velocity values are scaled by the gapwise average velocity at that radius, $v_r^* = 4\pi rhv_r/Q$. Folgar–Tucker model with $C_I = 0.001$, $\kappa = 1$, natural closure. Calculations by VerWeyst [Ver98].

The coupled solution, Fig. 6.11(b), is generally similar but there are some differences. Where the upper wall of the disk joins the sprue, the fibers are oriented in a way that increases the viscosity; this reduces the velocity there, causing a noticeable dip in the streamlines. The orientation pattern is similar to the decoupled case, though the crossflow region has a different shape and the shell layer on the lower mold wall starts developing sooner.

A quantitative comparison of these results appears in Fig. 6.12. Here the profiles of velocity and flow-direction orientation are shown at two radial locations: $r/h = 5$ is just inside the edge of Figs. 6.11, while $r/h = 50$ is farther out along the flow path. (The total thickness of the disk is $2h$ and the coordinate systems is shown in Fig. 4.19.)

At the first radius, $r/h = 5$, there is a slight asymmetry of the velocity profile for the coupled flow, though differences in the two velocity profiles are small at this radius. Both orientation profiles are asymmetric and have more noticeable differences, though these are small compared to the accuracy of most fiber orientation measurements.

At the larger radius, $r/h = 50$, both the velocity and orientation profiles are symmetric. The flow from $r/h = 5$ to this point has removed the earlier asymmetries. The orientation displays a classical core/shell pattern, with highly aligned fibers near the wall and strong crossflow alignment at the midplane[3]. The velocity profile for the coupled case is slightly more pointed than the decoupled case. This is because the fibers near the midplane have a small value of A_{11}, which leads to a smaller value of $\mathbb{A}_{1313}$ and a smaller viscosity at the midplane compared to the material near the walls. The effect is subtle, and has no discernable effect on the fiber orientation.

The differences in orientation for a coupled and decoupled calculation for this flow are small, and are confined to a region close to the center of the disk. For this geometry and for N_p values not much larger than 20, a decoupled calculation of flow and orientation would give acceptable results for almost any purpose.

6.6.2 General Criteria for Decoupling

One can make a good theoretical argument that flow and orientation may be decoupled, even with orientation-dependent rheology, if the fiber suspension is flowing in a sufficiently narrow gap [Tuc91]. Many injection molded and compression molded parts are designed with thin shell-like geometries. This makes efficient use of the material and allows the part to be cooled quickly, which keeps the molding cycle time low. Figures 4.16 and 4.18 are examples of this type of geometry, though the analysis applies to parts that have more complex in-plane shapes and to parts that are gently curved.

Let the 1–2 plane be the midplane of the part so that x_3 is the thickness coordinate. The geometry is characterized by a total thickness $2h$ and an in-plane dimension L. Define the parameter ϵ as

$$\epsilon \equiv \frac{h}{L} \tag{6.46}$$

We will consider mold cavities that are geometrically thin, $\epsilon \ll 1$.

To determine general criteria for decoupling, one writes the full equations for coupled flow and estimates the relative magnitudes of all terms in the equation. Then, for example, when one term is smaller than others by a factor of ϵ or even ϵ^2, it can be neglected. This provides a systematic way to simplify the governing equations.

To apply the same technique to the constitutive equation (6.29), we need to estimate the magnitudes of the components of $\mathbb{A}$. For this, we assume that the orientation is nearly planar in the 1–2 plane, as it is in most molded parts. The typical magnitude of the out-of-plane orientation is described by a parameter δ, and a good way to estimate this is

$$\delta = \sqrt{A_{33}^{\mathrm{SSF}}} \tag{6.47}$$

[3] Nodal values of the orientation components in this Galerkin finite element calculation can occasionally take on non-physical values, e.g., $A_{11} > 1$ or $A_{11} < 0$, as seen in these figures. These are the correct solutions to the finite element equations, so they must be allowed in the solution process. If they are corrected to physical values before being used in the closure approximation and constitutive equation, they do not pollute the rest of the solution [Ver98, VT02].

where A_{33}^{SSF} is the out-of-plane orientation component for steady state in simple shear flow[4]. Then, for example, we expect $\mathbb{A}_{1313}$ to have a magnitude of approximately δ^2.

Following through with the analysis reveals four different flow regimes [Tuc91]:

Regime I. Decoupled lubrication flow: $N_p\delta^2 \ll 1$ and $\delta \geq \epsilon$

This is the regime where a decoupled solution is just as accurate as a coupled solution. The equations of motion in the 1 and 2 directions simplify to

$$\frac{\partial p}{\partial x_1} = \frac{\partial \tau_{13}}{\partial x_3} \tag{6.48}$$

$$\frac{\partial p}{\partial x_2} = \frac{\partial \tau_{23}}{\partial x_3} \tag{6.49}$$

and the 3-direction equation shows that pressure p does not depend on x_3. From these equations, the flow is driven by the pressure gradient in the 1–2 plane, and resisted by the gapwise shear stresses τ_{13} and τ_{23}.

Also, one can approximate the important components of the rate of deformation as

$$D_{13} \simeq \frac{1}{2}\frac{\partial v_1}{\partial x_3} \tag{6.50}$$

$$D_{23} \simeq \frac{1}{2}\frac{\partial v_2}{\partial x_3} \tag{6.51}$$

Because the orientation is very flat (δ is small) and N_p is not too large ($N_p\delta^2 \ll 1$), the fiber-suspension constitutive equation for the important stress components reduces to

$$\tau_{13} = 2\eta_I D_{13} = \eta_I \frac{\partial v_1}{\partial x_3} \tag{6.52}$$

$$\tau_{23} = 2\eta_I D_{23} = \eta_I \frac{\partial v_2}{\partial x_3} \tag{6.53}$$

Substituting these results back into Eqns. (6.48) and (6.49) gives a pair of equations that contains only velocity, pressure and η_I. Neither the fiber orientation tensor $\mathbb{A}$ nor the particle number N_p appear, so the velocity distribution can be found without knowing the fiber orientation state.

This regime extends the classical lubrication approximation for flow in a narrow gap to fiber suspensions, hence the name. The simplifications of Eqns. (6.48–6.51) for generalized Newtonian fluids and non-isothermal flows, sometimes called the *generalized Hele-Shaw* model [HS80], formed the basis of injection molding simulations for many years. This approach is still used in so-called 2.5-D or *midplane* simulations.

Regime II. Coupled lubrication flow: $N_p\delta^2 \geq 1$ and $\epsilon \ll \delta \ll 1$

If $N_p\delta^2$ is too large for Regime I to apply, a simplification is still available if $\epsilon \ll \delta \ll 1$. Equations (6.48–6.51) still apply, but the relevant stress components are now given by

$$\tau_{13} = 2\eta_I \left(D_{13} + 2N_p\left[\mathbb{A}_{1313}D_{13} + \mathbb{A}_{1323}D_{23}\right]\right) \tag{6.54}$$

$$\tau_{23} = 2\eta_I \left(D_{23} + 2N_p\left[\mathbb{A}_{2313}D_{13} + \mathbb{A}_{2323}D_{23}\right]\right) \tag{6.55}$$

4 The original paper [Tuc91] suggested $\delta = C_I^{1/3}$; however, this was based on the quadratic closure for simplicity. Since we now have ARD models that can adjust A_{33} independently, Eqn. (6.47) is more appropriate.

Now N_p and $\mathbb{A}$ appear, so the flow and orientation are coupled and must be solved simultaneously. However, only the shear stresses τ_{13} and τ_{23} affect the velocity solution; in-plane stretching motions and the extensional viscosity of the suspension are not important. While this regime is coupled, the solution could be carried out in a much more simplified form than the general 3-D equations.

Regime III. General narrow gap flow: $N_p\delta^2 \geq 1$ and $\delta \sim \epsilon$

In this parameter range the flow and orientation are coupled, there are no significant simplifications, and the full 3-D equations should be solved.

Regime IV. Plug flow with shear boundary layer: $N_p\epsilon^2 \gg 1$ and $\delta \ll \epsilon$

In this regime the fiber orientation is flat, $\delta \ll \epsilon$. This might reasonably be the case, for example, with sheet molding compound (SMC), where the fiber length is much greater than the gap height. Then, if N_p is large enough, one can argue that the gapwise velocity profile is very flat, similar to a lubricated squeezing flow, with shearing near the wall confined to a thin boundary layer. This idea offers some theoretical support to the observation that SMC materials typically exhibit a flat velocity profile across the cavity thickness [BC85]. However, in practice, thin resin-rich layers at the surfaces of the molding compound, and low viscosity near the mold surfaces when a cold charge is placed in a hot mold, are also important factors.

As a general rule, one can expect a decoupled flow solution to give useful fiber orientation predictions whenever the flow is in a geometrically thin mold cavity where the flow length is large compared to the cavity thickness. Decoupled solutions are less reliable near sudden contractions or expansions in the cross-section, or in parts that have a "blocky" geometry where the dimensions are similar in all directions.

6.7 Orientation Dependent Isotropic Viscosity Model

A recent alternative to the fiber suspension rheological model of Eqn. (6.29) uses a local isotropic viscosity that best approximates the stress response of the suspension to the local flow. That is, instead of the full anisotropic fiber suspension constitutive equation,

$$\boldsymbol{\tau} = 2\eta_I \left[\mathbf{D} + N_p \mathbb{A} : \mathbf{D}\right] \tag{6.56}$$

this model uses an isotropic constitutive equation,

$$\boldsymbol{\tau} = 2\eta^{\text{IISO}}\mathbf{D} \tag{6.57}$$

where the viscosity η^{IISO} depends on the fiber orientation state and $\mathbf{D}$. Here IISO stands for *informed isotropic* viscosity, since the viscosity value is informed by the local orientation state. The IISO model was developed by Favaloro et al. [Fav17, FTP18, TF19], and we use their name here. The same model was developed independently by Li and Luyé [LL19], who termed η^{IISO} the *optimal scalar viscosity*.

To get the best match between the isotropic and anisotropic model, recall from the energy balance equation (6.45) that the rate of energy dissipation due to viscosity is $\boldsymbol{\tau} : \mathbf{D}$. The IISO viscosity is found by matching this dissipation rate for the IISO model to the dissipation rate for the full anisotropic model. Using Eqns. (6.56) and (6.57), this gives[5]

$$2\eta^{\text{IISO}}\mathbf{D}:\mathbf{D} = 2\eta_I\mathbf{D}:\mathbf{D} + 2\eta_I N_p \mathbf{D}:\mathbb{A}:\mathbf{D} \tag{6.58}$$

Algebraic rearrangement then gives the basic result:

$$\eta^{\text{IISO}} = \eta_I\left[1 + N_p\frac{\mathbf{D}:\mathbb{A}:\mathbf{D}}{\mathbf{D}:\mathbf{D}}\right] \tag{6.59}$$

Recall that the scalar strain rate is $\dot{\gamma} = \sqrt{2\mathbf{D}:\mathbf{D}}$ (Eqn. (4.14)). Using this to define a dimensionless rate-of-deformation tensor $\mathbf{D}^* \equiv \mathbf{D}/\dot{\gamma}$, we can rewrite Eqn. (6.59) as

$$\eta^{\text{IISO}} = \eta_I\left[1 + 2N_p\mathbf{D}^*:\mathbb{A}:\mathbf{D}^*\right] \tag{6.60}$$

Tseng et al. [TF19] call $K_S \equiv \mathbf{D}^*:\mathbb{A}:\mathbf{D}^*$ the stretching kernel, and they use a dimensionless Trouton ratio $R_T \equiv 2N_p$, so they write Eqn. (6.60) as

$$\eta^{\text{IISO}} = \eta_I\left[1 + R_T K_S\right] \tag{6.61}$$

All of the forms (6.59–6.61) are equivalent, as long as N_p is independent of shear rate. Tseng at al. do make R_T a function of shear rate, though we will not pursue that here.

The IISO model is attractive because it can be implemented in a mold filling simulation [TF19, LL19] in a fairly straightforward manner. At each time step the current fiber orientation is used to find the viscosity at each point in the flow, and the continuity and momentum equations are solved to find the velocity at that time. Those velocities are used to project the orientation to the next time step, solving the fiber orientation model, and the cycle repeats. In contrast, implementing the full fiber suspension model (6.56) requires major changes to the flow solution, and one can easily encounter numerical difficulties as N_p increases [TS19].

The IISO model exactly reproduces the shear stress response of the full model in simple shear flow, Eqn. (6.32). However, because it is isotropic, it gives normal-stress differences in simple shear that are zero, rather than the normal-stress response of Eqn. (6.33). The IISO model does match the stress response of the full model in planar extensional flow, Eqn. (6.38).

The IISO model is accurate in channel flows where the gapwise shear stresses dominate the momentum balance. It may do a better job than the decoupled-flow approximation at capturing the core thickness near the entrance of an injection mold [TF19, LL19]. This is because it assigns a higher viscosity to the core fluid near the mold entrance, where the core fibers are not yet highly aligned. This, in turn, flattens the velocity profile near the midplane, which reduces the growth of flow-direction alignment near the core. The IISO model has little effect on the shell orientation in typical molded parts, since material near the shell quickly acquires the steady-state orientation for simple shear flow.

The IISO model is only partially successful at representing strong anisotropic effects like those shown in Section 6.5. The change in flow pattern at the entrance to a contraction using IISO can be similar to Fig. 6.8(b) [TF19], though at a much larger value of N_p than the experiment. For the anisotropic squeeze flow in Fig. 6.10, if the squeeze flow is lubricated then the IISO model predicts that the charge remains circular. This misses the anisotropy of the flow. An

[5] $\mathbf{D}:\mathbb{A}:\mathbf{D} = D_{ij}\mathbb{A}_{jikl}D_{lk}$ is a scalar product, as is $\mathbf{D}:\mathbf{D} = D_{ij}D_{ji}$.

elliptical flow front can be predicted by IISO for a transfer molding flow where the suspension flows radially outward in a narrow gap, like a center-gated disk; however, the anisotropic effects are not as strong with the IISO model as they are with the full model [Fav18].

The IISO model offers some ability to capture the effect of fiber orientation changes on rheology and flow, so it can improve some aspects of fiber orientation prediction. It is, however, an incomplete representation of the anisotropic nature of fiber suspension rheology, so it will necessarily fail to predict features of the flow that are driven by strong anisotropy. As a result, it fits in between the decoupled approximation, which ignores the fibers altogether when modeling flow, and the full anisotropic suspension model, which takes fibers into account but can also make the numerical solution of flow problems difficult.

6.8 Transient Rheological Experiments

An alternative to the steady-shear experiments discussed in Section 6.2 is to measure the transient rheological properties of a fiber suspension. This is usually done in shear flow, but experiments in planar elongation and mixed shear/elongation are also possible.

Experiments of this type provide a way to explore the connection between fiber orientation and stress, and they offer one of the few routes to measure anisotropic rheological parameters such as η_I and N_p. They also offer a possible way to determine orientation model parameters such as C_I and κ, though this approach has seen limited success.

6.8.1 Flow Geometries

The basic experiment subjects a sample having a defined, repeatable fiber orientation state to a simple shear flow at constant rate $\dot{\gamma}$, and measures the resulting stresses τ_{13}, N_1 and N_2 as functions of time (or strain). Instruments that have been used to do this for fiber reinforced injection molding compounds include:

- Parallel disk rheometer [SAC04, WOT08]. A disk-shaped sample is confined between two solid disks, and one of the disks is rotated to create shear flow. The shear stress is determined from the measured torque, and the normal stress difference $N_1 - N_2$ is determined from the axial force. The shear strain and shear rate in the sample are non-uniform, varying from zero at the axis of rotation to maximum at the rim of the disks. This makes transient experiments more difficult to interpret.
- Cone and plate rheometer. The sample is confined between a flat disk and a shallow cone, one of which rotates to produce the shearing motion. The shear rate and shear strain in the sample are uniform; however, the gap height goes to zero at the axis of rotation so there is always a concern about fiber motion being constrained by the walls. Wall effects can be reduced by drilling a hole in the sample (the *cone and donut* geometry [EBWVG09b, EBWVG09a]) or, to a lesser extent, using a truncated cone [PGBC20]. The torque on the rotating disk gives the shear stress, and the axial force gives the first normal stress difference N_1.

- Sliding plate rheometer [OAE+11, CWB15]. A rectangular sample with constant thickness is held between two parallel plates. Shearing is created by sliding one plate relative to the other, and a shear-stress transducer in the face of one plate measures τ_{13}. It is also possible to measure N_1 with this device, though normal-stress measurements for fiber filled polymers have not been reported.

The gap height of a shear rheometer compared to the fiber length affects the repeatability of the experiments. Bibbo et al. [BDA85] recommend a gap height of at least 1.2 times the fiber length for model suspensions, while Eberle et al. [EBWVG09a] found that a gap height of at least 2.2 times L_w was required. This implies that cone and plate devices are useful for short fiber materials, while a sliding plate rheometer with a large gap height is needed for long fiber materials.

Measurements have also been performed in lubricated squeeze flow (planar elongation) [LB16] and non-lubricated squeeze flow (mixed shear and elongation) [LWB17, LWB21].

6.8.2 Initial Fiber Orientation State

In transient rheology experiments, it is essential to have a repeatable, and preferably known, initial fiber orientation state. While early experiments tended to assume that the initial orientation was isotropic, there is no reliable way to create isotropic orientation. Techniques that have been used to create repeatable initial orientation include:

- Reverse shear flow. In many rheometers the sample can be pre-sheared in the reverse direction by a large amount, often $\gamma = -1{,}000$ in rotational rheometers and $\gamma = -100$ in a sliding plate rheometer. This should establish the steady-state orientation for simple shear flow in the reverse direction. Then forward shearing can begin, and a repeatable transient response of the suspension can be measured [SCMA04, SCG+04]. Often a rest time is inserted between the two movements so that any viscoelastic stresses in the matrix can relax out.
- Injection molding. The initial sample can be cut from an injection molded plaque or disk that has the desired thickness. This can be placed in the rheometer, heated, and squeezed slightly to conform to the rheometer surfaces. The initial orientation will be non-uniform across the thickness, owing to the typical core/shell structure of injection molded parts, but it will at least be repeatable [CBW16, PGBC20].
- Extrusion followed by compression molding. Strands of the polymer/fiber mixture can be extruded, and pieces of the extrudate cut to convenient lengths. A layer of these can be arranged in a compression mold, either aligned in a single direction or randomly oriented in the plane of the mold, then heated and molded into a solid sample [SBO20]. Multiple layers can be combined when a thicker sample is needed. Aligning the extrudates tends to produce a sample with high fiber alignment, and samples of this type usually use a crossflow initial orientation.

For any flow geometry, the best practice is to measure the initial orientation state experimentally. Some experiments also measure the orientation as a function of strain to complement the stress measurements. This is done by using multiple samples, halting the experiment at different levels of shear strain, solidifying the sample, and measuring the orientation state at each strain level.

6.8.3 Comparing Models to Data

Once reliable data has been obtained, one can attempt to fit a model to the data. Often both a fiber orientation model and a rheological model are fitted simultaneously.

Figure 6.13 shows an example. This data is for a short fiber reinforced thermoplastic, and used a parallel disk geometry. The initial orientation state was created by shearing the material in the reverse direction up to $\gamma_R = -1{,}000$ (γ_R is the shear strain at the outer radius of the disks), then the flow was reversed and the data shown in the figure was recorded. The solid lines in the figure were determined by adjusting the orientation parameters C_I and κ from the Folgar–Tucker/RSC model and the rheological parameters η_I and N_p from Eqn. (6.29) to fit the shear stress data only. The initial orientation state was calculated from the reverse flow using the parameters of the orientation model. The curves in Fig. 6.13 are calculated to account for the variation in orientation and stress with radius in this device [WOT08].

The fit to the shear stress data in Fig. 6.13(a) is remarkably good, except for a short period at the start of the experiment. This initial period is probably affected by a viscoelastic response in the PBT matrix, which is not accounted for in the rheological model. This may also be a period where the fiber orientation model is not accurate.

The model greatly over-predicts the steady-state normal stress difference in Fig. 6.13(b). The model does capture the qualitative features of the data, predicting a negative normal stress difference at the start of the experiment, followed by a transient rise to a positive and steady normal stress difference at large strains.

Another important feature of Fig. 6.13 is that a small value of the RSC parameter, $\kappa = 0.112$, is required to fit the data. Many similar experiments also display slow orientation kinetics in simple shear, and one must use an SRF, RSC or RPR model with adjustable κ if the model is to come close to the data.

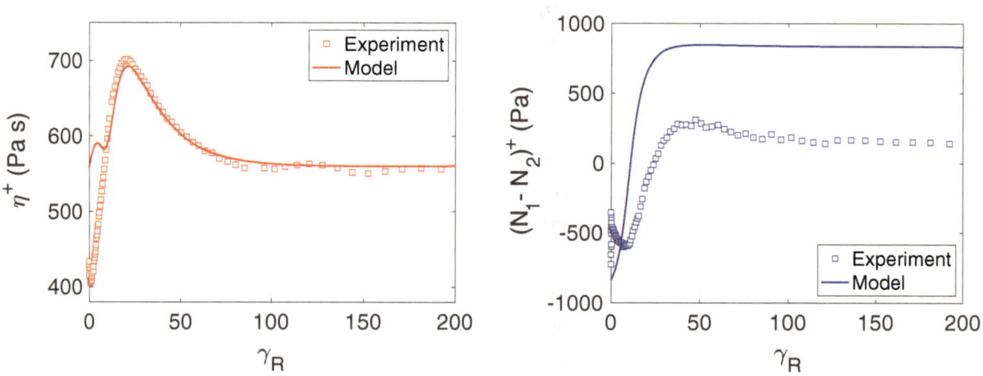

Figure 6.13 Rheological measurements in the startup of simple shear flow for PBT with 30% by weight short glass fibers. Parallel disk rheometer. γ_R is the shear strain at the outer radius. Initial orientation from reverse flow of $\gamma_R = -1{,}000$. Lines are the Folgar–Tucker/RSC model with $C_I = 0.0112$, $\kappa = 0.112$, $\eta_I = 55.7$ Pa s, $N_p = 123$ and the natural closure. Experimental data from [WOT08].

While this particular experiment has produced reasonable values of C_I and κ, rheological stress measurements may not be the best way to obtain orientation parameters. Ortman et al. [OBWA12] used orientation parameters from rheology experiments [OBWW12] to predict orientation in a center gated disk with 30% by weight long glass fiber/PP, and found a good match to the molding data. In contrast, Meyer et al. [MHB14] found that orientation parameters from stress data of a short glass fiber/PP composite over-predicted the flow-direction orientation in an end-gated plaque, and better predictions were obtained by fitting the parameters to molding data near the gate. A recent study [KLC+20b] found good agreement between molding simulations and orientation measurements in a commercial part using parameters determined with a sliding plate rheometer. However, these parameters were fitted to transient *orientation* data from the rheometer, rather than to the stress data.

At this time, the best way to determine parameters for a fiber orientation model is to fit fiber orientation measurements, either from rheometry experiments or from molded samples. Rheometry experiments offer a way to do this with controlled deformation, something that is not available when fitting molding data. Offsetting this, the maximum shear rate in most rheometers is less than $10\,\mathrm{s}^{-1}$, which is several orders of magnitude lower than typical shear rates in injection molding.

6.8.4 Insights from Transient Rheology Experiments

While transient rheological experiments have not yet provided an easy way to fit orientation parameters, they have provided useful insights into the orientation and stress behavior of fiber suspensions. These include the following:

- Shear stress and normal stress transients confirm the qualitative prediction of Eqn. (6.29) for fiber suspension rheology.
- Slow transients in the shear and normal stress confirm the presence of slow orientation kinetics and the need for RSC or RPR type orientation models.
- Within the limits of experimental equipment, fiber orientation evolves with strain, independent of strain rate (e.g., [CWB15]). This strain-time scaling is consistent with the orientation models in Chapt. 5.
- The stress response of long fiber materials ($L_w > 1\,\mathrm{mm}$) is more complex than for short fiber materials, and Eqn. (6.29) does not fully represent the stress response of long fiber composites [CBW16].

6.9 Chapter Summary

Fibers affect the rheology of a fluid polymer matrix in much the same way that they affect the elastic properties of a solid polymer. The rheology of a fiber suspension is closely connected to the fiber orientation, so in general the solutions for fluid velocity and fiber orientation are *coupled*, and the two problems must be solved simultaneously. This requires a constitutive equation that gives the stress in the fiber suspension as a function of **D**, the orientation state and other parameters like the matrix viscosity and fiber volume fraction.

When a fiber follows Jeffery's orientation equation for $\xi = 1$, the relative velocity between the fiber and the fluid is parallel to the fiber axis, and is proportional to the fluid stretching rate in the fiber direction. This generates a tension or compression in the fiber that resists the stretching motion. This stress in the fiber direction is the main contribution of fibers to suspension viscosity.

When the fiber contribution is averaged over all orientations, the resulting stress is controlled by the fourth-order orientation tensor, and is proportional to the product $\mathbb{A} : \mathbf{D}$. Since $\mathbb{A}$ is usually determined from the second-order tensor using a closure approximation, any rheological model for a fiber suspension requires an accurate closure. The natural/ORE and IBOF forms that are recommended for orientation modeling also give useful results for rheological predictions.

A full orientation-dependent rheological model can be used to solve coupled flow-orientation problems. The results are a good match to experiments where coupling greatly affects the flow pattern, e.g., flow in an axisymmetric contraction. However, for many flow geometries, coupling has little effect on the flow pattern. This is particularly true for flow in narrow gaps, which occurs frequently with injection and compression molded parts. In these cases a *decoupled calculation* can be used, where the flow field is solved without using any orientation data and the results are used to predict orientation. Most CAE software packages use the decoupled approach.

Rheological experiments using a capillary or slit-die rheometer can measure the steady shear viscosity of a fiber suspension at shear rates that are typical of injection mold filling. This type of data is useful for injection molding simulations. Transient rheological experiments, using parallel disk, cone-and-plate or sliding plate instruments, are limited to low shear rates, but reveal time-dependent (or strain-dependent) behavior that is related to the transient development of fiber orientation. Models for fiber suspension rheology reproduce these transients qualitatively, for both shear and normal stress. However, complete quantitative agreement is difficult to achieve, especially for LFT materials. This may be due to phenomena that are not in the model, such as fiber bending, stresses due to fiber-fiber interactions, or confinement of fibers by the rheometer walls. Transient rheological experiments do exhibit the slow orientation kinetics discussed in Section 5.7, supporting the need for that type of orientation model.

7 Fiber Length Degradation during Processing

Composite properties depend on fiber length as well as on orientation, and fiber length can change during processing. Fiber length degradation is particularly important for long fiber injection molding materials, where the pellets start with fibers 10–12 mm long but molded parts will have average fiber lengths in the 1–3 mm range. For these materials it is useful to understand how fibers break, and to be able to predict the length distribution in molded parts.

The fiber length model used in most CAE software [PT09] is discussed in this chapter. To develop this model, the first step is to write a conservation law that governs changes in the fiber length distribution. We then examine the mechanics of fiber breakage, and develop an expression for the breakage rate. The properties of this model are explored, including the effects of the model parameters, and model predictions are compared to experiments.

7.1 Model Development

7.1.1 Conservation of Fiber Length

For this chapter the discrete fiber lengths for the fiber length distribution will be

$$\ell_i = i\,\Delta\ell \tag{7.1}$$

where $\Delta\ell$ is a convenient length increment. The index i runs from 1 to n, where ℓ_n is the largest fiber length that could be present. For a long fiber injection molding compound ℓ_n would be the fiber length in the pellets.

Also for this chapter, subscripts i, j and k will run from 1 to n, and repeated indices will *not* imply summation.

Researchers tend to use a small length increment, say $\Delta\ell = 0.1\,\text{mm}$, which can give $n \geq 100$. Larger values of $\Delta\ell$ may be used in commercial software to reduce computation time, though as we will see, there are limits on how large $\Delta\ell$ can be.

We will develop the model in terms of the number-based fiber length distribution N_i. It is helpful to think of a small volume of material containing many fibers, with each N_i value representing the number of fibers in the volume with length ℓ_i. The total length of all the fibers in the volume is

$$L_t = \sum_{i=1}^{n} N_i \ell_i \tag{7.2}$$

Each N_i will change as fibers break, but the total length L_t will not. We want to write an equation for the rate of change of N_i based on this concept.

Consider a fiber that breaks into two shorter pieces. Call that fiber a *parent* and the resulting shorter pieces *children*. Define the *breakage rate* P_i such that $P_i \Delta t$ is the probability that a parent fiber of length ℓ_i will break during the time interval Δt. Each parent fiber length can have a different breakage rate. We will develop a detailed expression for P_i in Section 7.1.2, but for now all we need is its definition.

Each breakage event creates children, and we must account for them too. Define the *child fiber generation rate* R_{ji} such that $R_{ji}\Delta t$ is the probability that a child of length ℓ_j is generated by breaking a parent fiber of length ℓ_i during the time interval Δt. The first index of R_{ji} gives the child length and the second index gives the parent length.

We assume that all breakage events involve a single parent fiber breaking into two child fibers, so the two child lengths must sum to the parent length. If a parent of length ℓ_i breaks into children with lengths ℓ_j and ℓ_k, we must have $\ell_j + \ell_k = \ell_i$. Since the fiber lengths are numbered as in Eqn. (7.1), we can write this as $j + k = i$.

If parent i breaks to create child k, it must also create a child $j = i - k$. The probabilities of creating these two child fibers must be the same, so the child generation rates must satisfy

$$R_{ki} = R_{(i-k)i} \tag{7.3}$$

For example, if $\Delta\ell = 1$ mm and we break a parent that is 10 mm long, then the probability of producing a 6 mm child equals the probability of producing a 4 mm child, $R_{6,10} = R_{4,10}$.

Since two children are created each time a parent fiber breaks, the number of children created by breaking fibers of a given length ℓ_i is twice the number of breaks. This relates the child generation rate to the breakage rate by

$$\sum_{j=1}^{i-1} R_{ji} = 2P_i \tag{7.4}$$

The upper limit on this sum reflects the fact that the child length ℓ_j must be shorter than the parent length ℓ_i.

We are now ready to write a conservation equation for the fiber length distribution. Consider fibers of length ℓ_i. The number of these fibers that break during any Δt will be the product of the number of fibers that are present N_i and the probability that each of those fibers breaks $P_i \Delta t$. Thus, the contribution to dN_i/dt of parent fibers breaking will be $-P_i N_i$. The negative sign appears because breaking parents of length ℓ_i causes N_i to decrease.

N_i can also increase if fibers longer than ℓ_i break to create children of length ℓ_i. The contribution to dN_i/dt from parents of length ℓ_k breaking to create children of length ℓ_i is $R_{ik}N_k$. Any fiber longer than ℓ_i is a possible parent, so we must sum this contribution over all parent lengths $k > i$ to get the total child contribution to dN_i/dt.

Combining the decrease in N_i due to breakage and the increase due to creation of children gives a conservation equation for N_i:

$$\frac{dN_i}{dt} = -P_i N_i + \sum_{k=i+1}^{n} R_{ik} N_k \tag{7.5}$$

This is the basic governing equation for fiber length. It shows how each N_i changes as parent fibers break and create child fibers. The lower limit on the sum appears because a child fiber can never be longer than its parent.

Equation (7.5) also ensures the conservation of total fiber length. If one differentiates the expression for total length, Eqn. (7.2), then substitutes Eqn. (7.5) for dN_i/dt and uses Eqns. (7.3) and (7.4), one can show that $dL_t/dt = 0$ and total fiber length is conserved. This is only true if the shortest fiber length in the calculation $\ell_1 = \Delta\ell$ has a zero breakage rate, $P_1 = 0$. If this is not true then total length is lost over time, because the shortest fibers break but their children are not accounted for. To avoid this problem, $\Delta\ell$ should be smaller than the shortest fiber that can break.

7.1.2 Breakage Rate and Unbreakable Length

We now need an expression for the breakage rate P_i in terms of physical parameters such as shear stress and fiber length.

We saw in Section 6.3 that fluid motion relative to a fiber is primarily parallel to the fiber axis. This relative velocity is proportional to $\mathbf{D}:\mathbf{pp}$, which is the stretching rate of the fluid along the fiber direction $\mathbf{p}$. Drag forces from the fluid cause either tension or compression along the fiber axis, depending on whether the sign of $\mathbf{D}:\mathbf{pp}$ is positive or negative.

Tensile forces that arise in processing flows are never large enough to break the fibers in tension. However, a slender fiber loaded in compression can fail by buckling at a much smaller stress than the tensile strength, and the required forces can readily occur in processing. In Fig. 6.3, fiber (c) is in tension and cannot be broken, while fiber (a) is in compression and could potentially buckle. Thus, we will focus on compressive loads in the fibers, and compare these loads to the critical load for buckling.

A useful estimate for the force in a fiber appears in Dinh and Armstrong's fiber suspension model [DA84], which gives the magnitude of the compressive force at the fiber center as

$$f_i(\mathbf{p}) = \frac{\zeta \eta_m \ell_i^2}{8}(-\mathbf{D}:\mathbf{pp}) \tag{7.6}$$

The subscript i in $f_i(\mathbf{p})$ indicates the fiber length ℓ_i, while ζ is a dimensionless drag coefficient and η_m is the viscosity of the polymer matrix. With this definition $f_i(\mathbf{p})$ is positive when $\mathbf{D}:\mathbf{pp} < 0$ and the fiber is in compression. Notice that f_i depends on the fiber length (longer fibers develop larger forces) and on the fiber orientation $\mathbf{p}$.

Buckling is a type of elastic instability. A perfectly straight fiber loaded in compression will not fail until the stress in the fiber reaches the compressive strength of the material. However, if the fiber bends slightly out of line, then it experiences a bending moment at its center that is proportional to the load and to the amount of misalignment. If the load is small enough, the fiber can support this bending moment with little additional deformation. If the load becomes too large then the fiber bends more to accommodate the bending stress, which increases the misalignment, which increases the bending moment, which further increases fiber bending. Then the fiber buckles.

Glass and carbon fibers are brittle, so they bend elastically until they reach their failure stress at some point along the fiber, at which point they break. Thus, we assume that each time a fiber buckles, it breaks.

From classical Euler–Bernoulli buckling theory, the critical compressive force to buckle a fiber that is simply loaded at its ends is

$$f_{\text{crit}} = \frac{\pi^3 E_f d_f^4}{64 \ell_i^2} \tag{7.7}$$

Here E_f is the elastic modulus of the fiber and d_f is the fiber diameter. Note that the fiber stiffness E_f appears here but the fiber tensile strength does not. This is because buckling is an elastic phenomenon.

We expect that a fiber will buckle and break if $f_i > f_{\text{crit}}$. Combining Eqns. (7.6) and (7.7), a fiber will buckle if

$$\frac{f_i(\mathbf{p})}{f_{\text{crit}}} = \frac{8 \zeta \eta_m \ell_i^4}{\pi^3 E_f d_f^4} (-\mathbf{D} : \mathbf{pp}) \geq 1 \tag{7.8}$$

To simplify this expression we define a dimensionless breakage parameter B_i as

$$B_i \equiv \frac{4 \zeta \eta_m \dot{\gamma} \ell_i^4}{\pi^3 E_f d_f^4} \tag{7.9}$$

and a dimensionless rate-of-deformation tensor $\mathbf{D}^*$,

$$\mathbf{D}^* \equiv \mathbf{D} / \dot{\gamma} \tag{7.10}$$

With these, the criterion for a fiber to break, Eqn. (7.8), becomes

$$B_i \left(-2\mathbf{D}^* : \mathbf{pp}\right) \geq 1 \tag{7.11}$$

$\mathbf{D}^*$ represents the "shape" of the rate-of-deformation tensor (e.g., shear vs. elongation), independent of the magnitude of $\mathbf{D}$. All other physical parameters (matrix viscosity, scalar strain rate, fiber length, fiber diameter, elastic modulus and fiber drag coefficient) combine to give B_i. Note that the shear stress $\eta_m \dot{\gamma}$ appears in the numerator of B_i, and B_i is proportional to the fiber aspect ratio ℓ_i / d_f raised to the fourth power.

To translate this into a breakage probability, we first assume that P_i is proportional to the shear rate $\dot{\gamma}$, and to the fraction of fibers of length ℓ_i that have orientations such that $f_i(\mathbf{p}) / f_{\text{crit}} > 1$. Call the constant of proportionality the *breakage coefficient* C_B, so that

$$P_i = C_B \dot{\gamma} \times \left(\text{fraction of fibers with } f_i(\mathbf{p}) / f_{\text{crit}} > 1\right) \tag{7.12}$$

Making P_i proportional to $\dot{\gamma}$ ensures that two flows at different strain rates but with the same value of B_i will produce the same amount of breakage for any given total shear strain. This is the same type of strain-time scaling we observe with fiber orientation. C_B is an empirical parameter whose value is determined by matching experiments.

The fraction of fibers for which $f_i(\mathbf{p}) / f_{\text{crit}} > 1$ depends on B_i and on the orientation distribution function. Figure 7.1 shows how $(-2\mathbf{D}^* : \mathbf{pp})$ varies with orientation for a simple shear flow. Fibers in the red zones are in compression while fibers in the blue zones are in tension. (Compare this figure to Fig. 6.3.)

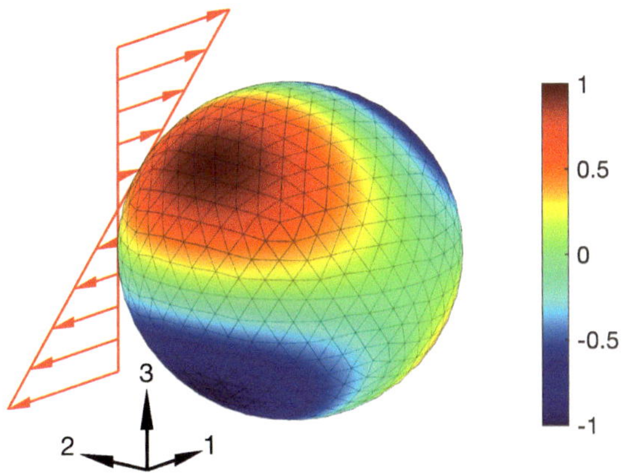

Figure 7.1 Dimensionless fluid contraction rate $(-2\mathbf{D}^* : \mathbf{pp})$ as a function of orientation for simple shear flow, $v_1 = \dot{\gamma} x_3$.

In simple shear flow the maximum value of $(-2\mathbf{D}^* : \mathbf{pp})$ is unity, so we need $B_i > 1$ for any of the red areas to have $B_i\,(-2\mathbf{D}^* : \mathbf{pp}) > 1$. The portion of the red areas where this is true increases as B_i increases beyond unity, and for $B_i \gg 1$ a fiber could buckle almost anywhere in the red and yellow areas.

To get the fraction of fibers that have $f_i(\mathbf{p})/f_{\text{crit}} > 1$ we also need the orientation distribution function. There is no simple formula for this, but Phelps et al. [PAKT13] used steady-state orientation distributions in simple shear flow for different C_I values to find the fraction of fibers that satisfied Eqn. (7.11) and could possibly buckle. The following empirical expression was a reasonable approximation to their data:

$$P_i = \begin{cases} 0 & \text{for} \quad B_i \le 1 \\ C_B \dot{\gamma}\left[1 - \exp(1 - B_i)\right] & \text{for} \quad B_i > 1 \end{cases} \tag{7.13}$$

In this model, fibers with $B_i \le 1$ will not break. Using Eqn. (7.9), this implies that fibers do not break if $\ell_i \le L_{\text{ub}}$, where L_{ub} is the maximum unbreakable fiber length for the current flow conditions and fiber properties, and

$$L_{\text{ub}} = d_f \left(\frac{\pi^3 E_f}{4\zeta \eta_m \dot{\gamma}} \right)^{1/4} \tag{7.14}$$

If we shear a fiber suspension for a long time at constant stress $\eta_m \dot{\gamma}$, all fibers should be broken to a length of L_{ub} or smaller. Thus, L_{ub} characterizes the long-time behavior of the fiber length model.

Using L_{ub} we can rewrite the breakage rate (7.13) to show the fiber length explicitly,

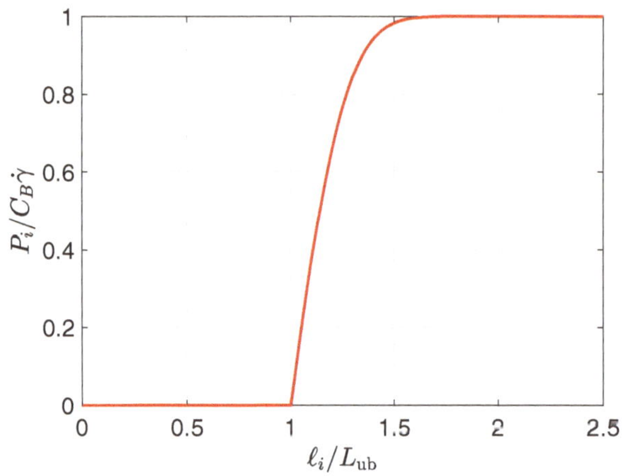

Figure 7.2 Breakage rate P_i as a function of fiber length ℓ_i, Eqn. (7.15).

$$P_i = \begin{cases} 0 & \text{for} \quad \ell_i \leq L_{\mathrm{ub}} \\ C_B\dot{\gamma}\left[1 - \exp\left(1 - \{\ell_i/L_{\mathrm{ub}}\}^4\right)\right] & \text{for} \quad \ell_i > L_{\mathrm{ub}} \end{cases} \tag{7.15}$$

This expression is used in commercial software, and is shown in Fig. 7.2.

In practice, fibers somewhat shorter than L_{ub} may break, either because they have a pre-existing curvature that makes them more susceptible to buckling, or because they are pushed out of line by forces from other fibers. Expressions for P_i that incorporate this possibility have been developed [DDV+13]. However, the predictions of the model are not very sensitive to the details of P_i, especially since the drag coefficient ζ (and thus L_{ub}) are adjusted to match experimental data.

Equation (7.14) shows how different material and process parameters affect fiber breakage by changing L_{ub}. Increasing the shear stress $\eta_m\dot{\gamma}$ decreases L_{ub} and breaks the fibers to shorter lengths, while increasing the fiber modulus E_f, perhaps by changing from glass to carbon fibers, decreases L_{ub} and allows longer fibers to survive the flow. However, both shear stress and fiber modulus are raised to the 1/4 power in the expression for L_{ub}, so large changes in these variables are needed to produce a significant change in the unbreakable length.

One might think that increasing the fiber diameter d_f would increase the final fiber length, and in the model this is true. However, the mechanical properties of a discontinuous fiber composite depend on the fiber aspect ratio ℓ/d_f, not on the fiber length by itself. Since L_{ub} is proportional to d_f, increasing the fiber diameter does not change the unbreakable fiber aspect ratio L_{ub}/d_f. Thus, increasing the fiber diameter should have little effect on mechanical properties.

7.1.3 Child Generation Rates

To complete the model we need an expression for the child generation rates R_{ik}. The center of the fiber is the most likely location for a break, so R_{ik} should be maximum at the center of

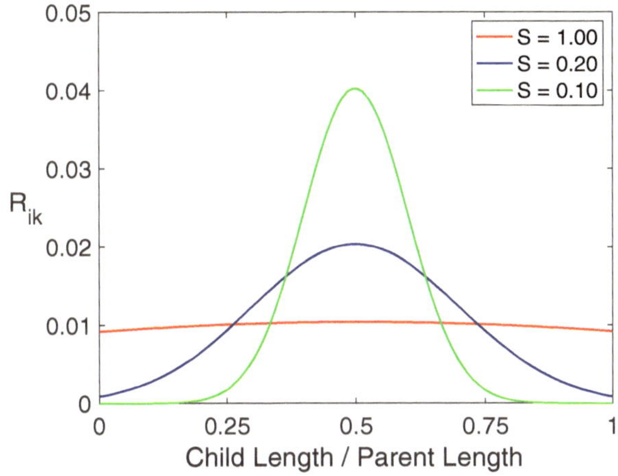

Figure 7.3 Child fiber generation rates R_{ik} for different values of the parameter S. Units for R_{ik} are arbitrary.

the parent fiber, $\ell_i = \ell_k/2$. Because there are forces on the fiber from fiber-fiber interactions, and because the fiber strength varies from point to point, some breaks could happen at other locations. From these requirements, it is reasonable to assume a Gaussian breakage profile along the parent fiber,

$$R_{ik} = \alpha_k G_{\text{PDF}}(\ell_i, \ell_k/2, S\ell_k) \tag{7.16}$$

Here $G_{\text{PDF}}(x, \bar{x}, \sigma)$ is the Gaussian probability density function for the variable x with mean $\bar{x}$ and standard deviation σ, and α_k is a normalization factor to be determined.

S is a dimensionless model parameter that controls the shape of the Gaussian breakage profile. A small value of S gives a high probability that the fiber will break near its center, while a large S value distributes the probable breakage points more evenly along the fiber length. Figure 7.3 shows some examples.

The normalization factor α_k is chosen so that R_{ik} satisfies Eqn. (7.4). Working out the details, we can write R_{ik} explicitly as

$$R_{ik} = \begin{cases} 2P_i \dfrac{G_{\text{PDF}}(\ell_i, \ell_k/2, S\ell_k)}{\sum_{j=1}^{k-1} G_{\text{PDF}}(\ell_j, \ell_k/2, S\ell_k)} & \text{for} \quad i < k \\ 0 & \text{for} \quad i \geq k \end{cases} \tag{7.17}$$

Substituting this expression and the breakage rate (7.15) into the conservation equation (7.5) gives a complete model for fiber breakage. One starts from an initial condition $N_i(t=0)$ and integrates Eqn. (7.5) forward in time to get $N_i(t)$. At any time step, the number-based distribution N_i can be used to find the weight-based distribution w_i and the average lengths L_n and L_w, as in Section 2.5. A function implementing this model, `solveFLDstar`, appears in the accompanying software.

7.2 Effects of Model Parameters

The fiber length model has three parameters, each taking a specific role.

The **breakage coefficient** C_B controls the rate of change of fiber length. In Eqn. (7.5) both P_i and R_{ik}, and thus dN_i/dt, are proportional to $C_B\dot{\gamma}$. Larger values of C_B produce faster changes in the fiber length without changing the shapes of the fiber length distribution; smaller values give the same changes at a slower rate. The time dependence of the model predictions is governed by $C_B\dot{\gamma}t$.

Fibers only break when they rotate into an orientation that is favorable for buckling (Fig. 7.1) and moving all fibers to these orientations requires large amounts of strain. Thus, we expect $C_B \ll 1$. While C_B may be related to the strain needed for a fiber to complete a Jeffery orbit [DDV+13], here we treat it as an adjustable parameter.

The **drag coefficient** ζ affects L_{ub}, the length below which fibers do not break. L_{ub} defines the length scale of the steady-state solution of the model. For calculations where the shear stress $\eta_m\dot{\gamma}$ is constant, L_{ub} is also constant. Then the length dependence of the model is controlled by ℓ/L_{ub}, the fiber length scaled by the unbreakable length. This also accounts for the values of shear stress, d_f and E_f, which contribute to L_{ub}.

We will see in a moment that it is important to choose $\Delta\ell$, the length increment used to discretize the fiber length distribution, so that $\Delta\ell \le L_{ub}$.

If Eqn. (7.6) is a good estimate for the force in a fiber, then the value of ζ should be around unity, say $0.2 \le \zeta \le 5$. Values somewhat outside this range would still be reasonable.

The **child generation parameter** S affects the shape of the fiber length distribution by controlling the distribution of child fibers produced when fibers break. An example appears below. Reasonable values lie in the range of $0.1 \le S \le 2$. Values smaller than 0.1 force all fibers to break very close to their midpoints and, if all fibers have the same initial length ℓ_0, then at later times the distribution only has lengths of ℓ_0, $\ell_0/2$, $\ell_0/4$, and so forth. This is unrealistic. When $S = 2$ any child length is almost equally likely, so increasing S beyond this value has a negligible affect on the results.

Results of an example calculation appear in Fig. 7.4. All fibers have an initial length of $15L_{ub}$, we use $S = 0.2$, and the effects of all other parameters are accounted for by using $C_B\dot{\gamma}t$ as the time variable and ℓ_i/L_{ub} as the length variable. The primary calculations give $N_i(t)$, and those results are used to compute the average lengths and the weight-based distribution at each time step.

The time dependence of L_n and L_w are plotted in Fig. 7.4(a). Both L_n and L_w equal $15L_{ub}$ at $t = 0$, since all fibers have that length. The average lengths decrease rapidly at first, then more slowly, with substantial changes occurring by $C_B\dot{\gamma}t = 3$. L_w decreases more slowly that L_n. At long times, $C_B\dot{\gamma}t \ge 10$, both lengths are very close to steady state. The final values of L_n and L_w are slightly below L_{ub}. This is because breaking a fiber longer than L_{ub} creates child fibers shorter than L_{ub}.

Weight-based fiber length distributions from this calculation are shown in Figs. 7.4(b–e). At $C_b\dot{\gamma}t = 1.5$ we see an early stage of fiber breakage, where more than 20% of the fiber weight still has the initial length and the remainder of the fibers are distributed across shorter lengths. By $C_B\dot{\gamma}t = 3$ the distribution has a definite peak near L_{ub} with a tail at greater lengths, similar to the experimental distributions in Figs. 2.10(b) and 3.17(b). Five percent by weight of the

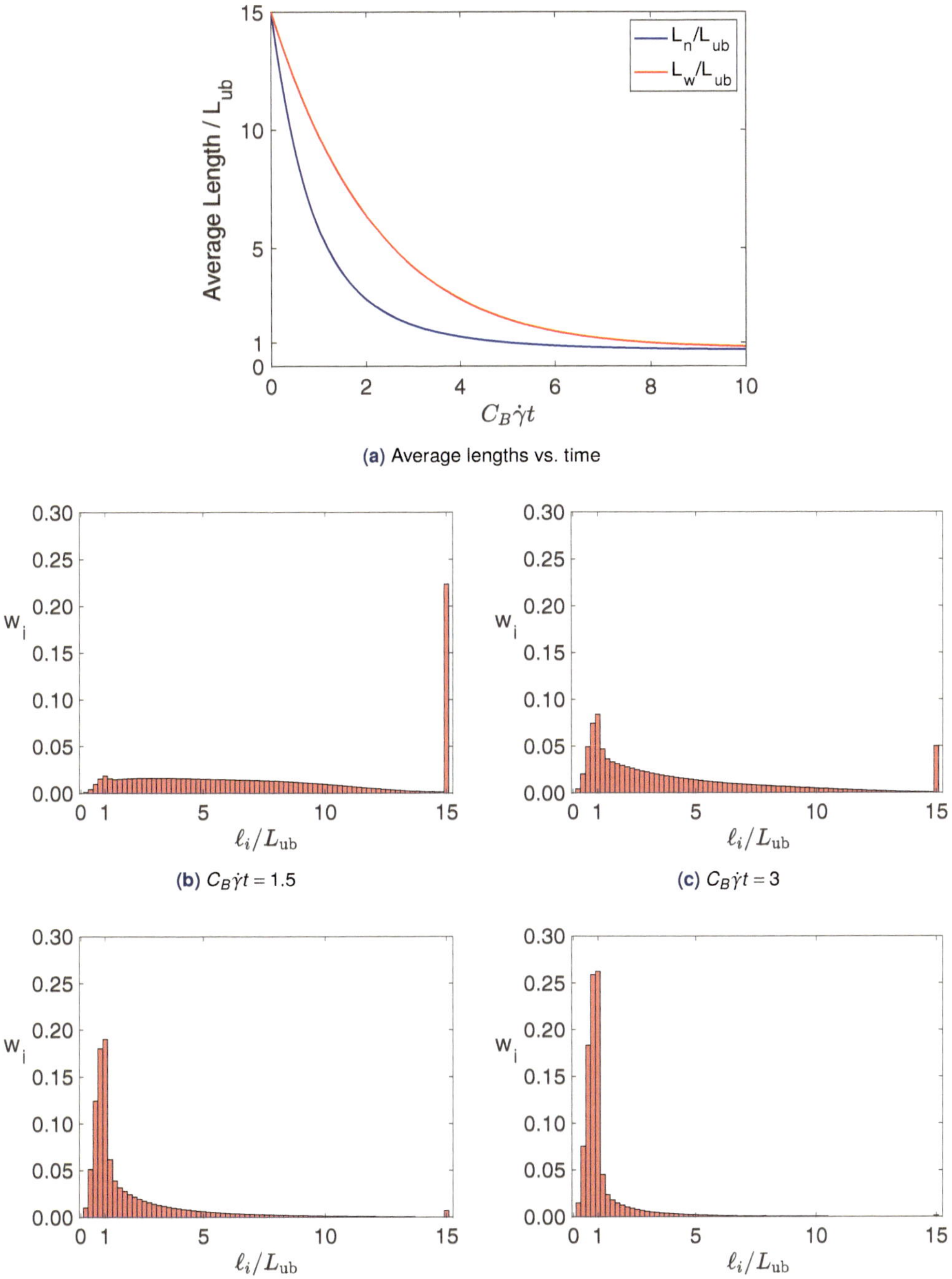

(a) Average lengths vs. time

(b) $C_B\dot{\gamma}t = 1.5$

(c) $C_B\dot{\gamma}t = 3$

(d) $C_B\dot{\gamma}t = 5$

(e) $C_B\dot{\gamma}t = 7$

Figure 7.4 Example results from the fiber length model using $\Delta\ell = 0.2L_{\text{ub}}$ and $S = 0.2$. In the initial condition all fibers have $\ell = 15L_{\text{ub}}$.

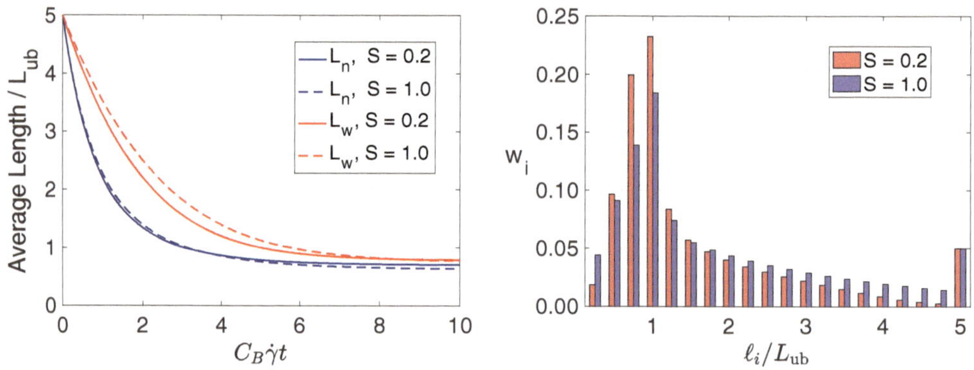

(a) Average lengths vs. time **(b)** Length distributions at $C_B\dot{\gamma}t = 3$

Figure 7.5 Effect of the parameter S on the fiber length model. $\Delta\ell = 0.25L_{ub}$ and the initial condition has all fibers with $\ell = 5L_{ub}$.

fibers still have the initial length. At $C_B\dot{\gamma}t = 5$ the peak near L_{ub} is larger, the tail at longer fiber lengths is smaller, and just a few fibers with the initial length are visible. By $C_B\dot{\gamma}t = 7$ there are no visible fibers with the initial length, and almost no fibers with $\ell_i/L_{ub} > 5$. While measures like L_n and L_w are useful metrics for fiber length, the fiber length model provides full information on the fiber length distribution.

The effect of changing the child generation parameter S is shown in Fig. 7.5. This figure compares results for two values $S = 0.2$ and $S = 1$ (see also Fig. 7.3). The time histories of L_n and L_w are similar, with L_w decreasing a bit more slowly for the larger value of S. The weight-based length distributions at $C_B\dot{\gamma}t = 3$ are similar as well. The larger S value distributes the breaks more evenly along parent fibers, so it has more fibers at longer lengths ($\ell_i/L_{ub} \geq 2$) and at short lengths ($\ell_i/L_{ub} \leq 0.5$) than the smaller value of S. The peak in the orientation distribution is at L_{ub} for both S values. A close look at Fig. 7.5(a) shows that $S = 1.0$ has slightly lower steady-state values of L_n and L_w than $S = 0.2$. This data shows that S can be used to fine-tune the shape of fiber length distribution, while the major effects of the model are reflected in C_B and L_{ub}.

In any numerical calculation one must discretize the fiber length distribution using a length increment $\Delta\ell$. Smaller values of $\Delta\ell$ give more accurate solutions of the model at the cost of increased computation time. In a mold filling simulation one must calculate the fiber length distribution at every node, so there is good reason to keep the number of fiber length variables per node as small as possible. Choosing $\Delta\ell$ involves striking a balance between discretization error and computation time.

Figure 7.6 explores this tradeoff. Three calculations were done using different values of $\Delta\ell$, with $S = 0.2$ and all other parameters held constant.

Figures 7.6(a) and (b) show that increasing $\Delta\ell$ from $0.1L_{ub}$ to $0.5L_{ub}$ has a small affect on L_n and L_w. The weight-based distributions at $C_B\dot{\gamma}t = 3$, Figs. 7.6(c) and (d), also look very similar to one another. Both of these distributions show 5% of the fiber weight still at the initial length, $\ell_i/L_{ub} = 10$. (The peaks at $\ell_i/L_{ub} = 1$ have different w_i values because each distribution is normalized so that $\sum w_i = 1$, and the smaller $\Delta\ell$ value has more length increments.) These two cases show that one loses little accuracy by using $\Delta\ell = 0.5L_{ub}$.

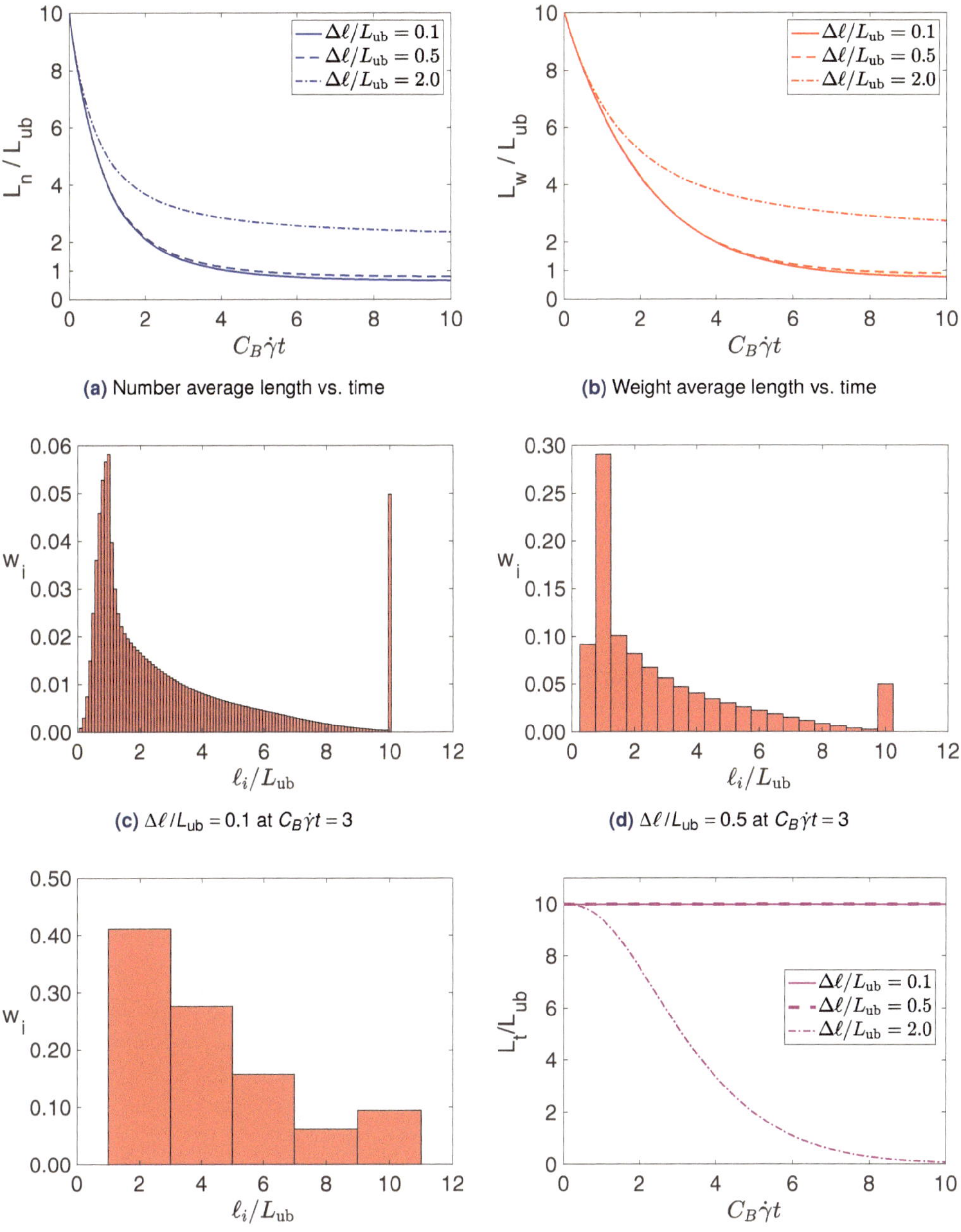

(a) Number average length vs. time

(b) Weight average length vs. time

(c) $\Delta\ell/L_{ub} = 0.1$ at $C_B\dot{\gamma}t = 3$

(d) $\Delta\ell/L_{ub} = 0.5$ at $C_B\dot{\gamma}t = 3$

(e) $\Delta\ell/L_{ub} = 2.0$ at $C_B\dot{\gamma}t = 3$

(f) Total fiber length vs. time

Figure 7.6 Effect of length increment $\Delta\ell$ on accuracy of fiber length calculations. The initial condition has all fibers with $\ell = 10L_{ub}$ and $S = 0.2$. Note the large errors in average length in (a) and (b) and the severe loss of total fiber length in (f) when $\Delta\ell > L_{ub}$.

The results for $\Delta\ell = 2L_{ub}$ are quite different. In Figs. 7.6(a) and (b) we see that both L_n and L_w have steady-state values of $2L_{ub}$, rather than slightly below L_{ub}. This is a major error. In fact, L_n and L_w will have incorrect steady-state values of $\Delta\ell$ in *any* calculation where $\Delta\ell > L_{ub}$. This is because the calculation has no way to represent fiber shorter than $\Delta\ell$, so the most the model can do is reduce all fibers to a length of $\ell_1 = \Delta\ell$.

For $\Delta\ell = 2L_{ub}$ the weight-based distribution function, Fig. 7.6(e), seems to have the right general shape, but the weight fraction of fibers at the initial length is nearly twice as large as it should be. This occurs because the calculation has another serious error: it loses total fiber length. Figure 7.6(f) shows the total fiber length L_t, Eqn. (7.2), for all three calculations. The two smaller values of $\Delta\ell$ preserve fiber length perfectly, with L_t always equal to its initial value. But when $\Delta\ell = 2L_{ub}$ the fiber length declines over time, and most of it has disappeared by $C_B\dot{\gamma}t = 10$.

This loss of L_t occurs because, when $\Delta\ell > L_{ub}$, the breakage rate P_1 for the shortest fibers in the calculation is not zero. As longer fibers break, children of length $\ell_1 = \Delta\ell$ are produced. When these fibers break, their lengths disappear from the calculation because there is no way to account for their children. This distorts the results in Fig. 7.6(e), since the weight-based distribution function is normalized to have $\sum w_i = 1$, but fibers with lengths less than $\Delta\ell$ are missing from the normalization.

In the breakage rate model of Eqn. (7.15), the breakage rate P_i is zero as long as $\ell_i \le L_{ub}$. Any calculation that has at least one fiber length meeting this criterion will conserve total fiber length. And, as this example shows, little accuracy is lost when the value of $\Delta\ell$ approaches L_{ub}.

In a fiber length calculation, always choose a length increment $\Delta\ell \le L_{ub}$.

7.3 Implementation in Mold Filling Software

To use the fiber length model in a mold filling simulation, we recognize that the left-hand side of Eqn. (7.5) is the time derivative following a material point, $\dot{N}_i$. We expand this derivative in terms of spatially fixed values as in Eqn. (5.8), so that Eqn. (7.5) becomes

$$\frac{\partial N_i}{\partial t} + v_1\frac{\partial N_i}{\partial x_1} + v_2\frac{\partial N_i}{\partial x_2} + v_3\frac{\partial N_i}{\partial x_3} = -P_i N_i + \sum_{k=i+1}^{n} R_{ik} N_k \tag{7.18}$$

From here, the same numerical techniques that are used to solve for nodal values of orientation tensor components are used to solve for the nodal N_i values. This requires solving for n variables at each node, where n is the number of fiber lengths in the discrete fiber length distribution.

7.3.1 Fiber Length in an End-Gated Strip

To illustrate the general features of the model, Figure 7.7 shows a fiber length solution for the end-gated strip geometry of Fig. 4.16. In this calculation the length L of the strip is relatively short, and a larger-than-normal value of $C_B = 0.2$ has been used to create a substantial change in the fiber length distribution over this short flow distance. A more realistic C_B value would give a similar picture, stretched horizontally over a longer flow length.

The calculation uses the isothermal Newtonian velocity solution, Eqn. (4.107) and Fig. 4.17. The value of $S = 0.2$ and the remaining problem parameters combine to give $L_{\text{ub}}^{\text{wall}}$, the unbreakable fiber length at the upper and lower mold walls. Rather than choose a value for $L_{\text{ub}}^{\text{wall}}$ we use this length to scale all fiber lengths.

In this flow, the shear rate $\dot{\gamma}$ varies linearly with the thickness-direction coordinate x_3 (see Eqn. (4.108)), and the matrix viscosity η_m is constant. If we define the scalar magnitude of the shear stress as $\tau = \eta_m \dot{\gamma}$, then

$$\tau = \tau_{\text{wall}} \frac{x_3}{h} \tag{7.19}$$

where τ_{wall} is the shear stress at the mold wall.

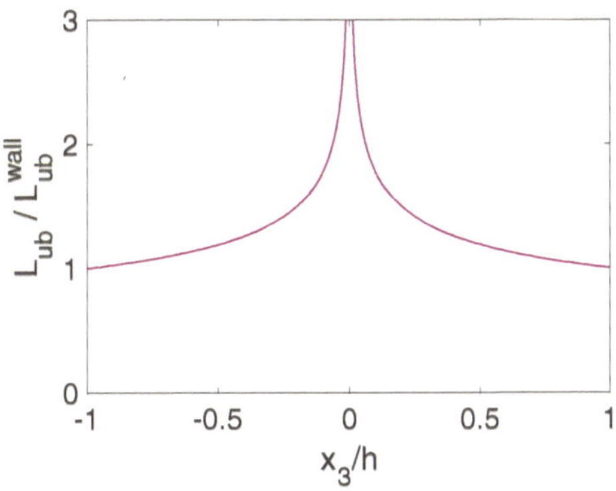

(a) L_{ub} vs. gapwise position

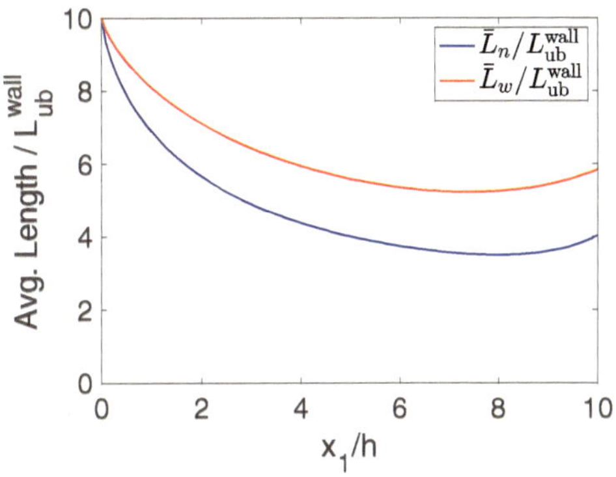

(b) Thickness-averaged L_n and L_w

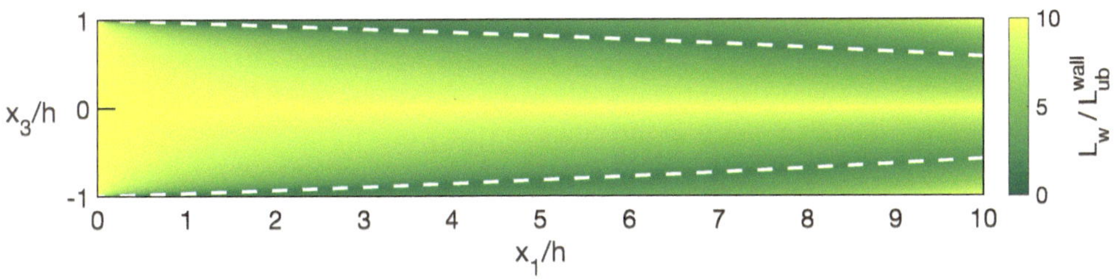

(c) Pointwise values of L_w

Figure 7.7 Calculated fiber lengths in an end-gated strip using the isothermal Newtonian velocity field, Eqn. (4.107). Cavity length $L = 10h$, $C_B = 0.2$, $S = 0.2$, $\Delta\ell = 0.2L_{\text{ub}}^{\text{wall}}$ and all fibers at the inlet have $\ell = 10L_{\text{ub}}^{\text{wall}}$. Dashed white lines in (c) separate points that have passed through the fountain flow from points arriving directly from the gate.

From Eqn. (7.14) this is related to the unbreakable length at the mold wall by

$$L_{\text{ub}}^{\text{wall}} = d_f \left(\frac{\pi^3 E_f}{4 \zeta \tau_{\text{wall}}} \right)^{1/4} \tag{7.20}$$

As we move away from the wall, the shear stress decreases and L_{ub} increases. Using Eqns. (7.14), (7.20) and (7.19), this dependence is

$$L_{\text{ub}} = L_{\text{ub}}^{\text{wall}} \left(\frac{\tau_{\text{wall}}}{\tau} \right)^{1/4} = L_{\text{ub}}^{\text{wall}} \left(\frac{x_3}{h} \right)^{-1/4} \tag{7.21}$$

This expression is graphed in Fig. 7.7(a). While L_{ub} increases as we move from the wall to the midplane, the 1/4 power dependence means this dependence is not very strong, and L_{ub} is only 3 times its wall value at $x_3/h = 0.01$.

Equation (7.19) applies for *any* pressure-driven flow between parallel plates (see, e.g., [HS80, DT01]), so Eqn. (7.21) and Fig. 7.7(a) are general results for the gapwise variation of L_{ub} in a thin mold cavity. They apply to non-Newtonian fluids, non-isothermal flows and any thin cavity geometry. In the current example τ_{wall} and $L_{\text{ub}}^{\text{wall}}$ are constant, but they would vary along the flow path for a non-isothermal flow or a different mold cavity shape.

As we saw in Chapt. 3, burn-off measurements of fiber length use a sample that extends across the full thickness of the part. To compare predictions to these measurements we must average the computed results across the part thickness as well. For this purpose we calculate the thickness-averaged values of L_n and L_w,

$$\overline{L}_n = \frac{1}{2h} \int_{-h}^{h} L_n(x_3)\, \mathrm{d}x_3 \tag{7.22}$$

$$\overline{L}_w = \frac{1}{2h} \int_{-h}^{h} L_w(x_3)\, \mathrm{d}x_3 \tag{7.23}$$

Figure 7.7(b) shows how these values vary with flow distance for our example. At the gate, $x_1 = 0$, the calculation assumes that all fibers have $\ell = 10 L_{\text{ub}}^{\text{wall}}$, so both $\overline{L}_n$ and $\overline{L}_w$ have this same value. Both $\overline{L}_n$ and $\overline{L}_w$ decrease with flow distance, with $\overline{L}_n$ taking a lower value than $\overline{L}_w$ at any given value of x_1, as expected.

A surprising feature of Fig. 7.7(b) is that the average fiber lengths begin to increase with flow distance starting around $x_1/h = 8$. We know that flow and deformation only cause fiber length to decrease, so how can this be?

The answer can be explained with the help of Fig. 7.7(c), which shows the point-by-point solution for L_w. This is a side view of the mold cavity. Material to the left of the dashed white lines lines flowed to its final position directly from the gate at a constant value of x_3. Near the walls the velocity is low and the shear rate is high, so material near the walls accumulates shear strain rapidly with flow distance, and the fiber length decreases quickly with x_1. Near the midplane, $x_3 = 0$, the velocity is high and the shear rate is low, so material near the midplane accumulates little shear strain for a given flow distance and the fiber length decreases slowly with x_1. Exactly on the midplane, the shear stress and shear rate are zero, so fibers there do not break at all.

The dashed white lines in Fig. 7.7(c) show the position at the end of filling for material that entered the mold cavity at $t = 0$. Material to the right of these lines entered the cavity close to the midplane, caught up to the flow front during filling, and was moved out toward the wall by the fountain flow at the flow front. (See Section 4.3.4.) The fiber length for this material degraded slowly while close to the midplane, and more rapidly once it was moved closer to the wall. The farther down the cavity we go, the more recently the material passed through the fountain flow, so fiber length *increases* with x_1 at constant x_3 for points to the right of the dashed white lines.

This is the source of the higher fiber lengths near $x_1/h = 10$ and $x_3/h = \pm 1$. When the results for all points are averaged across the thickness, $\overline{L}_n$ and $\overline{L}_w$ increase with x_1 toward the end of the mold cavity, as shown in Fig. 7.7(c).

7.3.2 Comparison with Experiments

A comparison between predictions of this model and an experiment [PAKT13] is shown in Figs. 7.8 and 7.9. The part is a center-gated disk (Fig. 4.18) with a 90 mm radius and a 3 mm total thickness. The material is polypropylene with 40% by weight of long glass fibers. The calculations use $\Delta\ell = 0.1\,\text{mm}$ and the longest fibers are 13 mm, so $n = 130$.

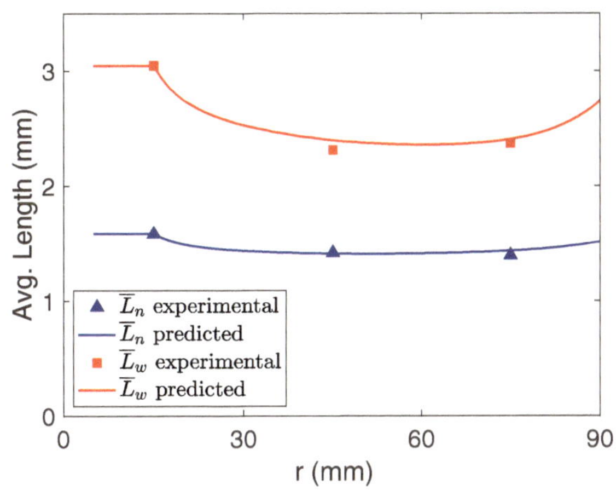

(a) Thickness averaged L_n and L_w

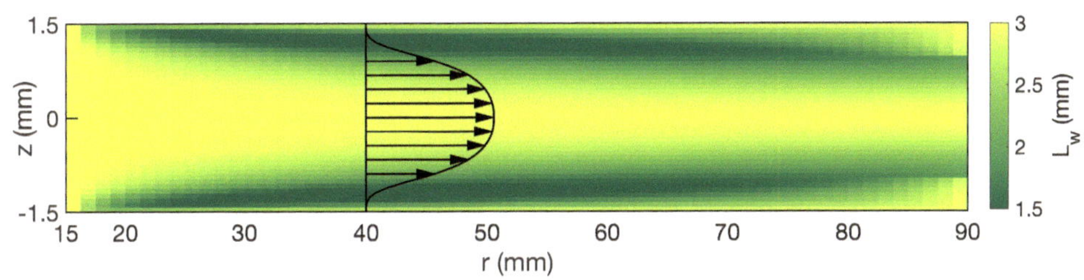

(b) Pointwise values of L_w

Figure 7.8 Average fiber length vs. position in a center-gated disk molded from PP with 40% by weight of long glass fibers. $\zeta = 0.55$, $C_B = 0.025$, $S = 1.0$. Experimental data and predictions from [PAKT13].

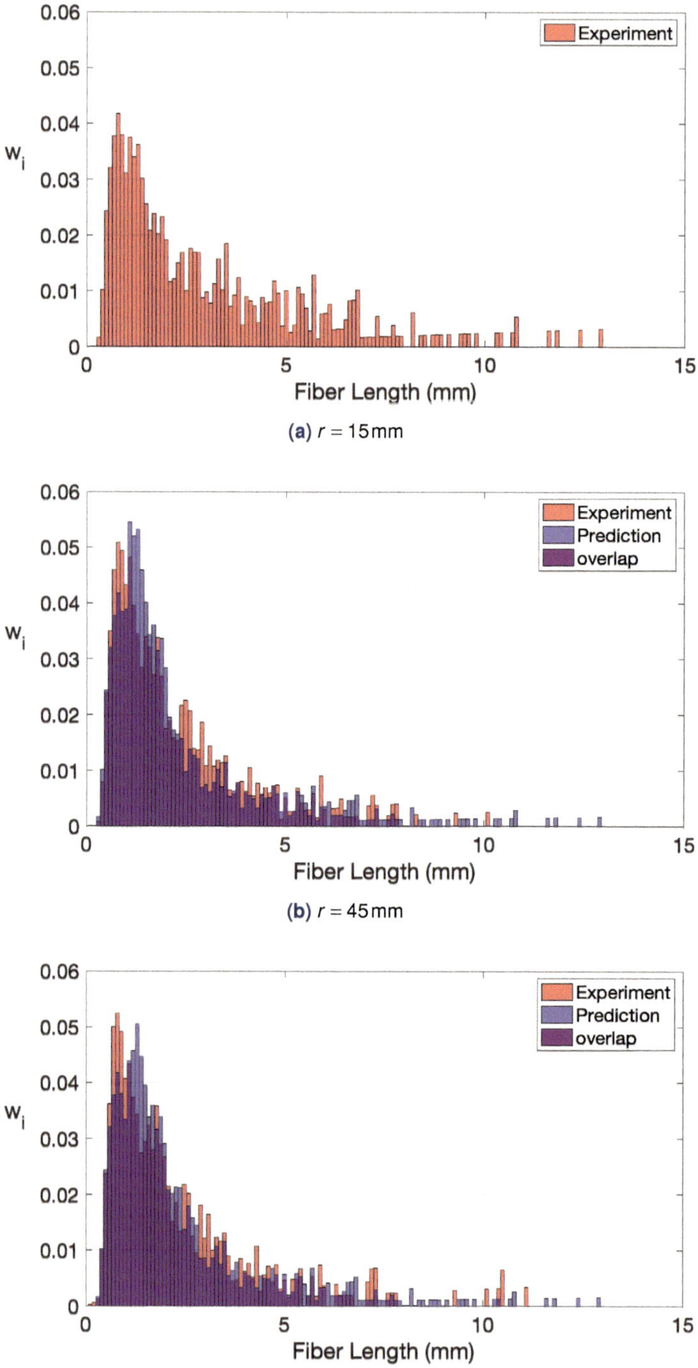

(a) $r = 15\,\text{mm}$

(b) $r = 45\,\text{mm}$

(c) $r = 75\,\text{mm}$

Figure 7.9 Weight-based fiber length distributions at three locations along the flow path in a center-gated disk molded from PP with 40% by weight of long glass fibers. Experimental data and predictions from [PAKT13].

Figure 7.8(a) compares measured values of $\overline{L}_n$ and $\overline{L}_w$, the fiber lengths averaged across the part thickness, with the model predictions. Data at the first experimental point, $r = 15\,\text{mm}$, is used as the initial condition for the calculations, so only the second and third points are predicted. The calculated values, which use $\zeta = 0.55$, $C_B = 0.025$ and $S = 1.0$, match the data quite well. These parameter values were chosen by trial and error to produce a good match to the data, and the values are all reasonable in terms of the theory. Both the experiments and predictions show an upturn in average length, especially for $\overline{L}_w$, near the end of the flow, due to fountain effects.

The predicted spatial distribution of L_w is shown in Fig. 7.8(b). This is a side view of the r–z plane of the disk, as in Fig. 4.109. The general pattern is the same as Fig. 7.7, with longer fibers at the center, shorter fibers near the walls, and longer fibers in the regions filled through the fountain flow. In this case layers of longer fibers near the wall are more distinct and are visible along most of the flow length. This occurs because the velocity calculation accounts for the non-Newtonian nature of the polymer and the increase in polymer viscosity due to cooling near the mold walls. This produces the type of velocity profile shown in Fig. 7.8(b), while Fig. 7.7 uses the parabolic velocity profile of a Newtonian fluid, Fig. 4.17. In the flow portion of this calculation, material that rolls onto the wall via the fountain flow soon freezes there, which stops the fiber length from changing. This preserves the thin yellow layers near the upper and low mold walls in the figure.

Figure 7.9 provides a more detailed look at the experimental and predicted fiber length distributions. The experimental data in Fig. 7.9(a), $r = 15\,\text{mm}$, was used as the inlet boundary condition for the fiber length distribution, so there is no prediction at this point. Figures 7.9(b) and (c) show both the experimental and predicted length distributions at the two downstream measurement points. In these figures, translucent blue bars for the predicted distributions have been overlaid on the experimental red bars, so overlap between the two appears as purple. Red and blue bars extending above the purple are areas of disagreement. All of these distributions are averaged through the part thickness.

The predicted distributions in Figs. 7.9(b) and (c) started from the initial condition of Fig. 7.9(a), so the noise in that figure, especially at the higher fiber lengths, carries over into the predictions. The match between the predicted and experimental distributions is good, though the experimental distribution has more fibers around 1 and 3 mm long, and the predictions have more fibers around 2 mm long. These errors offset one another in the calculation of $\overline{L}_n$ and $\overline{L}_w$. These differences in the length distributions would cause only a tiny change in the predicted mechanical properties of the composite.

7.4 Chapter Summary

Fibers can break during processing, and this is a concern for long fiber injection molding materials. The mechanism of breakage is buckling, when fluid drag forces place fibers in compression with enough force that fibers bend out of line, buckle and break.

A fiber length model can predict the fiber length distribution at each point in a molding. The model combines a conservation equation for fiber length, an expression for the rate at which parent fibers break, and a distribution for the lengths of the resulting child fibers.

The probability that a fiber will break depends on the its length, and the model predicts an unbreakable length L_{ub}, below which fibers will not break. L_{ub} increases as the flow stress $\eta_m \dot{\gamma}$ decreases, so lower stresses help preserve the fiber length during processing. This dependence is weak, $L_{\text{ub}} \propto (\eta_m \dot{\gamma})^{-1/4}$, so large reductions in stress are needed to achieve substantial increases in final fiber length.

The fiber length model has three parameters: a breakage coefficient C_B that controls the overall breakage rate, a drag coefficient ζ that affects L_{ub}, and a parameter S that affects the distribution of child fiber lengths. These are phenomenological parameters that must be selected to match experimental data. The values from Section 7.3.2 are a good starting point.

For injection molding in simple geometries, the model predicts better preservation of fiber length near the midplane, where the flow stress and shear rate are low, and shorter fiber lengths near the walls, where the flow stress and shear rate are high. There can also be a layer of longer fibers close to the mold walls, created by the fountain flow at the flow front.

Predictions of the model can match experimental data well. Comparisons with experiments primarily use thickness-averaged length distributions and average lengths, because it is difficult to measure changes in the fiber length distribution across the thickness of an injection molded part.

8 Mechanical Properties and Orientation

The reason for predicting fiber orientation and fiber length to begin with was to predict mechanical properties in the solid part. This chapter shows how fiber orientation and fiber length are related to mechanical properties.

We focus on linear elastic behavior, starting with a review of the elasticity of anisotropic materials. Here we define the stiffness tensor, and show how it is related to engineering properties such as Young's moduli and Poisson ratios in different directions. Different types of material symmetry are also explored.

We then discuss the concept of homogenization, where a heterogeneous material at the microscale (i.e., fibers in a matrix) is modeled as a homogeneous material with equivalent properties. For discontinuous fiber composites a two-step homogenization process is used. The first step is to find the properties of a composite with unidirectional fibers of uniform length, and the second step averages over the fiber orientation distribution and, if desired, the fiber length distribution. Options for each of these steps are discussed and tested. We also show how to treat the variation of orientation across the thickness of an injection molded part, which makes the flexural stiffness different from the tensile stiffness.

Thermal expansion, which is important to warping and dimensional stability, can be predicted once the composite stiffness is known, and we show how to do that. The elastic-plastic behavior of ductile matrix composites can also be modeled, but a detailed discussion is outside the scope of this book. This chapter does lay the foundation for modeling nonlinear properties, and references are provided for readers who want to explore this in more depth.

8.1 Anisotropic Elasticity

8.1.1 Stress and Strain

For structural analysis we need to relate the stresses in a material to its deformation.

For stress we use the total stress tensor $\boldsymbol{\sigma}$ from Section 6.1.1. While a solid composite changes volume very little when loaded, those changes are comparable in magnitude to the shearing and stretching deformations, so the extra stress tensor $\boldsymbol{\tau}$ is not used for elastic behavior.

To describe deformation, start with $\mathbf{u}$, the *displacement vector*, with $\mathbf{u}(\mathbf{x})$ representing the difference between the position of a material point in the deformed and undeformed configurations. In the notation of Section 4.1.2,

$$\mathbf{u} \equiv \mathbf{x} - \mathbf{x}' \tag{8.1}$$

If **u** does not vary with position **x**, then the material has translated without deforming, so deformation must be related to the way **u** changes with **x**. This suggests using derivatives like $\partial u_i/\partial x_j$. The full set of derivatives has the same form as the velocity gradient **L**, with **u** replacing **v**. Just as the symmetric part of **L** gave the rate-of-deformation tensor **D**, the symmetric part of $\partial u_i/\partial x_j$ is a measure of deformation. The resulting tensor $\boldsymbol{\varepsilon}$ is called the *small strain tensor*, or simply the strain tensor, and is defined such that

$$\varepsilon_{ij} = \frac{1}{2}\left(\frac{\partial u_i}{\partial x_j} + \frac{\partial u_j}{\partial x_i}\right) \tag{8.2}$$

The strain tensor is symmetric, $\varepsilon_{ij} = \varepsilon_{ji}$. Strains are small if each component of $\boldsymbol{\varepsilon}$ is $\ll 1$. Typically this means strains of a few percent or less.

The diagonal components of $\boldsymbol{\varepsilon}$ are normal strains, and for small strains they match the usual definition that normal strain equals change in length divided by original length.

The off-diagonal components of $\boldsymbol{\varepsilon}$ are shear strains. In many texts, shear strain is defined as the change in angle between two initially perpendicular material lines. For small deformations this leads to the following measures of shear strain:

$$\gamma_{12} = \frac{\partial u_1}{\partial x_2} + \frac{\partial u_2}{\partial x_1} \qquad \gamma_{23} = \frac{\partial u_2}{\partial x_3} + \frac{\partial u_3}{\partial x_2} \qquad \gamma_{31} = \frac{\partial u_3}{\partial x_1} + \frac{\partial u_1}{\partial x_3} \tag{8.3}$$

These differ from the off-diagonal components of $\boldsymbol{\varepsilon}$ by a factor of two. The shear strains ε_{12}, ε_{23} and ε_{31} are components of a tensor that has normal components ε_{11}, ε_{22} and ε_{33}, and they are sometimes called *tensor shear strains.* The γ_{ij}s are called *engineering shear strains.* In this book we use engineering shear strains only in defining the engineering constants. For all other purposes we use the small strain tensor $\boldsymbol{\varepsilon}$ and its components.

8.1.2 Stiffness and Compliance Tensors

A material behaves in a *linear elastic* manner if the components of stress σ_{ij} are linear functions of the strain components ε_{kl}. Most solids are linearly elastic for small deformations. The most general linear relationship between stress and strain is

$$\sigma_{ij} = \mathbb{C}_{ijkl}\,\varepsilon_{kl} \qquad \text{or} \qquad \boldsymbol{\sigma} = \mathbb{C} : \boldsymbol{\varepsilon} \tag{8.4}$$

The coefficients $\mathbb{C}_{ijkl}$ are components of a fourth-order tensor $\mathbb{C}$, the *stiffness tensor.* If we know the stiffness tensor of a material, we can predict its stress response to any strain.

As a fourth-order tensor, $\mathbb{C}$ has 81 components. However, not all of these components are independent. Because both $\boldsymbol{\sigma}$ and $\boldsymbol{\varepsilon}$ are symmetric, $\mathbb{C}$ must have the minor symmetries

$$\mathbb{C}_{ijkl} = \mathbb{C}_{jikl} \qquad \text{and} \qquad \mathbb{C}_{ijkl} = \mathbb{C}_{ijlk} \tag{8.5}$$

In addition, if the energy stored in the material by elastic deformation depends only on the strain and not on the loading path (and we always assume this), then the stiffness tensor also has the major symmetry

$$\mathbb{C}_{ijkl} = \mathbb{C}_{klij} \tag{8.6}$$

Together, Eqns. (8.5) and (8.6) are called the symmetries of an elasticity tensor. The fourth-order orientation tensor $\mathbb{A}$ has all of these symmetries, plus some additional ones.

We can invert Eqn. (8.4) and write strain as a function of stress,

$$\varepsilon_{ij} = \mathbb{S}_{ijkl}\,\sigma_{kl} \qquad \text{or} \qquad \boldsymbol{\varepsilon} = \mathbb{S} : \boldsymbol{\sigma} \tag{8.7}$$

$\mathbb{S}$ is the *compliance tensor*, and it contains the same information as $\mathbb{C}$.

The compliance tensor $\mathbb{S}$ is the inverse of the stiffness tensor $\mathbb{C}$. To write this relationship we need the symmetric fourth-order identity tensor $\mathbb{I}$, whose components are

$$\mathbb{I}_{ijkl} = \frac{1}{2}\left(\delta_{ik}\delta_{jl} + \delta_{il}\delta_{jk}\right) \tag{8.8}$$

This has the property that, for any symmetric second-order tensor like $\boldsymbol{\varepsilon}$,

$$\boldsymbol{\varepsilon} : \mathbb{I} = \mathbb{I} : \boldsymbol{\varepsilon} = \boldsymbol{\varepsilon} \tag{8.9}$$

and for any fourth-order tensor like $\mathbb{C}$ that has the two minor symmetries of Eqn. (8.5),

$$\mathbb{C} : \mathbb{I} = \mathbb{I} : \mathbb{C} = \mathbb{C} \tag{8.10}$$

In tensor notation, we can then write the relationship between $\mathbb{C}$ and $\mathbb{S}$ as

$$\mathbb{S} : \mathbb{C} = \mathbb{I} \qquad \text{or} \qquad \mathbb{S} = \mathbb{C}^{-1} \tag{8.11}$$

The compliance tensor also has the symmetries of an elasticity tensor, Eqns. (8.5) and (8.6).

8.1.3 Contracted Notation

The contracted notation introduced in Section 2.3.5.1 is useful for writing out the components of $\mathbb{C}$ and $\mathbb{S}$, and for performing calculations on stress-strain relationships like Eqn. (8.4).

The stress and strain tensors are contracted to 6×1 column vectors using the same rule as Eqn. (2.70),

$$\{\boldsymbol{\sigma}\} = \begin{Bmatrix} \sigma_{11} \\ \sigma_{22} \\ \sigma_{33} \\ \sigma_{23} \\ \sigma_{31} \\ \sigma_{12} \end{Bmatrix} = \begin{Bmatrix} \sigma_{1} \\ \sigma_{2} \\ \sigma_{3} \\ \sigma_{4} \\ \sigma_{5} \\ \sigma_{6} \end{Bmatrix} \qquad \text{and} \qquad \{\boldsymbol{\varepsilon}\} = \begin{Bmatrix} \varepsilon_{11} \\ \varepsilon_{22} \\ \varepsilon_{33} \\ \varepsilon_{23} \\ \varepsilon_{31} \\ \varepsilon_{12} \end{Bmatrix} = \begin{Bmatrix} \varepsilon_{1} \\ \varepsilon_{2} \\ \varepsilon_{3} \\ \varepsilon_{4} \\ \varepsilon_{5} \\ \varepsilon_{6} \end{Bmatrix} \tag{8.12}$$

The stiffness and compliance tensors are contracted using the rule of Eqn. (2.73). For stiffness this is

$$\begin{aligned}
[\mathbb{C}] &= \begin{bmatrix}
\mathbb{C}_{1111} & \mathbb{C}_{1122} & \mathbb{C}_{1133} & \mathbb{C}_{1123} & \mathbb{C}_{1131} & \mathbb{C}_{1112} \\
\mathbb{C}_{2211} & \mathbb{C}_{2222} & \mathbb{C}_{2233} & \mathbb{C}_{2223} & \mathbb{C}_{2231} & \mathbb{C}_{2212} \\
\mathbb{C}_{3311} & \mathbb{C}_{3322} & \mathbb{C}_{3333} & \mathbb{C}_{3323} & \mathbb{C}_{3331} & \mathbb{C}_{3312} \\
\mathbb{C}_{2311} & \mathbb{C}_{2322} & \mathbb{C}_{2333} & \mathbb{C}_{2323} & \mathbb{C}_{2331} & \mathbb{C}_{2312} \\
\mathbb{C}_{3111} & \mathbb{C}_{3122} & \mathbb{C}_{3133} & \mathbb{C}_{3123} & \mathbb{C}_{3131} & \mathbb{C}_{3112} \\
\mathbb{C}_{1211} & \mathbb{C}_{1222} & \mathbb{C}_{1233} & \mathbb{C}_{1223} & \mathbb{C}_{1231} & \mathbb{C}_{1212}
\end{bmatrix} \\
&= \begin{bmatrix}
\mathbb{C}_{11} & \mathbb{C}_{12} & \mathbb{C}_{13} & \mathbb{C}_{14} & \mathbb{C}_{15} & \mathbb{C}_{16} \\
\mathbb{C}_{21} & \mathbb{C}_{22} & \mathbb{C}_{23} & \mathbb{C}_{24} & \mathbb{C}_{25} & \mathbb{C}_{26} \\
\mathbb{C}_{31} & \mathbb{C}_{32} & \mathbb{C}_{33} & \mathbb{C}_{34} & \mathbb{C}_{35} & \mathbb{C}_{36} \\
\mathbb{C}_{41} & \mathbb{C}_{42} & \mathbb{C}_{43} & \mathbb{C}_{44} & \mathbb{C}_{45} & \mathbb{C}_{46} \\
\mathbb{C}_{51} & \mathbb{C}_{52} & \mathbb{C}_{53} & \mathbb{C}_{54} & \mathbb{C}_{55} & \mathbb{C}_{56} \\
\mathbb{C}_{61} & \mathbb{C}_{62} & \mathbb{C}_{63} & \mathbb{C}_{64} & \mathbb{C}_{65} & \mathbb{C}_{66}
\end{bmatrix}
\end{aligned} \tag{8.13}$$

The corresponding equation for compliance uses $\mathbb{S}$ in place of $\mathbb{C}$ in every location.

Our notation is that $\boldsymbol{\sigma}$, $\boldsymbol{\varepsilon}$, $\mathbb{C}$ and $\mathbb{S}$ are the full tensors, with tensor components σ_{ij}, ε_{ij}, $\mathbb{C}_{ijkl}$ and $\mathbb{S}_{ijkl}$. $\{\boldsymbol{\sigma}\}$ and $\{\boldsymbol{\varepsilon}\}$ are 6×1 column vectors with vector components σ_m and ε_m, and $[\mathbb{C}]$ and $[\mathbb{S}]$ are 6×6 matrices of the contracted stiffness and compliance tensors, with matrix components $\mathbb{C}_{mn}$ and $\mathbb{S}_{mn}$.

The contracted form of the fourth-order identity tensor $\mathbb{I}$ is *not* the 6×6 identity matrix. Instead, working out the components from Eqn. (8.8),

$$[\mathbb{I}] \equiv \begin{bmatrix} 1 & 0 & 0 & 0 & 0 & 0 \\ 0 & 1 & 0 & 0 & 0 & 0 \\ 0 & 0 & 1 & 0 & 0 & 0 \\ 0 & 0 & 0 & \frac{1}{2} & 0 & 0 \\ 0 & 0 & 0 & 0 & \frac{1}{2} & 0 \\ 0 & 0 & 0 & 0 & 0 & \frac{1}{2} \end{bmatrix} \tag{8.14}$$

It would be convenient to write Eqn. (8.4) in contracted form as $\{\boldsymbol{\sigma}\} = [\mathbb{C}]\{\boldsymbol{\varepsilon}\}$, but this would be incorrect because some of the 36 components in $[\mathbb{C}]$ are representing both themselves and others from the 81 tensor components of $\mathbb{C}_{ijkl}$. We can account for these "missing" components by defining a matrix $[\mathbb{R}]$ [Reu71, NF98],

$$[\mathbb{R}] \equiv \begin{bmatrix} 1 & 0 & 0 & 0 & 0 & 0 \\ 0 & 1 & 0 & 0 & 0 & 0 \\ 0 & 0 & 1 & 0 & 0 & 0 \\ 0 & 0 & 0 & 2 & 0 & 0 \\ 0 & 0 & 0 & 0 & 2 & 0 \\ 0 & 0 & 0 & 0 & 0 & 2 \end{bmatrix} . \tag{8.15}$$

Using this, the contracted form of the anisotropic stiffness equation (8.4) is

$$\{\boldsymbol{\sigma}\} = [\mathbb{C}]\,[\mathbb{R}]\,\{\boldsymbol{\varepsilon}\} \tag{8.16}$$

This allows us to perform tensor operations like $\mathbb{C} : \boldsymbol{\varepsilon}$ as matrix multiplications.

For the contracted notation used in this book, two other common operations are the double-dot product of two fourth-order tensors,

$$\mathbb{T} = \mathbb{U} : \mathbb{V} \qquad \text{contracts to} \qquad [\mathbb{T}] = [\mathbb{U}]\,[\mathbb{R}]\,[\mathbb{V}] \tag{8.17}$$

and the inverse of a fourth-order tensor,

$$\mathbb{C} = \mathbb{S}^{-1} \qquad \text{contracts to} \qquad [\mathbb{C}] = [\mathbb{I}]\,[\mathbb{S}]^{-1}\,[\mathbb{I}] \tag{8.18}$$

Our notation is that $\mathbb{S}^{-1}$ is the fourth-order tensor inverse of $\mathbb{S}$, while $[\mathbb{S}]^{-1}$ is the matrix inverse of $[\mathbb{S}]$. See Appendix A.4 for more details.

8.1.4 Material Symmetries

The matrix [$\mathbb{C}$] has 36 components. However, the major symmetry of Eqn. (8.6) show that this matrix is symmetric, $\mathbb{C}_{mn} = \mathbb{C}_{nm}$. This reduces the number of independent components to 21. If we know nothing else about a material, 21 stiffness components are needed to fully describe its stress-strain behavior. Such a material is said to be *generally anisotropic* or triclinic.

Many anisotropic substances have inherent symmetries in their structure, and the stiffness tensor will have the same symmetries. This allows us to simplify [$\mathbb{C}$] or [$\mathbb{S}$] and reduce the number of properties needed to characterize a material.

The first symmetry of interest is an **orthotropic** material. This has three mutually perpendicular planes of material symmetry. That is, a sample of the material could subjected to a test (say, a shear test), then flipped about a symmetry plane and re-tested, and the result would be the same. A composite reinforced with a plain-woven fabric is a good example. The material may have different properties in the warp and weft directions of the fabric, and will certainly have different properties in the thickness direction, but rotating the material 180° about either the warp or weft directions does not change the properties.

For an orthotropic material, many components of the stiffness tensor are zero. If the 1, 2 and 3 coordinate axes are normal to the symmetry planes, [$\mathbb{C}$] reduces to

$$[\mathbb{C}] = \begin{bmatrix} \mathbb{C}_{11} & \mathbb{C}_{12} & \mathbb{C}_{13} & 0 & 0 & 0 \\ \mathbb{C}_{12} & \mathbb{C}_{22} & \mathbb{C}_{23} & 0 & 0 & 0 \\ \mathbb{C}_{13} & \mathbb{C}_{23} & \mathbb{C}_{33} & 0 & 0 & 0 \\ 0 & 0 & 0 & \mathbb{C}_{44} & 0 & 0 \\ 0 & 0 & 0 & 0 & \mathbb{C}_{55} & 0 \\ 0 & 0 & 0 & 0 & 0 & \mathbb{C}_{66} \end{bmatrix} \tag{8.19}$$

The contracted compliance tensor [$\mathbb{S}$] has the same pattern of zero and non-zero components. An orthotropic material has nine independent elastic constants.

A second possibility is that the material has an axis of symmetry, with properties that are the same for any direction in the plane normal to this axis. The material is then said to be **transversely isotropic** about this axis. A continuous fiber composite with unidirectional fibers is transversely isotropic. This imposes additional relationships among the components of [$\mathbb{C}$]. If the 1 coordinate axis is aligned with the symmetry axis, then

$$[\mathbb{C}] = \begin{bmatrix} \mathbb{C}_{11} & \mathbb{C}_{12} & \mathbb{C}_{12} & 0 & 0 & 0 \\ \mathbb{C}_{12} & \mathbb{C}_{22} & \mathbb{C}_{23} & 0 & 0 & 0 \\ \mathbb{C}_{12} & \mathbb{C}_{23} & \mathbb{C}_{22} & 0 & 0 & 0 \\ 0 & 0 & 0 & (\mathbb{C}_{22} - \mathbb{C}_{23})/2 & 0 & 0 \\ 0 & 0 & 0 & 0 & \mathbb{C}_{66} & 0 \\ 0 & 0 & 0 & 0 & 0 & \mathbb{C}_{66} \end{bmatrix} \tag{8.20}$$

and [$\mathbb{S}$] follows the same pattern. A transversely isotropic material has 5 independent elastic constants.

Equations (8.19) and (8.20) apply only when the coordinate system is aligned with the material symmetry axes. (For a transversely isotropic material it is only necessary for the x_1 axis to be aligned with the symmetry axis.) We can transform $\mathbb{C}$ or $\mathbb{S}$ to other coordinate systems using the fourth-order tensor transformation rule as shown in Appendix A.3. However, in an arbitrary coordinate system, any of the components of $[\mathbb{C}]$ or $[\mathbb{S}]$ can be non-zero, even if the material is orthotropic or transversely isotropic.

A material whose properties are the same in every direction is **isotropic**. Ordinary metals, polymers and ceramics are usually treated as isotropic. Now the material has only two independent elastic constants, and in *any* coordinate system the stiffness matrix is

$$[\mathbb{C}] = \begin{bmatrix} \mathbb{C}_{11} & \mathbb{C}_{12} & \mathbb{C}_{12} & 0 & 0 & 0 \\ \mathbb{C}_{12} & \mathbb{C}_{11} & \mathbb{C}_{12} & 0 & 0 & 0 \\ \mathbb{C}_{12} & \mathbb{C}_{12} & \mathbb{C}_{11} & 0 & 0 & 0 \\ 0 & 0 & 0 & (\mathbb{C}_{11}-\mathbb{C}_{12})/2 & 0 & 0 \\ 0 & 0 & 0 & 0 & (\mathbb{C}_{11}-\mathbb{C}_{12})/2 & 0 \\ 0 & 0 & 0 & 0 & 0 & (\mathbb{C}_{11}-\mathbb{C}_{12})/2 \end{bmatrix} \quad (8.21)$$

8.1.5 Engineering or Technical Constants

The behavior of an isotropic material is usually described using properties like Young's modulus E, the Poisson ratio ν and the shear modulus G. The corresponding properties for an anisotropic material are called *engineering constants* or *technical constants*. We want to define these and relate them to the components of $\mathbb{C}$.

Each engineering constant is defined in terms of a mechanical test where a stress is applied and one or more strains are measured. In a tensile test, a single stress component σ_{11} is applied, with all other stress components zero, and the axial strain ε_{11} and one transverse strain ε_{22} are measured. Young's modulus is then determined as $E = \sigma_{11}/\varepsilon_{11}$ and the Poisson ratio is $\nu = -\varepsilon_{22}/\varepsilon_{11}$.

If the material is anisotropic then the Young's moduli and Poisson ratios will vary with the direction of the test. For an applied stress σ_{11}, the engineering constants of an anisotropic material are

$$E_{11} = \frac{\sigma_{11}}{\varepsilon_{11}} \qquad \nu_{12} = -\frac{\varepsilon_{22}}{\varepsilon_{11}} \qquad \nu_{13} = -\frac{\varepsilon_{33}}{\varepsilon_{11}} \quad (8.22)$$

In this notation E_{11} is the Young's modulus for an applied stress of σ_{11}, and ν_{12} is the Poisson ratio for a stress in the 1 direction and contraction in the 2 direction. A tensile test in the 2 direction would measure E_{22}, ν_{21} and ν_{23}, and a tensile test in the 3 direction would measure E_{33}, ν_{31} and ν_{32}. In general $\nu_{12} \neq \nu_{21}$.

To determine the shear modulus G for an isotropic material, a shear stress σ_{12} is applied and the shear strain is measured. The shear modulus is then $G = \sigma_{12}/\gamma_{12} = \sigma_{12}/(2\varepsilon_{12})$. For anisotropic materials, a shear experiment where σ_{12} is applied gives the shear modulus G_{12},

$$G_{12} = \frac{\sigma_{12}}{\gamma_{12}} = \frac{\sigma_{12}}{2\varepsilon_{12}} \quad (8.23)$$

Similarly, G_{23} is the shear modulus for σ_{23} loading and G_{31} is the shear modulus for σ_{31} loading. Because $\sigma_{ij} = \sigma_{ji}$ and $\varepsilon_{ij} = \varepsilon_{ji}$, there is no difference between G_{12} and G_{21}.

Many books on the mechanics of composite materials combine $[\mathbb{R}]$ and $\{\boldsymbol{\varepsilon}\}$ to form a different vector of strain components $\{\tilde{\boldsymbol{\varepsilon}}\}$ that we will call the engineering strain vector,

$$\{\tilde{\boldsymbol{\varepsilon}}\} \equiv [\mathbb{R}]\{\boldsymbol{\varepsilon}\} = \begin{Bmatrix} \varepsilon_{11} \\ \varepsilon_{22} \\ \varepsilon_{33} \\ 2\varepsilon_{23} \\ 2\varepsilon_{31} \\ 2\varepsilon_{12} \end{Bmatrix} = \begin{Bmatrix} \varepsilon_{11} \\ \varepsilon_{22} \\ \varepsilon_{33} \\ \gamma_{23} \\ \gamma_{31} \\ \gamma_{12} \end{Bmatrix} \tag{8.24}$$

$\{\tilde{\boldsymbol{\varepsilon}}\}$ uses the engineering shear strains for the shear components but retains the tensor shear strains for the normal components. This allows Eqn. (8.16) to be written as

$$\{\boldsymbol{\sigma}\} = [\mathbb{C}]\{\tilde{\boldsymbol{\varepsilon}}\} \tag{8.25}$$

The matrix inverse of this equation is

$$\{\tilde{\boldsymbol{\varepsilon}}\} = [\mathbb{C}]^{-1}\{\boldsymbol{\sigma}\} \tag{8.26}$$

The matrix inverse of $[\mathbb{C}]$ is, in many texts, called the compliance matrix or the compliance tensor. However, as Eqn. (8.18) shows, $[\mathbb{C}]^{-1}$ is *not* the contraction of $\mathbb{S}$ according to our rules for contracting fourth-order tensors. In this book we contract all tensors using the same rule, so we will simply call this matrix $[\mathbb{C}]^{-1}$.

Equation (8.26) is often used when defining the engineering constants, and we follow that pattern here. For each equation below, the expression in square brackets is $[\mathbb{C}]^{-1}$. One can translate between the stiffness tensor and the engineering constants by forming this matrix and taking its matrix inverse to get $[\mathbb{C}]$, or by taking the matrix inverse of a known $[\mathbb{C}]$ and using these equations to interpret the results in terms of engineering constants.

The engineering constants for an **orthotropic** material are related to the components of $[\mathbb{C}]^{-1}$ according to:

$$\begin{Bmatrix} \varepsilon_{11} \\ \varepsilon_{22} \\ \varepsilon_{33} \\ \gamma_{23} \\ \gamma_{31} \\ \gamma_{12} \end{Bmatrix} = \begin{bmatrix} \frac{1}{E_{11}} & -\frac{\nu_{21}}{E_{22}} & -\frac{\nu_{31}}{E_{33}} & 0 & 0 & 0 \\ -\frac{\nu_{12}}{E_{11}} & \frac{1}{E_{22}} & -\frac{\nu_{32}}{E_{33}} & 0 & 0 & 0 \\ -\frac{\nu_{13}}{E_{11}} & -\frac{\nu_{23}}{E_{22}} & \frac{1}{E_{33}} & 0 & 0 & 0 \\ 0 & 0 & 0 & \frac{1}{G_{23}} & 0 & 0 \\ 0 & 0 & 0 & 0 & \frac{1}{G_{31}} & 0 \\ 0 & 0 & 0 & 0 & 0 & \frac{1}{G_{12}} \end{bmatrix} \begin{Bmatrix} \sigma_{11} \\ \sigma_{22} \\ \sigma_{33} \\ \sigma_{23} \\ \sigma_{31} \\ \sigma_{12} \end{Bmatrix} \tag{8.27}$$

It appears that there are twelve engineering constants: three Young's moduli, six Poisson ratios and three shear moduli. However, the fact that $[\mathbb{C}]$and $[\mathbb{C}]^{-1}$ are symmetric means that only three of the Poisson ratios are independent. Using that symmetry, the Poisson ratios must satisfy

$$\frac{\nu_{12}}{E_{11}} = \frac{\nu_{21}}{E_{22}} \qquad \frac{\nu_{23}}{E_{22}} = \frac{\nu_{32}}{E_{33}} \qquad \frac{\nu_{31}}{E_{33}} = \frac{\nu_{13}}{E_{11}} \tag{8.28}$$

With these three relationships, an orthotropic material has nine independent engineering constants that give the nine independent components of $[\mathbb{C}]$ in Eqn. (8.19).

In the accompanying software[1], the function C2eng converts a stiffness tensor $\mathbb{C}$ to engineering constants based on Eqns. (8.27) and (8.28), while eng2C forms the orthotropic stiffness tensor $\mathbb{C}$ for a given set of engineering constants.

For a **transversely isotropic** material the properties in two of the principal directions are equal. Using the 1 direction as the symmetry axis and the 2–3 plane as the symmetry plane, a transversely isotropic material has the properties

$$\begin{Bmatrix} \varepsilon_{11} \\ \varepsilon_{22} \\ \varepsilon_{33} \\ \gamma_{23} \\ \gamma_{31} \\ \gamma_{12} \end{Bmatrix} = \begin{bmatrix} \frac{1}{E_{11}} & -\frac{\nu_{21}}{E_{22}} & -\frac{\nu_{21}}{E_{22}} & 0 & 0 & 0 \\ -\frac{\nu_{12}}{E_{11}} & \frac{1}{E_{22}} & -\frac{\nu_{23}}{E_{22}} & 0 & 0 & 0 \\ -\frac{\nu_{12}}{E_{11}} & -\frac{\nu_{23}}{E_{22}} & \frac{1}{E_{22}} & 0 & 0 & 0 \\ 0 & 0 & 0 & \frac{1}{G_{23}} & 0 & 0 \\ 0 & 0 & 0 & 0 & \frac{1}{G_{12}} & 0 \\ 0 & 0 & 0 & 0 & 0 & \frac{1}{G_{12}} \end{bmatrix} \begin{Bmatrix} \sigma_{11} \\ \sigma_{22} \\ \sigma_{33} \\ \sigma_{23} \\ \sigma_{31} \\ \sigma_{12} \end{Bmatrix} \tag{8.29}$$

In addition, because any direction in the 2–3 plane is equivalent to any other, the shear modulus in that plane is related to the transverse Young's modulus and Poisson ratio by

$$\frac{1}{G_{23}} = \frac{2(1+\nu_{23})}{E_{22}} \tag{8.30}$$

Equation (8.28) still relates the Poisson ratios, so five engineering constants suffice to describe the elastic properties of a transversely isotropic material. A convenient set is E_{11}, E_{22}, ν_{12}, ν_{23} and G_{12}.

Equations (8.27) and (8.29) apply when the coordinate axes are aligned with the material symmetry axes.

If a material is **isotropic** then the properties are equal in all directions. There is only one Young's modulus E, one Poisson ratio ν, one shear modulus G, and these are related to $[\mathbb{C}]^{-1}$ by

$$\begin{Bmatrix} \varepsilon_{11} \\ \varepsilon_{22} \\ \varepsilon_{33} \\ \gamma_{23} \\ \gamma_{31} \\ \gamma_{12} \end{Bmatrix} = \begin{bmatrix} \frac{1}{E} & -\frac{\nu}{E} & -\frac{\nu}{E} & 0 & 0 & 0 \\ -\frac{\nu}{E} & \frac{1}{E} & -\frac{\nu}{E} & 0 & 0 & 0 \\ -\frac{\nu}{E} & -\frac{\nu}{E} & \frac{1}{E} & 0 & 0 & 0 \\ 0 & 0 & 0 & \frac{1}{G} & 0 & 0 \\ 0 & 0 & 0 & 0 & \frac{1}{G} & 0 \\ 0 & 0 & 0 & 0 & 0 & \frac{1}{G} \end{bmatrix} \begin{Bmatrix} \sigma_{11} \\ \sigma_{22} \\ \sigma_{33} \\ \sigma_{23} \\ \sigma_{31} \\ \sigma_{12} \end{Bmatrix} \tag{8.31}$$

This equation applies in any coordinate system. In addition, the shear modulus is related to E and ν by

$$G = \frac{E}{2(1+\nu)} \tag{8.32}$$

so an isotropic material has only two independent elastic constants. The function iso2C in the accompanying software forms the stiffness tensor $\mathbb{C}$ for an isotropic material using E and ν.

[1] *http://github.com/charlestucker3/Fiber-Orientation-Tools*

We can now transform between engineering constants and the stiffness tensor for different types of anisotropic materials. In what follows we will use tensor notation in equations involving $\mathbb{C}$ and $\mathbb{S}$, while comparisons to experiments will use engineering constants. Numerical calculations are always carried out in contracted notation using matrix operations.

8.2 Homogenization

While the properties of a composite material vary with position on the length scale of the fibers, we routinely measure properties at the scale of a tensile test sample and use them in structural analysis. *Homogenization* is the process of modeling these large-scale, effective properties using small-scale information such as fiber and matrix properties, fiber volume fraction, fiber orientation and fiber length. Homogenization methods are also called *micromechanics models*.

8.2.1 General Principles

There are many approaches to homogenization, but most begin with the idea of a *representative volume element* or RVE. An RVE is a small volume that contains fibers and matrix in the same proportions as the composite, arranged in a way that is representative of the overall material.

Figure 8.1 shows several possible RVEs for a composite with unidirectional fibers of uniform length. When RVEs are constructed explicitly, they can either use a periodic arrangement of fibers like Figs. 8.1(a) and (b), or random arrangements like Figs. 8.1(c) and (d). RVEs can also include other microstructural features, such as a distribution of fiber orientation or fiber length.

Most mechanical property predictions use mean-field theories, in which case the RVE is a conceptual tool and is never specified explicitly. Alternately, one can construct three-dimensional RVEs and analyze them numerically, for instance using finite element stress analysis, as discussed in Section 8.3.4.

The concept of an RVE allows us to say precisely what is meant by the average properties of a composite. To do this, let the volume of the RVE be V, with V_f being the portion of that volume occupied by the fibers and V_m the portion occupied by the matrix. The fiber volume fraction is $\phi_f = V_f / V$ and the matrix volume fraction is $\phi_m = V_m / V$. We assume for now that there are no other phases, so $\phi_m + \phi_f = 1$.

To make predictions we must also know the fiber and matrix properties. For linear elastic analysis these are described by the stiffness tensors for each phase. Using superscripts f and m to denote the fiber and matrix respectively,

$$\boldsymbol{\sigma}^f = \mathbb{C}^f : \boldsymbol{\varepsilon}^f \quad \text{and} \quad \boldsymbol{\sigma}^m = \mathbb{C}^m : \boldsymbol{\varepsilon}^m \tag{8.33}$$

Now imagine that we deform the RVE by applying some displacements to the boundary. This will induce stresses and strains in the RVE which, because the RVE is non-homogeneous, will be complicated functions of position $\mathbf{x}$. It is challenging to solve for the stress distribution $\boldsymbol{\sigma}(\mathbf{x})$

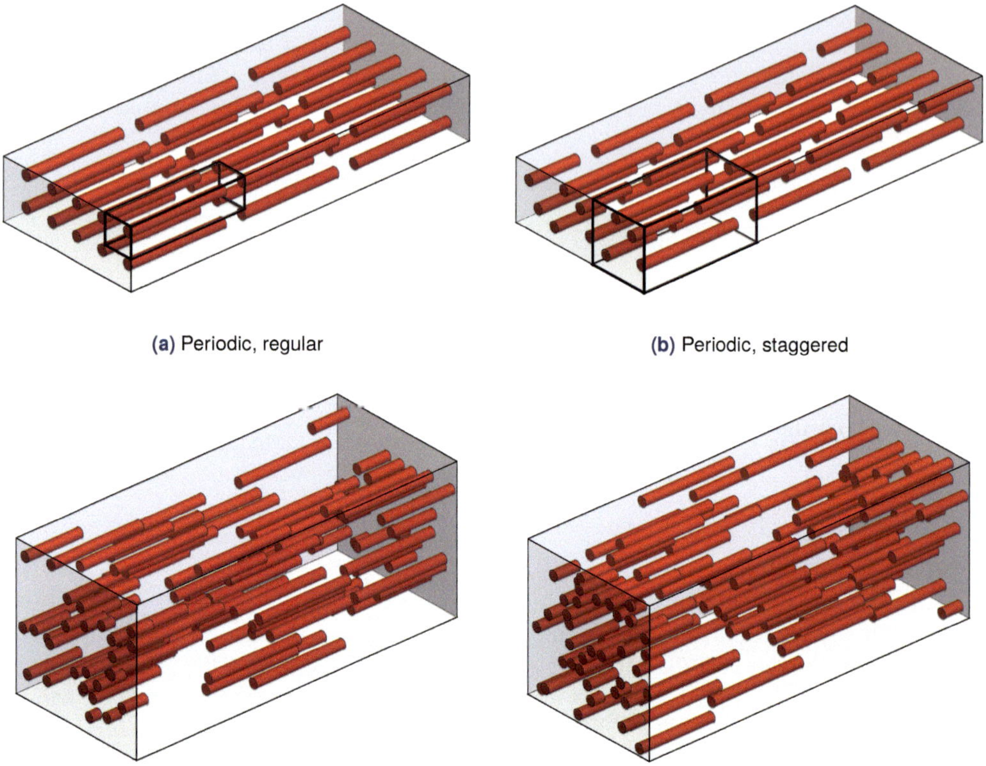

(a) Periodic, regular

(b) Periodic, staggered

(c) Random

(d) Random

Figure 8.1 Representative volume elements for a composite with unidirectional fibers of uniform length. The heavy lines in (a) and (b) indicate unit cells. The RVEs in (c) and (d) are periodic in the fiber direction, so fibers that exit the box on one face re-enter the box on the opposite face.

and the strain distribution $\boldsymbol{\varepsilon}(\mathbf{x})$ at the micro-scale in an RVE. Fortunately, for many purposes we can use average values.

Define the volume-average stress and volume-average strain in the RVE as

$$\overline{\boldsymbol{\sigma}} = \frac{1}{V}\int_V \boldsymbol{\sigma}(\mathbf{x})\,\mathrm{d}V \qquad \text{and} \qquad \overline{\boldsymbol{\varepsilon}} = \frac{1}{V}\int_V \boldsymbol{\varepsilon}(\mathbf{x})\,\mathrm{d}V \tag{8.34}$$

The effective (homogenized) stiffness of the RVE is the stiffness tensor $\mathbb{C}$ that relates these two quantities,

$$\overline{\boldsymbol{\sigma}} = \mathbb{C} : \overline{\boldsymbol{\varepsilon}} \tag{8.35}$$

This is the formal definition for the homogenized stiffness of a composite material.

From this point forward we use $\mathbb{C}$ with no superscript to denote the homogenized stiffness of the composite. Other stiffness tensors like $\mathbb{C}^f$ and $\mathbb{C}^m$ will be distinguished by superscripts.

In developing solutions for $\mathbb{C}$ it is helpful to define the average stress and strain in the fibers and in the matrix:

$$\overline{\boldsymbol{\sigma}}^f = \frac{1}{V_f}\int_{V_f} \boldsymbol{\sigma}^f(\mathbf{x})\,\mathrm{d}V_f \qquad \overline{\boldsymbol{\varepsilon}}^f = \frac{1}{V_f}\int_{V_f} \boldsymbol{\varepsilon}^f(\mathbf{x})\,\mathrm{d}V_f \tag{8.36}$$

$$\overline{\boldsymbol{\sigma}}^m = \frac{1}{V_m}\int_{V_m} \boldsymbol{\sigma}^m(\mathbf{x})\,\mathrm{d}V_m \qquad \overline{\boldsymbol{\varepsilon}}^m = \frac{1}{V_m}\int_{V_m} \boldsymbol{\varepsilon}^m(\mathbf{x})\,\mathrm{d}V_m \tag{8.37}$$

Because $V_f + V_m = V$ and the fiber and matrix volume fractions are $\phi_f = V_f/V$ and $\phi_m = V_m/V$, these are connected to the overall averages by

$$\overline{\boldsymbol{\sigma}} = \phi_f\overline{\boldsymbol{\sigma}}^f + \phi_m\overline{\boldsymbol{\sigma}}^m \tag{8.38}$$

$$\overline{\boldsymbol{\varepsilon}} = \phi_f\overline{\boldsymbol{\varepsilon}}^f + \phi_m\overline{\boldsymbol{\varepsilon}}^m \tag{8.39}$$

Also, because the stress-strain relations (8.33) are linear, the average fiber and matrix stresses and strains obey the relationships

$$\overline{\boldsymbol{\sigma}}^f = \mathbb{C}^f : \overline{\boldsymbol{\varepsilon}}^f \qquad \text{and} \qquad \overline{\boldsymbol{\sigma}}^m = \mathbb{C}^m : \overline{\boldsymbol{\varepsilon}}^m \tag{8.40}$$

We now define a *strain concentration tensor* that relates the average strain in the fibers to the average strain in the matrix:

$$\overline{\boldsymbol{\varepsilon}}^f = \mathbb{B} : \overline{\boldsymbol{\varepsilon}}^m \tag{8.41}$$

$\mathbb{B}$ is a fourth-order tensor that has the same minor symmetries as $\mathbb{C}$, $\mathbb{B}_{ijkl} = \mathbb{B}_{jikl} = \mathbb{B}_{ijlk}$, but does not have the major symmetry, so in general $\mathbb{B}_{ijkl} \neq \mathbb{B}_{klij}$.

As shown in Appendix B.4.3, combining Eqns. (8.33–8.35) and (8.38–8.41) gives a basic equation for the stiffness of the RVE,

$$\mathbb{C} = \left[\phi_m\mathbb{C}^m + \phi_f\mathbb{C}^f : \mathbb{B}\right] : \left[\phi_m\mathbb{I} + \phi_f\mathbb{B}\right]^{-1} \tag{8.42}$$

If we can develop an expression for $\mathbb{B}$, this equation can be used to find $\mathbb{C}$. The engineering constants can then be determined using the results from Section 8.1.5.

As a simple example, if we assume that the fiber and matrix have the same average strain, $\overline{\boldsymbol{\varepsilon}}^f = \overline{\boldsymbol{\varepsilon}}^m = \overline{\boldsymbol{\varepsilon}}$, then $\mathbb{B} = \mathbb{I}$ and Eqn. (8.42) gives the composite stiffness as

$$\mathbb{C} = \phi_m\mathbb{C}^m + \phi_f\mathbb{C}^f \tag{8.43}$$

This formula is known as the *Voigt average*, and it gives an upper bound on $\mathbb{C}$. By upper bound we mean that some arrangement of the fiber and matrix phases within an RVE could give a stiffness tensor as large as Eqn. (8.43), but no arrangement can give a higher stiffness.

Another simple option is to assume that the fiber and matrix have the same average stress, $\overline{\boldsymbol{\sigma}}^f = \overline{\boldsymbol{\sigma}}^m = \overline{\boldsymbol{\sigma}}$. In this case the compliance tensor for the composite has a simple form,

$$\mathbb{S} = \phi_f\mathbb{S}^f + \phi_m\mathbb{S}^m \tag{8.44}$$

This is known as the *Reuss average*, and it gives a lower bound on $\mathbb{C} = \mathbb{S}^{-1}$. This corresponds to Eqn. (8.42) with $\mathbb{B} = \mathbb{S}^f : \mathbb{C}^m$.

For most composites the Voigt and Reuss bounds are quite far from one another, and they do not use any information about the shapes or orientations of the fibers. We will need some additional analysis to find practical estimates for $\mathbb{C}$.

8.2.2 Two-Step Homogenization

With discontinuous fiber composites, it is common to use two homogenization steps: a first step to account for fiber and matrix properties with unidirectional fibers of uniform length, and a second step to account for fiber orientation and fiber length distribution.

For simplicity, consider a composite where all the fibers have a uniform length. Figure 8.2 shows an RVE for this composite and the steps in the homogenization process:

Decomposition: Partition the RVE into *pseudo-grains*. Each pseudo-grain has the same ϕ_f and ϕ_m as the overall composite, and all the fibers in each pseudo-grain have the same orientation **p**. The relative sizes of the pseudo-grains are determined by the fiber orientation distribution function.

First step: Homogenize each pseudo-grain, finding the stiffness for each grain.

Second step: Homogenize the pseudo-grains to find the final composite stiffness.

If the composite has a distribution of fiber lengths, the pseudo-grains are constructed so that each one has uniform fiber length as well as unidirectional orientation. The pseudo-grain stiffness is then a function of both orientation **p** and length ℓ, and the second step homogenizes over both **p** and ℓ.

The next section discusses the first homogenization step and Section 8.4 discusses the second step.

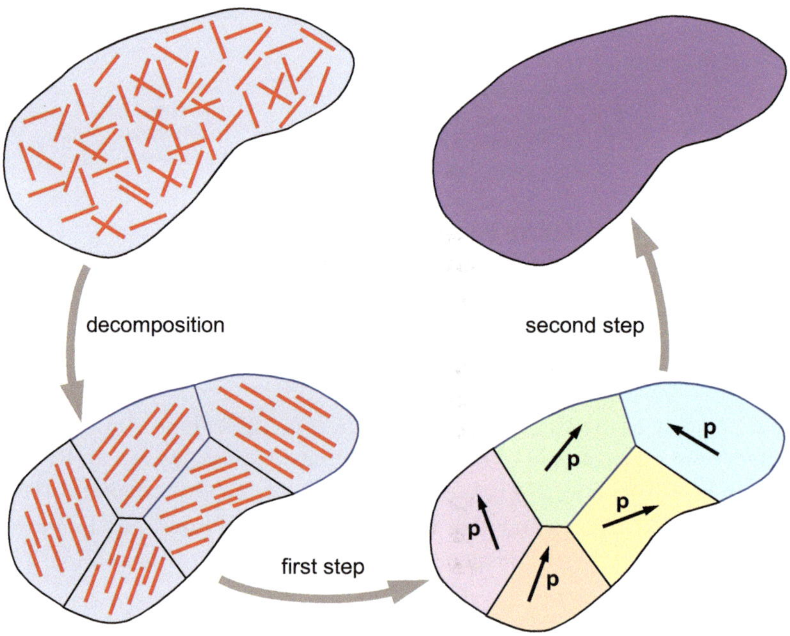

Figure 8.2 Schematic of the two-step homogenization process. After Lielens et al. [LPC+98].

8.3 Stiffness Estimates for Unidirectional Discontinuous Fiber Composites

A variety of approaches have been used to estimate the stiffness of a discontinuous fiber composite with unidirectional fibers [TL99]. The Eshelby/Mori–Tanaka model and the related Lielens/double-inclusion model are believed to be the most accurate [TL99, PFD04], and are widely used in commercial software. The Halpin–Tsai equations are also discussed here because they are reasonably accurate and simple to use. All three models are available in the accompanying software[2]. We evaluate them here by comparing their predictions to finite element calculations on RVEs.

While the examples here are for fiber reinforcement, the Eshelby/Mori–Tanaka model and the Lielens/double-inclusion model are equally suitable for disk-shaped reinforcements like flakes or platelets. In that case the orientation vector is normal to the face of the disk (Figs. 4.8(b) and (d)) and $a/b < 1$.

8.3.1 Eshelby/Mori–Tanaka Model

While our interest is in predicting the stiffness of discontinuous fiber composites, Eshelby [Esh57] developed an exact solution to a different problem, a *homogeneous ellipsoidal inclusion.* Consider a large body with the mechanical properties of the matrix material. A small portion of this body, ellipsoidal in shape, is the inclusion. Suppose that the inclusion undergoes a phase transformation or some other process that would cause it to strain by an amount $\boldsymbol{\varepsilon}^T$ if it were not embedded in the matrix. Because it is embedded, it will have a different strain $\boldsymbol{\varepsilon}^C$, and the matrix will also have a complex strain pattern. Eshelby showed that if the inclusion is ellipsoidal in shape, the strain $\boldsymbol{\varepsilon}^C$ in the inclusion is uniform, and is given by

$$\boldsymbol{\varepsilon}^C = \mathbb{E}^m : \boldsymbol{\varepsilon}^T \tag{8.45}$$

$\mathbb{E}^m$ is a fourth-order tensor called the *Eshelby tensor*, and for an isotropic matrix it depends only on the shape of the inclusion and the matrix Poisson ratio. Analytical expressions for the components of the Eshelby tensor have been worked out [Mur82, TA89], and the `eshtens` function in the accompanying software give $\mathbb{E}^m$ for a spheroid with any value of a/b. Eshelby's solution is intimately related to Jeffery's solution for a rigid ellipsoidal particle in a Newtonian fluid (e.g., [WT01]).

While the original Eshelby problem considers a homogeneous inclusion and a far-field strain of zero, one can form an equivalent problem for an *inhomogeneous* inclusion (i.e., a fiber embedded in the matrix) with an applied strain $\boldsymbol{\varepsilon}^0$ far from the fiber. The key result, derived in Appendix B.4.1, is that the average strain in the fiber is related to the far-field strain by

$$\overline{\boldsymbol{\varepsilon}}^f = \mathbb{H}^m : \boldsymbol{\varepsilon}^0 \tag{8.46}$$

with $\mathbb{H}^m$ given by

$$\mathbb{H}^m = \left[\mathbb{I} + \mathbb{E}^m : \mathbb{S}^m : \left(\mathbb{C}^f - \mathbb{C}^m\right)\right]^{-1} \tag{8.47}$$

[2] The functions are `mori`, `lielens` and `halpin`.

The superscripts on $\mathbb{H}^m$ and $\mathbb{E}^m$ indicate that they are calculated using the matrix material as the continuous phase.

If the composite is dilute, $\phi_f \rightarrow 0$, then the average strain $\overline{\boldsymbol{\varepsilon}}$ equals the far-field strain $\boldsymbol{\varepsilon}^0$. Equation (8.46) can then be used together with equations from Section 8.2.1 (Appendix B.4.2 shows the details) to find the composite stiffness,

$$\mathbb{C} = \mathbb{C}^m + \phi_f \left(\mathbb{C}^f - \mathbb{C}^m\right) : \mathbb{H}^m \tag{8.48}$$

This is a good model for a very dilute composite, say $\phi_f < 0.01$, with unidirectional fibers. The fiber and matrix properties are accounted for by $\mathbb{C}^f$ and $\mathbb{C}^m$, and the fiber aspect ratio is accounted for in the Eshelby tensor. For cylindrical fibers the ellipsoidal axis ratio is replaced by ℓ / d_f. However, Eqn. (8.48) is not accurate at practical volume fractions because it ignores interactions between each fiber and its neighbors.

Modeling the interactions between fibers in a composite is challenging, but a conjecture first made by Mori and Tanaka [MT73] is very helpful. The essence of the Mori–Tanaka assumption [Ben87] is that, for finite values of fiber volume fraction, the average strain in the fiber is given by

$$\overline{\boldsymbol{\varepsilon}}^f = \mathbb{H}^m : \overline{\boldsymbol{\varepsilon}}^m \tag{8.49}$$

That is, at finite volume fraction, a fiber responds to the average *matrix* strain in the same way that an isolated fiber responds to the far-field strain, Eqn. (8.46). This represents an averaged or mean-field interaction between fibers. In terms of the strain concentration tensor, Eqn. (8.41), the Mori–Tanaka assumption is that $\mathbb{B}^{\mathrm{MT}} = \mathbb{H}^m$, or

$$\mathbb{B}^{\mathrm{MT}} = \left[\mathbb{I} + \mathbb{E}^m : \mathbb{S}^m : \left(\mathbb{C}^f - \mathbb{C}^m\right)\right]^{-1} \tag{8.50}$$

Using this in Eqn. (8.42) gives the Mori–Tanaka prediction for composite stiffness.

The full set of elastic constants for a discontinuous fiber composite using the Eshelby/Mori–Tanaka model was first developed by Tandon and Weng [TW84], so this is sometimes called the Tandon and Weng model. The Eshelby/Mori–Tanaka model is also equivalent to several other models that were derived by different routes [TL99].

8.3.2 Lielens/Double-Inclusion Model

The Mori–Tanaka model matches the stiffness of the dilute model, Eqn. (8.48), as $\phi_f \rightarrow 0$, and for $\phi_f = 0$ both models give $\mathbb{C} = \mathbb{C}^m$. Thus, the Mori–Tanaka model includes the dilute model while capturing interactions between fibers as the volume fraction increases from zero.

One could also imagine a composite with a very high volume fraction of fibers, $\phi_f \rightarrow 1$. Such a material might best be modeled by treating the fibers as the continuous phase and the matrix as the dispersed phase. We can write the Mori–Tanaka model for this case by interchanging the f and m superscripts in Eqns. (8.49) and (8.47):

$$\overline{\boldsymbol{\varepsilon}}^m = \mathbb{H}^f : \overline{\boldsymbol{\varepsilon}}^f \qquad \text{where} \qquad \mathbb{H}^f = \left[\mathbb{I} + \mathbb{E}^f : \mathbb{S}^f : \left(\mathbb{C}^m - \mathbb{C}^f\right)\right]^{-1} \tag{8.51}$$

$\mathbb{E}^f$, the Eshelby tensor in this expression, is evaluated using the fiber properties.

Comparing this to Eqn. (8.41), the strain concentration tensor for this model is $\mathbb{B}^u = \left(\mathbb{H}^f\right)^{-1}$, or

$$\mathbb{B}^u = \left[\mathbb{I} + \mathbb{E}^f : \mathbb{S}^f : \left(\mathbb{C}^m - \mathbb{C}^f\right)\right] \tag{8.52}$$

We label this tensor with a superscript u because it matches an *upper bound* on composite stiffness from the Hashin–Shtrikman–Willis bounds [HS63, Wil77]. The standard Mori–Tanaka model, Eqn. (8.50), matches the corresponding lower bound. The Hashin–Shtrikman–Willis bounds can take reinforcement geometry into account, and are tighter than the Voigt and Reuss bounds.

Lielens et al. [LPC+98] suggested that the Mori–Tanaka/lower-bound model should be accurate at low volume fractions, and the upper-bound model should be accurate at very high volume fractions. To cover the entire range of volume fractions, they proposed interpolating between these two bounds using an inverse rule of mixtures on $\mathbb{B}$:

$$\mathbb{B}^{\text{Lielens}} = \left[(1-\xi)\left(\mathbb{B}^{\text{MT}}\right)^{-1} + \xi\left(\mathbb{B}^u\right)^{-1}\right]^{-1} \tag{8.53}$$

where the interpolating factor is

$$\xi = \frac{\phi_f + \phi_f^2}{2} \tag{8.54}$$

The Lielens model is also called the *double-inclusion model* [PFD04].

8.3.3 Halpin–Tsai Equations

Historically, the first calculations of the full set of elastic constants for an aligned discontinuous fiber composite used the Halpin–Tsai equations. These equations are still used in the literature, and they provide a convenient way to estimate engineering constants without the tensor operations of the preceding sections.

Halpin and Tsai originally found that a number of models for both continuous fiber composites and particle reinforced composites could be expressed in a common form [HK76]:

$$\frac{P}{P_m} = \frac{1+\zeta\eta\phi_f}{1-\eta\phi_f} \qquad \text{with} \qquad \eta = \frac{(P_f/P_m)-1}{(P_f/P_m)+\zeta} \tag{8.55}$$

Here P represents one of the elastic moduli of the composite while P_f and P_m are the corresponding properties of the fiber and the matrix. ζ is a dimensionless parameter that depends on the matrix Poisson ratio and the property being calculated. Note that the engineering constants are calculated directly.

To adapt this to discontinuous fiber composites, Halpin et al. [AHP69, Hal69] made the additional assumptions that the transverse and shear properties of a discontinuous fiber composite could be approximated by the formulae for a continuous fiber composite, and that the effect of fiber aspect ratio on E_{11} could be accounted for by choosing $\zeta = 2(\ell/d_f)$.

Table 8.1 Traditional Halpin–Tsai parameters for a discontinuous fiber composite with unidirectional fibers, used in Eqn. (8.55)

P	P_f	P_m	ζ	Comments
E_{11}	E_f	E_m	$2(\ell/d_f)$	Longitudinal Young's modulus
E_{22}	E_f	E_m	2	Transverse Young's modulus
G_{12}	G_f	G_m	1	Longitudinal shear modulus
G_{23}	G_f	G_m	$\frac{1+\nu_m}{3-\nu_m-4\nu_m^2}$	Transverse shear modulus
ν_{12}				Poisson ratio, $\nu_{12} = \phi_f \nu_f + \phi_m \nu_m$

The resulting combinations of P, P_f, P_m and ζ used to find the elastic properties of a unidirectional, discontinuous fiber composite are given in Table 8.1. ν_{12} is found using a rule-of-mixtures formula rather than Eqn. (8.55). The composite is transversely isotropic, so ν_{23} is found using G_{23} and E_{22} in Eqn. (8.30).

Some adjustments to the basic Halpin–Tsai form have been proposed [HK76, TL99], especially for fiber volume fractions close to the packing limit for unidirectional fibers, though none of these forms has been widely used for discontinuous fiber composites.

8.3.4 Numerical Analysis of an RVE

A very different way to obtain homogenized stiffness values is to explicitly construct one or more RVEs and use a numerical scheme like the finite element method to analyze the structure. This treats interactions between fibers explicitly, and can model realistic arrangements of the fibers. A disadvantage is that every combination of fiber properties, matrix properties, fiber aspect ratio and fiber volume fraction requires a new calculation.

The fiber arrangements can be periodic, like Figs. 8.1(a) and (b), or random like Figs. 8.1(c) and (d). For a periodic arrangement, the analysis can focus on a unit cell, like the black boxes in Figs. 8.1(a) and (b). This is efficient, but may not represent the true fiber arrangement in a composite. Random RVEs like Figs. 8.1(c) and (d) are more representative of real structures, but one must analyze a larger volume to obtain useful results. An RVE whose edges are 4 to 5 fiber lengths long is recommended [NDK⁺20]. Random RVEs are created by a statistical process, and each realization for a given set of parameters will be different from the next. Figures 8.1(c) and (d) are two realizations created by the same program, and the differences between them are apparent. The most accurate results for random RVEs are obtained by analyzing multiple realizations and averaging the results.

It is also possible to use micro-CT data and construct RVEs that exactly match a specific physical sample. Of course other physical samples from the same composite will have different fiber and matrix geometry, so the challenge of multiple realizations is still present.

Once the RVE has been chosen, a mesh or grid is generated, and the properties of each element are set either to the fiber properties or the matrix properties.

The next step is to impose boundary conditions and determine the components of the homogenized stiffness tensor $\mathbb{C}$. If the applied boundary conditions are consistent with a uniform strain $\boldsymbol{\varepsilon}^0$, then it is easy to show that the average strain in the RVE is $\overline{\boldsymbol{\varepsilon}} = \boldsymbol{\varepsilon}^0$ (e.g., [TL99]). For random RVEs one can choose a specific applied strain component to be non-zero, say ε^0_{11}, and set all other components of $\boldsymbol{\varepsilon}^0$ to zero. The boundaries of the RVE are then subjected to displacements consistent with this strain,

$$\mathbf{u} = \boldsymbol{\varepsilon}^0 \cdot \mathbf{x} \tag{8.56}$$

and the stresses are analyzed. The resulting average stresses $\overline{\boldsymbol{\sigma}}$ are determined, either by evaluating the constraint forces on each face of the RVE and dividing by the area of the face, or by averaging the element stresses over the entire volume. One column of the stiffness matrix $[\mathbb{C}]$ is then determined using Eqn. (8.35). Since all the strain components except ε^0_{11} are zero, this is

$$\begin{array}{lll} \mathbb{C}_{1111} = \overline{\sigma}_{11}/\varepsilon^0_{11} & \mathbb{C}_{2211} = \overline{\sigma}_{22}/\varepsilon^0_{11} & \mathbb{C}_{3311} = \overline{\sigma}_{33}/\varepsilon^0_{11} \\ \mathbb{C}_{2311} = \overline{\sigma}_{23}/\varepsilon^0_{11} & \mathbb{C}_{3111} = \overline{\sigma}_{31}/\varepsilon^0_{11} & \mathbb{C}_{1211} = \overline{\sigma}_{12}/\varepsilon^0_{11} \end{array} \tag{8.57}$$

The process is repeated for each strain component to build up the complete $\mathbb{C}$ tensor.

For periodic structures it is more efficient to apply periodic boundary conditions, so that the displacements on any face of the structure match the displacements on the periodic face, plus a difference corresponding to the applied strain.

One may also perform a numerical tensile test, where the RVE is stretched in one direction and allowed to contract freely in the perpendicular directions while maintaining periodicity in the boundary displacements. This effectively applies a single stress component, and the corresponding strain components are used to determine components of the compliance tensor $\mathbb{S}$ and/or the engineering constants.

8.3.5 Comparison and Recommendations

Figure 8.3 compares predictions of the mean-field models for E_{11} and E_{22} as a function of volume fraction. Traditional discontinuous fiber composites rarely have $\phi_f > 0.30$, but the full range of volume fraction is included here to illustrate some of the relationships between the theories. The dilute Eshelby model uses Eqn. (8.48), and the reverse dilute model uses the same equation with the roles of the fiber and matrix interchanged. The upper bound model uses $\mathbb{B}^u$ from Eqn. (8.52) in (8.42).

All models except reverse dilute give $E_{11} = E_{22} = E_m$ when $\phi_f = 0$, and all models except dilute Eshelby give $E_{11} = E_{22} = E_f$ when $\phi_f = 1$. The dependence of modulus on volume fraction is linear for the dilute and reverse dilute models, as it must be when there is no interaction between fibers.

At intermediate volume fractions, the upper bound model is always higher than the Mori–Tanaka/lower bound, as it should be. The distance between the upper and lower bounds is substantial, especially for E_{22} at higher volume fractions. The Mori–Tanaka model approaches the dilute model as $\phi_f \to 0$, and the upper bound approaches the reverse dilute model as $\phi_f \to 1$.

The Lielens model always falls between the upper and lower bounds, as it must. It also matches Mori–Tanaka at low volume fractions, and the upper bound at very high volume fractions.

The Halpin–Tsai model is quite close to the Mori–Tanaka results for this fiber aspect ratio, though we will see below that this is not always the case.

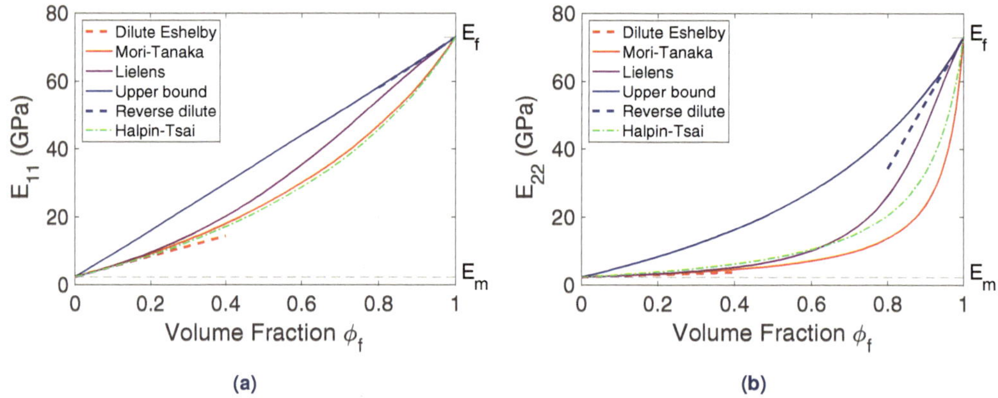

Figure 8.3 Predictions of stiffness as a function of fiber volume fraction for unidirectional discontinuous fiber composites. Fiber and matrix properties for glass fiber/polycarbonate from Table 8.2 with $\ell/d_f = 10$.

Note that the dilute, reverse dilute and upper bound models are not used in practice. They are included here to illustrate the relationships between other models.

A comparison of the Mori–Tanaka, Lielens and Halpin–Tsai equations to finite element calculations on periodic unit cells appears in Fig. 8.4. The properties are typical of a glass fiber/thermoplastic composite, with $E_f/E_m = 30$, $\nu_f = 0.2$, $\nu_m = 0.38$, and $\phi_f = 0.20$. Finite element calculations [TL99] were carried out for cylindrical fibers with $\ell/d_f = 1, 2, 4, 8, 12$ and 24, with one calculation at $\ell/d_f = 48$. The RVEs are three-dimensional, and use either a regular or a staggered arrangement of fibers as shown in Figs. 8.1(a) and (b). When viewed along the fiber direction, the fibers are either in a square or a hexagonal array. The analysis used periodic boundary conditions.

The fiber-direction modulus E_{11}, Fig. 8.4(a), is a strong function of aspect ratio, and rises in an S-shaped curve when ℓ/d_f is plotted on a logarithmic axis. For very large fiber aspect ratios, all models asymptote to a value that is close to the rule of mixtures, $E_{11} = \phi_m E_m + \phi_f E_f$, which gives $E_{11}/E_m = 6.80$ in this case. The finite element values of E_{11} for the hexagonal staggered array at low aspect ratios are probably unrealistically stiff, because the rules used for building the RVEs gave very close end-to-end fiber spacing in these cases. If we ignore that data set, the models all track the finite element data fairly well. There is little difference between the Mori–Tanaka and Lielens predictions at these volume fractions, and the Halpin–Tsai prediction appears to rise too slowly at high aspect ratios.

The transverse modulus E_{22} in Fig. 8.4(b) has a weak dependence on aspect ratio, though there is some variation that the Mori–Tanaka and Lielens models capture for some of the finite element geometries. The Halpin–Tsai equations ignore any aspect ratio dependence of E_{22}, and this is a reasonable approximation.

The Poisson ratios ν_{12} and ν_{23}, shown in Figs. 8.4(c) and (d), do show some modest dependence on aspect ratio, some of which is captured by the Mori–Tanaka and Lielens models. The Halpin–Tsai result for ν_{12} is reasonable, though the value of ν_{23} is over-predicted.

This set of finite element calculations did not include the shear moduli G_{12} and G_{23}, but the mean-field theory calculations are shown in Figs. 8.4(e) and (f). Both values are close to 1.5

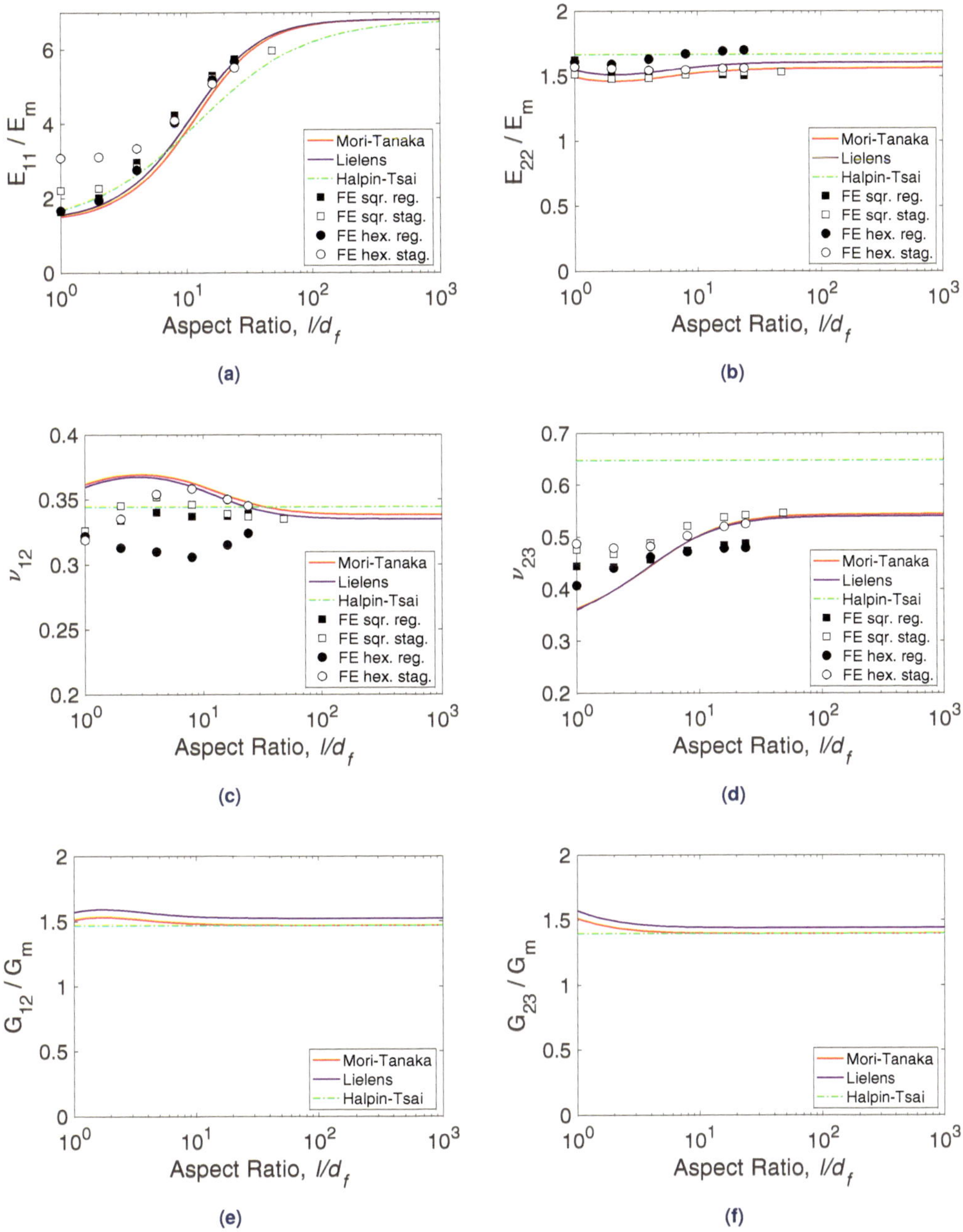

Figure 8.4 Predictions of stiffness as a function of fiber aspect ratio for unidirectional discontinuous fiber composites. $E_f/E_m = 30$, $\nu_f = 0.2$, $\nu_m = 0.38$, $\phi_f = 0.20$. Finite element (FE) data for periodic arrays of cylindrical fibers from [TL99].

times the matrix shear modulus G_m, and neither one has much dependence on aspect ratio for the Mori–Tanaka and Lielens theories.

Multiple tests of different analytical models against finite element data [TL99, HRG02, PFD04, GL14] also show that the Mori–Tanaka and Lielens models are the best mean-field theories for predicting the stiffness of a composite with unidirectional fibers. The two models give nearly equal results for volume fractions up to around 20%, while the Lielens model is more accurate at higher volume fractions. For typical injection molded composites, either model is suitable. These models are also recommended for composites with disk-shaped reinforcements.

8.4 Averaging Stiffness for Orientation and Length

8.4.1 Definitions and Notation

Once the properties of each pseudo-grain been determined, the next step is to homogenize the pseudo-grains. For this we need some new notation.

Let $\overline{\boldsymbol{\sigma}}^p(\mathbf{p},\ell)$ and $\overline{\boldsymbol{\varepsilon}}^p(\mathbf{p},\ell)$ be the volume-average stress and strain in a pseudo-grain with orientation $\mathbf{p}$ and fiber length ℓ. These are simple volume averages as in Eqn. (8.34), except that the average is over the pseudo-grain volume. The stiffness of a pseudo-grain $\mathbb{C}^p(\mathbf{p},\ell)$ is then defined from[3]

$$\overline{\boldsymbol{\sigma}}^p = \mathbb{C}^p : \overline{\boldsymbol{\varepsilon}}^p \tag{8.58}$$

which is the pseudo-grain version of Eqn. (8.35). $\mathbb{C}^p(\mathbf{p},\ell)$ is the stiffness tensor calculated in Section 8.3.

To connect pseudo-grain properties to the full RVE properties, first consider the case when the composite has a distribution of fiber orientation but uniform fiber length. If the orientation distribution were discrete, with a finite number of orientations $\mathbf{p}$, then the volume of each pseudo-grain would be proportional to the fraction of fibers with orientation $\mathbf{p}$. Then the volume average of stress or strain over the entire RVE would be the sum of the pseudo-grain averages, weighted by the fraction of fibers with each orientation. Generalizing this to a continuous distribution of orientation $\psi(\mathbf{p})$, the RVE averages are related to the pseudo-grain averages by

$$\overline{\boldsymbol{\sigma}} = \langle \overline{\boldsymbol{\sigma}}^p \rangle \quad \text{and} \quad \overline{\boldsymbol{\varepsilon}} = \langle \overline{\boldsymbol{\varepsilon}}^p \rangle \tag{8.59}$$

where the angle brackets denote an orientation average as in Eqn. (2.32),

$$\langle \overline{\boldsymbol{\sigma}}^p(\mathbf{p}) \rangle \equiv \oint \overline{\boldsymbol{\sigma}}^p(\mathbf{p})\, \psi(\mathbf{p})\, \mathrm{d}\mathbf{p} \quad \text{and} \quad \langle \overline{\boldsymbol{\varepsilon}}^p(\mathbf{p}) \rangle \equiv \oint \overline{\boldsymbol{\varepsilon}}^p(\mathbf{p})\, \psi(\mathbf{p})\, \mathrm{d}\mathbf{p} \tag{8.60}$$

The computation of orientation averages will be discussed in Section 8.4.3. When the fibers have a distribution of lengths, a similar average can be taken over the length distribution function. This will be discussed in Section 8.4.4.

[3] We will omit the arguments $(\mathbf{p},\ell)$ from pseudo-grain quantities, except where needed for clarity.

8.4.2 Types of Stiffness Averages

Our goal is to find the overall stiffness tensor $\mathbb{C}$ that relates $\overline{\boldsymbol{\sigma}}$ to $\overline{\boldsymbol{\varepsilon}}$ as in Eqn. (8.35). To start, take the orientation average of the pseudo-grain stiffness equation (8.58):

$$\langle \overline{\boldsymbol{\sigma}}^p \rangle = \langle \mathbb{C}^p : \overline{\boldsymbol{\varepsilon}}^p \rangle \tag{8.61}$$

From Eqn. (8.59) the left-hand side is $\overline{\boldsymbol{\sigma}}$, but we cannot immediately separate the two terms on the right-hand side because both of them depend on $\mathbf{p}$.

One option is to assume that the average pseudo-grain strain $\overline{\boldsymbol{\varepsilon}}^p$ is the same for all pseudo-grains. Mathematically,

$$\overline{\boldsymbol{\varepsilon}}^p(\mathbf{p}) = \overline{\boldsymbol{\varepsilon}} \qquad \forall\, \mathbf{p} \tag{8.62}$$

When this is substituted into Eqn. (8.61) we can bring $\overline{\boldsymbol{\varepsilon}}$ outside of the orientation average, because it does not depend on $\mathbf{p}$. This gives

$$\overline{\boldsymbol{\sigma}} = \langle \mathbb{C}^p \rangle : \overline{\boldsymbol{\varepsilon}} \tag{8.63}$$

or, comparing this to Eqn. (8.35),

$$\mathbb{C} = \langle \mathbb{C}^p \rangle \tag{8.64}$$

This is called the **Voigt average**, or sometimes the *aggregate average*. It says that the overall RVE stiffness can be calculated by taking the orientation average of the pseudo-grain stiffness $\mathbb{C}^p$. The Voigt average is the standard approach for the second homogenization step.

For later reference we substitute Eqn. (8.42) for $\mathbb{C}^p$ into this equation and write the Voigt stiffness average as

$$\mathbb{C} = \left\langle \left[\phi_m \mathbb{C}^m + \phi_f \mathbb{C}^{f,p} : \mathbb{B}^p\right] : \left[\phi_m \mathbb{I} + \phi_f \mathbb{B}^p\right]^{-1} \right\rangle \tag{8.65}$$

Here we have added a superscript p to $\mathbb{B}$ to emphasize that this is a pseudo-grain quantity that depends on the fiber orientation $\mathbf{p}$. Also, $\mathbb{C}^f$ can depend on $\mathbf{p}$ if the fiber material is anisotropic, which is true for carbon and aramid fibers. We assume that the matrix material is isotropic, so no additional superscript is needed on $\mathbb{C}^m$.

In addition to the Voigt average, there are other ways to homogenize the pseudo-grains. One can take a **Reuss average**, which assumes that the average stress is the same for all pseudo-grains, $\overline{\boldsymbol{\sigma}}^p(\mathbf{p}) = \overline{\boldsymbol{\sigma}}$. The result is that the overall compliance of the composite is the orientation average of the pseudo-grain compliances,

$$\mathbb{S} = \langle \mathbb{S}^p \rangle \qquad \text{or} \qquad \mathbb{C} = \left\langle \left(\mathbb{C}^p\right)^{-1} \right\rangle^{-1} \tag{8.66}$$

The Reuss average consistently under-predicts the overall stiffness [DGR07] and is rarely used.

One can also take a **Mori–Tanaka average** by assuming that the average *matrix* strain is the same for all pseudo-grains,

$$\overline{\boldsymbol{\varepsilon}}^{m,p}(\mathbf{p}) = \overline{\boldsymbol{\varepsilon}}^m \qquad \forall\, \mathbf{p} \tag{8.67}$$

After some derivation (e.g., [PFD04]), the overall stiffness is

$$\mathbb{C} = \left[\phi_m \mathbb{C}^m + \phi_f \left\langle \mathbb{C}^{f,p} : \mathbb{B}^p \right\rangle\right] : \left[\phi_m \mathbb{I} + \phi_f \left\langle \mathbb{B}^p \right\rangle\right]^{-1} \tag{8.68}$$

This equation has nearly the same form as the Voigt average, Eqn. (8.65), but here the tensors $\left(\mathbb{C}^{f,p} : \mathbb{B}^p\right)$ and $\mathbb{B}^p$ are averaged separately. The Mori–Tanaka average is sometimes referred to as a *dispersed average* [EMG94]. Equation (8.68) can also be obtained using a one-step homogenization on a composite with distributed orientation. While the Mori–Tanaka average is acceptable for two-phase composites, when extended to composites with three or more phases, it can give stiffness tensors that lack the major symmetry $\mathbb{C}_{ijkl} = \mathbb{C}_{klij}$, a result that is non-physical [BDC91, PFD04].

Comparisons both to experiments and to finite element analysis on full RVEs [LHG02, NDK+20] show that the Voigt average is the best choice for the second homogenization step. The Voigt average is the standard way to perform orientation averaging, and we will use it exclusively here.

8.4.3 Computing Orientation Averages

We now need to compute orientation averages like $\left\langle \mathbb{C}^p(\mathbf{p}) \right\rangle$. For linear elastic properties, $\mathbb{C}^p$ is the same for every pseudo-grain except for a coordinate transformation, and there is a simple way to calculate the average.

Because each pseudo-grain has aligned, axisymmetric fibers, it must be transversely isotropic around its $\mathbf{p}$ vector. As a result, it can be written in tensor notation as [AT87, CTYM90]

$$\begin{aligned}
\mathbb{C}^p_{ijkl}(\mathbf{p}) &= B_1\, p_i p_j p_k p_l + B_2^u\, p_i p_j \delta_{kl} + B_2^l\, p_k p_l \delta_{ij} \\
&\quad + B_3 \left(p_i p_k \delta_{jl} + p_i p_l \delta_{jk} + p_j p_l \delta_{ik} + p_j p_k \delta_{il}\right) \\
&\quad + B_4\, \delta_{ij}\delta_{kl} + B_5 \left(\delta_{ik}\delta_{jl} + \delta_{il}\delta_{jk}\right)
\end{aligned} \tag{8.69}$$

The B_is are invariants of $\mathbb{C}^p$, so they do not depend on the choice of coordinate system or on the orientation $\mathbf{p}$.

Software that implements any of the micromechanics models from Section 8.3 returns a single stiffness tensor in some reference configuration. Call this tensor $\hat{\mathbb{C}}^p$. If we define the reference configuration such that the 1 axis is the symmetry axis, then the B_is are related to the components of $\hat{\mathbb{C}}^p$ (written in contracted notation) by

$$B_1 = \hat{\mathbb{C}}^p_{11} + \hat{\mathbb{C}}^p_{22} - 2\hat{\mathbb{C}}^p_{12} - 4\hat{\mathbb{C}}^p_{66} \tag{8.70}$$

$$B_2^u = \hat{\mathbb{C}}^p_{12} - \hat{\mathbb{C}}^p_{23} \tag{8.71}$$

$$B_2^l = \hat{\mathbb{C}}^p_{21} - \hat{\mathbb{C}}^p_{23} \tag{8.72}$$

$$B_3 = \hat{\mathbb{C}}^p_{66} + \left(\hat{\mathbb{C}}^p_{23} - \hat{\mathbb{C}}^p_{22}\right)/2 \tag{8.73}$$

$$B_4 = \hat{\mathbb{C}}^p_{23} \tag{8.74}$$

$$B_5 = \left(\hat{\mathbb{C}}^p_{22} - \hat{\mathbb{C}}^p_{23}\right)/2 \tag{8.75}$$

Equations (8.69–8.75) hold for any transversely isotropic fourth-order tensor that has the minor symmetries $\mathbb{C}_{ijkl} = \mathbb{C}_{jikl} = \mathbb{C}_{ijlk}$, but they do not assume that the tensor has the major symmetry[4] $\mathbb{C}_{ijkl} = \mathbb{C}_{klij}$. If the tensor has the major symmetry (which $\hat{\mathbb{C}}^p$ does) then $B_2^l = B_2^u$ and there are five independent B_is, corresponding to the five independent elastic constants of a transversely isotropic material.

Taking the orientation average of both sides of Eqn. (8.69), we find

$$\begin{aligned}\left\langle \mathbb{C}^p_{ijkl}(\mathbf{p})\right\rangle &= B_1 \langle p_i p_j p_k p_l\rangle + B_2^u \langle p_i p_j \delta_{kl}\rangle + B_2^l \langle p_k p_l \delta_{ij}\rangle \\ &+ B_3 \langle p_i p_k \delta_{jl} + p_i p_l \delta_{jk} + p_j p_l \delta_{ik} + p_j p_k \delta_{il}\rangle \\ &+ B_4 \langle \delta_{ij}\delta_{kl}\rangle + B_5 \langle \delta_{ik}\delta_{jl} + \delta_{il}\delta_{jk}\rangle \end{aligned} \tag{8.76}$$

Since δ_{ij} is independent of $\mathbf{p}$ it can come outside the averaging brackets. Also, $\langle p_i p_j p_k p_l\rangle$ is the fourth-order orientation tensor $\mathbb{A}_{ijkl}$ and $\langle p_i p_j\rangle$ is the second-order orientation tensor A_{ij}. Equation (8.76) then simplifies to

$$\begin{aligned}\left\langle \mathbb{C}^p_{ijkl}\right\rangle &= B_1 \mathbb{A}_{ijkl} + B_2^u A_{ij}\delta_{kl} + B_2^l A_{kl}\delta_{ij} \\ &+ B_3 \left(A_{ik}\delta_{jl} + A_{il}\delta_{jk} + A_{jl}\delta_{ik} + A_{jk}\delta_{il}\right) \\ &+ B_4 \delta_{ij}\delta_{kl} + B_5 \left(\delta_{ik}\delta_{jl} + \delta_{il}\delta_{jk}\right) \end{aligned} \tag{8.77}$$

To compute an orientation average of $\mathbb{C}^p$, start with $\hat{\mathbb{C}}^p$ in the reference configuration, calculate the B_i values from Eqns. (8.70–8.75), then combine those values with the orientation tensors in this equation to find $\left\langle \mathbb{C}^p_{ijkl}\right\rangle$. The function `oravg4` in the accompanying software performs these operations.

Equation (8.77) expresses an important result[5]:

The orientation average for stiffness can be computed using only the second-order and fourth-order orientation tensors.

If we are calculating linear elastic properties, we only need the orientation tensors; reconstructing the full orientation distribution function is not required.

Of course, when orientation is predicted from a mold filling simulation, only the second-order tensor $\mathbf{A}$ is calculated directly; the fourth-order tensor $\mathbb{A}$ is found using a closure approximation. This has an important implication:

Predictions of linear elastic properties are sensitive to the choice of closure approximation.

Fortunately, the closure approximations that are most accurate for modeling flow-induced orientation are also accurate when predicting mechanical properties (e.g., [DGR07]).

---

4 The $\mathbb{B}^p$ tensor does not have this major symmetry, but it can still be averaged using this procedure, e.g., if computing a Mori–Tanaka orientation average.

5 This applies to any type of orientation average: Voigt, Reuss or Mori–Tanaka

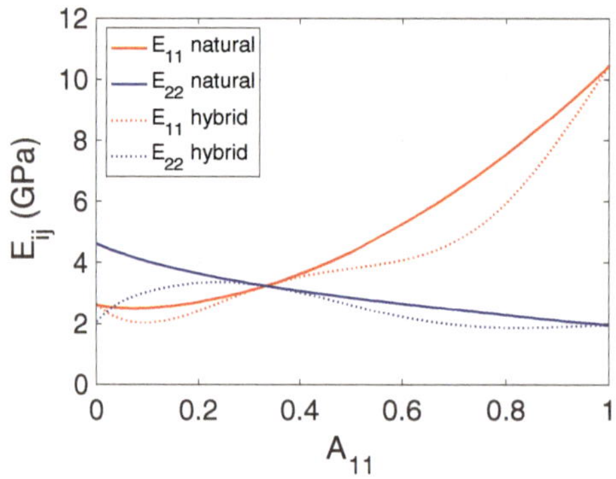

(a) Transversely isotropic orientation

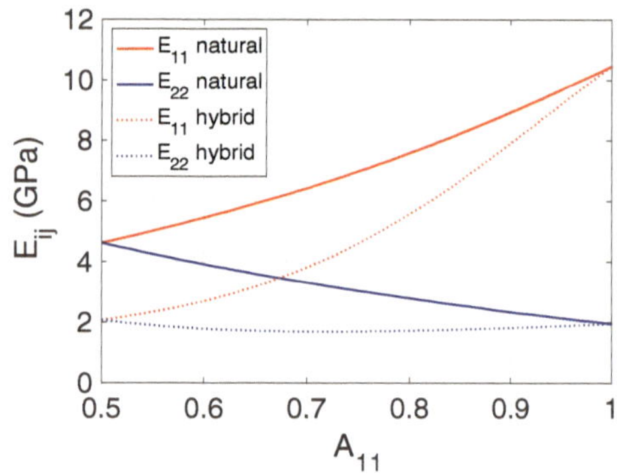

(b) Planar orientation

Figure 8.5 Stiffness vs. orientation using Voigt averaging with the natural and hybrid closures. (a) Transversely isotropic orientation states, $A_{22} = A_{33} = (1 - A_{11})/2$ and all other $A_{ij} = 0$. (b) Planar orientation states, $A_{22} = 1 - A_{11}$ and all other $A_{ij} = 0$. Calculations for 40% by weight glass fiber in PP with properties from Table 8.2 and $\ell/d_f = 100$.

Table 8.2 Constituent mechanical properties used in example calculations of stiffness and thermal expansion

Material	E (GPa)	ν	ρ (g/cm^3)	α (10^{-6} K^{-1})
E-glass fiber	73	0.21	2.55	4.9
Polypropylene (PP)	1.325	0.42	0.92	
Polycarbonate (PC)	2.24	0.38	1.20	65

Two examples are shown in Fig. 8.5. Here the Mori–Tanaka model for unidirectional orientation has been used to find $\hat{\mathbb{C}}^p$ for a polypropylene matrix with 40% by weight of long glass fibers, $\ell/d_f = 100$. The fiber and matrix properties are given in Table 8.2.

Two types of orientation are studied. Figure 8.5(a) shows transversely isotropic orientation states, where A_{11} is the orientation in the symmetry direction and $A_{22} = A_{33}$. $A_{11} = 0$ in this plot represents random-in-plane orientation in the 2–3 plane, so $E_{22} = E_{33} > E_{11}$ for that case. $A_{11} = 1/3$ gives isotropic orientation, and the curves for E_{11} and E_{22} cross at that point. $A_{11} = 1$ is unidirectional orientation, so E_{11} attains it maximum value and E_{22} has its minimum value. The natural closure gives a smooth and reasonable interpolation between these points. The dotted lines show predictions using the hybrid closure, Eqns. (5.37) and (5.38). These look odd, and are certainly not representative of real composites. This highlights the role of the closure approximation in mechanical property predictions.

Properties for planar orientation states are shown in Fig. 8.5(b). The fibers lie in the 1–2 plane, so $A_{33} = 0$ in all cases. $A_{11} = 0.5$ is random-in-plane orientation, so $E_{11} = E_{22}$ at that point. As A_{11} increases, E_{11} increases and E_{22} decreases, reaching the same points for unidirectional orientation as in Fig. 8.5(a). Results for the natural closure are reasonable, while the hybrid closure gives very low modulus values for most of the orientation range.

Results for the IBOF closure (not shown) are nearly indistinguishable from the natural closure results in these calculations. One should always use the natural, IBOF or ORW closure, or something equally accurate, when predicting mechanical properties.

Orientation tensors can also be used to average any second-order symmetric tensor $\mathbf{T}(\mathbf{p})$ that is transversely isotropic about the $\mathbf{p}$ axis. These tensors have the form

$$T_{ij}(\mathbf{p}) = a_1\, p_i p_j + a_2\, \delta_{ij} \tag{8.78}$$

If the reference-configuration tensor $\hat{\mathbf{T}}$ has the 1 axis as the symmetry axis, then the invariants are

$$a_1 = \hat{T}_{11} - \hat{T}_{22} \tag{8.79}$$

$$a_2 = \hat{T}_{22} \tag{8.80}$$

and the orientation average for a second-order tensor is

$$\langle T_{ij}(\mathbf{p}) \rangle = a_1 A_{ij} + a_2 \delta_{ij} \tag{8.81}$$

The function `oravg2` in the accompanying software computes this average.

8.4.4 Averaging over Fiber Length

A composite with a distribution of fiber lengths can be modeled in one of two ways. The first option is to calculate pseudo-grain properties for each fiber length in the first homogenization step, then average over both the length distribution and the orientation distribution in the second step. The second option is to choose a single, representative fiber length for the first homogenization step.

To set up the first option, assume that we know the weight-based length distribution w_i for a set of discrete fiber lengths ℓ_i as in Section 2.5, and that the weight-based distribution has been normalized so that $\sum_i w_i = 1$. Also assume that the length and orientation distributions are separable, Eqn. (2.113), so that the fibers of any length have the same orientation distribution as the fibers of any other length[6].

Now $\overline{\boldsymbol{\sigma}}^p$ and $\overline{\boldsymbol{\varepsilon}}^p$ are functions of both $\mathbf{p}$ and ℓ_i, as is $\mathbb{C}^p$, and Eqns. (8.59) are easily extended to average over both length and orientation,

$$\overline{\boldsymbol{\sigma}} = \left\langle \sum_i w_i\, \overline{\boldsymbol{\sigma}}^p(\mathbf{p}, \ell_i) \right\rangle \quad \text{and} \quad \overline{\boldsymbol{\varepsilon}} = \left\langle \sum_i w_i\, \overline{\boldsymbol{\varepsilon}}^p(\mathbf{p}, \ell_i) \right\rangle \tag{8.82}$$

The weight-based distribution w_i is used here rather than the number-based distribution N_i, because the volume of the pseudo-grain for each fiber length should be proportional to the volume of those fibers in the composite.

In the first homogenization step we must calculate a separate pseudo-grain stiffness for each ℓ_i in the length distribution. Call these tensors $\hat{\mathbb{C}}^p(\ell_i)$, since they are calculated in the reference configuration.

6 This assumption can easily be relaxed, but one rarely has the necessary microstructural information.

Using the Voigt assumption (8.62) that all pseudo-grains have the same average strain, the Voigt average for stiffness is then

$$\mathbb{C} = \left\langle \sum_i w_i \,\hat{\mathbb{C}}^p(\mathbf{p}, \ell_i) \right\rangle \tag{8.83}$$

That is:

We can model the effects of fiber length by first averaging the stiffness tensor over all lengths using the weight-based length distribution, and then averaging over orientation.

The second option is to use a single, representative fiber length in the first homogenization step, and average over just the orientation in the second step. The number-average and weight-average lengths L_n and L_w are good candidates for this representative length, though other options have been tried as well. Hine et al. [HRG02] tested a variety of options against finite element RVE simulations using an experimental fiber length distribution. They recommend the number-average length L_n as the best single length for predicting stiffness.

A comparison of the two approaches is shown in Table 8.3. Here we have used the fiber length distributions from Fig. 3.17 (30% by weight short glass in polypropylene) and Fig. 2.10 (40% by weight long glass in polypropylene), together with the constituent properties in Table 8.2, to calculate the stiffness of a unidirectional composite. The full Voigt average over fiber length, Eqn. (8.83), as well as single-length calculations using L_n and L_w, were performed.

The results in Table 8.3 show that only E_{11} is sensitive to the treatment of fiber length. This is consistent with Fig. 8.4. The other elastic constants are different between the short fiber and long fiber materials because of the difference in fiber volume fraction.

For the short fiber material, E_{11} using L_n is significantly below the Voigt average, which is probably the most accurate result, while E_{11} using L_w is higher then the Voigt average. In contrast, for the long fiber material the value of E_{11} using L_n is within 1% of the Voigt average. In this case all three predictions are similar. Looking at Fig. 8.4(a), we can see that L_n and L_w for the short fiber composite fall on the steeply rising part of the curve of E_{11} vs. ℓ / d_f, so this calcula-

Table 8.3 Predicted stiffness for unidirectional composites using different treatments of the fiber length distribution. Fiber length distributions are given in previous figures as indicated. d_f = 0.010 mm for the short fibers and 0.017 mm for the long fibers.

	E_{11} (GPa)	E_{22} (GPa)	ν_{12}	G_{12} (GPa)
30 wt.% short glass fiber/PP, Fig. 3.17				
Voigt average using w_i	6.73	1.69	0.22	0.71
$L_n/d_f = 14.3$	5.80	1.68	0.22	0.71
$L_w/d_f = 26.1$	7.99	1.68	0.21	0.71
40 wt.% long glass fiber/PP, Fig. 2.10				
Voigt average using w_i	14.48	1.86	0.21	0.80
$L_n/d_f = 93$	14.59	1.86	0.21	0.80
$L_w/d_f = 152$	14.95	1.86	0.21	0.80

tion is very sensitive to the treatment of fiber length. In contrast, L_n and L_w for the long fiber material fall on the upper, flat part of the curve, where the sensitivity to fiber aspect ratio is small.

The recommended practice for treating fiber length is to use a Voigt average over the length distribution, Eqn. (8.83), if the required data is available. If a single length is used, then the number-average length L_n is the best choice, though this may affect the accuracy of the predictions.

8.4.5 Modeling the Layered Structure of Injection Moldings

Injection molded parts have a layered orientation structure, and this must also be taken into account when predicting mechanical properties, especially the flexural stiffness.

To see why, consider the data for two injection molded parts shown in Fig. 8.6. Here the stiffness has been calculated at each point across the thickness where orientation was measured. The first material has a polycarbonate (PC) matrix with 30% by weight of short glass fibers (Fig. 5.16) and the second has a polypropylene (PP) matrix with 40% by weight of long glass fibers (Fig. 3.8). The Mori–Tanaka model, a Voigt average for orientation, the natural closure and constituent properties from Table 8.2 were used. The short fiber properties use $\ell / d_f = 20$ and the long fiber properties use $\ell / d_f = 100$.

For both samples the profile of E_{11} across the thickness mirrors the profile of A_{11}, while E_{22} follows the profile of A_{22} and E_{33} follows the profile of A_{33}. The short fiber/PC composite has thick shell layers with strong flow alignment, so its shell layers have a high E_{11} and low E_{22}. The long fiber/PP composite has a thick core layer, much lower flow-direction alignment in the shell and a distinct skin layer. As a result, its E_{11} and E_{22} values are closer to one another across the part thickness.

These stiffness variations across the thickness can be averaged to give an overall stiffness for the sample. However, while all layers are equally important for tensile stiffness, the outer layers are more important for flexural (bending) stiffness. Flexural properties are particularly important for injection molded parts, since they play a key role in warpage.

If the mold filling simulation uses a three-dimensional mesh with multiple elements across the thickness, and if this mesh carries over into the structural analysis, then the effects of thickness-direction property variations on tensile and bending behavior are accounted for automatically. This requires that $\mathbb{C}$ be calculated for every element in the mesh, or possibly at multiple integration points within each element.

Alternately, if the mold filling simulation uses a midplane analysis, where each element is a triangle or quadrilateral with a given thickness, then a third homogenization step is needed. In this step the material is modeled as a laminate of layers, with a layer for every location across the part thickness where orientation is calculated. The stiffness of each layer is computed using the two-step homogenization process described above, then the laminate stiffness is calculated using classical lamination theory. This theory assumes that all layers of the laminate are in plane stress ($\sigma_{13} = \sigma_{23} = \sigma_{33} = 0$, when the 3 axis is the thickness direction), and the laminate deforms such that material lines that were straight and perpendicular to the midplane of the laminate ($x_3 = 0$) remain straight and perpendicular. This calculation is implemented in the function `Clayer2laminate` in the accompanying software, and details are available in

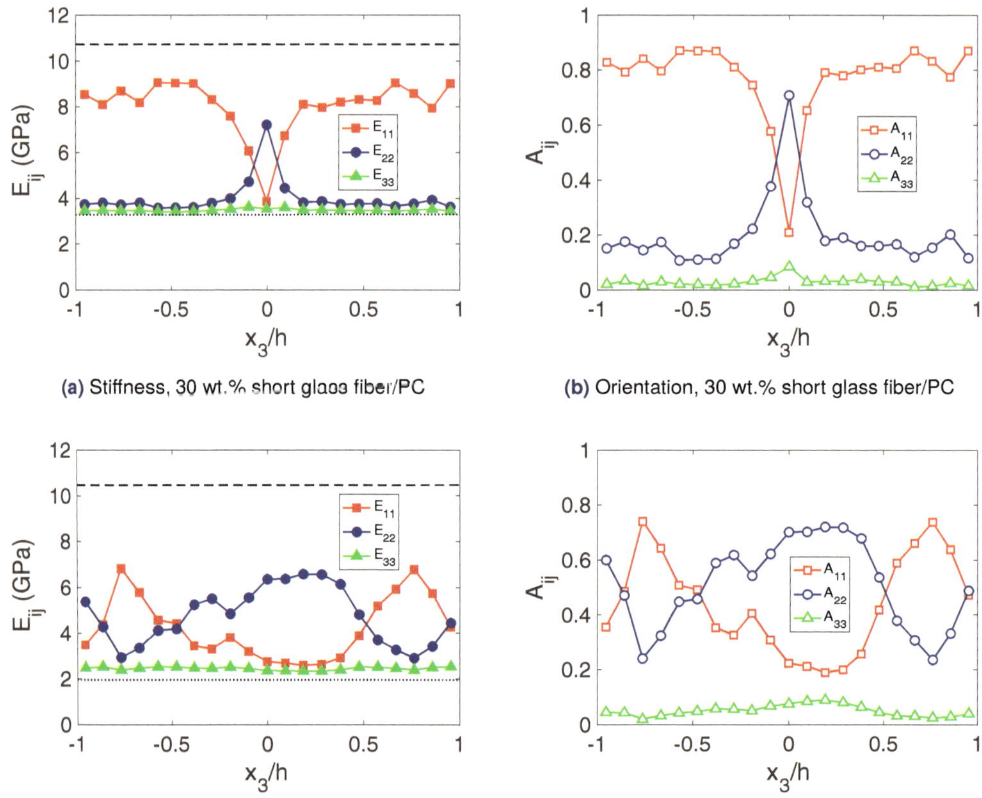

(a) Stiffness, 30 wt.% short glass fiber/PC

(b) Orientation, 30 wt.% short glass fiber/PC

(c) Stiffness, 40 wt.% long glass fiber/PP

(d) Orientation, 40 wt.% long glass fiber/PP

Figure 8.6 Profiles of predicted stiffness (left) and measured orientation (right) across the thickness of two injection molded samples. The broken black lines in (a) and (c) represent E_{11} and E_{22} for the same composites with unidirectional fibers. Short fiber/PC orientation from Fig. 5.16; long fiber/PP orientation from Fig. 3.8. Mori–Tanaka model and the natural closure. Fiber and matrix properties from Table 8.2.

many books (e.g., [Jon99, DI06, Gib16]). The result is a complete description of both in-plane and bending stiffness for the layered structure. For a mold filling simulation, this treatment is applied to every element in the mesh.

Table 8.4 shows the resulting tensile and flexural moduli in the 1 and 2 directions for the two composites in Fig. 8.6. The short fiber/PC composite, with its thick, flow-aligned shell layers, is much stiffer in the flow direction than in the crossflow direction ($E_{11} > E_{22}$) for both tension and bending. In contrast, the long fiber/PP composite is more nearly isotropic, with smaller differences between E_{11} and E_{22}. This is due to its thick, crossflow-aligned core layer and to low flow-direction alignment in the shell. When loaded in tension this composite is stiffer in the 2 direction than the 1 direction, but in bending it is stiffer in the 1 direction. This occurs because the core and shell are equally important in tension, but the outer layers, which have some flow-direction alignment, are more important in bending. It is import to perform a third homogenization step when orientation varies across the thickness.

Table 8.4 Predicted tensile and flexural moduli for the two injection molded samples in Fig. 8.6

	E_{11} (GPa)	E_{22} (GPa)
30 wt.% short glass fiber/PC		
Tensile	8.04	4.02
Flexural	8.51	3.75
40 wt.% long glass fiber/PP		
Tensile	4.26	4.79
Flexural	4.94	4.13

8.5 Comparison to Experiments

The ultimate test for any mechanical property prediction is to compare it to experiments on real materials. Two comparisons for injection molded composites are shown in Tables 8.5 and 8.6. In both cases the Mori–Tanaka model was used with Voigt averaging.

Table 8.5 shows data for a short fiber/polyarylamide composite [DGR07]. This was a 1 mm thick plate which showed little evidence of a core/shell structure, so the plate was treated as a single layer. Experiments were conducted at two temperatures to obtain two different values of matrix stiffness. A single fiber aspect ratio was used. Different treatments of the fourth-order orientation tensor were explored, and here we show results using the measured $\mathbb{A}$ and results using the ORF closure [CT95] to calculate $\mathbb{A}$. ORF is an orthotropic fitted closure that is close to the natural and IBOF closures.

This sample has strong flow-direction alignment ($A_{11} = 0.793$) and the properties reflect this, with E_{11} larger than E_{22}. This anisotropy is more pronounced at the higher temperature, where the matrix is softer and the fibers have more influence on the stiffness. Predictions at both temperatures are close to the experimental results, though all predicted values are slightly higher than the experiments. Prediction errors for the room-temperature data are less than 8%, which is satisfactory. Errors in predicting the stiffness at the higher temperature are as large as 16%, though the trends are captured correctly.

Table 8.5 Experimental and predicted tensile moduli for a short glass fiber/polyarylamide composite [DGR07]. Predictions use either the measured fourth-order tensor or the ORF closure [CT95]. Mori–Tanaka/Voigt model, $\phi_f = 0.165$, $\ell / d_f = 25$.

Sample	Experimental (GPa)	Measured $\mathbb{A}$ (GPa)	ORF $\mathbb{A}$ (GPa)
Room temperature			
E_{11}	11.80	12.18	12.28
E_{22}	7.22	7.76	7.79
120 °C			
E_{11}	3.74	4.18	4.24
E_{22}	1.26	1.45	1.46

Table 8.6 Experimental and predicted tensile moduli for 40% by weight long glass fiber/polypropylene injection molded composites [NBH+08]. Experimental values are the mean ± one standard deviation. Mori–Tanaka/Voigt model using measured orientation, measured length distribution and laminate analysis.

Sample	Experimental (GPa)	Predicted (GPa)
Center-gated disk		
E_{11}	5.32±0.12	5.55
E_{22}	7.52±0.02	7.91
ISO plaque		
E_{11}	6.06±0.18	6.02
E_{22}	7.22±0.48	6.94

Significantly, the predictions using the ORF closure are very close to the predictions using the measured fourth-order tensor, with differences between the two predictions less than 1.5% in all cases. This supports the idea that the best closure approximations provide useful mechanical property predictions, and that any errors introduced by the closure are smaller than other approximations in the property models.

A comparison for long fiber/polypropylene composites appears in Table 8.6. Two different mold geometries were tested, and both moldings were 3 mm thick. For these calculations the full fiber length distribution was used, as well as the measured orientation, and the laminated plate analysis described in Section 8.4.5 was used to account for the variation of orientation across the part thickness.

Both long fiber moldings are stiffer in the crossflow direction than in the flow direction, $E_{22} > E_{11}$. This anisotropy is due to a thick, crossflow-aligned core, and is strongest in the center-gated disk, with its diverging radial flow. The agreement between the predicted and experimental stiffness for both samples is excellent here, with the largest error just over 5%. The predictions for the ISO plaque even fall within the standard deviation of the experiments.

As these examples show, the models in this chapter can quantitatively represent the effect of fiber orientation and length on stiffness. When comparing predictions to experiments, agreement within ±5% is considered excellent, and agreement within ±10% is acceptable.

8.6 Thermal Expansion

Thermal expansion behavior is also affected by fiber orientation. Glass and carbon fibers have lower thermal expansion coefficients than polymer matrices, so adding fibers tends to reduce thermal expansion. This not only affects the performance of the molded part in service, it can also drive warping of the part as it cools from the mold temperature to room temperature. Not surprisingly, a good deal of effort has gone into modeling thermal expansion of discontinuous fiber composites.

8.6.1 Definitions

For an anisotropic material, the dimensional changes associated with temperature change are described by a *thermal expansion tensor* $\boldsymbol{\alpha}$. If the state of zero strain is defined at a reference temperature T_0 and the material is allowed to expand or contract freely, then the thermal strain due to a temperature change $\Delta T \equiv T - T_0$ is

$$\boldsymbol{\varepsilon}_{\text{thermal}} = \boldsymbol{\alpha}\, \Delta T \tag{8.84}$$

The strain tensor $\boldsymbol{\varepsilon}$ is symmetric, so $\boldsymbol{\alpha}$ must be symmetric too, $\alpha_{ij} = \alpha_{ji}$. Multiplying $\boldsymbol{\alpha}$ by a temperature change gives strain, which is dimensionless, so typical units for $\boldsymbol{\alpha}$ are $10^{-6}\,\text{K}^{-1}$.

If a material is isotropic, its thermal expansion tensor is isotropic as well:

$$\boldsymbol{\alpha} = \alpha\,\mathbf{I} = \begin{bmatrix} \alpha & 0 & 0 \\ 0 & \alpha & 0 \\ 0 & 0 & \alpha \end{bmatrix} \qquad \text{isotropic} \tag{8.85}$$

The scalar quantity α is the familiar coefficient of linear thermal expansion.

A transversely isotropic material has, in its principal material axes, a longitudinal expansion coefficient α_L and a transverse expansion coefficient α_T, so that

$$\boldsymbol{\alpha} = (\alpha_L - \alpha_T)\,\mathbf{pp} + \alpha_T\,\mathbf{I} = \begin{bmatrix} \alpha_L & 0 & 0 \\ 0 & \alpha_T & 0 \\ 0 & 0 & \alpha_T \end{bmatrix} \qquad \text{transversely isotropic} \tag{8.86}$$

The matrix expression applies when the symmetry axis $\mathbf{p}$ is aligned with the 1 coordinate axis, and the tensor expression is valid in any axes.

If a material is not isotropic, the off-diagonal components of the thermal expansion tensor can be non-zero, unless the coordinate system is aligned with the eigenvectors of $\boldsymbol{\alpha}$.

For combined thermal and mechanical loading, the stiffness tensor from Eqn. (8.4) multiplies the mechanical part of the strain, which is the difference between the total strain and the thermal strain. This gives a stress-strain-temperature relationship of

$$\boldsymbol{\sigma} = \mathbb{C} : (\boldsymbol{\varepsilon} - \boldsymbol{\alpha}\,\Delta T) \tag{8.87}$$

This can also be written as

$$\boldsymbol{\sigma} = \mathbb{C} : \boldsymbol{\varepsilon} - \boldsymbol{\beta}\,\Delta T \qquad \text{with} \qquad \boldsymbol{\beta} \equiv \mathbb{C} : \boldsymbol{\alpha} \tag{8.88}$$

$\boldsymbol{\beta}$ is a second-order tensor called the *thermal stress tensor;* multiplying it by $-\Delta T$ gives the thermal portion of the stress.

8.6.2 First Homogenization Step

We would like to predict $\boldsymbol{\alpha}$ or $\boldsymbol{\beta}$ as a function of the fiber and matrix properties, fiber orientation and fiber length. The two-step homogenization approach discussed in Section 8.2.2 is used for thermal expansion as well.

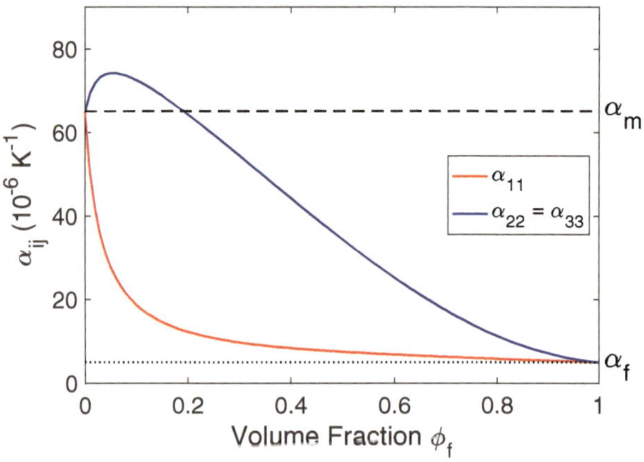

Figure 8.7 Predictions of thermal expansion vs. fiber volume fraction for a unidirectional composite. Lielens model for stiffness. Polycarbonate with glass fibers, $\ell/d_f = 100$, other properties in Table 8.2.

For the first homogenization step, a variety of models have been proposed (e.g., [Hal69, TT85, PFD04]). However, Levin [Lev67] showed that, for two-phase composites, once the stiffness tensor $\mathbb{C}^p$ of a pseudo-grain has been predicted, the thermal stress tensor of the pseudo-grain $\boldsymbol{\beta}^p$ can be calculated directly. Levin's result was cast into a form that is convenient for discontinuous fiber composites by Dray et al. [DGR07]:

$$\boldsymbol{\beta}^p = \left[\mathbb{C}^f - \mathbb{C}^p\right] : \left[\mathbb{C}^f - \mathbb{C}^m\right]^{-1} : \boldsymbol{\beta}^m + \left[\mathbb{C}^m - \mathbb{C}^p\right] : \left[\mathbb{C}^m - \mathbb{C}^f\right]^{-1} : \boldsymbol{\beta}^f \tag{8.89}$$

A derivation is given in Appendix B.4.4. This equation gives the thermal stress tensor that corresponds to *any* stiffness tensor $\mathbb{C}^p$. If $\mathbb{C}^p$ was predicted using a Mori–Tanaka model, it gives the Mori–Tanaka estimate for $\boldsymbol{\beta}^p$. Note that the fiber could be anisotropic, both in stiffness and thermal expansion, though the matrix is assumed to be isotropic.

Figure 8.7 shows calculations for a undirectional composite using the Lielens stiffness model together with Eqn. (8.89). The properties correspond to polycarbonate with glass fibers, $\ell/d_f = 100$, with other properties given in Table 8.2. As with most discontinuous fiber composites, the fibers have a much smaller thermal expansion coefficient than the matrix.

In Fig. 8.7, α_{11}, the expansion coefficient in the fiber direction, decreases rapidly as volume fraction increases, because the fibers tend to constrain the expansion of the matrix in the fiber direction. However, the transverse expansion coefficient α_{22} initially *increases* with fiber volume fraction. In this case, the transverse expansion of the composite is greater than the matrix expansion coefficient α_m up to $\phi_f = 0.2$. In this range the matrix is trying to expand volumetrically and, since the fibers restrict its ability to expand in the 1 direction, there is enhanced expansion transverse to the fibers.

At higher volume fractions the volumetric effect of the fibers dominates, and the transverse expansion coefficient decreases. As $\phi_f \to 1$ both expansion coefficients go to the fiber value, α_f. Of course, the practical range of ϕ_f for discontinuous fiber composites is on the left side of this figure.

8.6.3 Second Homogenization Step

To develop an equation for the second homogenization step, write Eqn. (8.88) for the pseudo-grain,

$$\overline{\boldsymbol{\sigma}}^p = \mathbb{C}^p : \overline{\boldsymbol{\varepsilon}}^p - \boldsymbol{\beta}^p \, \Delta T \tag{8.90}$$

Then take the orientation average of this equation as in Section 8.4.2, using the Voigt assumption that all pseudo-grains have the same average strain, Eqn. (8.62). This gives

$$\overline{\boldsymbol{\sigma}} = \langle \mathbb{C}^p \rangle : \overline{\boldsymbol{\varepsilon}} - \langle \boldsymbol{\beta}^p \rangle \, \Delta T \tag{8.91}$$

Comparing this to Eqn. (8.88), the Voigt average for the thermal stress tensor is simply the average of the pseudo-grain $\boldsymbol{\beta}^p$s:

$$\boldsymbol{\beta} = \langle \boldsymbol{\beta}^p \rangle \tag{8.92}$$

If the fiber length distribution is also being modeled, this average is

$$\boldsymbol{\beta} = \left\langle \sum_i w_i \, \hat{\boldsymbol{\beta}}^p (\mathbf{p}, \ell_i) \right\rangle \tag{8.93}$$

where $\hat{\boldsymbol{\beta}}^p$ is the pseudo-grain thermal stress tensor in a reference configuration.

Once $\boldsymbol{\beta}$ has been found, the overall thermal expansion coefficient can be calculated by inverting the second part of Eqn. (8.88),

$$\boldsymbol{\alpha} = \mathbb{C}^{-1} : \boldsymbol{\beta} \tag{8.94}$$

Since $\boldsymbol{\beta}$ is a second-order tensor, Eqns. (8.80) and (8.81) are used to calculate the orientation average in (8.92). Only the second-order orientation tensor is required for this step, so the closure approximation does not influence $\boldsymbol{\beta}$, though it does influence $\boldsymbol{\alpha}$ through $\mathbb{C}$.

Calculations showing the effect of fiber orientation on thermal expansion appear in Fig. 8.8. The composite is glass fiber/polycarbonate as in Fig. 8.7, with $\ell / d_f = 100$ and 30% by weight fibers or $\phi_f = 0.168$.

Figure 8.8(a) shows transversely isotropic orientation states with the 1 axis as the symmetry axis. For $A_{11} = 0$ the orientation is random in the 2–3 plane. The 1 direction is the out-of-plane direction, and α_{11} is much greater than α_m. This effect is similar to the transverse thermal expansion of the unidirectional composite in Fig. 8.7, but now the fibers constrain the thermal expansion for all directions in the 2–3 plane, so the expansion in the 1 direction is very large.

As A_{11} increases we see α_{11} decreasing and $\alpha_{22} = \alpha_{33}$ increasing. The two curves cross at $A_{11} = 1/3$, which is isotropic orientation. As the fibers become more aligned in the 1 direction, a low value of α_{11} is achieved, while α_{22} approaches or even exceeds the expansion coefficient of the matrix.

Planar orientation states are shown in Fig. 8.8(b). $A_{11} = 0.5$ is random-in-plane orientation, so $\alpha_{11} = \alpha_{22}$, and the out-of-plane coefficient α_{33} is very large. As A_{11} increases we see α_{11} decreasing and α_{22} increasing. The out-of-plane expansion α_{33} decreases slowly, eventually meeting the α_{22} curve when $A_{11} = 1$ and the fibers are all aligned in the 1 direction.

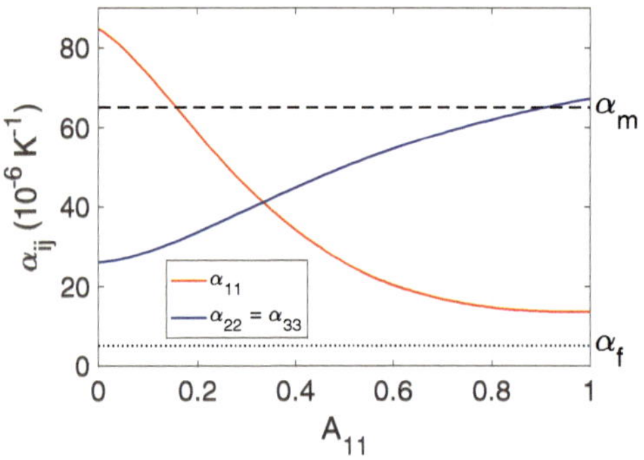

(a) Transversely isotropic orientation

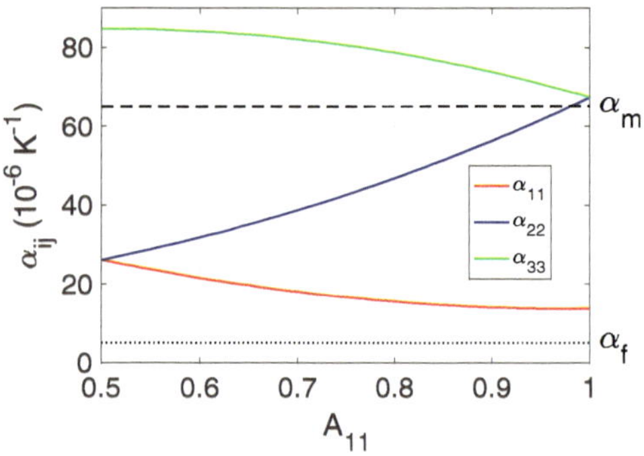

(b) Planar orientation

Figure 8.8 Predictions of thermal expansion vs. fiber orientation. Lielens model for stiffness, Voigt averaging and the natural closure. (a) Transversely isotropic orientation states, $A_{22} = A_{33} = (1 - A_{11})/2$ and all other $A_{ij} = 0$. (b) Planar orientation states, $A_{22} = 1 - A_{11}$ and all other $A_{ij} = 0$. PC with 30% by weight glass fiber, properties from Table 8.2 and $\ell/d_f = 100$.

As long as the fibers have lower thermal expansion than the matrix, we can expect the following trends:

Thermal expansion will be lower in directions with greater fiber orientation. However, thermal expansion in directions with low fiber orientation can be greater than the expansion of the matrix alone.

8.7 Nonlinear Stress-Strain Behavior

The models for stiffness and thermal expansion discussed above are for linear material behavior, where stress is proportional to strain and/or temperature change. This is appropriate for small, reversible deformations, and for the initial, linear portion of a material's stress-strain curve. However, to treat yielding, plastic deformation, and the path to failure, one must analyze nonlinear behavior.

Homogenization models have been developed for nonlinear stress-strain behavior. These models build on the ideas discussed in this chapter but are more complex, so only a brief summary will be given here. Readers wanting to know more about this type of modeling can start with the papers of Doghri and Tinel [DT05, DT06], and with a review of recent developments [SAS20].

The two-step homogenization process is once again used. In the first homogenization step, the deformation of the material is analyzed in small increments. These increments are often called time steps, though the material response depends on deformation rather than on time. Thus, for a time step Δt the strain will change by $\Delta\boldsymbol{\varepsilon}$ and the stress by $\Delta\boldsymbol{\sigma}$.

For each time step, the *matrix* stress-strain response within a pseudo-grain is approximated by a linear relationship,

$$\Delta\boldsymbol{\sigma}^m(\mathbf{x},t) = \widetilde{\mathbb{C}}^m(t) : \Delta\boldsymbol{\varepsilon}^m(\mathbf{x},t) \tag{8.95}$$

In this expression $\widetilde{\mathbb{C}}^m(t)$ is a *reference tangent stiffness.* For very small deformations it equals the matrix stiffness tensor $\mathbb{C}^m$, but as the matrix yields and deforms plastically, $\widetilde{\mathbb{C}}^m$ will decrease. A similar treatment can be used for the fiber material, though in most polymer composites the fibers remain within their elastic limit up to failure, so the linear elastic stiffness $\mathbb{C}^f$ can be used for the fibers.

A key idea of this approximation is that $\widetilde{\mathbb{C}}^m$ depends on time (i.e., it will change as the material strains), but it does not depend on position $\mathbf{x}$. With this assumption, any homogenization scheme for the linear stiffness of a unidirectional composite can be applied to the time step, with $\widetilde{\mathbb{C}}^m$ taking the place of $\mathbb{C}^m$. The Mori–Tanaka and Lielens/double-inclusion models are commonly used.

The analysis then proceeds time step by time step. However, there are multiple issues that still must be addressed in the calculations. These include choosing a plasticity model for the matrix elastic-plastic behavior, and deciding how $\widetilde{\mathbb{C}}^m$ will be chosen at each time step. Once the material begins to yield $\widetilde{\mathbb{C}}^m$ becomes anisotropic, making the computation of the Eshelby tensor more complex. It is common to identify an isotropic portion of $\widetilde{\mathbb{C}}^m$ and use that in computing the Eshelby tensor, but to use the full anisotropic $\widetilde{\mathbb{C}}^m$ everywhere else [DO03]. Also, one typically needs the average matrix strain $\overline{\boldsymbol{\varepsilon}}^m$ in the pseudo-grain to determine the reference tangent stiffness at each increment, but this is not known until the increment has been solved. Thus, an iterative equation-solving procedure is needed within each time step.

Different options for the first homogenization step are compared to finite element calculations on an RVE by Pierard et al. [PGS+07], and the best models are identified.

The second homogenization step proceeds as before, and commonly uses a Voigt average. For each time step a total strain increment $\Delta\overline{\boldsymbol{\varepsilon}}$ is imposed, and the total stress increment is computed as the orientation average of the pseudo-grain stresses,

$$\Delta\overline{\boldsymbol{\sigma}}(t) = \left\langle \Delta\overline{\boldsymbol{\sigma}}^p(t) \right\rangle \tag{8.96}$$

However, because the stress response of each pseudo-grain is not a linear function of strain, the averaging equations from Section 8.4.3 cannot be used. Instead, the full orientation distribution function $\psi(\mathbf{p})$ must be reconstructed or otherwise approximated. The usual approach is to choose n pseudo-grain orientations $\mathbf{p}^i$, $i = 1$ to n, and identify a weighting factor f_i for each grain. One method for this is given in [DT06], and another option is shown in the accompanying software[7]. The set of $(\mathbf{p}^i, f_i)$ values are then a discrete representation of $\psi(\mathbf{p})$. It is common for the reconstructed distribution function to match the second-order and fourth-order orientation tensors, so the $(\mathbf{p}^i, f_i)$ values satisfy

$$\mathbf{A} = \sum_{i=1}^{n} f_i \mathbf{p}^i \mathbf{p}^i \tag{8.97}$$

$$\mathbb{A} = \sum_{i=1}^{n} f_i \mathbf{p}^i \mathbf{p}^i \mathbf{p}^i \mathbf{p}^i \tag{8.98}$$

7 *https://github.com/charlestucker3/Fiber-Orientation-Tools*

Orientation averages are then computed as

$$\langle \overline{\boldsymbol{\sigma}}^p \rangle = \sum_{i=1}^{n} f_i \, \overline{\boldsymbol{\sigma}}^p(\mathbf{p}^i) \tag{8.99}$$

While predictions of nonlinear stress-strain behavior have more approximations than linear models, they still do a good job of predicting material behavior. For example, Nguyen et al. [NKF$^+$09] show good agreement to experiments on long fiber/PP injection molded samples.

Predicting the failure of a discontinuous fiber composite is more difficult. Stress-based failure criteria for composites have been generalized to discontinuous fiber materials with a distribution of orientations [VB99, NKF$^+$09], and the elastic-plastic models described above have been extended to incorporate progressive damage at the pseudo-grain level [KDA$^+$11, KDBD15]. However, no generally applicable scheme for strength prediction has emerged. This may be because failure in these materials can initiate from regions of high hydrostatic stress, which occur where the end of one fiber is very close to another fiber. The combination of *in-situ* micro-CT scanning and finite element analysis on the actual microscale geometry is helping elucidate these mechanisms [HAXS20, HS21].

8.8 Chapter Summary

Discontinuous fiber composites are anisotropic because fibers influence the composite properties more along their axes than in other directions. Composites are stiffer in directions with high fiber alignment and more compliant in directions with low fiber alignment.

The linear elastic properties of an anisotropic material are described by its stiffness tensor $\mathbb{C}$. Technical constants such as the Young's moduli, Poisson ratios and shear moduli in different directions can be determined from the stiffness tensor. Alternately, the stiffness tensor can be determined from a complete set of engineering constants. If a material exhibits symmetries in its elastic behavior (e.g., orthotropic, transversely isotropic, isotropic), its stiffness tensor will have fewer independent components.

Homogenization determines the properties of homogeneous material that is equivalent to a heterogeneous composite. For discontinuous fiber composites a two-step homogenization process is used. The composite is decomposed into pseudo-grains, each with a unidirectional orientation and a uniform fiber length. The first homogenization step analyzes the pseudo-grain properties, then the second step accounts for the fiber orientation distribution and, if desired, the fiber length distribution.

In the first homogenization step, mean-field theories provide analytical formulas for the stiffness of a unidirectional composite. The best mean-field theories are the Mori–Tanaka model and the Lielens/double-inclusion model. Pseudo-grain properties can also be determined by performing a numerical stress analysis (e.g., using the finite element method) on a representative volume element. All of the engineering constants of a unidirectional composite are affected by the fiber volume fraction and by the fiber and matrix properties. However, only the Young's modulus in the fiber direction E_{11} is strongly influenced by the fiber aspect ratio ℓ / d_f.

For the second homogenization step a Voigt average is standard. This assumes that the average strain in each pseudo-grain equals the overall average strain in the composite. As a result, the composite stiffness is simply the orientation average of the pseudo-grain stiffnesses. This average can be computed using only the second-order and fourth-order orientation tensors; the full fiber orientation distribution function is not required. If the fourth-order tensor is determined by a closure approximation, the choice of closure affects the stiffness results. The best closures for fiber orientation prediction (natural, IBOF) are quite accurate and should be used for mechanical property prediction.

Thermal expansion of an anisotropic material is described by a second-order thermal expansion tensor $\boldsymbol{\alpha}$ or, equivalently, a thermal stress tensor $\boldsymbol{\beta}$. Two-step homogenization is also used to predict thermal expansion. For the first step, $\boldsymbol{\beta}$ for a pseudo-grain can be determined from the composite stiffness $\mathbb{C}$ and the fiber and matrix properties, with no additional modeling. Using a Voigt average for the second step, $\boldsymbol{\beta}$ for the composite is simply the orientation average of the pseudo-grain $\boldsymbol{\beta}$s. Fibers reduce thermal expansion along their length, but at volume fractions typical of discontinuous fiber composites they can cause thermal expansion to increase in the direction transverse to the fibers.

Nonlinear properties can also be modeled, again using two-step homogenization. In a nonlinear calculation the orientation tensors alone do not provide sufficient information, and an approximate orientation distribution function $\psi(\mathbf{p})$ must be reconstructed to do the orientation averaging. Useful models exist for elastic-plastic behavior, and can predict the nonlinear stress-strain curve of a ductile composite. However, determining the stress at failure is more challenging, and there are no widely accepted models for predicting strength.

9 Current and Future Trends

While many aspects of fiber orientation modeling are well established, the models are not perfect and research on them continues. In recent years the CAE software industry has readily incorporated new or improved models into their software. Readers can expect this trend to continue, with new features being added on a regular basis.

This chapter covers two areas of modeling from recent research. The first, fiber migration, is now appearing in commercial software. The second, direct simulation of multiple fibers, may affect the way that fiber orientation and length are modeled in the future. The coverage of each topic is introductory, and more details can be found in the references.

9.1 Fiber Migration

9.1.1 Volume Fraction Variations in Injection Molded Parts

Up to this point we have assumed that the fiber volume fraction ϕ_f has the same value everywhere. However, multiple studies show significant variations in fiber volume fraction from one location to another within a molded part [TA93, VGWB$^+$12, GO18]. For injection molded composites these variations take two forms:

- A thickness-direction variation, where fiber volume fraction is higher near the midplane and lower near the mold walls. This is similar to the core/shell structure of fiber orientation.
- A flow-direction variation, where fibers are more concentrated at the last locations to fill, and may have lower-than-average concentration near the gate.

The magnitudes of these variations increase with average fiber volume fraction up to some point, then decrease at very high volume fractions.

Figure 9.1 shows examples of both types of variation. This data is from a long rectangular plaque, 2.85 mm thick and gated at one end, molded from polypropylene with 40% by weight long glass fibers [GO18]. The geometry is like Fig. 4.16 but with a point gate. The full study tested samples from 5 to 60 weight percent, and variations in volume fraction were the largest in the range of 30–50 weight percent.

The thickness-direction variation in ϕ_f is shown in Fig. 9.1 (a). Here, a 10 mm × 15 mm sample was cut from the center of the plaque and analyzed using micro-CT. A thresholding operation separated fiber voxels from matrix voxels, allowing the fiber volume fraction to be determined in any region of the sample. The CT image was divided into slices across the thickness, much like a fiber orientation analysis, to find ϕ_f as a function of x_3, the thickness-direction coordinate.

Figure 9.1 (a) shows a strong peak in volume fraction at the midplane, with a relatively flat profile of ϕ_f closer to the upper and lower mold walls. The highest fiber volume fraction is 1.5

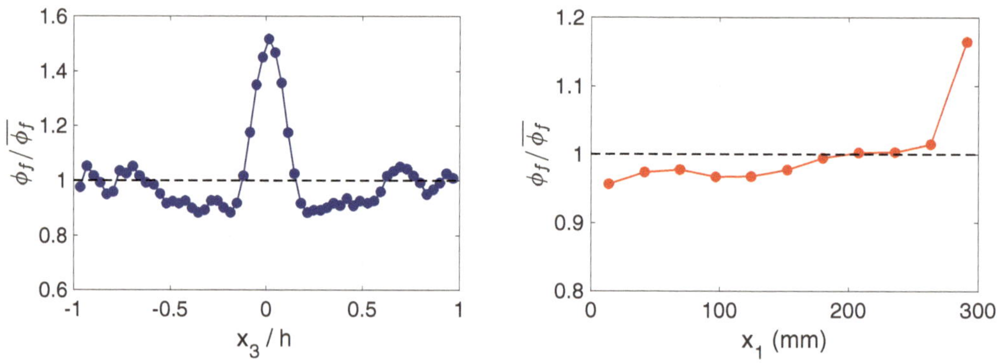

(a) Thickness direction (b) Flow direction

Figure 9.1 Variations of local fiber volume fraction in an injection molded plaque with 40% by weight long glass fiber in PP. (a) Local volume fraction across the thickness direction at the center of the plaque. (b) Thickness-averaged ϕ_f as a function of flow distance x_1. The final point is at the end of the plaque. ϕ_f values are normalized by the average value in the plaque $\overline{\phi_f}$. Data from [GO18].

times the nominal value, and the lowest is 0.9 times the nominal. This is a substantial effect that will influence the mechanical behavior of the material. In these experiments the midplane peak in ϕ_f develops soon after the gate, and persists until close to the end of flow, where flow-direction variations dominate.

The flow-direction data in Fig. 9.1(b) was collected by cutting the plaque into 25 mm × 13 mm rectangular pieces, weighing each piece, then burning off the matrix and weighing the fibers. The resulting fiber weight fraction is easily converted to fiber volume fraction, and each data point is an average volume fraction across the thickness of the part.

The most striking feature of Fig. 9.1(b) is that the volume fraction at the end of the plaque is higher than the nominal value. Measurements on short shots show that this region of high fiber volume fraction develops at the flow front soon after filling starts, and moves along with the flow front during filling. Along the rest of the plaque length the variation is small, and the thickness-averaged volume fraction is within a few percent of the overall average. The area nearest the gate seems to have lower-than-average volume fraction, though this variation is not as large as at the far end of the plaque.

According to Jeffery's equation, the fibers translate with the fluid. This is true for fibers in dilute suspensions. However, most discontinuous fiber composites are far from dilute and, as the data above show, the fibers move somewhat differently, migrating relative to the surrounding fluid. This relative motion only occurs when the suspension is deforming, so it is referred to as *shear-induced migration.*

9.1.2 Shear-Induced Migration Phenomena

Shear-induced migration has been studied extensively for suspensions of spherical particles in Newtonian fluids, e.g. [LA87, ATG+91, PAB+92, KHL94]. Experiments on fiber suspensions [MBA+94, SSBG18] show that fiber suspensions behave similarly. From these studies, shear-induced particle migration involves two competing phenomena:

- Particles migrate from regions of high shear rate toward regions of low shear rate.
- Particles migrate from regions of high volume fraction toward regions of low volume fraction.

The first phenomenon explains the tendency of fibers to collect near the midplane. For mold filling in a narrow gap, the shear rate is zero at the midplane and increases toward the wall (see Fig. 4.17). Fibers are attracted to the low-shear-rate region near the midplane, and begin to collect there.

The second phenomenon explains why the fibers do not *all* end up on the midplane in a molded part. Once the fiber volume fraction increases near the midplane, this second phenomenon operates and pushes fibers back toward the mold walls. If the flow continues long enough, a steady-state profile of volume fraction is reached where these two phenomena balance one another, and the migration rate goes to zero.

These studies have also provided information about the *rate* of particle migration. First, the migration rate is proportional to the shear rate $\dot{\gamma}$. This means that shear-induced migration follows the same strain-time scaling as fiber orientation (Section 4.2.6). As a result, changes in ϕ_f depend on the total strain, but not on the rate at which that strain was accumulated.

Second, the viscosity of the matrix fluid η_m does not affect the migration rate, provided it does not vary with position. This is because all forces on the particles are driven by stresses in the fluid, both forces that induce migration and forces that resist migration. Changing the fluid viscosity changes both types of forces equally, so the migration rate remains unchanged.

Third, larger particles migrate faster than smaller particles. The rate of migration for spherical particles is proportional to a^2, where a is the radius of the spheres. Particle size does not affect the steady-state profile of volume fraction, only the rate at which it is achieved.

Fourth, the overall velocity field is affected by changes in particle volume fraction. Suspension viscosity increases with particle volume fraction, so locations that develop higher volume fraction due to migration will have higher viscosity. This decreases the local shear rate, which reduces the tendency of fibers to migrate toward those regions. For example, in an injection molding flow where the particles concentrate near the midplane, particle migration will make the velocity profile flatter near the midplane and steeper near the walls. In general, the velocity solution and the solution for particle volume fraction are *coupled*, much like the flow/fiber orientation solutions in Chapt. 6.

Finally, both experiments and theory [CLWL21] show that

- Flow-direction variations in ϕ_f result from shear-induced migration acting in combination with the fountain effect.

The mechanism for this is sketched in Fig. 9.2. As the material enters the mold cavity, shear-induced migration causes particles to collect near the midplane. As discussed in Section 4.3.4, the velocity at the midplane is higher than the gapwise average velocity, so this particle-rich material catches up to the flow front, enters the fountain flow, and spreads across the entire thickness of the cavity. This produces a high particle concentration across the entire part thickness at the flow front. For any location in the mold, once the flow front has moved past, shear-induced migration pushes particles near the wall back toward the midplane, re-establishing

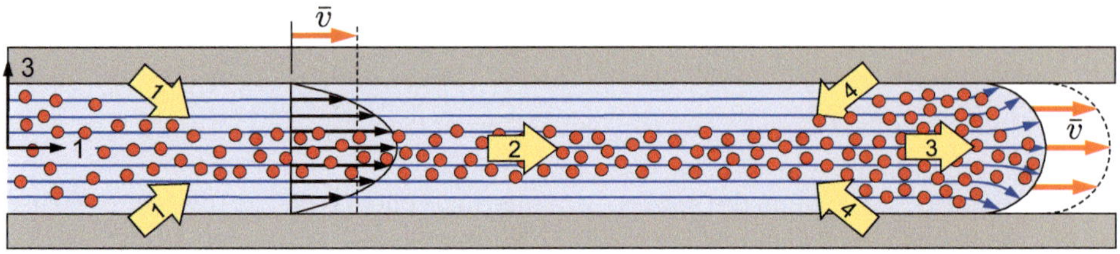

Figure 9.2 Mechanism for creating flow-direction variations in particle volume fraction. (1) Shear-induced migration drives particles toward the midplane. (2) Particles on the midplane move faster than the average velocity $\bar{v}$, catching up to the flow front. (3) At the flow front, the fountain flow spreads the particles across the gap height. (4) After the flow front passes, shear-induced migration drives particles back toward the midplane.

the thickness-direction profile of ϕ_f. Thus, we do not need any mechanism other than shear-induced migration to explain flow-direction variations in ϕ_f in injection molded parts.

9.1.3 Modeling Shear-Induced Migration

Quantitative models for particle migration have been developed for spherical particles, and these provide the basis for molding migration in fiber reinforced composites. Here we present the basic structure of the model, which gives an equation for the fiber volume fraction as a function of position and time.

To start, we need a variable that describes the particle flux due to migration. Particles migrate if their velocity is different from the overall fluid motion, so we want a flux relative to the average motion of the suspension. Using $\mathbf{v}$ for the average suspension velocity (the fluid and the particle phases combined) and $\mathbf{v}_p$ for the average velocity of the particles, and retaining ϕ_f as the particle volume fraction, the *migration flux vector* $\mathbf{j}_\perp$ relative to the fluid motion is defined as

$$\mathbf{j}_\perp \equiv \phi_f \left(\mathbf{v}_p - \mathbf{v}\right) \tag{9.1}$$

Next, we write a mass balance equation for the particle phase. Consider a small volume fixed in space through which the suspension flows. The concentration of particles inside the volume can only increase if there are more particles migrating into the volume than migrating out. Allowing for both the average motion of the suspension and the migration flux, we find

$$\frac{\partial \phi_f}{\partial t} + \mathbf{v} \cdot \nabla \phi_f = -\nabla \cdot \mathbf{j}_\perp \quad \text{or} \quad \frac{\partial \phi_f}{\partial t} + v_i \frac{\partial \phi_f}{\partial x_i} = -\frac{\partial j_{\perp,i}}{\partial x_i} \tag{9.2}$$

All particle migration models must obey this equation. The full suspension (particles and fluid together) must also obey the general mass balance equation (6.43). Subtracting Eqn. (9.2) from that equation gives a mass balance for the fluid phase, so a separate fluid-phase mass balance equation is not needed.

The next task is to develop a relationship for the migration flux $\mathbf{j}_\perp$. The first model for migration of spherical particles was the *diffusive flux model* [LA87, PAB+92]. This model focused on the kinematics of particle-particle interactions and used scaling arguments to write the expression for the migration flux. An alternate approach is the *suspension balance model* (SBM) [NB94, MB99]. Here the migration flux is determined by a force balance on the particles, so the suspension balance model is more closely tied to the rheology of the suspension. Both models accurately predict experimental observations for spherical-particle suspensions in controlled shear flows [PAB+92, MB99, SBA04]. Recent practice has centered on the suspension balance model, so we present it here.

In the suspension balance model the constitutive equation for the suspension is written as the sum of the average stress in the fluid (which is Newtonian) and a contribution $\mathbf{\Sigma}^p$ from the particles:

$$\boldsymbol{\tau} = 2\eta_m \mathbf{D} + \mathbf{\Sigma}^p \tag{9.3}$$

Next, a momentum balance is written for the particle phase. Assuming that inertia and body forces are negligible, if $\mathbf{\Sigma}^p$ varies with position (i.e., $\nabla \cdot \mathbf{\Sigma}^p \neq 0$) then there is a net force on the particles associated with $\mathbf{\Sigma}^p$. This force could come from particle-particle interactions or from spatial variations in shear rate or viscosity, and it causes the particles to migrate. As the particles migrate they experience a drag force proportional to the difference between their velocity $\mathbf{v}_p$ and the suspension velocity $\mathbf{v}$. They acquire a velocity $\mathbf{v}_p$ where this drag force balances the driving force from $\nabla \cdot \mathbf{\Sigma}^p$.

Using a formula for the drag force based on the settling of spheres in dense suspensions, Morris and Boulay [MB99] wrote the migration flux as

$$\mathbf{j}_\perp = \frac{2a^2}{9\eta_m} f_h(\phi_f)\, \nabla \cdot \mathbf{\Sigma}^p \qquad \text{or} \qquad j_{\perp,i} = \frac{2a^2}{9\eta_m} f_h(\phi_f)\, \frac{\partial \Sigma^p_{ji}}{\partial x_j} \tag{9.4}$$

In this expression $f_h(\phi_f)$, called the sedimentation hindrance function, is a dimensionless function that describes how the particle drag force increases with volume fraction. For spherical particles it is written as [MM06]

$$f_h(\phi_f) = \left(1 - \frac{\phi_f}{\phi_{\max}}\right)\left(1 - \phi_f\right)^{\alpha-1} \tag{9.5}$$

where α is an empirical constant, usually $2 \le \alpha \le 4$. $\phi_{\max}$ is the maximum particle volume fraction, which for spherical particles is the packing limit, $\phi_{\max} = 0.68$. $\phi_{\max}$ plays an important role in flow-induced migration models, since ϕ_f can never be greater than $\phi_{\max}$. The inclusion of $\phi_{\max}$ in Eqn. (9.5) guarantees this, since $\mathbf{j}_\perp$ will go to zero as $\phi_f \to \phi_{\max}$.

Equation (9.4) will only predict migration across streamlines if the suspension exhibits normal stresses in simple shear flow. To capture this, Morris and Boulay proposed a constitutive equation for the particle stress:

$$\mathbf{\Sigma}^p = 2\eta_p(\phi_f)\,\mathbf{D} - \eta_n(\phi_f)\,\dot{\gamma}\,\widehat{\mathbf{Q}} \tag{9.6}$$

Here $\eta_p(\phi_f)$ and $\eta_n(\phi_f)$ are the particle contributions to the shear viscosity and the normal stresses, respectively. For spherical particles, Morris and Boulay use

$$\frac{\eta_p}{\eta_m} = 2.5\,\phi_{\max}\frac{(\phi_f/\phi_{\max})}{(1-\phi_f/\phi_{\max})} + K_s \frac{(\phi_f/\phi_{\max})^2}{(1-\phi_f/\phi_{\max})^2} \tag{9.7}$$

$$\frac{\eta_n}{\eta_m} = K_n \frac{(\phi_f/\phi_{\max})^2}{(1-\phi_f/\phi_{\max})^2} \tag{9.8}$$

Values of $K_s = 0.1$ and $K_n = 0.75$ provided a good fit to experimental data for spherical particles.

Returning to Eqn. (9.6), the first term is a Newtonian contribution that, when used in Eqn. (9.3), increases the Newtonian viscosity of the suspension. The second term in Eqn. (9.6) gives normal stresses in simple shear flow that are proportional to the shear rate $\dot{\gamma}$. $\widehat{\mathbf{Q}}$ is a dimensionless tensor that determines the relative magnitude of the normal stresses. In a coordinate system where 1 is the flow direction and 2 is the direction in which velocity changes, $\widehat{\mathbf{Q}}$ is given by

$$\widehat{\mathbf{Q}} = \begin{bmatrix} 1 & 0 & 0 \\ 0 & \hat{\lambda}_2 & 0 \\ 0 & 0 & \hat{\lambda}_3 \end{bmatrix} \tag{9.9}$$

where $\hat{\lambda}_2$ and $\hat{\lambda}_3$ are additional, dimensionless model parameters. Typical values are $\hat{\lambda}_2 = 0.8$ and $\hat{\lambda}_3 = 0.5$.

To get a better idea of how the suspension balance model works, consider a 1–3 shearing flow where v_1 depends only on x_3 and $v_2 = v_3 = 0$. The shear rate $\dot{\gamma} = \partial v_1 / \partial x_3$ could depend on x_3 as well, but the matrix fluid viscosity η_m is constant.

Equation (9.6) then gives the particle contribution to the stress as

$$\mathbf{\Sigma}^p = \dot{\gamma} \begin{bmatrix} -\eta_n & -\eta_p & 0 \\ -\eta_p & -\hat{\lambda}_3 \eta_n & 0 \\ 0 & 0 & -\hat{\lambda}_2 \eta_n \end{bmatrix} \tag{9.10}$$

The 3-direction component of $\nabla \cdot \mathbf{\Sigma}^p$ for this flow, which will drive migration across the streamlines, is $\partial \Sigma^p_{33} / \partial x_3$. Using this in Eqn. (9.4), the migration flux across the streamlines is

$$j_{\perp,3} = -\frac{2\hat{\lambda}_2 a^2}{9\eta_m} f_h(\phi_f) \left[\eta_n(\phi_f) \underline{\frac{\partial \dot{\gamma}}{\partial x_3}} + \dot{\gamma} \underline{\frac{\partial \eta_n}{\partial \phi_f} \frac{\partial \phi_f}{\partial x_3}} \right] \tag{9.11}$$

Here we have used $\partial \eta_n / \partial x_3 = \left(\partial \eta_n / \partial \phi_f \right) \left(\partial \phi_f / \partial x_3 \right)$, since η_n varies with volume fraction.

The underlined terms in Eqn. (9.11) show that the suspension balance model exhibits the two phenomena of shear-induced migration from Section 9.1.2: the first term creates a flux from high shear rate to low and the second term creates a flux from high volume fraction to low. The flux is proportional to a^2 and, because η_n is proportional to η_m in (9.7) and (9.8), the matrix viscosity cancels out and does not affect the flux.

Substituting $\mathbf{j}_\perp$ from Eqn. (9.4) into Eqn. (9.2) gives an equation for ϕ_f as a function of position and time. This becomes the primary equation that is solved to find $\phi_f(\mathbf{x}, t)$. A boundary condition on ϕ_f is needed at the inlet, while at the mold walls there is no migration normal to the wall, $\mathbf{j}_\perp \cdot \mathbf{n} = 0$. As mentioned above, the solution for $\phi_f(\mathbf{x}, t)$ is coupled with the equations governing the velocity $\mathbf{v}(\mathbf{x}, t)$ because viscosity depends on ϕ_f.

The suspension balance model as originally developed and written here applies only to simple shear flows. More general versions of the model have been developed for two-dimensional motion, and are appropriate for numerical calculations of complex flow problems [FMB⁺02, MSM09].

9.1.4 Application to Injection Molding

Shear-induced migration models have been implemented in commercial injection molding codes (e.g., [TCH18a]) and the qualitative predictions are correct, though quantitative comparisons between predictions and experiments are limited. When applying these models to fiber suspensions, important question are how to choose the values of parameters like a, $\phi_{\max}$, K_s, K_n and α, and whether to replace the functions $f_h(\phi_f)$, $\eta_p(\phi_f)$ and $\eta_n(\phi_f)$ with something related to fiber suspension rheology.

There are two goals for modeling shear-induced migration in fiber suspensions: predicting the migration rate and predicting the magnitude of the peak in ϕ_f.

The particle radius a affects the migration rate so it is important in transient predictions. Interestingly, it does not affect steady-state profiles of ϕ_f. Mondy et al. [MBA⁺94] measured transient profiles of ϕ_f in a large-gap Couette flow, and found that longer fibers migrated more quickly that shorter fibers at the same d_f and ϕ_f. The rate of diffusion was roughly proportional to the volume of an individual fiber to approximately the 2/3 power. This suggests modeling fibers with an effective sphere radius a_e that makes the sphere volume $(4/3)\pi a_e^3$ proportional to the volume of a single fiber $(\pi/4)\ell d_f^2$:

$$a_e \propto \frac{d_f}{2}\left(\frac{3}{4}\frac{\ell}{d_f}\right)^{1/3} \tag{9.12}$$

Several researchers have used this expression as an equality for suspensions [FPTZ00, TCH18a, PGBC19], though Mondy et al. only suggest proportionality.

Precise modeling of migration rates in fiber suspensions is challenging, as shown in experiments by Perumal et al. [PGBC19]. In these experiments a polyamide-6 reinforced with 33% by weight short glass fibers was injection molded in a center-gated disk mold like Fig. 4.18. A smaller disk was cut from the center of this molding and placed in a parallel-disk rheometer. There it was re-melted, then sheared by rotating the upper disk as shown in Fig. 9.3(a). The shearing was halted at a given value of shear strain, the sample was allowed to cool and solidify, and a small piece from near the outer radius was removed and analyzed using micro-CT. This provided both orientation and volume fraction profiles across the thickness.

The as-molded sample had a peak of fiber volume fraction near the midplane, shown in Fig. 9.3(b), with a profile similar to the plaque sample in Fig. 9.1(a). In a parallel-disk rheometer the shear stress $\tau_{\theta z}$ is uniform across the gap at a given radius, so the shear rate is nearly uniform as well. With a constant shear rate, the fibers should migrate from regions of high volume fraction toward regions of low volume fraction, so the peak in ϕ_f should diminish, and eventually disappear. This is exactly what happens, as shown by the experimental data in Figs. 9.3(c) and (d). This decay in the peak of ϕ_f is driven by the gradient in ϕ_f (the second term in brackets in Eqn. (9.11)), and the term involving the gradient in $\dot{\gamma}$ (the first term in brackets) has almost no role in this experiment.

The curves in Figs. 9.3(c) and (d) are predictions using the suspension balance model for spherical particles, as described in Section 9.1.3. The effective particle radius a_e was used as an adjustable parameter. Using $a_e = 65\,\mu\text{m}$ (the solid line) gives an excellent fit to the data at $\gamma = 120$, but the peak does not flatten fast enough to match the data at $\gamma = 240$. In contrast, doubling the radius to $a_e = 130\,\mu\text{m}$ matches the data at $\gamma = 240$ very well, but the peak in ϕ_f decays too quickly to match the data at $\gamma = 120$. For this suspension the right-hand side of Eqn. (9.12) equals $31.6\,\mu\text{m}$, so these values of a_e have the right order of magnitude.

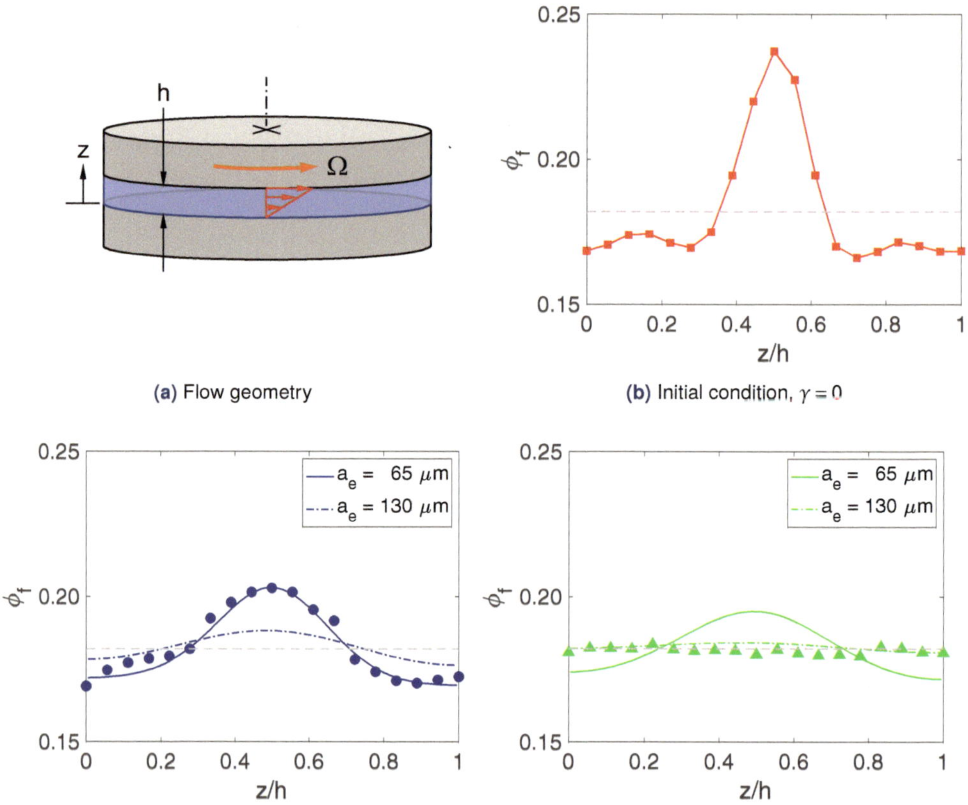

(a) Flow geometry

(b) Initial condition, $\gamma = 0$

(c) $\gamma = 120$

(d) $\gamma = 240$

Figure 9.3 Shear-induced fiber migration in PA6 with 33 wt.% short glass fibers. (a) The sample is sheared between parallel disks. (b)–(d) compare measured fiber volume fractions (points) across the thickness for different shear strains γ to predictions of the suspension balance model for two different values of effective sphere radius a_e. Experimental data from [PGBC19].

This comparison of theory to experiment suggests that particle migration rates in fiber suspensions are more complex than in suspensions of spherical particles. Migration rates for fibers may be more sensitive to volume fraction, and they may depend on the fiber orientation state.

In this particular experiment fiber orientation changes as the fibers migrate. The initial sample has a typical core/shell orientation: fibers near the midplane are oriented mainly in the θ direction while fibers near the walls are oriented mainly in the radial direction. As the sample is sheared in the parallel-disk rheometer, the shell fibers rotate toward the θ direction. Whether this affects the migration rates is not known. Perumal et al. [PGBC19] modified the SBM by using an expression for $\mathbf{\Sigma}^p$ based on fiber suspension rheology similar to Eqn. (6.29), which does introduce some orientation dependence in $\mathbf{j}_\perp$, but still could not fit all of their experimental data with a single value of a_e.

The second goal for injection molding is to accurately predict the height and width of the peak in volume fraction on the midplane of the part. In the SBM this height is influenced by $\phi_{\max}$, but the correct choice of $\phi_{\max}$ for a fiber suspension is not obvious. If fibers are randomly oriented in space and do not bend, then their maximum volume fraction is inversely proportional to their aspect ratio [EG86],

$$\phi_{\max} \simeq \frac{5.3\,d_f}{\ell} \tag{9.13}$$

This leads to low limits on volume fraction for longer fibers, e.g., $\phi_f \leq 0.053$ for $\ell/d_f = 100$. However, if the fibers are at least partially aligned they pack much more densely, and modest amounts of fiber bending also allow denser packing. In fact, most LFT materials have volume fractions well above the limit of Eqn. (9.13).

Strednak et al. [SSBG18] studied shear-induced migration of fiber suspensions in pipe flow, and found that the steady-state degree of segregation increased as the product $\overline{\phi}_f \ell / d_f$ increased. Here $\overline{\phi}_f$ is the overall average volume fraction. However, the centerline concentration of fibers did not go to the same value of ϕ_f for all experiments, but depended on average concentration and fiber aspect ratio.

Tseng et al. [TCH18a] implemented the suspension balance model in a 3-D injection mold filling simulation and simulated a center-gated disk molded from a PBT with 30% by weight short glass fibers. They compared their predictions to data from Vélez-Garcia et al. [VGWB+12, VG12]. This molding shows noticeable shear-induced migration at some locations, and the calculations reproduce the general trends of the data, which is encouraging. Unfortunately, the ϕ_f data is derived from polished cross-sections, and is very noisy compared to the micro-CT data in Figs. 9.1(a) and 9.3. As a result, one cannot draw strong conclusions about the details of the fiber migration model from this comparison.

Overall, the general form of the spherical-particle migration models works well for injection molded composites. However, there is room for improvement in many of the details of the model. Additional research, both theoretical and experimental, is needed to refine the models and develop robust methods to set the values of model parameters for any given material.

9.1.5 Fiber/Matrix Separation in SMC and GMT

A very different mechanism can cause *fiber/matrix separation* in materials like SMC and GMT, which are typically processed by compression molding. In these materials the fibers are long compared to the cavity thickness, are initially arranged in bundles or strands, and have significant flexibility. The strands are initially oriented almost parallel to the mold surface, and are more or less randomly oriented in that plane.

As this mat of fibers is compressed, fibers are pressed against the mold walls, especially where multiple fibers cross one another. The friction between the fibers and the mold surface can retard the movement of the fibers relative to the resin. In extreme cases (e.g., a GMT charge with high fiber volume fraction and a very slow closing speed) the fibers do not move laterally at all and the resin squeezes out, like water from a sponge. More commonly, the fibers move more slowly than the resin, so the area under the initial charge ends up with a higher fiber volume fraction while the last places to fill have a lower fiber volume fraction.

In addition, long fibers must bend to flow into ribs or around sharp corners. A fiber with unfavorable orientation can bridge the entrance to a rib, inhibiting other fibers from entering. This can lead to a reduced fiber volume fraction in a rib [KWTO17]. In extreme cases there may be no fibers at all at the top of the rib.

These forms of fiber/matrix separation are driven by fiber-wall interactions, and the size and flexibility of fibers plays a major role. These phenomena are not accounted for in the orientation and migration models discussed so far. A few models have been proposed for this type of fiber/matrix separation in flat or gently curved parts, balancing the drag forces on fibers from the fluid against fiber-wall friction [PGAC+17, KW20]. These models can readily be included in commercial software, but may not be accurate at a sharp corner or rib entrance. Direct fiber simulations, where each individual fiber is modeled explicitly, can be used to model fibers near ribs and corners. We discuss those simulations in the next section.

9.2 Direct Fiber Simulations

9.2.1 Overview

The models of fiber orientation discussed in Chapt. 5 are all phenomenological. They modify Jeffery's equation to make the model behave like a suspension of interacting fibers, but they are not derived from a fundamental analysis of fiber motion. As a result, the parameters of these models can only be determined by fitting experimental data.

A different approach is to calculate directly the motion of each individual fiber, and to use multiple fibers so that phenomena like fiber-fiber interactions are represented. Recall from Section 4.2 that Jeffery's equation is based on a force balance and moment balance for a single fiber. Fibers in a concentrated suspension must also obey Newton's laws of motion. If the forces and moments acting on them can be accurately represented, the resulting motions of the fibers should be accurate as well. This is the idea behind *direct fiber simulations* (DFS). Because these calculations apply Newton's laws to each individual fiber, they are also referred to as *mechanistic models*[1].

Figure 9.4 shows two configurations of multiple fibers from one DFS calculation. The fibers in Fig. 9.4(a) are the initial state, and there are more than 1,000 fibers whose centroids lie in a simulation box. Fluid fills the space between the fibers but is not shown. This suspension is concentrated, and fiber-fiber interactions are very likely. In the simulation, the box is sheared and the influence of neighboring fibers outside the box is represented using periodic boundary conditions—an approach much like the numerical simulation of RVEs in Chapt. 8.

Over time each fiber moves and reorients, and the fiber configuration changes. The same fibers after a shear strain of $\dot{\gamma}t = 100$ are shown in Fig. 9.4(b). These fibers are generally aligned in the flow direction, as we would expect. However, they are not fully aligned. This is a realistic behavior that has been captured without using any empirical coefficients like C_I. Quantitative data for this calculation appears in Fig. 9.7.

A key advantage of direct fiber simulations is that they can include physical phenomena that are not represented in the models from previous chapters. The phenomena include:

- Fiber-fiber interactions, with or without friction,
- Fiber-wall interactions, including the behavior of fibers confined in a narrow gap,

1 Other names include direct bundle simulation, the discrete element method (DEM), direct numerical simulation (DNS), particle level simulation (PLS) and particle simulation method (PSM).

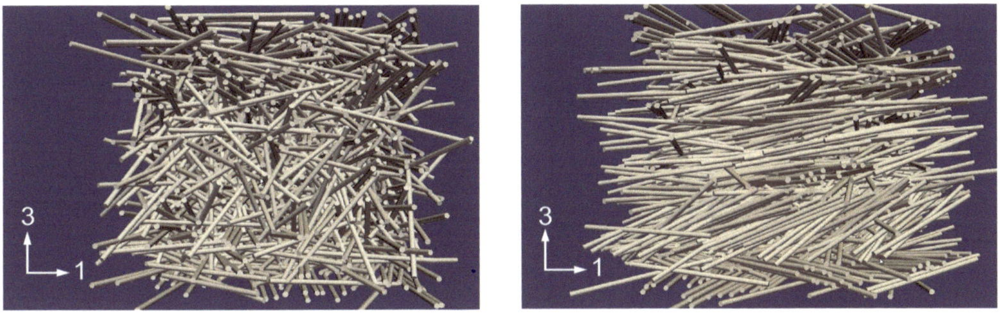

(a) Initial fiber configuration (b) After shearing by $\dot{\gamma}t = 100$

Figure 9.4 Fibers in the representative volume element of a direct fiber simulation. The fibers in (a) are subjected to a simple shear deformation, $v_1 = \dot{\gamma}x_3$, to create (b). Images by Dr. Armin Kech and Dr. Susanne Kugler, Robert Bosch GmbH.

- Fiber flexibility,
- Fiber breakage, and
- Fiber/matrix separation.

The main disadvantage of direct fiber simulations is that they require a large amount of computation, so the number of fibers that can be modeled is limited. The earliest calculations treated a few dozen fibers, though it is currently possible to model thousands of fibers. Run times for a single case can range from hours to days.

9.2.2 Simulation Methods

A complete treatment of direct fiber simulation methods is beyond the scope of this book, and readers can consult the references in [KKCO20] for details. The general framework is the same for most codes, and we summarize that here.

9.2.2.1 Representing the Fibers

The first step is to choose how fibers will be represented in the calculation. If only rigid fibers are of interest, then a rigid cylinder like Fig. 9.5(a) can be used. Some codes treat each fiber as a chain of spherical beads, Fig. 9.5(b), or a series of linked cylindrical segments, Fig. 9.5(c). These approaches are more general, and offer the option to include fiber flexibility. This is done by allowing the segments (either spheres or cylinders) to bend at the joints where they are connected, as in Fig. 9.5(d). These joints are spring-like, and provide a torque that tries to straighten the joint. This torque is proportional to the bend angle. The spring constant for the joint is matched to the bending stiffness of the fiber and to the length of the fiber segments. Most recent calculations use cylindrical segments, like Figs. 9.5(c) and (d), for computational efficiency. The number of segments per fiber is chosen to represent bent fiber shapes without unduly increasing computation time.

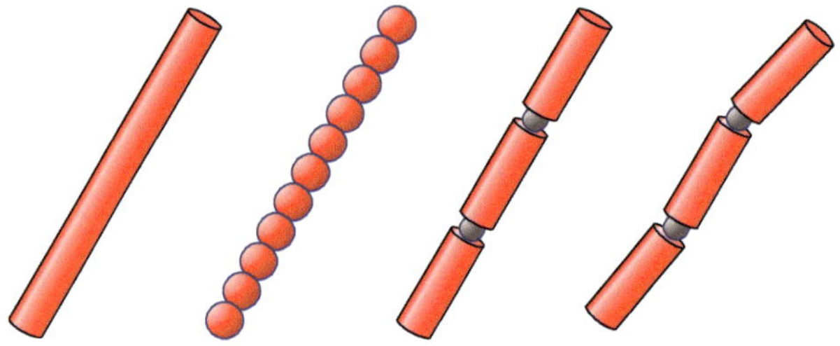

Figure 9.5 A cylindrical fiber (a) can be modeled either as a chain of spherical beads (b) or a chain of cylindrical segments (c). Both (b) and (c) can model a flexible fiber by bending at the joints (d).

9.2.2.2 Initial Configuration

Next, an initial configuration of fibers is generated. One must specify the position and orientation of each fiber, and fibers may not pass through one another. Computational approaches for initial packing of rigid fibers are discussed, e.g., by [AERT13, Sch17]. If the fibers are straight, rigid and randomly oriented, then they have a maximum volume fraction given by Eqn. (9.13). Materials like LFTs often exceed that limit, so the fibers must either bend or align locally in the initial configuration. If fiber bending is allowed, one can generate fibers that initially overlap, and use the simulation code to remove the overlaps [SBO20]. Alternately, one can generate a precursor configuration at a volume fraction that obeys the rigid-fiber packing limit, then use the simulation code to compress it to the desired volume fraction [GGOS20].

Once the initial configuration is set, the simulation proceeds in a series of time steps. At each time step, the forces and moments on each fiber segment are calculated. For each fiber this includes hydrodynamic forces and moments from the overall fluid motion, interaction forces from other fibers, and forces from the walls. For discontinuous fiber composites, the effects of inertia and gravity are usually negligible, though these can be simulated.

9.2.2.3 Hydrodynamic Forces

Hydrodynamic forces from the overall fluid motion are modeled using the underlying solution of Jeffery's equation, or something similar. Fluid rotation and deformation exert a torque on each fiber segment, depending on its orientation. If these are the only forces acting on a rigid fiber, then its resulting motion should be the Jeffery motion. A useful test for a simulation code is to model a single, rigid fiber in simple shear flow, and check that the calculated motion matches the path and period of the Jeffery orbit for the fiber's aspect ratio.

The development of Jeffery's equation, and all of the subsequent models in Chapt. 5, assumed that the unperturbed fluid velocity varies linearly along the length of the fiber, so the fiber rotation was governed by the local values of $\mathbf{D}$ and $\mathbf{W}$ (see Eqn. (6.16)). For longer fibers this may not be the case, but this is easily handled in a direct fiber simulation: each segment of the fiber experiences the velocity and velocity gradient at its position.

9.2.2.4 Interaction Forces

Fiber-fiber interaction forces can take one of several forms, depending on the distance between the fibers. Consider a fiber numbered i and a potentially interacting fiber numbered j, as shown in Fig. 9.6. At each time step the centroid positions $\mathbf{x}_i$ and $\mathbf{x}_j$ and orientations $\mathbf{p}_i$ and $\mathbf{p}_j$ are known. This information can be used to determine the unit vector normal to both fibers $\mathbf{n}_{ij}$ and the minimum distance between their centerlines h_{ij}. One also determines s_i and s_j, the distances from each centroid to the point of closest approach along the fiber centerlines. Equations are given, e.g., by [YKD94, AERT13]. If both s_i and s_j lie within the lengths of their respective fibers, then fibers i and j can have a short-range interaction.

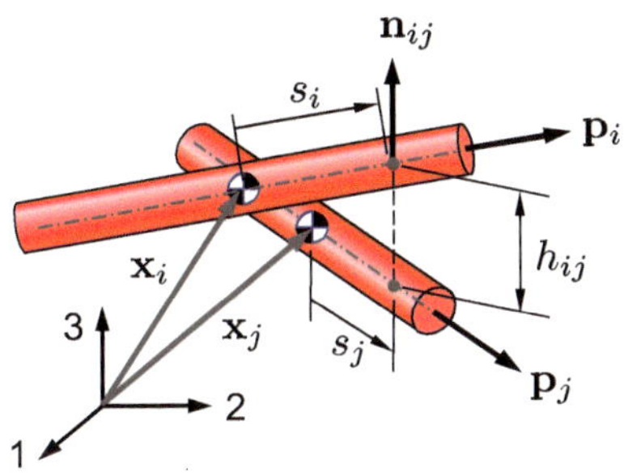

Figure 9.6 Nomenclature for determining the shortest distance between two fiber segments.

If the distance h_{ij} equals the fiber diameter d_f then the two fibers are in mechanical contact. They cannot penetrate one another, so any motion trying to bring the fibers closer together will be resisted by a contact force than maintains $h_{ij} = d_f$. This is usually handled with a penalty scheme that adds a highly nonlinear spring between the two fibers. The force in this spring increases exponentially as the gap between the fiber surfaces goes to zero [YM95, SSK00]. The spring is activated when the gap between the fibers falls below a critical value that is small compared to d_f.

Many codes neglect friction, in which case the contact force acts only in the direction of $\mathbf{n}_{ij}$. It is possible to include Coulomb friction in the simulation [SSK00, SBO20].

If the fibers are close together but not in mechanical contact, they can still interact through short-range hydrodynamic forces. When two fibers are approaching one another ($\mathrm{d}h_{ij}/\mathrm{d}t < 0$), matrix fluid is squeezed out of the gap between them. This increases the local fluid pressure, providing a force that resists the approach of the fibers. If the distance between the fibers is increasing, this interaction resists the separation. Lubrication theory can be used to analyze the fluid flow in the gap between the fibers, providing a formula for the interaction force [YKD94]. The simplest formula loses accuracy when the fibers are nearly parallel or the contact point is near the end of a fiber, but models that more accurately treat those cases have been developed [KBP+21]. The force due to a short-range hydrodynamic interaction also acts normal to both fibers, in the direction of $\mathbf{n}_{ij}$. Tangential hydrodynamic forces are much smaller than normal forces, and are usually neglected.

The strength of a short-range hydrodynamic interaction decreases as the gap between the fibers grows, so short-range interactions are neglected when the distance between the fibers exceeds some critical value, e.g., $h_{ij} > 1.5d_f$ [YM95].

Fibers can also interact when the distance between them is large compared to d_f. However, these long-range hydrodynamic interactions are expensive to compute, and are usually thought to be unimportant in concentrated suspensions. For any given fiber, the closest neighbors tend to screen the long-range interactions from fibers that are farther away. While it is possible to include long-range hydrodynamic interactions [PTFTZ02], they are usually ignored.

Fiber-wall interactions are modeled in the same way as fiber-fiber interactions: a short-range hydrodynamic force when the distance between the fiber and the wall is small, transitioning to a contact force (and possibly a tangential force due to friction) when the fiber touches the wall.

9.2.2.5 Solution and Time Stepping

Once all the forces and moments have been determined, equations are assembled for the total force and total moment on each fiber. These equations can be solved for the motion $\dot{\mathbf{x}}_i$ and $\dot{\mathbf{p}}_i$ for each fiber. Using these, the position and orientation of each fiber are updated for a step size Δt, and the time step is complete. Since the position and orientation of every fiber are known, one can easily calculate orientation tensors or other microstructural parameters at each time step.

Most codes use a decoupled approach, where the fluid velocity field is pre-determined and is unaffected by the presence of the fibers. It is possible to couple the fluid-flow solution and the motion of the fibers, at least at the length scale of a fiber segment, and a few simulations have done this [LU08]. In a coupled calculation the fluid velocity solution is updated at each time step as well.

This time-step cycle is repeated until the desired total time is reached. Because fibers can come into contact with small rotations, the time step size Δt must be small. This, together with the amount of computation required for each time step, means that calculation times can range from hours to days.

A variety of numerical approaches are used to accelerate the calculation, so the details of any particular code may differ from the description above.

9.2.3 Applications

Direct fiber simulations can be used in several ways. Here we discuss three popular applications: parameter identification, full-part simulation and creation of new orientation models.

9.2.3.1 Parameter Identification

One of the earliest uses of direct fiber simulations was to predict the interaction coefficient C_I. Orientation models have since acquired additional parameters, so this is now called parameter identification.

An example result is shown in Fig. 9.7. Here, more than 1,000 fibers were initially generated in a rectangular box (Fig. 9.4(a)), using a specified initial orientation tensor. The box was then subjected to simple shear deformation, using periodic boundary conditions to represent interactions with fibers outside the box. At each time step the orientation tensor components were

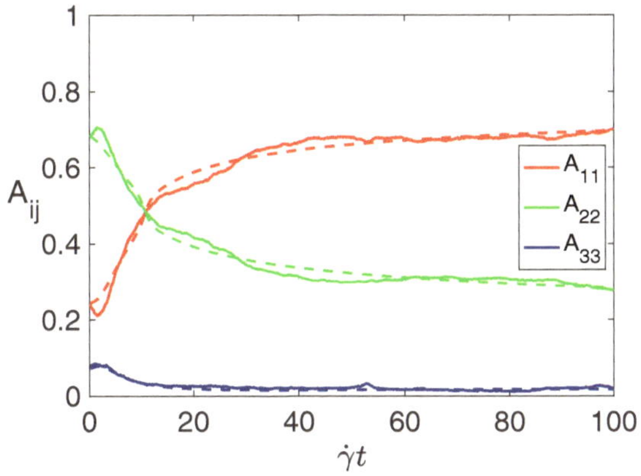

Figure 9.7 Results for a direct fiber simulation of simple shear flow (solid lines), and a tensor model fitted to the data (broken lines) using MRD/RSC with $C_I = 0.0168$, $D_2 = 0.5$, $D_3 = 0.0471$ and $\kappa = 0.3251$ and the natural closure. DFS data by Dr. Armin Kech and Dr. Susanne Kugler, Robert Bosch GmbH.

calculated, giving the time history in Fig. 9.7. The fibers at $\dot{\gamma}t = 100$ are shown in Fig. 9.4(b). For additional information about the calculation, see [KKO20].

The initial orientation is fairly flat in the 1–2 plane and biased toward the 2 direction. Shearing gradually reorients the fibers in the 1 direction, while A_{33} remains small. At $\dot{\gamma}t = 100$ the orientation is leveling out, but has not quite reached a steady state.

To identify model parameters, the initial orientation tensor from the DFS data was evolved using an MRD/RSC model and the natural closure. This calculation was combined with an optimization algorithm, which adjusted the model parameters to find values that best fit the DFS data. The resulting parameters for Fig. 9.7 are $C_I = 0.0168$, $D_2 = 0.5$, $D_3 = 0.0471$ and $\kappa = 0.3251$. Predictions using these values are shown in the figure as broken lines. The tensor-based model fits the DFS orientation well through the entire transient. Similar fits can be obtained with a pARD/RSC model [KKO20]. This shows that tensor models can do an excellent job of representing the behavior of concentrated fiber suspensions.

Comparisons between DFS models and experiments on injection molding compounds have been reported by Kugler et al. [KLC+20a] and Simon et al. [SBO20]. Both groups used a sliding plate rheometer to provide controlled simple shear flow, with different schemes for creating a repeatable initial orientation state. Direct fiber simulations were performed, with an initial fiber configuration matched closely to the experiment. The DFS results did a good job of matching the steady-state orientation for large shear strains. This was achieved without any adjustable parameters, showing that DFS results can produce useful values of C_I and any ARD parameters.

However, in both of these studies the direct fiber simulation had much faster orientation kinetics than the experiment. The experimental results were consistent with an RSC or SRF model with $\kappa = 0.25$ to 0.30, while the DFS results were similar to Jeffery kinetics. The reason for this discrepancy is not known. Further research is needed before reliable values of κ can be determined using direct fiber simulations.

Another topic that can be explored using direct fiber simulations is **confinement** of fibers by the mold walls. This occurs when fibers are longer than the height of the gap in which they flow. Confinement restricts the orientations the fiber can experience and alters the orientation dynamics, even for single fibers [PSAC+16, SACCK16].

The effect of confinement on a concentrated suspension has been studied using a direct fiber simulation [MPS+16]. A confined suspension was simulated using solid walls at the top and bottom of the simulation box, with periodic boundary conditions on the sides. The corresponding unconfined case used periodic boundary conditions on all sides. The fibers were rigid, with $\phi_f = 0.115$ and $\ell/d_f = 20$, and the total gap height for the confined case was 0.6ℓ.

Results for both the confined and unconfined suspensions were well fit by a Folgar–Tucker model and the IBOF closure. However, the confined suspension had lower steady-state alignment and was fit by a higher interaction coefficient than the unconfined case ($A_{11} = 0.80$, $C_I = 0.006$ confined; $A_{11} = 0.95$, $C_I = 0.0015$ unconfined). Even this moderate degree of confinement created a significant difference in orientation parameters.

This suggests that the degree of confinement is important when reporting data or evaluating results. While the Folgar–Tucker model and its variants continue to be useful, even for highly confined composites like SMC [PPG+20], the parameter values may depend on the degree of confinement.

Fiber breakage can also be modeled using direct fiber simulation, by allowing segments to separate when they are bent beyond a critical radius of curvature. A detailed numerical study [SIS19] shows that DFS models match the trends observed in experiments: fiber length degradation increases with increasing viscosity, shear rate and fiber volume fraction.

This study also showed that fiber-fiber interactions were important in fiber breakage. If a fiber is substantially bent by interaction forces, it becomes more susceptible to being broken by fluid-mechanical forces. This explains why fiber breakage rates increase with fiber volume fraction. Fibers near a wall also break more readily than fibers away from a wall, and the simulation predicted a layer with decreased L_w near the wall in simple shear flow. This would be interesting to probe in experiments.

DFS breakage predictions were compared to experiments in a Couette rheometer by Chang et al. [CBC+20]. The simulations provided a good match to the time history of L_n and L_w. This opens a path either to identify parameters for the fiber breakage model from Chapt. 7 using direct fiber simulations, or develop improved models for the breakage rate and child generation rate.

While direct fiber simulations cannot fully replace physical experiments when identifying model parameters, there has been substantial progress toward that goal. One can expect direct fiber simulations to play an increasingly important role in parameter identification in the future.

9.2.3.2 Full-Part Simulations

Tensor-based orientation models were originally developed to avoid calculating the orientation of every fiber in a molded part. However, with direct fiber simulation, it is now possible to do precisely that. Several research groups have reported direct fiber calculations for full molded parts, or for selected portions of a part. The computation times are long, and various simplifications are used to keep the computing time reasonable.

Sasayama et al. [SSKM20] simulated the injection of an SFT material into a center-gated disk mold. A conventional mold filling code was used to generate the velocity distribution at each time step, and these velocities drove the direct fiber simulation. The entire sprue was included in the DFS calculation, but only a narrow strip of the disk, extending from the sprue to the edge, was modeled. The visual agreement between simulation results and micro-CT images from molded parts is remarkable, and the quantitative agreement between DFS and experiment was good as well. This suggests that direct fiber simulations can be effective in predicting orientation in injection molded composites.

Full-part simulations for compression molded parts have been performed for SMC [MSB+20, THO20] and GMT [KKT20]. In these materials the fibers are initially in bundles or strands of a few hundred filaments. If the bundles remain intact as the material is molded, then one can simulate bundles rather than individual fibers, greatly reducing the size of the calculation [MSB+20].

Another way to reduce computation is to simplify the model for fiber motion. This is the approach taken in a commercial code, 3D TIMON CompositePRESS [NS14]. The fibers are composed of cylindrical segments and can bend freely at the joints. Fiber-fiber interactions are neglected. The motion of a fiber at each time step is computed by first assuming that each joint moves with the fluid velocity at that point, then adjusting the joint positions to maintain the length of each fiber segment. If the fiber is straight and the velocity field is linear, this model should behave like a Jeffery fiber with $\xi = 1$. However, the fiber will bend when the velocity varies nonlinearly along its length. This DFS scheme has been applied to SFT and LFT injection moldings, with encouraging comparisons to experiments [NS14].

Using this code, Teuwsen et al. [THO20] further simplified the calculation by modeling only a few representative fibers in each bundle. There are no fiber-fiber interactions in this simulation, so the fibers in a bundle can separate from one another, though this rarely occurred.

Kuhn et al. [KKT20] used a full mechanistic fiber simulation with fiber-fiber and fiber-wall interactions, but reduced computation by modeling only a portion of the part. They compared these calculations to the 3D TIMON code, and to predictions using standard tensor-based models, all of which simulated the entire molding. Meyer et al. [MSB+20] also compared their direct bundle simulation to tensor-based orientation predictions.

Orientations predicted in these DFS calculations compare well to micro-CT measurements on compression molded parts. The DFS models captured the tendency of fibers to align parallel to the edges of the part, as well as the complex orientation pattern near a weld line. Conventional tensor-based models also do a good job of predicting orientation away from the mold edges, but cannot represent the details of orientation at the edges or near a weld line.

Direct fiber simulations of full parts can also predict fiber/matrix separation, since the fibers and matrix have different velocities in these models. Meyer et al. [MSB+20] found that their simulation for SMC predicted resin-rich regions in the correct locations, but over-predicted the degree of separation. Kuhn et al. [KKT20] also found good agreement with the fiber/matrix separation in a rib using a full mechanistic simulation. 3D TIMON was less accurate in predicting separation, since it did not include fiber-fiber interactions.

These comparisons between experiments and direct fiber simulations are encouraging. Computation times ranged from one to six days, so these codes are not ready for everyday CAE use. However, we now have the ability to model individual fibers in realistic part geometries, and this could become a practical tool in the future.

9.2.3.3 New Orientation Models

A very different use for direct fiber simulations is to create new models for fiber orientation, replacing completely the models from Chapt. 5. Within a mold filling simulation, the fiber orientation model is simply a function that returns $\dot{\mathbf{A}}$, the rate of change of the orientation tensor, for any given $\mathbf{A}$, $\mathbf{D}$ and $\mathbf{W}$. While that function could use Jeffery-type equations, it could use any other method as well.

This has led researchers to apply techniques from machine learning to develop new orientation models. Results from a set of direct fiber simulations are used as offline training data, and the machine learning tools convert that into an orientation model. This approach does not require a closure approximation, nor any fitting parameters like C_I. Of course, a machine-learning model is only as good as the DFS data it learns from. A few preliminary studies have been reported [SAAC+18, YAMG+20], and additional work will be needed to develop practical tools for orientation analysis.

9.3 Chapter Summary

This chapter has discussed two recent developments: models of fiber migration and direct fiber simulation (DFS).

Shear-induced particle migration causes the fiber volume fraction ϕ_f to vary with position in injection molded SFT and LFT materials. Fiber volume fraction tends to be higher near the midplane of the part and at the last locations to fill. The suspension balance model can represent shear-induced migration for fibers, and is appearing in commercial software. There is still a need for additional controlled experiments and for a reliable way to identify model parameters.

In SMC and GMT materials, fiber/matrix separation is caused by friction between the fibers and the mold walls. For flat or gently curved parts this can be predicted with a continuum model. Separation may also be severe at sharp bends in the part or near the entrances to ribs. Predicting this usually requires a direct fiber simulation.

Direct fiber simulations have become an important tool for understanding fiber orientation. DFS codes can explore phenomena and reveal details that are not present in tensor-based orientation models. These include fiber-fiber and fiber-wall interactions, fiber flexibility, fiber breakage and fiber/matrix separation. Run times for most DFS codes range from hours to days, and calculations are usually restricted to a few thousand fibers.

One use of DFS calculations is to identify parameters for a fiber orientation model. Calculated rotary diffusion parameters like C_I are a good match to experiments, but slow-kinetics parameters like κ may not be.

When fibers are confined between mold walls, calculations suggest they will be less aligned at steady state, and so need different rotary diffusion parameters. One should pay attention to the relative values of fiber length and gap height when interpreting data and selecting model parameters.

Direct fiber simulations have been performed for entire molded parts, and for selected portions parts. Compression molded materials like SMC and GMT are particularly suitable for whole-part simulation, since the DFS code can track fiber bundles rather than individual fibers. At least one commercial code can perform DFS calculations as part of a mold filling simulation, using a simplified model of flexible fibers.

A set of DFS calculations could potentially be converted into a new model for fiber orientation, without any parameter fitting or closure approximations, using the tools of machine learning.

Tensor-based orientation predictions have become a standard CAE tool for injection and compression molded composites. Additional microstructural variables like the fiber length distribution and fiber volume fraction can now be modeled as well. However, there is still room for improvement. We can expect continued development of the models in this book, or perhaps their replacement by better models, as computing power, the ability to probe microstructure experimentally, and our understanding of fiber orientation increase.

A Vectors and Tensors

Models for fiber orientation make extensive use of vectors and tensors. This appendix defines vector and tensor operations like $\mathbf{D} \cdot \mathbf{A}$ and $\mathbb{A} : \mathbf{D}$ and shows how to calculate them numerically in MATLAB. The three forms of notation used in this book are explained and connected: direct notation (e.g., $\mathbf{A}$ and $\mathbb{A}$), index notation (A_{ij} and $\mathbb{A}_{ijkl}$) and contracted notation ($\{\mathbf{A}\}$ and $[\mathbb{A}]$). We treat only Cartesian coordinates here, except for the last section, which gives $\mathbf{D}$ and $\mathbf{W}$ in cylindrical coordinates.

In this appendix, s is an arbitrary scalar, $\mathbf{u}$ and $\mathbf{v}$ represent arbitrary vectors, $\mathbf{T}$ and $\mathbf{S}$ are arbitrary second-order tensors and $\mathbb{F}$ and $\mathbb{G}$ are arbitrary fourth-order tensors.

A.1 Definitions and Notation

A.1.1 Scalars and Vectors

A physical quantity that has magnitude but no direction is a *scalar*. Examples are temperature and pressure. Scalars are represented by light-face italic symbols like T and p.

If a physical quantity has both magnitude and direction, like force, then it is a *vector*. Vectors are represented by boldface lowercase symbols like $\mathbf{f}$.

To use a vector in numerical calculations we must express it in a coordinate system. A Cartesian coordinate system is defined by an origin and three orthogonal *unit vectors*, $\hat{\boldsymbol{\delta}}_1$, $\hat{\boldsymbol{\delta}}_2$ and $\hat{\boldsymbol{\delta}}_3$. Any vector $\mathbf{v}$ can then be described by moving its tail to the origin and projecting its length onto each coordinate axis. The *components* of the vector, v_1, v_2 and v_3, are the distances along the coordinate axes from the origin to the projection points. Figure A.1 shows a vector $\mathbf{v}$ in the 1–2 plane together with the unit vectors $\hat{\boldsymbol{\delta}}_1$ and $\hat{\boldsymbol{\delta}}_2$ and the vector's components v_1 and v_2.

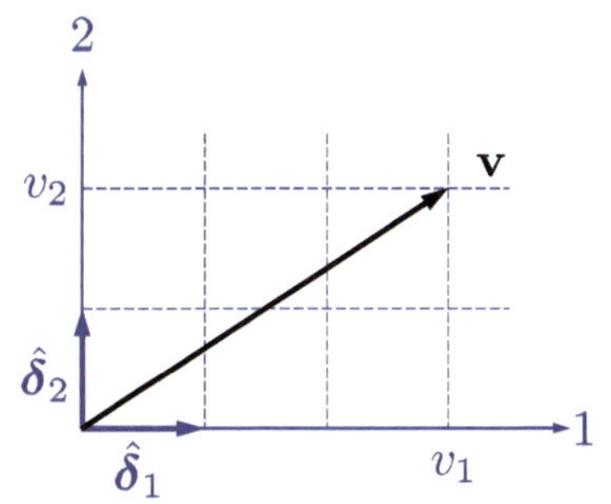

Figure A.1 Unit vectors $\hat{\boldsymbol{\delta}}_1$ and $\hat{\boldsymbol{\delta}}_2$ in a Cartesian coordinate system and their relationship to a vector $\mathbf{v} = 3\hat{\boldsymbol{\delta}}_1 + 2\hat{\boldsymbol{\delta}}_2$.

We can write any vector as a sum of products of its components (which are scalar) and the unit vectors:

$$\mathbf{v} = v_1 \hat{\boldsymbol{\delta}}_1 + v_2 \hat{\boldsymbol{\delta}}_2 + v_3 \hat{\boldsymbol{\delta}}_3 \tag{A.1}$$

This book uses 1-2-3 to label vector components in Cartesian coordinates, though some texts use x, y and z.

A.1.2 Tensors

While a vector associates a scalar component with each coordinate direction, some physical quantities associate a scalar component with each *ordered pair* of coordinate directions. These quantities are *second-order tensors*, or simply tensors. Stress and rate of deformation are examples. Second-order tensors are denoted by boldface capital letters like **T** or, in a few cases, by boldface lowercase Greek letters like $\boldsymbol{\sigma}$.

Extending the pattern of Eqn. (A.1), a tensor can be written in terms of its components as

$$\begin{aligned}\mathbf{T} = {} & T_{11}\, \hat{\boldsymbol{\delta}}_1\hat{\boldsymbol{\delta}}_1 + T_{12}\, \hat{\boldsymbol{\delta}}_1\hat{\boldsymbol{\delta}}_2 + T_{13}\, \hat{\boldsymbol{\delta}}_1\hat{\boldsymbol{\delta}}_3 \\ & + T_{21}\, \hat{\boldsymbol{\delta}}_2\hat{\boldsymbol{\delta}}_1 + T_{22}\, \hat{\boldsymbol{\delta}}_2\hat{\boldsymbol{\delta}}_2 + T_{23}\, \hat{\boldsymbol{\delta}}_2\hat{\boldsymbol{\delta}}_3 \\ & + T_{31}\, \hat{\boldsymbol{\delta}}_3\hat{\boldsymbol{\delta}}_1 + T_{32}\, \hat{\boldsymbol{\delta}}_3\hat{\boldsymbol{\delta}}_2 + T_{33}\, \hat{\boldsymbol{\delta}}_3\hat{\boldsymbol{\delta}}_3\end{aligned} \tag{A.2}$$

Each component of a second-order tensor has two subscripts, corresponding to the two unit vectors it multiplies. The products of two unit vectors are called *unit dyads* and the order of multiplication matters: $\hat{\boldsymbol{\delta}}_1\hat{\boldsymbol{\delta}}_2$ is not the same as $\hat{\boldsymbol{\delta}}_2\hat{\boldsymbol{\delta}}_1$.

We often write a tensor as the matrix of its components,

$$\mathbf{T} = \begin{bmatrix} T_{11} & T_{12} & T_{13} \\ T_{21} & T_{22} & T_{13} \\ T_{31} & T_{32} & T_{33} \end{bmatrix} \tag{A.3}$$

Other physical quantities, like stiffness, are tensors of higher order. A fourth-order tensor associates a scalar component with each ordered product of four unit vectors, and is denoted by a blackboard bold symbol like $\mathbb{C}$. Writing this out in detail for an arbitrary fourth-order tensor $\mathbb{F}$,

$$\begin{aligned}\mathbb{F} = {} & \mathbb{F}_{1111}\, \hat{\boldsymbol{\delta}}_1\hat{\boldsymbol{\delta}}_1\hat{\boldsymbol{\delta}}_1\hat{\boldsymbol{\delta}}_1 + \mathbb{F}_{1112}\, \hat{\boldsymbol{\delta}}_1\hat{\boldsymbol{\delta}}_1\hat{\boldsymbol{\delta}}_1\hat{\boldsymbol{\delta}}_2 + \mathbb{F}_{1113}\, \hat{\boldsymbol{\delta}}_1\hat{\boldsymbol{\delta}}_1\hat{\boldsymbol{\delta}}_1\hat{\boldsymbol{\delta}}_3 \\ & + \mathbb{F}_{1121}\, \hat{\boldsymbol{\delta}}_1\hat{\boldsymbol{\delta}}_1\hat{\boldsymbol{\delta}}_2\hat{\boldsymbol{\delta}}_1 + \mathbb{F}_{1122}\, \hat{\boldsymbol{\delta}}_1\hat{\boldsymbol{\delta}}_1\hat{\boldsymbol{\delta}}_2\hat{\boldsymbol{\delta}}_2 + \mathbb{F}_{1123}\, \hat{\boldsymbol{\delta}}_1\hat{\boldsymbol{\delta}}_1\hat{\boldsymbol{\delta}}_2\hat{\boldsymbol{\delta}}_3 \\ & + \mathbb{F}_{1131}\, \hat{\boldsymbol{\delta}}_1\hat{\boldsymbol{\delta}}_1\hat{\boldsymbol{\delta}}_3\hat{\boldsymbol{\delta}}_1 + \mathbb{F}_{1132}\, \hat{\boldsymbol{\delta}}_1\hat{\boldsymbol{\delta}}_1\hat{\boldsymbol{\delta}}_3\hat{\boldsymbol{\delta}}_2 + \mathbb{F}_{1133}\, \hat{\boldsymbol{\delta}}_1\hat{\boldsymbol{\delta}}_1\hat{\boldsymbol{\delta}}_3\hat{\boldsymbol{\delta}}_3 \\ & + \ldots \qquad \text{(81 total terms)}\end{aligned} \tag{A.4}$$

Each component of a fourth-order tensor has four indices, corresponding to the four unit vectors it multiplies, and each index can take on the values 1, 2 or 3. In three dimensions a fourth-order tensor has 81 components.

Tensors of any other order can be defined in the same way. In general, a tensor of order n associates a scalar component with each ordered product of n unit vectors. A scalar is a 0th order tensor and a vector is a first-order tensor.

A.1.3 Index Notation

Index notation provides a convenient way to write expressions involving the components of vectors and tensors. The numerical indices of vector and tensor components are replaced by symbols like i, j, k or l, which can take on the values 1, 2 or 3.

There are two rules in index notation:

1. **A repeated index in a product implies summation**. Suppose we wanted to calculate the dot product of two vectors, $\mathbf{u}\cdot\mathbf{v} = u_1 v_1 + u_2 v_2 + u_3 v_3 = \sum_{i=1}^{3} u_i v_i$. In index notation we can shorten this to $\mathbf{u}\cdot\mathbf{v} = u_i v_i$. The appearance of index i twice on the right-hand side tells us to sum over all values of i.

   Similarly, in $S_{ij} T_{ji}$ both the i and the j are repeated, so this expression is computed as $\sum_{i=1}^{3}\sum_{j=1}^{3} S_{ij} T_{ji}$.
2. **A free index implies multiple scalar equations**. In the equation $u_i = T_{ij} v_j$ the index j is repeated in a product, but i is not. A non-repeated index like i in this example is called a free index, and it can be interpreted in two ways. Either this expression gives three scalar equations ($i = 1$, 2 or 3) for the three components of the vector $\mathbf{u}$, or we can multiply each of those scalar equations by the corresponding $\hat{\boldsymbol{\delta}}_i$ and sum them to obtain a single vector equation.

   If an expression has two free indices, the result is a second-order tensor. If it has four free indices, it evaluates to a fourth-order tensor.

Using the second rule, we can equate expressions in direct notation and index notation:

$$\mathbf{v} = v_i \tag{A.5}$$

$$\mathbf{T} = T_{ij} \tag{A.6}$$

$$\mathbb{F} = \mathbb{F}_{ijkl} \tag{A.7}$$

We will use this format to show how expressions in direct notation are evaluated using index notation.

A.1.4 Identity Tensors and the Kronecker Delta

An important second-order tensor is the *identity tensor* or *unit tensor* $\mathbf{I}$. This is the tensor that has ones on the diagonal and zeros everywhere else:

$$\mathbf{I} = \begin{bmatrix} 1 & 0 & 0 \\ 0 & 1 & 0 \\ 0 & 0 & 1 \end{bmatrix} \tag{A.8}$$

In index notation the components of $\mathbf{I}$ are written as δ_{ij}, where δ_{ij} is the *Kronecker delta*,

$$\delta_{ij} = \begin{cases} 1 & \text{if} \quad i = j \\ 0 & \text{if} \quad i \neq j \end{cases} \tag{A.9}$$

The components of the symmetric fourth-order identity tensor $\mathbb{I}$ are

$$\mathbb{I}_{ijkl} = \frac{1}{2}\left(\delta_{ik}\delta_{jl} + \delta_{il}\delta_{jk}\right) \tag{A.10}$$

A.1.5 Transpose and Symmetries

The transpose of a second-order tensor $\mathbf{T}$ is denoted by $\mathbf{T}^T$, and is found by interchanging the order of the indices in $\mathbf{T}$. In matrix form this is the same as a matrix transpose,

$$\mathbf{T}^T = \begin{bmatrix} T_{11} & T_{21} & T_{31} \\ T_{12} & T_{22} & T_{32} \\ T_{13} & T_{23} & T_{33} \end{bmatrix} \tag{A.11}$$

In index notation the transpose is

$$\left(T_{ij}\right)^T = T_{ji} \tag{A.12}$$

A second-order tensor that is equal to its transpose is *symmetric*,

$$\mathbf{T} = \mathbf{T}^T \quad \text{or} \quad T_{ij} = T_{ji} \qquad \text{symmetric} \tag{A.13}$$

and the matrix of its components is a symmetric matrix.

A tensor that is equal to the negative of its transpose is *anti-symmetric* or *skew symmetric*:

$$\mathbf{T} = -\mathbf{T}^T \quad \text{or} \quad T_{ij} = -T_{ji} \qquad \text{anti-symmetric} \tag{A.14}$$

The diagonal components of an anti-symmetric tensor must all equal zero.

Fourth-order tensors are symmetric (or not) with respect to specific pairs of indices. A fourth-order tensor is symmetric with respect to its first two indices if

$$\mathbb{F}_{ijkl} = \mathbb{F}_{jikl} \tag{A.15}$$

and symmetric with respect to its last two indices if

$$\mathbb{F}_{ijkl} = \mathbb{F}_{ijlk} \tag{A.16}$$

Together, these two are called the *minor symmetries*. A fourth-order tensor can also have the *major symmetry*, in which case the first two indices can be exchanged with the last two:

$$\mathbb{F}_{ijkl} = \mathbb{F}_{klij} \tag{A.17}$$

A.1.6 The Trace of a Tensor

The trace of a second-order tensor is the sum of its diagonal components:

$$\mathrm{tr}\,\mathbf{T} = T_{11} + T_{22} + T_{33} = T_{ii} \tag{A.18}$$

A.2 Algebraic Operations

A.2.1 Addition and Subtraction

Vectors and tensors are added (or subtracted) by adding (or subtracting) their components,

$$\mathbf{u} + \mathbf{v} = u_i + v_i \tag{A.19}$$

$$\mathbf{S} + \mathbf{T} = S_{ij} + T_{ij} \tag{A.20}$$

$$\mathbb{F} + \mathbb{G} = \mathbb{F}_{ijkl} + \mathbb{G}_{ijkl} \tag{A.21}$$

A.2.2 Multiplication

When a vector or tensor is **multiplied by a scalar** s, multiply each component by s:

$$s\mathbf{v} = s v_i \tag{A.22}$$
$$s\mathbf{T} = s T_{ij} \tag{A.23}$$
$$s\mathbb{F} = s\mathbb{F}_{ijkl} \tag{A.24}$$

The **dot product of two vectors** is a scalar, and is computed as

$$\mathbf{u}\cdot\mathbf{v} = u_i v_i \tag{A.25}$$

There are two ways to compute the **dot product of a vector and a tensor**; both produce a vector:

$$\mathbf{T}\cdot\mathbf{v} = T_{ij} v_j \tag{A.26}$$
$$\mathbf{v}\cdot\mathbf{T} = v_i T_{ij} = \mathbf{T}^T\cdot\mathbf{v} \tag{A.27}$$

The **dot product of two second-order tensors** is also a second-order tensor:

$$\mathbf{T}\cdot\mathbf{S} = T_{ik} S_{kj} \tag{A.28}$$

If both tensors are stored in 3×3 matrix form, this is the same as a matrix multiplication.

In all of the cases above, the dot collects the nearest index on either side of the dot and sums over all values of that index.

Vector and tensor dot products are not commutative, so in general

$$\mathbf{S}\cdot\mathbf{T} \neq \mathbf{T}\cdot\mathbf{S} \tag{A.29}$$

Taking the dot product of a vector or second-order tensor with the second-order identity tensor returns the original quantity:

$$\mathbf{I}\cdot\mathbf{v} = \mathbf{v}\cdot\mathbf{I} = \mathbf{v} \tag{A.30}$$
$$\mathbf{I}\cdot\mathbf{T} = \mathbf{T}\cdot\mathbf{I} = \mathbf{T} \tag{A.31}$$

A **double-dot product** collects the two indices on either side of the double-dot and sums over all values of both indices.

$$\mathbf{S}:\mathbf{T} = S_{ij} T_{ji} \tag{A.32}$$
$$\mathbb{F}:\mathbf{T} = \mathbb{F}_{ijkl} T_{lk} \tag{A.33}$$
$$\mathbb{F}:\mathbb{G} = \mathbb{F}_{ijkl}\mathbb{G}_{lkmn} \tag{A.34}$$

The double-dot product of two second-order tensors is a scalar, the double-dot product of a fourth-order and a second-order tensor is a new second-order tensor, and double-dotting two fourth-order tensors gives a new fourth-order tensor.

If a fourth-order tensor has the two minor symmetries of Eqns. (A.15) and (A.16), taking a double-dot product with the fourth-order identity tensor returns the original tensor,

$$\mathbb{F}:\mathbb{I} = \mathbb{I}:\mathbb{F} = \mathbb{F} \tag{A.35}$$

The **cross product of two vectors** is a vector,

$$\mathbf{u} \times \mathbf{v} = \epsilon_{ijk} u_j v_k \tag{A.36}$$

Here ϵ_{ijk} is the *permutation symbol* or Levi-Civita symbol,

$$\epsilon_{ijk} = \begin{cases} 1 & \text{if } ijk = 123 \text{ or } 231 \text{ or } 312 \\ -1 & \text{if } ijk = 321 \text{ or } 132 \text{ or } 213 \\ 0 & \text{if any of } i,\ j \text{ or } k \text{ are the same} \end{cases} \tag{A.37}$$

The **dyadic product** of two vectors is a second-order tensor:

$$\mathbf{uv} = u_i v_j \tag{A.38}$$

Some texts write the dyadic product of two vectors as $\mathbf{u} \otimes \mathbf{v}$.

The dyadic product of four vectors is a fourth-order tensor,

$$\mathbf{uvwx} = u_i v_j w_k x_l \tag{A.39}$$

Some texts write this as $\mathbf{u} \otimes \mathbf{v} \otimes \mathbf{w} \otimes \mathbf{x}$.

A.2.3 Magnitude of a Vector

The magnitude of a vector is its length, and is independent of the vector's direction. This is denoted by $|\mathbf{v}|$ and is computed as

$$|\mathbf{v}| = \sqrt{\mathbf{v} \cdot \mathbf{v}} = \sqrt{v_1^2 + v_2^2 + v_3^2} \tag{A.40}$$

A.2.4 Tensor Inverse

The inverse of a second-order tensor, denoted by $\mathbf{T}^{-1}$, is also a second-order tensor. The dot product of a tensor and its inverse produces the identity tensor,

$$\mathbf{T} \cdot \mathbf{T}^{-1} = \mathbf{I} \tag{A.41}$$

When $\mathbf{T}$ is written in 3×3 matrix form, $\mathbf{T}^{-1}$ is the matrix inverse of $\mathbf{T}$.

Similarly, the double-dot product of a fourth-order tensor and its inverse gives the fourth-order identity tensor,

$$\mathbb{F} : \mathbb{F}^{-1} = \mathbb{I} \tag{A.42}$$

Equation (A.65) shows how to compute the inverse of a fourth-order tensor.

A.3 Coordinate Transformation

While a set of coordinate axes is needed to specify the components of a vector or tensor, the vector or tensor could just as well be described by its components in a different coordinate system. If we know the components in one coordinate system and we also know how that system is related to a second coordinate system, we can compute the components in the second coordinate system. This is known as *coordinate transformation.*

Figure A.2 shows an example for a vector in two dimensions. We know the blue 1–2 coordinate system with unit vectors $\hat{\boldsymbol{\delta}}_1$ and $\hat{\boldsymbol{\delta}}_2$, and we know the components of vector $\mathbf{v}$ in this system, v_1 and v_2. We wish to find the components v'_1 and v'_2 in the $1'$–$2'$ coordinate system, shown in red. The unit vectors in this second system are $\hat{\boldsymbol{\delta}}'_1$ and $\hat{\boldsymbol{\delta}}'_2$.

The relationship between the two coordinate systems is determined by the relative orientations of their unit vectors. This information is described by a 3 × 3 *transformation matrix* [**Q**], whose components are defined such that

$$Q_{ik} = \hat{\boldsymbol{\delta}}'_i \cdot \hat{\boldsymbol{\delta}}_k \qquad \text{(A.43)}$$

Note that the first index of Q_{ik} refers to one of the new coordinate axes, $1'$, $2'$ or $3'$, while the second index refers to the original coordinate axes, 1, 2 or 3.

The components in the new (primed) coordinate system are then given by

$$v'_i = Q_{ik} v_k \qquad \text{(A.44)}$$

$$T'_{ij} = Q_{ik} Q_{jl} T_{kl} \qquad \text{(A.45)}$$

$$\mathbb{F}'_{ijkl} = Q_{im} Q_{jn} Q_{ko} Q_{lp} \mathbb{F}_{mnop} \qquad \text{(A.46)}$$

[**Q**] is an orthogonal matrix, which means that

$$[\mathbf{Q}][\mathbf{Q}]^T = \mathbf{I} \qquad \text{(A.47)}$$

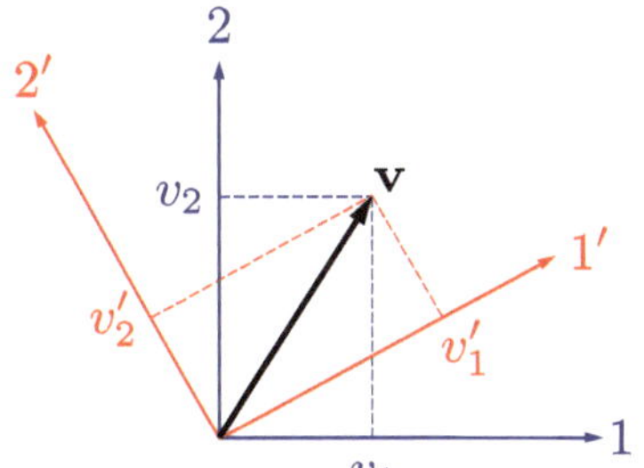

Figure A.2 A two-dimensional vector **v** and its components in two coordinate systems.

A.4 Contracted Notation

A.4.1 Definitions

The best way to display fourth-order tensors and use them in computations is *contracted notation.* There are several versions of contracted notation, and interested readers should consult Nadeau and Ferrari [NF98][1].

We could write the nine components of any second-order tensor as a 9×1 column vector,

$$\{\mathbf{T}\} = \begin{Bmatrix} T_{11} \\ T_{22} \\ T_{33} \\ T_{23} \\ T_{31} \\ T_{12} \\ T_{32} \\ T_{13} \\ T_{21} \end{Bmatrix} \tag{A.48}$$

Using the same indexing for rows and columns, we could write any fourth-order tensor as a 9×9 matrix of its components,

$$[\mathbb{F}] = \begin{bmatrix}
\mathbb{F}_{1111} & \mathbb{F}_{1122} & \mathbb{F}_{1133} & \mathbb{F}_{1123} & \mathbb{F}_{1131} & \mathbb{F}_{1112} & \mathbb{F}_{1132} & \mathbb{F}_{1113} & \mathbb{F}_{1121} \\
\mathbb{F}_{2211} & \mathbb{F}_{2222} & \mathbb{F}_{2233} & \mathbb{F}_{2223} & \mathbb{F}_{2231} & \mathbb{F}_{2212} & \mathbb{F}_{2232} & \mathbb{F}_{2213} & \mathbb{F}_{2221} \\
\mathbb{F}_{3311} & \mathbb{F}_{3322} & \mathbb{F}_{3333} & \mathbb{F}_{3323} & \mathbb{F}_{3331} & \mathbb{F}_{3312} & \mathbb{F}_{3332} & \mathbb{F}_{3313} & \mathbb{F}_{3321} \\
\mathbb{F}_{2311} & \mathbb{F}_{2322} & \mathbb{F}_{2333} & \mathbb{F}_{2323} & \mathbb{F}_{2331} & \mathbb{F}_{2312} & \mathbb{F}_{2332} & \mathbb{F}_{2313} & \mathbb{F}_{2321} \\
\mathbb{F}_{3111} & \mathbb{F}_{3122} & \mathbb{F}_{3133} & \mathbb{F}_{3123} & \mathbb{F}_{3131} & \mathbb{F}_{3112} & \mathbb{F}_{3132} & \mathbb{F}_{3113} & \mathbb{F}_{3121} \\
\mathbb{F}_{1211} & \mathbb{F}_{1222} & \mathbb{F}_{1233} & \mathbb{F}_{1223} & \mathbb{F}_{1231} & \mathbb{F}_{1212} & \mathbb{F}_{1232} & \mathbb{F}_{1213} & \mathbb{F}_{1221} \\
\mathbb{F}_{3211} & \mathbb{F}_{3222} & \mathbb{F}_{3233} & \mathbb{F}_{3223} & \mathbb{F}_{3231} & \mathbb{F}_{3212} & \mathbb{F}_{3232} & \mathbb{F}_{3213} & \mathbb{F}_{3221} \\
\mathbb{F}_{1311} & \mathbb{F}_{1322} & \mathbb{F}_{1333} & \mathbb{F}_{1323} & \mathbb{F}_{1331} & \mathbb{F}_{1312} & \mathbb{F}_{1332} & \mathbb{F}_{1313} & \mathbb{F}_{1321} \\
\mathbb{F}_{2111} & \mathbb{F}_{2122} & \mathbb{F}_{2133} & \mathbb{F}_{2123} & \mathbb{F}_{2131} & \mathbb{F}_{2112} & \mathbb{F}_{2132} & \mathbb{F}_{2113} & \mathbb{F}_{2121}
\end{bmatrix} \tag{A.49}$$

However, most second-order tensors of interest are symmetric, so the last three rows in Eqn. (A.48) duplicate other entries. Similarly, most fourth-order tensors of interest have the minor symmetries, Eqns. (A.15) and (A.16), so the last three rows and the last three columns of Eqn. (A.49) are redundant.

[1] Nadeau and Ferrari's $\eta = 1$ form is used in this book. Their $\eta = \sqrt{2}$ form is used by some research groups, e.g., [KGFB21].

If we consider only tensors with these symmetries, we can eliminate the redundant rows and columns without loss of information. Second-order tensors are then contracted to 6×1 column vectors,

$$\{\mathbf{T}\} = \begin{Bmatrix} T_{11} \\ T_{22} \\ T_{33} \\ T_{23} \\ T_{31} \\ T_{12} \end{Bmatrix} = \begin{Bmatrix} T_1 \\ T_2 \\ T_3 \\ T_4 \\ T_5 \\ T_6 \end{Bmatrix} \tag{A.50}$$

and fourth-order tensors are contracted to 6 × 6 matrices,

$$\begin{aligned}[\mathbb{F}] &= \begin{bmatrix} \mathbb{F}_{1111} & \mathbb{F}_{1122} & \mathbb{F}_{1133} & \mathbb{F}_{1123} & \mathbb{F}_{1131} & \mathbb{F}_{1112} \\ \mathbb{F}_{2211} & \mathbb{F}_{2222} & \mathbb{F}_{2233} & \mathbb{F}_{2223} & \mathbb{F}_{2231} & \mathbb{F}_{2212} \\ \mathbb{F}_{3311} & \mathbb{F}_{3322} & \mathbb{F}_{3333} & \mathbb{F}_{3323} & \mathbb{F}_{3331} & \mathbb{F}_{3312} \\ \mathbb{F}_{2311} & \mathbb{F}_{2322} & \mathbb{F}_{2333} & \mathbb{F}_{2323} & \mathbb{F}_{2331} & \mathbb{F}_{2312} \\ \mathbb{F}_{3111} & \mathbb{F}_{3122} & \mathbb{F}_{3133} & \mathbb{F}_{3123} & \mathbb{F}_{3131} & \mathbb{F}_{3112} \\ \mathbb{F}_{1211} & \mathbb{F}_{1222} & \mathbb{F}_{1233} & \mathbb{F}_{1223} & \mathbb{F}_{1231} & \mathbb{F}_{1212} \end{bmatrix} \\ &= \begin{bmatrix} \mathbb{F}_{11} & \mathbb{F}_{12} & \mathbb{F}_{13} & \mathbb{F}_{14} & \mathbb{F}_{15} & \mathbb{F}_{16} \\ \mathbb{F}_{21} & \mathbb{F}_{22} & \mathbb{F}_{23} & \mathbb{F}_{24} & \mathbb{F}_{25} & \mathbb{F}_{26} \\ \mathbb{F}_{31} & \mathbb{F}_{32} & \mathbb{F}_{33} & \mathbb{F}_{34} & \mathbb{F}_{35} & \mathbb{F}_{36} \\ \mathbb{F}_{41} & \mathbb{F}_{42} & \mathbb{F}_{43} & \mathbb{F}_{44} & \mathbb{F}_{45} & \mathbb{F}_{46} \\ \mathbb{F}_{51} & \mathbb{F}_{52} & \mathbb{F}_{53} & \mathbb{F}_{54} & \mathbb{F}_{55} & \mathbb{F}_{56} \\ \mathbb{F}_{61} & \mathbb{F}_{62} & \mathbb{F}_{63} & \mathbb{F}_{64} & \mathbb{F}_{65} & \mathbb{F}_{66} \end{bmatrix}\end{aligned} \tag{A.51}$$

Our notation is that $\mathbf{T}$ and $\mathbb{F}$ are the full tensors, with tensor components $\mathbf{T}_{ij}$ and $\mathbb{F}_{ijkl}$, while $\{\mathbf{T}\}$ and $[\mathbb{F}]$ are the contracted matrices with matrix components T_m and $\mathbb{F}_{mn}$. Table 2.1 gives the correspondence between tensor indices like ij or kl and contracted indices like m or n.

In this section we will write the equivalence between tensors and their contracted forms as

$$\mathbf{T} = \{\mathbf{T}\} \tag{A.52}$$

$$\mathbb{F} = [\mathbb{F}] \tag{A.53}$$

A.4.2 Identity Tensor

Using Eqn. (A.10), the contraction of the fourth-order identity tensor is

$$[\mathbb{I}] = \begin{bmatrix} 1 & 0 & 0 & 0 & 0 & 0 \\ 0 & 1 & 0 & 0 & 0 & 0 \\ 0 & 0 & 1 & 0 & 0 & 0 \\ 0 & 0 & 0 & \frac{1}{2} & 0 & 0 \\ 0 & 0 & 0 & 0 & \frac{1}{2} & 0 \\ 0 & 0 & 0 & 0 & 0 & \frac{1}{2} \end{bmatrix} \tag{A.54}$$

A.4.3 Operations

To add (or subtract) two tensors in contracted notation, add (or subtract) their matrices,

$$\mathbf{S}+\mathbf{T}=\{\mathbf{S}\}+\{\mathbf{T}\} \tag{A.55}$$

$$\mathbb{F}+\mathbb{G}=[\mathbb{F}]+[\mathbb{G}] \tag{A.56}$$

To multiply a tensor by a scalar, multiply its contracted form by the scalar,

$$s\mathbf{T}=s\{\mathbf{T}\} \tag{A.57}$$

$$s\mathbb{F}=s[\mathbb{F}] \tag{A.58}$$

Taking double-dot products in contracted notation requires an extra step to ensure that all components missing from the contracted form are properly accounted for. We know that in tensor multiplication $\mathbb{I}:\mathbf{T}=\mathbf{T}$, but if we simply take the matrix-vector product of the contractions $[\mathbb{I}]$ and $\{\mathbf{T}\}$ we obtain something different:

$$[\mathbb{I}]\{\mathbf{T}\}=\begin{bmatrix}1&0&0&0&0&0\\0&1&0&0&0&0\\0&0&1&0&0&0\\0&0&0&\frac{1}{2}&0&0\\0&0&0&0&\frac{1}{2}&0\\0&0&0&0&0&\frac{1}{2}\end{bmatrix}\begin{Bmatrix}T_{11}\\T_{22}\\T_{33}\\T_{23}\\T_{31}\\T_{12}\end{Bmatrix}=\begin{Bmatrix}T_{11}\\T_{22}\\T_{33}\\T_{23}/2\\T_{31}/2\\T_{12}/2\end{Bmatrix}\neq\{\mathbf{T}\} \tag{A.59}$$

To get the correct answer we need to account for the missing terms by inserting an additional matrix in the multiplication. Define the matrix $[\mathbb{R}]$ as the matrix inverse of $[\mathbb{I}]$,

$$\mathbb{R}\equiv\begin{bmatrix}1&0&0&0&0&0\\0&1&0&0&0&0\\0&0&1&0&0&0\\0&0&0&2&0&0\\0&0&0&0&2&0\\0&0&0&0&0&2\end{bmatrix} \tag{A.60}$$

Inserting $[\mathbb{R}]$ after $[\mathbb{I}]$ then gives the correct result,

$$[\mathbb{I}][\mathbb{R}]\{\mathbf{T}\}=\begin{bmatrix}1&0&0&0&0&0\\0&1&0&0&0&0\\0&0&1&0&0&0\\0&0&0&\frac{1}{2}&0&0\\0&0&0&0&\frac{1}{2}&0\\0&0&0&0&0&\frac{1}{2}\end{bmatrix}\begin{bmatrix}1&0&0&0&0&0\\0&1&0&0&0&0\\0&0&1&0&0&0\\0&0&0&2&0&0\\0&0&0&0&2&0\\0&0&0&0&0&2\end{bmatrix}\begin{Bmatrix}T_{11}\\T_{22}\\T_{33}\\T_{23}\\T_{31}\\T_{12}\end{Bmatrix}=\begin{Bmatrix}T_{11}\\T_{22}\\T_{33}\\T_{23}\\T_{31}\\T_{12}\end{Bmatrix}=\{\mathbf{T}\} \tag{A.61}$$

Based on this, the general rules for computing a double-dot product in contracted notation are:

$$\mathbb{F}:\mathbf{T}=[\mathbb{F}][\mathbb{R}]\{\mathbf{T}\} \tag{A.62}$$

$$\mathbb{F}:\mathbb{G}=[\mathbb{F}][\mathbb{R}][\mathbb{G}] \tag{A.63}$$

In contracted-notation computations, $[\mathbb{R}]$ should be inserted in the matrix product whenever a double-dot occurs in direct notation.

A.4.4 Fourth-Order Inverse

One must also be careful when computing tensor inverses. Using Eqn. (A.63) to write Eqn. (A.42) in contracted notation gives

$$[\mathbb{F}]\,[\mathbb{R}]\,[\mathbb{F}^{-1}] = [\mathbb{I}] \tag{A.64}$$

Treating this as a matrix equation, we can pre-multiply both sides by $[\mathbb{F}]^{-1}$ and then by $[\mathbb{R}]^{-1}$. Then, recalling that $[\mathbb{R}]^{-1} = [\mathbb{I}]$, the formula for computing a fourth-order inverse in contracted form is

$$\left[\mathbb{F}^{-1}\right] = \left[\mathbb{I}\right]\left[\mathbb{F}\right]^{-1}\left[\mathbb{I}\right] \tag{A.65}$$

In this expression, $[\mathbb{F}]^{-1}$ is the matrix inverse of $[\mathbb{F}]$, while $[\mathbb{F}^{-1}]$ is the contracted form of the tensor $\mathbb{F}^{-1}$.

A.4.5 Coordinate Transformation

We want to calculate coordinate transformations, Eqns. (A.45) and (A.46), in contracted notation. To do this (e.g., [NF98]), define a 6 × 6 transformation matrix $[\mathbb{Q}]$:

$$[\mathbb{Q}] = [\mathbb{Q}^a] + [\mathbb{Q}^b] \tag{A.66}$$

where

$$[\mathbb{Q}^a] = \begin{bmatrix} Q_{11}Q_{11} & Q_{12}Q_{12} & Q_{13}Q_{13} & Q_{12}Q_{13} & Q_{13}Q_{11} & Q_{11}Q_{12} \\ Q_{21}Q_{21} & Q_{22}Q_{22} & Q_{23}Q_{23} & Q_{22}Q_{23} & Q_{23}Q_{21} & Q_{21}Q_{22} \\ Q_{31}Q_{31} & Q_{32}Q_{32} & Q_{33}Q_{33} & Q_{32}Q_{33} & Q_{33}Q_{31} & Q_{31}Q_{32} \\ Q_{21}Q_{31} & Q_{22}Q_{32} & Q_{23}Q_{33} & Q_{22}Q_{33} & Q_{23}Q_{31} & Q_{21}Q_{32} \\ Q_{31}Q_{11} & Q_{32}Q_{12} & Q_{33}Q_{13} & Q_{32}Q_{13} & Q_{33}Q_{11} & Q_{31}Q_{12} \\ Q_{11}Q_{21} & Q_{12}Q_{22} & Q_{13}Q_{23} & Q_{12}Q_{23} & Q_{13}Q_{21} & Q_{11}Q_{22} \end{bmatrix} \tag{A.67}$$

and

$$[\mathbb{Q}^b] = \begin{bmatrix} 0 & 0 & 0 & Q_{13}Q_{12} & Q_{11}Q_{13} & Q_{12}Q_{11} \\ 0 & 0 & 0 & Q_{23}Q_{22} & Q_{21}Q_{23} & Q_{22}Q_{21} \\ 0 & 0 & 0 & Q_{33}Q_{32} & Q_{31}Q_{33} & Q_{32}Q_{31} \\ 0 & 0 & 0 & Q_{23}Q_{32} & Q_{21}Q_{33} & Q_{22}Q_{31} \\ 0 & 0 & 0 & Q_{33}Q_{12} & Q_{31}Q_{13} & Q_{32}Q_{11} \\ 0 & 0 & 0 & Q_{13}Q_{22} & Q_{11}Q_{23} & Q_{12}Q_{21} \end{bmatrix} \tag{A.68}$$

Coordinate transformations for second– and fourth-order tensors in contracted form are then

$$\left\{\mathbf{T}'\right\} = [\mathbb{Q}]\,\{\mathbf{T}\} \tag{A.69}$$

$$\left[\mathbb{F}'\right] = [\mathbb{Q}]\,[\mathbb{F}]\,[\mathbb{Q}]^T \tag{A.70}$$

A.5 Calculations in MATLAB

A.5.1 Vectors and Second-Order Tensors

For numerical calculations in MATLAB, a useful convention is to store vectors as 3 × 1 matrices (column vectors) and second-order tensors as 3 × 3 matrices. This is consistent with tensor notation, and the dot products in Eqns. (A.25) and (A.26) become standard matrix multiplications.

Table A.1 shows the MATLAB equivalents for operations involving vectors and second-order tensors. The functions `eye`, `trace`, `dot`, `cross`, `norm` and `inv` are standard in MATLAB.

Table A.1 MATLAB equivalents of vector and tensor operations using 3 × 1 storage for vectors `u` and `v` and 3 × 3 storage for second-order tensors `S` and `T`. `Q` is the 3 × 3 transformation matrix defined in Eqn. (A.43).

Operation	Tensor Notation	MATLAB Equivalent
identity tensor (3D)	$\mathbf{I}$	`eye(3)`
tensor transpose	$\mathbf{T}^T$	`T'`
tensor trace	$\text{tr}\mathbf{T}$	`trace(T)`
vector addition	$\mathbf{u}+\mathbf{v}$	`u + v`
tensor addition	$\mathbf{S}+\mathbf{T}$	`S + T`
scalar times a vector	$s\mathbf{v}$	`s * v`
scalar times a tensor	$s\mathbf{T}$	`s * T`
vector dot product	$\mathbf{u}\cdot\mathbf{v}$	`u' * v` **or** `dot(u,v)`
tensor-vector dot product	$\mathbf{T}\cdot\mathbf{v}$	`T * v`
vector-tensor dot product	$\mathbf{v}\cdot\mathbf{T}$	`T' * v`
tensor dot product	$\mathbf{S}\cdot\mathbf{T}$	`S * T`
tensor double-dot product	$\mathbf{S}:\mathbf{T}$	`trace(S * T)`
vector cross product	$\mathbf{u}\times\mathbf{v}$	`cross(u, v)`
dyadic product	$\mathbf{uv}$	`u * v'`
vector magnitude	$\lvert\mathbf{v}\rvert$	`norm(v)`
tensor inverse	$\mathbf{T}^{-1}$	`inv(T)`
vector coordinate transformation	$v'_i = Q_{ik} v_k$	`vprime = Q * v`
tensor coordinate transformation	$T'_{ij} = Q_{ik} Q_{jl} T_{kl}$	`Tprime = Q * T * Q'`

A.5.2 Fourth-Order Tensors

Fourth-order tensors are always stored in contracted form as 6 × 6 matrices. When a fourth-order tensor operates on a second-order tensor, the latter must first be converted to contracted form as a 6 × 1 column vector. In some cases (e.g., mechanical property calculations) it may be convenient to keep all second-order tensors in column-vector format.

Table A.2 shows operations in contracted notation and their MATLAB equivalents. The functions `tens2vec`, `vec2tens` and `inv4` are available in the accompanying software[2]. `inv` is the standard MATLAB function for a matrix inverse.

Table A.2 MATLAB equivalents of vector and tensor operations in contracted notation. Second-order tensors **S** and **T** are stored either as 3 × 3 matrices `S` and `T`, or as 6 × 1 column vectors `Sv` and `Tv`. Fourth-order tensors $\mathbb{F}$ and $\mathbb{G}$ are stored as 6 × 6 matrices `F4` and `G4`. `Q4` is the 6 × 6 transformation matrix defined in Eqns. (A.66–A.68).

Operation	Tensor Notation	MATLAB Equivalent
fourth-order identity tensor	$\mathbb{I}$	`I4 = diag([1,1,1,0.5,0.5,0.5])`
matrix for fourth-order products	$\mathbb{R}$	`R4 = diag([1,1,1,2,2,2])`
contract a second-order tensor	$\{\mathbf{T}\} \leftarrow \mathbf{T}$	`Tv = tens2vec(T)`
expand a second-order tensor	$\mathbf{T} \leftarrow \{\mathbf{T}\}$	`T = vec2tens(Tv)`
second-order addition	$\mathbf{S}+\mathbf{T}$	`Sv + Tv`
fourth-order addition	$\mathbb{F}+\mathbb{G}$	`F4 + G4`
scalar times a second-order tensor	$s\mathbf{T}$	`s * Tv`
scalar times a fourth-order tensor	$s\mathbb{F}$	`s * F4`
multiply fourth-order and second-order	$\mathbb{F}:\mathbf{T}$	`F4 * R4 * Tv`
multiply two fourth-order tensors	$\mathbb{F}:\mathbb{G}$	`F4 * R4 * G4`
fourth-order tensor inverse	$\mathbb{F}^{-1}$	`I4 * inv(F4) * I4` **or** `inv4(F4)`
coordinate transformation, second-order	$T'_{ij} = Q_{ik}Q_{jl}T_{kl}$	`TvPrime = Q4 * Tv`
coordinate transformation, fourth-order	$Q_{im}Q_{jn}Q_{ko}Q_{lp}\mathbb{F}_{mnop}$	`F4prime = Q4 * F4 * Q4'`

[2] *https://github.com/charlestucker3/Fiber-Orientation-Tools*

A.6 D and W in Cylindrical Coordinates

Expressions for the rate-of-deformation tensor **D** and the vorticity tensor **W** are more complicated in (r, θ, z) cylindrical coordinates than in Cartesian coordinates. This is because in a cylindrical coordinate system the directions of the unit vectors change as the θ coordinate changes. The following expressions are adapted from [Tuc89] and [DT01].

Rate of deformation:

$$D_{rr} = \frac{\partial v_r}{\partial r} \tag{A.71}$$

$$D_{\theta\theta} = \frac{1}{r}\frac{\partial v_\theta}{\partial \theta} + \frac{v_r}{r} \tag{A.72}$$

$$D_{zz} = \frac{\partial v_z}{\partial z} \tag{A.73}$$

$$D_{r\theta} = D_{\theta r} = \frac{1}{2}\left[r\frac{\partial}{\partial r}\left(\frac{v_\theta}{r}\right) + \frac{1}{r}\frac{\partial v_r}{\partial \theta} \right] \tag{A.74}$$

$$D_{\theta z} = D_{z\theta} = \frac{1}{2}\left[\frac{\partial v_\theta}{\partial z} + \frac{1}{r}\frac{\partial v_z}{\partial \theta} \right] \tag{A.75}$$

$$D_{zr} = D_{rz} = \frac{1}{2}\left[\frac{\partial v_z}{\partial r} + \frac{\partial v_r}{\partial z} \right] \tag{A.76}$$

Vorticity:

$$W_{r\theta} = -W_{\theta r} = \frac{1}{2}\left[\frac{1}{r}\frac{\partial v_r}{\partial \theta} - \frac{1}{r}\frac{\partial}{\partial r}(r v_\theta) \right] \tag{A.77}$$

$$W_{\theta z} = -W_{z\theta} = \frac{1}{2}\left[\frac{\partial v_\theta}{\partial z} - \frac{1}{r}\frac{\partial v_z}{\partial \theta} \right] \tag{A.78}$$

$$W_{zr} = -W_{rz} = \frac{1}{2}\left[\frac{\partial v_z}{\partial r} - \frac{\partial v_r}{\partial z} \right] \tag{A.79}$$

B Derivations

B.1 Jeffery's Equation for Planar Orientation and Planar Flow

For planar orientation in the 1–2 plane, the orientation vector and its derivative depends only on the angle ϕ and its derivative:

$$\mathbf{p} = \begin{Bmatrix} \cos\phi \\ \sin\phi \\ 0 \end{Bmatrix} \qquad \text{and} \qquad \dot{\mathbf{p}} = \dot{\phi} \begin{Bmatrix} -\sin\phi \\ \cos\phi \\ 0 \end{Bmatrix} \tag{B.1}$$

For planar flow at constant volume, the rate-of-deformation and vorticity tensors are determined by D_{11}, D_{12} and W_{12}:

$$\mathbf{D} = \begin{bmatrix} D_{11} & D_{12} & 0 \\ D_{12} & -D_{11} & 0 \\ 0 & 0 & 0 \end{bmatrix} \qquad \text{and} \qquad \mathbf{W} = \begin{bmatrix} 0 & W_{12} & 0 \\ -W_{12} & 0 & 0 \\ 0 & 0 & 0 \end{bmatrix} \tag{B.2}$$

Write out Jeffery's equation in matrix form, as in Eqn. (4.93), and substitute the expressions above, but do not expand the product $(\mathbf{D}:\mathbf{pp})$. This gives

$$\begin{aligned} \dot{\phi} \begin{Bmatrix} -\sin\phi \\ \cos\phi \\ 0 \end{Bmatrix} = & \begin{bmatrix} 0 & W_{12} & 0 \\ -W_{12} & 0 & 0 \\ 0 & 0 & 0 \end{bmatrix} \begin{Bmatrix} \cos\phi \\ \sin\phi \\ 0 \end{Bmatrix} \\ & + \xi \left(\begin{bmatrix} D_{11} & D_{12} & 0 \\ D_{12} & -D_{11} & 0 \\ 0 & 0 & 0 \end{bmatrix} \begin{Bmatrix} \cos\phi \\ \sin\phi \\ 0 \end{Bmatrix} - (\mathbf{D}:\mathbf{pp}) \begin{Bmatrix} \cos\phi \\ \sin\phi \\ 0 \end{Bmatrix} \right) \end{aligned} \tag{B.3}$$

The third line of this equation is identically zero and can be ignored. Multiply the first line by $-\sin\phi$ and the second line by $\cos\phi$ and add the equations. This gives

$$\begin{aligned} \dot{\phi}\left(\sin^2\phi + \cos^2\phi\right) = & -W_{12}\left(\sin^2\phi + \cos^2\phi\right) - \xi D_{11}\left(\sin\phi\cos\phi + \sin\phi\cos\phi\right) \\ & + \xi D_{12}\left(-\sin^2\phi + \cos^2\phi\right) + (\mathbf{D}:\mathbf{pp})\left(\sin\phi\cos\phi - \sin\phi\cos\phi\right) \end{aligned} \tag{B.4}$$

The $(\mathbf{D}:\mathbf{pp})$ terms cancel. Using standard trigonometric identities we obtain Eqn. (4.88), which is the desired result:

$$\dot{\phi} = -W_{12} - \xi D_{11}\sin(2\phi) + \xi D_{12}\cos(2\phi) \tag{B.5}$$

B.2 Tensor Form of the Folgar–Tucker Model

Here we derive the Folgar–Tucker equation for $\dot{\mathbf{A}}$, Eqn. (5.33).

B.2.1 Preliminaries

We will use spherical coordinates, for which the unit vectors are $\hat{\boldsymbol{\delta}}_r$, $\hat{\boldsymbol{\delta}}_\theta$ and $\hat{\boldsymbol{\delta}}_\phi$. In this coordinate system the orientation vector is the same as the unit vector in the r direction,

$$\mathbf{p} = \hat{\boldsymbol{\delta}}_r \tag{B.6}$$

and the identity tensor can be written as

$$\mathbf{I} = \hat{\boldsymbol{\delta}}_r\hat{\boldsymbol{\delta}}_r + \hat{\boldsymbol{\delta}}_\theta\hat{\boldsymbol{\delta}}_\theta + \hat{\boldsymbol{\delta}}_\phi\hat{\boldsymbol{\delta}}_\phi \tag{B.7}$$

In spherical coordinates the unit vectors change direction as θ or ϕ change. We will need the derivatives of the unit vectors with respect to these coordinates, which are (e.g., [BAH87])

$$\frac{\partial \hat{\boldsymbol{\delta}}_r}{\partial r} = 0 \qquad \frac{\partial \hat{\boldsymbol{\delta}}_\theta}{\partial r} = 0 \qquad \frac{\partial \hat{\boldsymbol{\delta}}_\phi}{\partial r} = 0 \tag{B.8}$$

$$\frac{\partial \hat{\boldsymbol{\delta}}_r}{\partial \theta} = \hat{\boldsymbol{\delta}}_\theta \qquad \frac{\partial \hat{\boldsymbol{\delta}}_\theta}{\partial \theta} = -\hat{\boldsymbol{\delta}}_r \qquad \frac{\partial \hat{\boldsymbol{\delta}}_\phi}{\partial \theta} = 0 \tag{B.9}$$

$$\frac{\partial \hat{\boldsymbol{\delta}}_r}{\partial \phi} = S\hat{\boldsymbol{\delta}}_\phi \qquad \frac{\partial \hat{\boldsymbol{\delta}}_\theta}{\partial \phi} = C\hat{\boldsymbol{\delta}}_\phi \qquad \frac{\partial \hat{\boldsymbol{\delta}}_\phi}{\partial \phi} = -S\hat{\boldsymbol{\delta}}_r - C\hat{\boldsymbol{\delta}}_\theta \tag{B.10}$$

We have used $S = \sin\theta$ and $C = \cos\theta$ here for conciseness.

Recall from Eqn. (5.19) that the gradient operator on the surface of the unit sphere is

$$\nabla_s = \hat{\boldsymbol{\delta}}_\theta \frac{\partial}{\partial \theta} + \hat{\boldsymbol{\delta}}_\phi \frac{1}{\sin\theta}\frac{\partial}{\partial \phi} \tag{B.11}$$

The Laplacian on the surface of the unit sphere, $\nabla_s^2 = \nabla_s \cdot \nabla_s$, can be derived from Eqn. (B.11) using Eqns. (B.9) and (B.10). The result is

$$\nabla_s^2 = \frac{1}{S}\frac{\partial}{\partial \theta}\left(S\frac{\partial}{\partial \theta}\right) + \frac{1}{S^2}\frac{\partial^2}{\partial \phi^2} \tag{B.12}$$

We will need the Laplacian of $\mathbf{pp}$. Using the above expression as well as Eqns. (B.6) and (B.7), this is

$$\begin{aligned}
\nabla_s^2(\mathbf{pp}) &= \frac{1}{S}\frac{\partial}{\partial \theta}\left[S\frac{\partial \hat{\boldsymbol{\delta}}_r}{\partial \theta}\hat{\boldsymbol{\delta}}_r + S\hat{\boldsymbol{\delta}}_r\frac{\partial \hat{\boldsymbol{\delta}}_r}{\partial \theta}\right] + \frac{1}{S^2}\frac{\partial}{\partial \phi}\left[\frac{\partial \hat{\boldsymbol{\delta}}_r}{\partial \phi}\hat{\boldsymbol{\delta}}_r + \hat{\boldsymbol{\delta}}_r\frac{\partial \hat{\boldsymbol{\delta}}_r}{\partial \phi}\right] \\
&= \frac{1}{S}\frac{\partial}{\partial \theta}\left[S(\hat{\boldsymbol{\delta}}_\theta\hat{\boldsymbol{\delta}}_r + \hat{\boldsymbol{\delta}}_r\hat{\boldsymbol{\delta}}_\theta)\right] + \frac{1}{S}\frac{\partial}{\partial \phi}\left[\hat{\boldsymbol{\delta}}_\phi\hat{\boldsymbol{\delta}}_r + \hat{\boldsymbol{\delta}}_r\hat{\boldsymbol{\delta}}_\phi\right] \\
&= -4\hat{\boldsymbol{\delta}}_r\hat{\boldsymbol{\delta}}_r + 2\hat{\boldsymbol{\delta}}_\theta\hat{\boldsymbol{\delta}}_\theta + 2\hat{\boldsymbol{\delta}}_\phi\hat{\boldsymbol{\delta}}_\phi \\
&= 2(\hat{\boldsymbol{\delta}}_r\hat{\boldsymbol{\delta}}_r + \hat{\boldsymbol{\delta}}_\theta\hat{\boldsymbol{\delta}}_\theta + \hat{\boldsymbol{\delta}}_\phi\hat{\boldsymbol{\delta}}_\phi) - 6\hat{\boldsymbol{\delta}}_r\hat{\boldsymbol{\delta}}_r \\
&= 2\mathbf{I} - 6\mathbf{pp}
\end{aligned} \tag{B.13}$$

Finally, let $\mathbf{T}(\mathbf{p})$ be any tensor quantity whose derivatives exist on the unit sphere. Because the unit sphere is a closed surface, the divergence theorem gives

$$\oint \nabla_s \cdot \mathbf{T}\, d\mathbf{p} = 0 \tag{B.14}$$

B.2.2 Separating the Jeffery and Diffusive Contributions

The definition of the second-order orientation tensor is

$$\mathbf{A}(t) = \oint \psi(\mathbf{p}, t)\,\mathbf{pp}\,\mathrm{d}\mathbf{p} \tag{B.15}$$

In this derivation we treat **p** as a coordinate that indexes points on the unit sphere, and assume that **A** changes with time only because $\psi(\mathbf{p})$ changes with time. Then, taking the material derivative of (B.15) gives

$$\dot{\mathbf{A}} = \frac{\mathrm{D}}{\mathrm{D}t}\left[\oint \psi(\mathbf{p}, t)\,\mathbf{pp}\,\mathrm{d}\mathbf{p}\right] = \oint \frac{\mathrm{D}\psi}{\mathrm{D}t}\,\mathbf{pp}\,\mathrm{d}\mathbf{p} \tag{B.16}$$

The Folgar–Tucker model assumes that fiber motion is the sum of a Jeffery part and a diffusive part. For the distribution function solution this is

$$\frac{\mathrm{D}\psi}{\mathrm{D}t} = -\nabla_s \cdot \left(\mathbf{q}^J + \mathbf{q}^d\right) \tag{B.17}$$

where $\mathbf{q}^J$ is the Jeffery contribution to the probability flux (5.15) and $\mathbf{q}^d$ is the diffusive contribution (5.29). Substituting Eqn. (B.17) into (B.16) shows that the material derivative of the orientation tensor is the sum of two parts: one from the Jeffery motion and one from diffusion.

$$\dot{\mathbf{A}} = \underbrace{\oint -(\nabla_s \cdot \mathbf{q}^J)\,\mathbf{pp}\,\mathrm{d}\mathbf{p}}_{\text{Jeffery}} + \underbrace{\oint -(\nabla_s \cdot \mathbf{q}^d)\,\mathbf{pp}\,\mathrm{d}\mathbf{p}}_{\text{diffusive}} = \dot{\mathbf{A}}^J + \dot{\mathbf{A}}^d \tag{B.18}$$

The Jeffery part, Eqn. (5.6), was derived in Section 5.1, so we only need to derive the diffusive part $\dot{\mathbf{A}}^d$.

B.2.3 Diffusive Contribution to $\dot{\mathbf{A}}$

The diffusive contribution to $\dot{\mathbf{A}}$ is

$$\dot{\mathbf{A}}^d = -\oint (\nabla_s \cdot \mathbf{q}^d)\,\mathbf{pp}\,\mathrm{d}\mathbf{p} \tag{B.19}$$

Integrating this by parts gives

$$\dot{\mathbf{A}}^d = -\oint \nabla_s \cdot (\mathbf{q}^d\mathbf{pp})\,\mathrm{d}\mathbf{p} + \oint \mathbf{q}^d \cdot \nabla_s(\mathbf{pp})\,\mathrm{d}\mathbf{p} \tag{B.20}$$

The first term on the right-hand side is zero from Eqn. (B.14). Substituting $\mathbf{q}^d = -C_I\dot{\gamma}\nabla_s\psi$ from Eqn. (5.29) into the second term gives

$$\dot{\mathbf{A}}^d = -\oint C_I\dot{\gamma}\nabla_s\psi \cdot \nabla_s(\mathbf{pp})\,\mathrm{d}\mathbf{p} \tag{B.21}$$

$C_I\dot{\gamma}$ does not depend on **p**, so it can be brought outside the integral. Integrating by parts then gives

$$\dot{\mathbf{A}}^d = -C_I\dot{\gamma}\oint \nabla_s \cdot [\nabla_s \cdot (\psi\mathbf{pp})]\,\mathrm{d}\mathbf{p} + C_I\dot{\gamma}\oint \psi\nabla_s \cdot \nabla_s(\mathbf{pp})\,\mathrm{d}\mathbf{p} \tag{B.22}$$

Again the first term on the right-hand side is zero from Eqn. (B.14), while the second term contains the orientation average of the Laplacian of **pp**.

$$\dot{\mathbf{A}}^d = C_I\dot{\gamma}\oint \psi\nabla_s^2(\mathbf{pp})\,\mathrm{d}\mathbf{p} = C_I\dot{\gamma}\left\langle \nabla_s^2(\mathbf{pp})\right\rangle \tag{B.23}$$

Inserting Eqn. (B.13) gives the final expression for the diffusive term,

$$\begin{aligned}\dot{\mathbf{A}}^d &= C_I\dot{\gamma}\,\langle 2\mathbf{I} - 6\mathbf{pp}\rangle \\ &= 2C_I\dot{\gamma}\,(\mathbf{I} - 3\mathbf{A})\end{aligned} \tag{B.24}$$

Adding this to the Jeffery contribution to $\dot{\mathbf{A}}$, Eqn. (5.6), gives the Folgar–Tucker expression for $\dot{\mathbf{A}}$, Eqn. (5.33).

B.3 Steady-State Orientation of RSC and RPR Models

We wish to show algebraically that RSC and RPR models have the same steady-state orientation as the standard model, provided that **D**, **W**, the rotary diffusion parameters and the closure approximation are the same.

To show this, it is helpful to decompose the second-order orientation tensor into its eigenvalues λ_i and eigenvectors $\mathbf{e}_i$,

$$\mathbf{A} = \lambda_1\mathbf{e}_1\mathbf{e}_1 + \lambda_2\mathbf{e}_2\mathbf{e}_2 + \lambda_3\mathbf{e}_3\mathbf{e}_3 \tag{B.25}$$

The Folgar–Tucker expression for $\dot{\mathbf{A}}$ can be used to develop expressions for the derivatives of the λ_is and $\mathbf{e}_i$s (see Section 5.7.3). These are Eqns. (5.100) and (5.101), which we repeat here:

$$\dot{\lambda}_i = 2\xi(\lambda_i\,\mathbf{D}:\mathbf{e}_i\mathbf{e}_i - \mathbf{e}_i\mathbf{e}_i:\mathbb{A}:\mathbf{D}) + 2C_I\dot{\gamma}(1-3\lambda_i) \qquad \text{(no sum on } i\text{)} \tag{B.26}$$

$$\dot{\mathbf{e}}_1 = \mathbf{W}\cdot\mathbf{e}_1 + \xi\left(\frac{\lambda_1+\lambda_2-2\bar{\mathbb{A}}_{1212}}{\lambda_1-\lambda_2}\right)\mathbf{D}:\mathbf{e}_1\mathbf{e}_2\mathbf{e}_2 + \xi\left(\frac{\lambda_1+\lambda_3-2\bar{\mathbb{A}}_{1313}}{\lambda_1-\lambda_3}\right)\mathbf{D}:\mathbf{e}_1\mathbf{e}_3\mathbf{e}_3 \tag{B.27}$$

Equations for $\dot{\mathbf{e}}_2$ and $\dot{\mathbf{e}}_3$ are found by permuting the indices in Eqn. (B.27).

In these expressions the fourth-order tensor $\mathbb{A}$ comes from a closure approximation, so it is fully determined by the eigenvalues λ_i and eigenvectors $\mathbf{e}_i$. Thus, we have a closed set of equations that can be integrated to find the λ_is and $\mathbf{e}_i$s as a function of time.

To convert this standard model to an RSC model, multiply the expression for $\dot{\lambda}_i$ by κ but leave the expressions for the $\dot{\mathbf{e}}_i$s unchanged. The RSC version of Eqn. (B.26) then reads

$$\dot{\lambda}_i^{\mathrm{RSC}} = 2\kappa\xi(\lambda_i\,\mathbf{D}:\mathbf{e}_i\mathbf{e}_i - \mathbf{e}_i\mathbf{e}_i:\mathbb{A}:\mathbf{D}) + 2\kappa C_I\dot{\gamma}(1-3\lambda_i) \tag{B.28}$$

Integrating this together with Eqn. (B.27) and its permutations gives the RSC histories for the λ_is and $\mathbf{e}_i$s.

For either model, steady state is reached when $\dot{\lambda}_i = 0$ and $\dot{\mathbf{e}}_i = 0$. Setting the right-hand sides of either Eqns. (B.26) and (B.27) (for the standard model) or Eqns. (B.28) and (B.27) (for the RSC model) equal to zero gives algebraic equations for the steady-state values of λ_i and $\mathbf{e}_i$.

With the right-hand side of the RSC equation (B.28) set equal to zero, dividing both sides by κ gives the steady-state version of the standard equation (B.26). The eigenvector equations are the same for both models, so the steady-state solutions for the standard and RSC models satisfy the same equations. As a result, the standard and RSC models always have the same steady-state solution, and this solution is independent of κ.

The RPR model calculates $\dot{\mathbf{\Lambda}}$ differently, but the model is still developed using $\dot{\lambda}_i^{\mathrm{RPR}} = \kappa \dot{\lambda}_i^{\mathrm{std}}$, so the same result applies. ARD models have slightly different forms of $\dot{\lambda}_i$ (Eqn. (5.109)), but follow the same argument.

B.4 Mechanical Property Equations

B.4.1 Eshelby Strain Concentration Tensor

We wish to derive Eqn. (8.47) for the tensor $\mathbb{H}^m$, which relates fiber strain to average strain for a dilute composite,

$$\overline{\boldsymbol{\varepsilon}}^f = \mathbb{H}^m : \overline{\boldsymbol{\varepsilon}} \tag{B.29}$$

Start with Eshelby's inclusion problem. A large body has the properties of the matrix $\mathbb{C}^m$. A small, ellipsoidal portion of that body, the *inclusion*, undergoes a phase transformation or some similar process that would induce a strain $\boldsymbol{\varepsilon}^T$ with no stress if the inclusion were free to deform. Because the inclusion is part of the larger body, it and the surrounding material instead acquire a strain $\boldsymbol{\varepsilon}^C(\mathbf{x})$. Outside the inclusion this strain depends on position $\mathbf{x}$, and $\boldsymbol{\varepsilon}^C \to 0$ as $\mathbf{x} \to \infty$. Within the inclusion $\boldsymbol{\varepsilon}^C$ is uniform and is given by

$$\boldsymbol{\varepsilon}^C = \mathbb{E}^m : \boldsymbol{\varepsilon}^T \tag{B.30}$$

Now apply an additional strain $\boldsymbol{\varepsilon}^A$ at infinity. Since the body has uniform stiffness, the additional strain is $\boldsymbol{\varepsilon}^A$ everywhere. Within the inclusion the total strain is $\boldsymbol{\varepsilon}^A + \boldsymbol{\varepsilon}^C$, but the portion $\boldsymbol{\varepsilon}^T$ does not contribute to the stress, so the stress in the inclusion is

$$\boldsymbol{\sigma}^I = \mathbb{C}^m : \left(\boldsymbol{\varepsilon}^A + \boldsymbol{\varepsilon}^C - \boldsymbol{\varepsilon}^T\right) \tag{B.31}$$

Next, consider a problem where the ellipsoidal volume has no transformation strain $\boldsymbol{\varepsilon}^T$ but has a different stiffness $\mathbb{C}^f$. This volume is usually called the *inhomogeneity*, but here we will call it the fiber. When the entire body is strained by $\boldsymbol{\varepsilon}^A$ at infinity, the fiber deforms less than the matrix and this introduces a perturbation in the strain field $\boldsymbol{\varepsilon}^C(\mathbf{x})$. The stress in the fiber is

$$\overline{\boldsymbol{\sigma}}^f = \mathbb{C}^f : \left(\boldsymbol{\varepsilon}^A + \boldsymbol{\varepsilon}^C\right) \tag{B.32}$$

We want to find a value of the transformation strain $\boldsymbol{\varepsilon}^T$ that makes the inclusion problem equivalent to this one by making the stresses in the ellipsoidal regions equal, $\boldsymbol{\sigma}^I = \overline{\boldsymbol{\sigma}}^f$.

In the notation of Section 8.3, $\boldsymbol{\varepsilon}^A = \overline{\boldsymbol{\varepsilon}}$ and

$$\overline{\boldsymbol{\varepsilon}}^f = \boldsymbol{\varepsilon}^A + \boldsymbol{\varepsilon}^C \tag{B.33}$$

Set the right-hand sides of Eqns. (B.31) and (B.32) equal and substitute this expression, then rearrange to find $\boldsymbol{\varepsilon}^T$:

$$\begin{aligned} \mathbb{C}^f : \overline{\boldsymbol{\varepsilon}}^f &= \mathbb{C}^m : \left(\overline{\boldsymbol{\varepsilon}}^f - \boldsymbol{\varepsilon}^T\right) \\ \left(\mathbb{C}^f - \mathbb{C}^m\right) : \overline{\boldsymbol{\varepsilon}}^f &= -\mathbb{C}^m : \boldsymbol{\varepsilon}^T \\ \boldsymbol{\varepsilon}^T &= -\mathbb{S}^m : \left(\mathbb{C}^f - \mathbb{C}^m\right) : \overline{\boldsymbol{\varepsilon}}^f \end{aligned} \tag{B.34}$$

Here we have used $\mathbb{S}^m = (\mathbb{C}^m)^{-1}$. Use (B.34) in Eshelby's relationship, Eqn. (B.30),

$$\boldsymbol{\varepsilon}^C = -\mathbb{E}^m : \mathbb{S}^m : \left(\mathbb{C}^f - \mathbb{C}^m\right) : \overline{\boldsymbol{\varepsilon}}^f \tag{B.35}$$

Substitute this into Eqn. (B.33), also using $\boldsymbol{\varepsilon}^A = \overline{\boldsymbol{\varepsilon}}$, and rearrange:

$$\begin{aligned} \overline{\boldsymbol{\varepsilon}}^f &= \overline{\boldsymbol{\varepsilon}} - \mathbb{E}^m : \mathbb{S}^m : \left(\mathbb{C}^f - \mathbb{C}^m\right) : \overline{\boldsymbol{\varepsilon}}^f \\ \left[\mathbb{I} + \mathbb{E}^m : \mathbb{S}^m : \left(\mathbb{C}^f - \mathbb{C}^m\right)\right] : \overline{\boldsymbol{\varepsilon}}^f &= \overline{\boldsymbol{\varepsilon}} \\ \overline{\boldsymbol{\varepsilon}}^f &= \left[\mathbb{I} + \mathbb{E}^m : \mathbb{S}^m : \left(\mathbb{C}^f - \mathbb{C}^m\right)\right]^{-1} : \overline{\boldsymbol{\varepsilon}} \end{aligned} \tag{B.36}$$

Comparing the last line to Eqn. (B.29) gives the final result,

$$\mathbb{H}^m = \left[\mathbb{I} + \mathbb{E}^m : \mathbb{S}^m : \left(\mathbb{C}^f - \mathbb{C}^m\right)\right]^{-1} \tag{B.37}$$

B.4.2 Eshelby Dilute Stiffness Model

To derive Eqn. (8.48) for the stiffness of a dilute composite, start with Eqn. (8.38) and substitute the fiber and matrix constitutive equations (8.40):

$$\begin{aligned} \overline{\boldsymbol{\sigma}} &= \phi_m \overline{\boldsymbol{\sigma}}^m + \phi_f \overline{\boldsymbol{\sigma}}^f \\ &= \phi_m \mathbb{C}^m : \overline{\boldsymbol{\varepsilon}}^m + \phi_f \mathbb{C}^f : \overline{\boldsymbol{\varepsilon}}^f \end{aligned} \tag{B.38}$$

Since $\overline{\boldsymbol{\varepsilon}} = \phi_m \overline{\boldsymbol{\varepsilon}}^m + \phi_f \overline{\boldsymbol{\varepsilon}}^f$, we can substitute $\phi_m \overline{\boldsymbol{\varepsilon}}^m = \overline{\boldsymbol{\varepsilon}} - \phi_f \overline{\boldsymbol{\varepsilon}}^f$ and obtain

$$\overline{\boldsymbol{\sigma}} = \mathbb{C}^m : \overline{\boldsymbol{\varepsilon}} + \phi_f \left(\mathbb{C}^f - \mathbb{C}^m\right) : \overline{\boldsymbol{\varepsilon}}^f \tag{B.39}$$

Now use Eqn. (B.29) to replace $\overline{\boldsymbol{\varepsilon}}^f$:

$$\overline{\boldsymbol{\sigma}} = \mathbb{C}^m : \overline{\boldsymbol{\varepsilon}} + \phi_f \left(\mathbb{C}^f - \mathbb{C}^m\right) : \mathbb{H}^m : \overline{\boldsymbol{\varepsilon}} \tag{B.40}$$

Since $\overline{\boldsymbol{\sigma}} = \mathbb{C} : \overline{\boldsymbol{\varepsilon}}$, this gives

$$\mathbb{C} = \mathbb{C}^m + \phi_f \left(\mathbb{C}^f - \mathbb{C}^m\right) : \mathbb{H}^m \tag{B.41}$$

which is Eqn. (8.48).

B.4.3 General Stiffness Equation

To derive Eqn. (8.42) start from Eqn. (B.38), which applies to any two-phase composite, and substitute $\overline{\boldsymbol{\varepsilon}}^f = \mathbb{B} : \overline{\boldsymbol{\varepsilon}}^m$.

$$\overline{\boldsymbol{\sigma}} = \left(\phi_m \mathbb{C}^m + \phi_f \mathbb{C}^f : \mathbb{B}\right) : \overline{\boldsymbol{\varepsilon}}^m \tag{B.42}$$

Make the same substitution into $\overline{\boldsymbol{\varepsilon}} = \phi_m \overline{\boldsymbol{\varepsilon}}^m + \phi_f \overline{\boldsymbol{\varepsilon}}^f$ and rearrange to get $\overline{\boldsymbol{\varepsilon}}^m$

$$\begin{aligned} \phi_m \overline{\boldsymbol{\varepsilon}}^m + \phi_f \mathbb{B} : \overline{\boldsymbol{\varepsilon}}^m &= \overline{\boldsymbol{\varepsilon}} \\ \left[\phi_m \mathbb{I} + \phi_f \mathbb{B}\right] : \overline{\boldsymbol{\varepsilon}}^m &= \overline{\boldsymbol{\varepsilon}} \\ \overline{\boldsymbol{\varepsilon}}^m &= \left[\phi_m \mathbb{I} + \phi_f \mathbb{B}\right]^{-1} : \overline{\boldsymbol{\varepsilon}} \end{aligned} \tag{B.43}$$

Use this in Eqn. (B.42),

$$\overline{\boldsymbol{\sigma}} = \left(\phi_m \mathbb{C}^m + \phi_f \mathbb{C}^f : \mathbb{B}\right) : \left[\phi_m \mathbb{I} + \phi_f \mathbb{B}\right]^{-1} : \overline{\boldsymbol{\varepsilon}} \tag{B.44}$$

Comparing this to $\overline{\boldsymbol{\sigma}} = \mathbb{C} : \overline{\boldsymbol{\varepsilon}}$ gives the general equation for composite stiffness in terms of $\mathbb{B}$, which is the desired result:

$$\mathbb{C} = \left(\phi_m \mathbb{C}^m + \phi_f \mathbb{C}^f : \mathbb{B}\right) : \left[\phi_m \mathbb{I} + \phi_f \mathbb{B}\right]^{-1} \tag{B.45}$$

B.4.4 General Equation for Thermal Stress

To derive the general expression for the thermal stress tensor $\boldsymbol{\beta}$, Eqn. (8.89), consider an RVE subjected to two loadings, A and B. We will analyze each loading separately, then combine them. Assume that the homogenized stiffness $\mathbb{C}$ has already been determined.

Loading A has a non-zero temperature change ΔT^A and displacement boundary conditions consistent with an average strain $\overline{\boldsymbol{\varepsilon}}^A$. This strain is chosen such that the average strains in the fiber and matrix are equal, $\overline{\boldsymbol{\varepsilon}}^{f,A} = \overline{\boldsymbol{\varepsilon}}^{m,A} = \overline{\boldsymbol{\varepsilon}}^A$, and the average stresses are also equal, $\overline{\boldsymbol{\sigma}}^{f,A} = \overline{\boldsymbol{\sigma}}^{m,A} = \overline{\boldsymbol{\sigma}}^A$.

The strain $\overline{\boldsymbol{\varepsilon}}^A$ that achieves this is found by equating the average stresses in the fibers and matrix, substituting the stress/strain/temperature relationships (8.88) for each phase, and rearranging:

$$\begin{aligned} \overline{\boldsymbol{\sigma}}^{f,A} &= \overline{\boldsymbol{\sigma}}^{m,A} \\ \mathbb{C}^f : \overline{\boldsymbol{\varepsilon}}^A - \boldsymbol{\beta}^f : \Delta T^A &= \mathbb{C}^m : \overline{\boldsymbol{\varepsilon}}^A - \boldsymbol{\beta}^m : \Delta T^A \\ \left(\mathbb{C}^f - \mathbb{C}^m\right) : \overline{\boldsymbol{\varepsilon}}^A &= \left(\boldsymbol{\beta}^f - \boldsymbol{\beta}^m\right) \Delta T^A \\ \overline{\boldsymbol{\varepsilon}}^A &= \left(\mathbb{C}^f - \mathbb{C}^m\right)^{-1} : \left(\boldsymbol{\beta}^f - \boldsymbol{\beta}^m\right) \Delta T^A \end{aligned} \tag{B.46}$$

Loading B has no temperature change, $\Delta T^B = 0$, and displacement boundary conditions consistent with an average strain of $\overline{\boldsymbol{\varepsilon}}^B = -\overline{\boldsymbol{\varepsilon}}^A$. The average stress for loading B is

$$\overline{\boldsymbol{\sigma}}^B = \mathbb{C} : \overline{\boldsymbol{\varepsilon}}^B = -\mathbb{C} : \overline{\boldsymbol{\varepsilon}}^A \tag{B.47}$$

Now combine the two loadings. The combined temperature change is $\Delta T = \Delta T^A + \Delta T^B = \Delta T^A$ and the combined average strain is $\overline{\boldsymbol{\varepsilon}} = \overline{\boldsymbol{\varepsilon}}^A + \overline{\boldsymbol{\varepsilon}}^B = 0$. The combined average stress is

$$\begin{aligned}\overline{\boldsymbol{\sigma}} &= \overline{\boldsymbol{\sigma}}^A + \overline{\boldsymbol{\sigma}}^B \\ &= \left(\phi_m \overline{\boldsymbol{\sigma}}^{m,A} + \phi_f \overline{\boldsymbol{\sigma}}^{f,A}\right) + \overline{\boldsymbol{\sigma}}^B\end{aligned} \tag{B.48}$$

Use the expressions from the second line of Eqn. (B.46) for $\overline{\boldsymbol{\sigma}}^{f,A}$ and $\overline{\boldsymbol{\sigma}}^{m,A}$ and substitute Eqn. (B.47) for $\overline{\boldsymbol{\sigma}}^B$. This gives

$$\overline{\boldsymbol{\sigma}} = \phi_m \left(\mathbb{C}^m : \overline{\boldsymbol{\varepsilon}}^A - \boldsymbol{\beta}^m : \Delta T^A\right) + \phi_f \left(\mathbb{C}^f : \overline{\boldsymbol{\varepsilon}}^A - \boldsymbol{\beta}^f : \Delta T^A\right) - \mathbb{C} : \overline{\boldsymbol{\varepsilon}}^A \tag{B.49}$$

On the right-hand side, collect terms that multiply $\overline{\boldsymbol{\varepsilon}}^A$. Also using $\Delta T = \Delta T^A$, we have

$$\overline{\boldsymbol{\sigma}} = -\left(\phi_m \boldsymbol{\beta}^m + \phi_f \boldsymbol{\beta}^f\right) \Delta T + \left(\phi_m \mathbb{C}^m + \phi_f \mathbb{C}^f - \mathbb{C}\right) : \overline{\boldsymbol{\varepsilon}}^A \tag{B.50}$$

Use Eqn. (B.46) to replace $\overline{\boldsymbol{\varepsilon}}^A$:

$$\overline{\boldsymbol{\sigma}} = -\left(\phi_m \boldsymbol{\beta}^m + \phi_f \boldsymbol{\beta}^f\right) \Delta T + \left(\phi_m \mathbb{C}^m + \phi_f \mathbb{C}^f - \mathbb{C}\right) : \left(\mathbb{C}^f - \mathbb{C}^m\right)^{-1} : \left(\boldsymbol{\beta}^f - \boldsymbol{\beta}^m\right) \Delta T \tag{B.51}$$

The homogenized stress will be $\overline{\boldsymbol{\sigma}} = \mathbb{C} : \overline{\boldsymbol{\varepsilon}} - \boldsymbol{\beta} \Delta T$. Because $\overline{\boldsymbol{\varepsilon}} = 0$ in this case, $\boldsymbol{\beta} = -\overline{\boldsymbol{\sigma}} / \Delta T$. Comparing this to Eqn. (B.51) gives

$$\boldsymbol{\beta} = \phi_m \boldsymbol{\beta}^m + \phi_f \boldsymbol{\beta}^f - \left(\phi_m \mathbb{C}^m + \phi_f \mathbb{C}^f - \mathbb{C}\right) : \left(\mathbb{C}^f - \mathbb{C}^m\right)^{-1} : \left(\boldsymbol{\beta}^f - \boldsymbol{\beta}^m\right) \tag{B.52}$$

This is a useful basic result for the thermal stress tensor. Equivalent expressions that include strain concentration tensors have been reported by Lielens et al. [LPC+98] and Pierard et al., [PFD04], and this appendix follows their derivation scheme.

Obtaining Eqn. (8.89) is now a matter of algebraically manipulating the right-hand side of Eqn. (B.52). Pre-multiply the first two terms by $\mathbb{I} = (\mathbb{C}^f - \mathbb{C}^m) : (\mathbb{C}^f - \mathbb{C}^m)^{-1}$ and distribute the last product over $\left(\boldsymbol{\beta}^f - \boldsymbol{\beta}^m\right)$:

$$\begin{aligned}\boldsymbol{\beta} = {} & \phi_m \left(\mathbb{C}^f - \mathbb{C}^m\right) : \left(\mathbb{C}^f - \mathbb{C}^m\right)^{-1} : \boldsymbol{\beta}^m + \phi_f \left(\mathbb{C}^f - \mathbb{C}^m\right) : \left(\mathbb{C}^f - \mathbb{C}^m\right)^{-1} : \boldsymbol{\beta}^f \\ & - \left(\phi_m \mathbb{C}^m + \phi_f \mathbb{C}^f - \mathbb{C}\right) : \left(\mathbb{C}^f - \mathbb{C}^m\right)^{-1} : \boldsymbol{\beta}^f + \left(\phi_m \mathbb{C}^m + \phi_f \mathbb{C}^f - \mathbb{C}\right) : \left(\mathbb{C}^f - \mathbb{C}^m\right)^{-1} : \boldsymbol{\beta}^m\end{aligned} \tag{B.53}$$

Group the terms that multiply $\boldsymbol{\beta}^m$ and those that multiply $\boldsymbol{\beta}^f$:

$$\begin{aligned}\boldsymbol{\beta} = {} & \left(\phi_m \mathbb{C}^f - \phi_m \mathbb{C}^m + \phi_m \mathbb{C}^m + \phi_f \mathbb{C}^f - \mathbb{C}\right) : \left(\mathbb{C}^f - \mathbb{C}^m\right)^{-1} : \boldsymbol{\beta}^m \\ & + \left(\phi_f \mathbb{C}^f - \phi_f \mathbb{C}^m - \phi_m \mathbb{C}^m - \phi_f \mathbb{C}^f + \mathbb{C}\right) : \left(\mathbb{C}^f - \mathbb{C}^m\right)^{-1} : \boldsymbol{\beta}^f\end{aligned} \tag{B.54}$$

Some terms cancel and others combine using $\phi_m + \phi_f = 1$:

$$\boldsymbol{\beta} = \left(\mathbb{C}^f - \mathbb{C}\right) : \left(\mathbb{C}^f - \mathbb{C}^m\right)^{-1} : \boldsymbol{\beta}^m + \left(-\mathbb{C}^m + \mathbb{C}\right) : \left(\mathbb{C}^f - \mathbb{C}^m\right)^{-1} : \boldsymbol{\beta}^f \tag{B.55}$$

In the second term, factor (-1) from the first expression in parentheses and use it to multiply the second expression in parentheses. This gives the final result,

$$\boldsymbol{\beta} = \left(\mathbb{C}^f - \mathbb{C}\right) : \left(\mathbb{C}^f - \mathbb{C}^m\right)^{-1} : \boldsymbol{\beta}^m + \left(\mathbb{C}^m - \mathbb{C}\right) : \left(\mathbb{C}^m - \mathbb{C}^f\right)^{-1} : \boldsymbol{\beta}^f \tag{B.56}$$

To apply this equation to a pseudo-grain, replace $\mathbb{C}$ by the pseudo-grain stiffness $\mathbb{C}^p$. This gives Eqn. (8.89).

Symbols and Abbreviations

In this book lightface italic symbols are scalars, lowercase bold symbols are vectors, uppercase bold symbols (and a few bold lowercase Greek letters) are second-order tensors, and blackboard bold symbols are fourth-order tensors. Abbreviations use roman letters. When i, j, k, l, m or n appear as subscripts they represent indices, and can take on any appropriate integer value. In the list below, numbers in parentheses indicate an equation where the symbol is defined.

Roman Letters

a, b, c	Semi-axis lengths of an ellipsoidal particle (4.83)
a_1, a_2	Invariants of a transversely isotropic second-order tensor (8.78)
ARD	Anisotropic rotary diffusion model, Section 5.6
$\mathbf{A}, A_{ij}$	Second-order orientation tensor (2.33)
$\{\mathbf{A}\}, A_m$	Contracted form of second-order orientation tensor (2.70)
$\bar{\mathbf{A}}, \bar{A}_{ij}$	Second-order orientation tensor in its principal axis system (2.57)
$\hat{\mathbf{A}}$	Target orientation tensor for ARD models (5.76)
$\mathbf{A}^{(n)}, A_{ij}^{(n)}$	Number-average orientation tensor (2.110)
$\mathbf{A}^{(w)}, A_{ij}^{(w)}$	Weight-average orientation tensor (2.111)
$\mathbb{A}, \mathbb{A}_{ijkl}$	Fourth-order orientation tensor (2.66)
$\bar{\mathbb{A}}, \bar{\mathbb{A}}_{ijkl}$	Fourth-order orientation tensor in the principal axes of the second-order tensor (5.43)
$[\mathbb{A}], \mathbb{A}_{mn}$	Contracted form of fourth-order orientation tensor (2.73)
$[\bar{\mathbb{A}}], \bar{\mathbb{A}}_{mn}$	Fourth-order orientation tensor in the principal axes of the second-order tensor, contracted form (5.43)
$\mathscr{A}, \mathscr{A}_{ijklmn}$	Sixth-order orientation tensor (2.98)
B_i	Breakage parameter for fibers of length ℓ_i (7.9)
$B_1, B_2^u, B_2^l, B_3, B_4, B_5$	Invariants of a transversely isotropic fourth-order tensor (8.69)
$\mathbf{B}, B_{ij}$	Green deformation tensor (Finger tensor) (4.70)
$\mathbf{B}^{-1}, B_{ij}^{-1}$	Piola deformation tensor (Almansi tensor) (4.68)
$\mathbb{B}$	Strain concentration tensor (8.41)
CAE	Computer-aided engineering
C_B	Breakage coefficient in fiber length model (7.12)
C_I	Folgar–Tucker interaction coefficient (5.29)
$\hat{C}_I$	Effective interaction coefficient for ARD models (5.76)
C_M	Parameter in the iARD model (5.87)
$\mathbf{C}, C_{ij}$	Rotary diffusion tensor in ARD models (5.75)
$\mathbb{C}, \mathbb{C}_{ijkl}$	Elastic stiffness tensor (8.4); homogenized stiffness tensor (8.35)
$[\mathbb{C}], \mathbb{C}_{mn}$	Contracted form of the stiffness tensor (8.13)

d_f	Fiber diameter
D_1, D_2, D_3	Parameters in the MRD model (5.89)
DFS	Direct fiber simulation, Section 9.2
D_r	Rotary diffusivity for planar orientation (5.25)
$\mathbf{D}, D_{ij}$	Rate-of-deformation tensor (4.5)
$\mathbf{D}^*, D^*_{ij}$	Dimensionless rate-of-deformation tensor (5.84)
$\hat{\mathbf{D}}$	Effective rate-of-deformation tensor for ARD models (5.76)
$\mathbf{D}^r$	Surface rotary diffusion tensor (5.81)
$\mathbf{e}_1, \mathbf{e}_2, \mathbf{e}_3$	Eigenvectors of the second-order orientation tensor (2.60)
E	Young's modulus, isotropic material (8.31)
E_f	Young's modulus of the fibers
E_m	Young's modulus of the matrix
E_{11}, E_{22}, E_{33}	Young's moduli, anisotropic material (8.22)
EBOF	Eigenvalue-based orthotropic fitted closure, Section 5.4.2.2
$\mathbb{E}^m$	Eshelby tensor with the matrix as the continuous phase (8.45)
F_k	Weighting factor for the kth fiber on a planar section (3.3)
$\mathbf{F}, F_{ij}$	Deformation gradient tensor (4.55)
$\mathbf{F}^{-1}, F^{-1}_{ij}$	Inverse deformation gradient tensor (4.56)
G	Shear modulus, isotropic material (8.31)
G_{12}, G_{23}, G_{31}	Shear moduli, anisotropic material (8.23)
GMT	Glass-mat thermoplastic composite
GNF	Generalized Newtonian fluid (6.6)
h	Half-thickness of mold cavity, Figs. 4.16 and 4.18
$-\dot{h}$	Mold closing velocity in squeezing flows, Fig. 4.27
h_{ij}	Shortest distance between the centerlines of fibers i and j, Fig. 9.6
$\mathbf{H}, H_{ij}$	Hessian matrix of a digital image (3.20)
$\mathbb{H}^m$	Strain concentration tensor for dilute Eshelby model (8.46)
$I(\mathbf{x})$	Image intensity at position $\mathbf{x}$, Section 3.2.2.1
I_D, II_D, III_D	Scalar invariants of $\mathbf{D}$ (4.11–4.13)
I_1, I_2, I_3	Alternate invariants of $\mathbf{D}$ (4.15–4.17) or $\mathbf{A}$ (5.55–5.57)
iARD	Improved anisotropic rotary diffusion model (5.87)
IBOF	Invariant-based orthotropic fitted closure, Section 5.4.2.1
IISO	Informed isotropic viscosity model (6.59)
$\mathbf{I}$	Second-order identity tensor (A.8)
$\mathbb{I}, \mathbb{I}_{ijkl}$	Symmetric fourth-order identity tensor (A.10)
$[\mathbb{I}]$	Contracted form of the fourth-order identity tensor (A.54)
ℓ	Fiber length
ℓ_i	Discrete values of fiber length in a fiber length distribution (7.1)
$\Delta\ell$	Fiber length increment in fiber breakage model (7.1)
L_n	Number-average fiber length (2.106)
L_w	Weight-average fiber length (2.107)
$\overline{L}_n, \overline{L}_w$	Thickness-averaged fiber lengths (7.22) and (7.23)
L_{ub}	Maximum unbreakable fiber length (7.14)
LFT	Long fiber/thermoplastic composite

$\mathbf{L}$, L_{ij}	Velocity gradient tensor (4.3)
$\mathbb{L}$, $\mathbb{L}_{ijkl}$	Fourth-order tensor formed from eigenvalues and eigenvectors of $\mathbf{A}$ (5.102)
m	Minor axis length of a fiber's ellipse on a section plane, Fig. 3.2; in Chapt. 6, parameter in the power-law fluid model (6.7)
M	Major axis length of a fiber's ellipse on a section plane, Fig. 3.2
M_z	Flow-type parameter (5.119)
MRD	Moldflow rotary diffusion model (5.89)
$\mathbb{M}$, $\mathbb{M}_{ijkl}$	Fourth-order tensor formed from eigenvectors of $\mathbf{A}$ (5.102)
n	Number of histogram bins in discrete fiber length distribution (Chapts. 2 and 7); in Chapt. 6, power-law index (6.7)
$\mathbf{n}$, n_i	Unit vector normal to a surface (6.1)
$\mathbf{n}_{ij}$	Unit vector normal to fibers i and j, Fig. 9.6
NLSF	Non-lubricated squeeze flow, Section 4.3.6.2
N_i	Number-based fiber length distribution, proportional to the number of fibers with length ℓ_i, Section 2.5
N_p	Particle number in model for fiber suspension stress (6.29)
N_s	Shear number in model for fiber suspension stress (6.21)
N_1, N_2	First and second normal stress differences (6.10) and (6.11)
p	Hydrostatic pressure (6.3)
$\mathbf{p}$, p_i	Unit vector describing instantaneous orientation of a fiber, Fig. 2.1
$\mathbf{p}'$, p'_i	Orientation vector in the reference configuration, Fig. 4.15
P_i	Breakage rate for fibers of length ℓ_i (7.5)
pARD	Principal-axis anisotropic rotary diffusion model (5.90)
$q(\phi)$	Probability flux in planar orientation, Fig. 5.2
$\mathbf{q}$	Probability flux in three-dimensional orientation (5.14)
$\mathbf{Q}$	Coordinate transformation matrix (A.43)
$[\mathbb{Q}]$	Coordinate transformation matrix for contracted notation (A.66–A.68)
r	Radial coordinate in cylindrical coordinate system, Fig. 4.19
r_e	Equivalent ellipsoidal axis ratio for a cylindrical fiber (4.95)
R_{ik}	Rate at which child fibers of length ℓ_i are generated by breaking parent fibers of length ℓ_k (7.5)
R_T	Trouton ratio (6.13)
RPR	Retarding principal rate model (5.107)
RSC	Reduced-strain closure model (5.103)
RVE	Representative volume element, Section 8.2.1
$[\mathbb{R}]$	Matrix used in contracted notation multiplications (A.60)
s	Distance along a fiber, measured from its centroid, Fig. 6.2
S	Child generation parameter in the fiber breakage model (7.17)
SBM	Suspension balance model, Section 9.1.3
SFT	Short fiber/thermoplastic composite
SMC	Sheet molding compound
SRF	Strain reduction factor model (5.93)
$\mathbf{S}$, S_{ij}	Structure tensor of an image (3.19)

$\mathbb{S}$, $\mathbb{S}_{ijkl}$	Elastic compliance tensor (8.7); homogenized compliance tensor
$\mathbf{t}$, t_i	Surface traction vector (6.1)
T	Period of Jeffery orbit (4.100); in Chapts. 5 and 8, temperature
ΔT	Temperature change (8.84)
$\mathbf{u}$, u_i	Displacement vector (8.1)
$\overline{v}$	Gapwise-average velocity (4.107)
v_r	Radial velocity in cylindrical coordinate system, Fig. 4.109
$\mathbf{v}$, v_i	Velocity vector
w_i	Weight-based fiber length distribution, proportional to the weight of fibers with length ℓ_i (2.104)
$\mathbf{W}$, W_{ij}	Vorticity tensor (4.9)
$\mathbf{x}$, x_i	Position vector; in Section 4.1.2, the instantaneous position of a material point
$\mathbf{x}'$, x_i'	Position of a material point in the reference configuration, Section 4.1.2
$\mathbf{x}_c$	Position of a fiber centroid, Fig. 6.2
z	Axial coordinate in cylindrical coordinate system, Fig. 4.19

Greek Letters

α	Thermal expansion coefficient, isotropic material (8.85); in Section 5.7.3, parameter in the RPR model
$\boldsymbol{\alpha}$, α_{ij}	Thermal expansion tensor (8.84)
β	Objective flow-type parameter (5.112)
$\boldsymbol{\beta}$, β_{ij}	Thermal stress tensor (8.88)
γ	Shear strain (4.60)
$\dot{\gamma}$	Scalar magnitude of the rate of deformation (4.14)
γ_{ij}	Engineering shear strains ($i \neq j$) (8.3)
δ	Parameter describing out-of-plane orientation in a thin cavity (6.47)
δ_{ij}	Kronecker delta (A.9)
$\hat{\boldsymbol{\delta}}_1, \hat{\boldsymbol{\delta}}_2, \hat{\boldsymbol{\delta}}_3$	Unit vectors in a Cartesian coordinate system, Fig. A.1
$\hat{\boldsymbol{\delta}}_r, \hat{\boldsymbol{\delta}}_\theta, \hat{\boldsymbol{\delta}}_\phi$	Unit vectors in a spherical coordinate system
ϵ	Parameter describing slenderness of a mold cavity (6.46)
ϵ_{ijk}	Permutation symbol (A.37)
$\boldsymbol{\varepsilon}$, ε_{ij}	Small strain tensor (8.2)
$\overline{\boldsymbol{\varepsilon}}$, $\overline{\varepsilon}_{ij}$	Volume-average strain in an RVE (8.34)
ζ	Fluid-fiber drag coefficient in fiber length model (7.14); in Chapt. 8, parameter in Halpin–Tsai equations, Table 8.1
η	Viscosity of a generalized Newtonian fluid (6.6); in Chapt. 8, parameter in the Halpin–Tsai equations (8.55)
$\overline{\eta}$	Extensional viscosity (6.12)
η_I	Isotropic viscosity parameter in model for fiber suspension stress (6.29)
η_m	Matrix viscosity (7.14)
θ	Angular coordinate in spherical coordinates, Fig. 2.1
κ	Parameter in the SRF, RSC and RPR models, Section 5.7
$\lambda_1, \lambda_2, \lambda_3$	Eigenvalues of the second-order orientation tensor (2.60)

$\mathbf{\Lambda}$	Diagonal matrix of eigenvalues (2.57)
μ	Viscosity of a Newtonian fluid (6.5)
ν	Poisson ratio, isotropic material (8.31)
ν_{ij}	Poisson ratio, anisotropic material ($i \neq j$) (8.22)
ξ	Particle shape factor in Jeffery's equation (4.86); in Chapt. 8, interpolating factor in Lielens model (8.54)
$\boldsymbol{\sigma}, \sigma_{ij}$	Total stress tensor (Cauchy stress) (6.1)
$\overline{\boldsymbol{\sigma}}, \overline{\sigma}_{ij}$	Volume-average stress in an RVE (8.34)
$\boldsymbol{\tau}, \tau_{ij}$	Extra stress tensor (6.4)
ϕ	Angular coordinate in spherical coordinates, Fig. 2.1
ϕ_f	Fiber volume fraction
ϕ_m	Matrix volume fraction
$\phi_{\max}$	Maximum particle volume fraction
$\psi(\mathbf{p}), \psi(\theta,\phi)$	Orientation distribution function for three-dimensional orientation (2.10)
$\psi_\phi(\phi)$	Orientation distribution function for planar orientation (2.3)
ψ^{AMI}	Objective flow-type parameter (5.116)
Ω	Angular velocity for rigid-body rotational motion (Chapt. 4); in Chapt. 5, parameter in the pARD model

Operators

$\mathrm{D}/\mathrm{D}t$	Material derivative (5.8)
$P(\ldots)$	Probability of an event
$S(\ldots)$	Fully symmetric part of a fourth-order tensor (5.59)
tr	Trace of a tensor (A.18)
∇	Gradient operator in space (5.17)
∇_s	Gradient operator on the surface of a unit sphere (5.19)
$\langle\ldots\rangle$	Orientation average (2.32)

References

[AERT13] A. I. Abd El-Rahman and C. L. Tucker. Mechanics of random discontinuous long-fiber thermoplastics—Part 1: Generation and characterization of initial geometry. *J. Appl. Mech.*, 80(5), 07 2013. 051007.

[AHP69] J. E. Ashton, J. C. Halpin, and P. H. Petit. *Primer on Composite Materials: Analysis.* Technomic, Stamford, Conn., 1969.

[Ari62] R. Aris. *Vectors, Tensors, and the Basic Equations of Fluid Mechanics.* Prentice-Hall, Englewood Cliffs, NJ, 1962.

[AS18] R. F. Agyei and M. D. Sangid. A supervised iterative approach to 3D microstructure reconstruction from acquired tomographic data of heterogeneous fibrous systems. *Compos. Struct.*, 206:234–246, 2018.

[Ast79] G. Astarita. Objective and generally applicable criteria for flow classification. *J. Non-Newt. Fluid Mech.*, 6(1):69–76, 1979.

[AT87] S. G. Advani and C. L. Tucker. The use of tensors to describe and predict fiber orientation in short fiber composites. *J. Rheol.*, 31:751–784, 1987.

[AT90] S. G. Advani and C. L. Tucker. Closure approximations for three-dimensional structure tensors. *J. Rheol.*, 34(3):367–386, 1990.

[ATG+91] J. R. Abbott, N. Tetlow, A. L. Graham, S. A. Altobelli, E. Fukushima, L. A. Mondy, and T. S. Stephens. Experimental observations of particle migration in concentrated suspensions: Couette flow. *J. Rheol.*, 35(5):773–795, 1991.

[Aut19] Autodesk, Inc. Autodesk Moldflow Insight 2019, Fiber Orientation Solver Validation. *https://damassets.autodesk.net/content/dam/autodesk/www/pdfs/Moldflow2019_Fiber_Orientation_validation.pdf*, 2019.

[BAH87] R. B. Bird, R. C. Armstrong, and O. Hassager. *Dynamics of Polymeric Liquids, Vol. 1, Fluid Mechanics.* John Wiley & Sons, New York, 2nd edition, 1987.

[Bay91] R. S. Bay. *Fiber Orientation in Injection Molded Composites: A Comparison of Theory and Experiment.* PhD thesis, University of Illinois at Urbana-Champaign, 1991.

[BC85] M. R. Barone and D. A. Caulk. Kinematics of flow of SMC. *Polym. Compos.*, 6:105–109, 1985.

[BDA85] M. A. Bibbo, S. M. Dinh, and R. C. Armstrong. Shear flow properties of semiconcentrated fiber suspensions. *J. Rheol.*, 29(6):905–929, 1985.

[BDC91] Y. Benveniste, G. Dvorak, and T. Chen. On diagonal and elastic symmetry of the approximate effective stiffness tensor of heterogeneous media. *J. Mech. Phys. Solids*, 39(7):927–946, 1991.

[BDD+19] T. Baranowski, D. Dobrovolskij, K. Dremel, A. Hölzing, G. Lohfink, K. Schladitz, and S. Zabler. Local fiber orientation from X-ray region-of-interest computed tomography of large fiber reinforced composite components. *Compos. Sci. Tech.*, 183:107786, 2019.

[Ben87] Y. Benveniste. A new approach to the application of Mori-Tanaka's theory in composite materials. *Mech. Mater.*, 6:147–157, 1987.

[BO20] A. Bechara and T. Osswald. Measuring fiber length in the core and shell regions of injection molded long fiber-reinforced thermoplastic plaques. *J. Compos. Sci.*, 4(3), 2020.

[Bre62] F. P. Bretherton. The motion of rigid particles in a shear flow at low Reynolds number. *J. Fluid Mech.*, 14:284–304, 1962.

[BT92a] R. S. Bay and C. L. Tucker. Fiber orientation in simple injection moldings. Part I: Theory and numerical methods. *Polym. Compos.*, 13(4):317–331, 1992.

[BT92b] R. S. Bay and C. L. Tucker. Fiber orientation in simple injection moldings. Part II: Experimental results. *Polym. Compos.*, 13(4):332–342, 1992.

[BT92c] R. S. Bay and C. L. Tucker. Stereological measurement and error estimates for three-dimensional fiber orientation. *Polym. Eng. Sci.*, 32(4):332–341, February 1992.

[Buz08] T. M. Buzug. *Computed Tomography: From Photon Statistics to Modern Cone-Beam CT.* Springer, Berlin, 2008.

[BYR+17] A. Bakharev, H. Yu, S. Ray, R. Speight, and J. Wang. Using new anisotropic rotational diffusion model to improve prediction of short fibers in thermoplastic injection molding. In *ANTEC 2017.* Society of Plastics Engineers, May 2017.

[CAD95] A. Clarke, G. Archenhold, and N. Davidson. A novel technique for determining the 3D spatial distribution of glass fibres in polymer composites. *Compos. Sci. Tech.*, 55(1):75–91, 1995.

[CBC+20] T.-C. Chang, A. Bechara, H. Celik, D. Brands, A. Yanev, and T. Osswald. Validation of fiber breakage in simple shear flow with direct fiber simulation. *J. Compos. Sci.*, 4(3):134, 2020.

[CBM87] D. J. Coyle, J. W. Blake, and C. W. Macosko. The kinematics of fountain flow in mold-filling. *AIChE J.*, 33(7):1168–1177, 1987.

[CBW16] M. J. Cieslinski, D. G. Baird, and P. Wapperom. Obtaining repeatable initial fiber orientation for the transient rheology of fiber suspensions in simple shear flow. *J. Rheol.*, 60(1):161–174, 2016.

[Che84] C.-Y. Chen. *Predictions of Monotonic Tensile Stress-Strain Curves of Short Fiber Reinforced Polymer Composites.* PhD thesis, University of Illinois at Urbana-Champaign, Urbana, IL, 1984.

[CK01] D. H. Chung and T. H. Kwon. Improved model of orthotropic closure approximation for flow-induced fiber orientation. *Polym. Compos.*, 22:636–649, 2001.

[CK02] D. H. Chung and T. H. Kwon. Invariant-based optimal fitting closure approximation for the numerical prediction of flow-induced fiber orientation. *J. Rheol.*, 46:169–194, 2002.

[CLWL21] Y. Chen, R. Luo, L. Wang, and S. Lee. Self-similarity in particle accumulation on the advancing meniscus. *J. Fluid Mech.*, 925:A10, 2021.

[CMZ90] J. Cheng and I. Manas-Zloczower. Flow field characterization in a Banbury mixer. *Int. Polym. Process.*, 5:178–183, 1990.

[Cox71] R. G. Cox. The motion of long slender bodies in a viscous fluid. Part 2. Shear flow. *J. Fluid Mech.*, 45(4):625–657, 1971.

[CT84] C.-Y. Chen and C. L. Tucker. Mechanical property predictions for short fiber/brittle matrix composites. *J. Reinf. Plast. Compos.*, 3:120–129, 1984.

[CT95] J. S. Cintra and C. L. Tucker. Orthotropic closure approximations for flow-induced fiber orientation. *J. Rheol.*, 39:1095–1122, 1995.

[CTYM90] C. W. Camacho, C. L. Tucker, S. Yalvaç, and R. L. McGee. Stiffness and thermal expansion predictions for hybrid short fiber composites. *Polym. Compos.*, 11(4):229–239, 1990.

[CWB15] M. J. Cieslinski, P. Wapperom, and D. G. Baird. Influence of fiber concentration on the startup of shear flow behavior of long fiber suspensions. *J. Non-Newt. Fluid Mech.*, 222:163–170, 2015.

[CWB19a] H. Chen, P. Wapperom, and D. G. Baird. A model incorporating the effects of flow type on fiber orientation for flows with mixed flow kinematics. *J. Rheol.*, 63(3):455–464, 2019.

[CWB19b] H. Chen, P. Wapperom, and D. G. Baird. The use of flow type dependent strain reduction factor to improve fiber orientation predictions for an injection molded center-gated disk. *Phys. Fluids*, 31:123105, 2019.

[DA84] S. M. Dinh and R. C. Armstrong. A rheological equation of state for semiconcentrated fiber suspensions. *J. Rheol.*, 28(3):207–227, 1984.

[dBGlR17] A. du Plessis, C. Broeckhoven, A. Guelpa, and S. G. le Roux. Laboratory x-ray micro-computed tomography: a user guideline for biological samples. *GigaScience*, 6(6):1–11, 2017.

[DCA97] N. C. Davidson, A. R. Clarke, and G. Archenhold. Large-area, high-resolution image analysis of composite materials. *J. Microscopy*, 185(2):233–242, 1997.

[DDV$^+$13] A. Durin, P. De Micheli, J. Ville, F. Inceoglu, R. Valette, and B. Vergnes. A matricial approach of fibre breakage in twin-screw extrusion of glass fibres reinforced thermoplastics. *Composites A*, 48:47–56, 2013.

[DGR07] D. Dray, P. Gilormini, and G. Régnier. Comparison of several closure approximations for evaluating the thermoelastic properties of an injection molded short-fiber composite. *Compos. Sci. Tech.*, 67(7):1601–1610, 2007.

[DI06] I. M. Daniel and O. Ishai. *Engineering Mechanics of Composite Materials.* Oxford University Press, New York, 2nd edition, 2006.

[DO03] I. Doghri and A. Ouaar. Homogenization of two-phase elasto-plastic composite materials and structures: Study of tangent operators, cyclic plasticity and numerical algorithms. *Int. J. Solids Struct.*, 40(7):1681–1712, 2003.

[Doi81] M. Doi. Molecular dynamics and rheological properties of concentrated solutions of rodlike polymers in isotropic and liquid crystalline phases. *J. Polym. Sci.: Polym. Phys. Ed.*, 19:229–243, 1981.

[DSF$^+$18] B. R. Denos, D. E. Sommer, A. J. Favaloro, R. B. Pipes, and W. B. Avery. Fiber orientation measurement from mesoscale CT scans of prepreg platelet molded composites. *Composites A*, 114:241–249, 2018.

[DT01] J. A. Dantzig and C. L. Tucker. *Modeling in Materials Processing.* Cambridge University Press, Cambridge, 2001.

[DT05] I. Doghri and L. Tinel. Micromechanical modeling and computation of elasto-plastic materials reinforced with distributed-orientation fibers. *Int. J. Plasticity*, 21(10):1919–1940, 2005.

[DT06] I. Doghri and L. Tinel. Micromechanics of inelastic composites with misaligned inclusions: Numerical treatment of orientation. *Comput. Methods Appl. Mech. Engrg.*, 195(13):1387–1406, 2006.

[EBWVG09a] A. P. R. Eberle, D. G. Baird, P. Wapperom, and G. M. Vélez-García. Obtaining reliable transient rheological data on concentrated short fiber suspensions using a rotational rheometer. *J. Rheol.*, 53(5):1049–1068, 2009.

[EBWVG09b] A. P. R. Eberle, D. G. Baird, P. Wapperom, and G. M. Vélez-García. Using transient shear rheology to determine material parameters in fiber suspension theory. *J. Rheol.*, 53(3):685–705, 2009.

[EC01] C. Eberhardt and A. Clarke. Fibre-orientation measurements in short-glass-fibre composites. Part I. Automated, high-angular-resolution measurement by confocal microscopy. *Compos. Sci. Tech.*, 61(12):1389–1400, 2001.

[EC02] C. N. Eberhardt and A. R. Clarke. Automated reconstruction of curvilinear fibres from 3D datasets acquired by X-ray microtomography. *J. Microscopy*, 206(1):41–53, 2002.

[ECV$^+$01] C. Eberhardt, A. Clarke, M. Vincent, T. Giroud, and S. Flouret. Fibre-orientation measurements in short-glass-fibre composites. Part II. A quantitative error estimate of the 2D image analysis technique. *Compos. Sci. Tech.*, 61(13):1961–1974, 2001.

[EG86] K. E. Evans and A. G. Gibson. Prediction of the maximum packing fraction achievable in randomly oriented short-fibre composites. *Compos. Sci. Tech.*, 25:149–162, 1986.

[EMG94] R. F. Eduljee, R. L. McCullough, and J. W. Gillespie. The influence of aggregated and dispersed textures on the elastic properties of discontinuous-fiber composites. *Compos. Sci. Tech.*, 50:381–391, 1994.

[Esh57] J. D. Eshelby. The determination of the elastic field of an ellipsoidal inclusion, and related problems. *Proc. Roy. Soc.*, A241(1226):376–396, 1957.

[ETM97] K. A. Ericsson, S. Toll, and J.-A. E. Månson. The two-way interaction between anisotropic flow and fiber orientation in squeeze flow. *J. Rheol.*, 41(3):491–511, 1997.

[EVGBW10] A. P. Eberle, G. M. Vélez-García, D. G. Baird, and P. Wapperom. Fiber orientation kinetics of a concentrated short glass fiber suspension in startup of simple shear flow. *J. Non-Newt. Fluid Mech.*, 165(3):110–119, 2010.

[Fav17] A. J. Favaloro. *Rheological Behavior and Manufacturing Simulation of Prepreg Platelet Molding Systems.* PhD thesis, Purdue University, 2017.

[Fav18] A. J. Favaloro. Unpublished research. 2018.

[FE88] G. Fischer and P. Eyerer. Measuring spatial orientation of short fiber reinforced thermoplastics by image analysis. *Polym. Compos.*, 9(4):297–304, 1988.

[FF85] S. Fakirov and C. Fakirova. Direct determination of the orientation of short glass fibers in an injection-molded poly(ethylene terephthalate) system. *Polym. Compos.*, 6(1):41–46, 1985.

[FHH⁺08] J. Férec, M. Heniche, M. C. Heuzey, G. Ausias, and P. J. Carreau. Numerical solution of the Fokker-Planck equation for fiber suspensions: Application to the Folgar-Tucker-Lipscomb model. *J. Non-Newt. Fluid Mech.*, 155:20–29, 2008.

[FHPG⁺09] J. Férec, M.-C. Heuzey, J. Pérez-González, L. de Vargas, G. Ausias, and P. J. Carreau. Investigation of the rheological properties of short glass fiber-filled polypropylene in extensional flow. *Rheol. Acta*, 48(1):59–72, 2009.

[FMAA20] J. Férec, D. Mezi, S. G. Advani, and G. Ausias. Axisymmetric flow simulations of fiber suspensions as described by 3D probability distribution function. *J. Non-Newt. Fluid Mech.*, 284:104367, 2020.

[FMB⁺02] Z. Fang, A. A. Mammoli, J. F. Brady, M. S. Ingber, L. A. Mondy, and A. L. Graham. Flow-aligned tensor models for suspension flows. *Int. J. Mult. Flow*, 28(1):137–166, 2002.

[FPTZ00] X.-J. Fan, N. Phan-Thien, and R. Zheng. Simulation of fibre suspension flow with shear-induced migration. *J. Non-Newt. Fluid Mech.*, 90:47–63, 2000.

[FT84] F. Folgar and C. L. Tucker. Orientation behavior of fibers in concentrated suspensions. *J. Reinf. Plast. Compos.*, 3:98–119, 1984.

[FT19] A. J. Favaloro and C. L. Tucker. Analysis of anisotropic rotary diffusion models for fiber orientation. *Composites A*, 126:105605, 2019.

[FTP18] A. J. Favaloro, H.-C. Tseng, and R. B. Pipes. A new anisotropic viscous constitutive model for composites molding simulation. *Composites A*, 115:112–122, 2018.

[GBY⁺18] S. Goris, T. Back, A. Yanev, D. Brands, D. Drummer, and T. A. Osswald. A novel fiber length measurement technique for discontinuous fiber-reinforced composites: A comparative study with existing methods. *Polym. Compos.*, 39(11):4058–4070, 2018.

[GGOS20] U. N. Gandhi, S. Goris, T. A. Osswald, and Y.-Y. Song. *Discontinous Fiber-Reinforced Composites: Fundamentals and Applications.* Hanser, Munich, 2020.

[Gib16] R. F. Gibson. *Principles of Composite Material Mechanics.* CRC Press, Boca Raton, Florida, 4th edition, 2016.

[GL14] E. Ghossein and M. Lévesque. A comprehensive validation of analytical homogenization models: The case of ellipsoidal particles reinforced composites. *Mech. Mater.*, 75:135–150, 2014.

[GO18] S. Goris and T. A. Osswald. Process-induced fiber matrix separation in long fiber-reinforced thermoplastics. *Composites A*, 105:321–333, 2018.

[GSKS16] U. Gandhi, D. B. Sebastian, V. Kunc, and Y. Song. Method to measure orientation of discontinuous fiber embedded in the polymer matrix from computerized tomography scan data. *J. Thermoplast. Compos. Mater.*, 29(12):1696–1709, 2016.

[Hal69] J. C. Halpin. Stiffness and expansion estimates for oriented short fiber composites. *J. Compos. Mater.*, 3:732–734, 1969.

[Han62] G. L. Hand. A theory of anisotropic fluids. *J. Fluid Mech.*, 13:33–64, 1962.

[HAXS19] I. Hanhan, R. Agyei, X. Xiao, and M. D. Sangid. Comparing non-destructive 3D X-ray computed tomography with destructive optical microscopy for microstructural characterization of fiber reinforced composites. *Compos. Sci. Tech.*, 184:107843, 2019.

[HAXS20] I. Hanhan, R. F. Agyei, X. Xiao, and M. D. Sangid. Predicting microstructural void nucleation in discontinuous fiber composites through coupled in-situ X-ray tomography experiments and simulations. *Scientific Reports*, 10(1):3564, 2020.

[HC92] C. A. Hieber and H. H. Chiang. Shear-rate-dependence modeling of polymer melt viscosity. *Polym. Eng. Sci.*, 32(14):931–938, 1992.

[Her09] G. T. Herman. *Fundamentals of Computerized Tomography: Image Reconstruction from Projections*. Springer, Berlin, 2nd edition, 2009.

[HK76] J. C. Halpin and J. L. Kardos. The Halpin-Tsai equations: A review. *Polym. Eng. Sci.*, 16:344–352, 1976.

[HP75] J. B. Harris and J. F. T. Pittman. Equivalent ellipsoidal axis ratios of slender rod-like particles. *J. Colloid Interf. Sci.*, 50(2):280–282, 1975.

[HRG02] P. J. Hine, H. Rudolf Lusti, and A. A. Gusev. Numerical simulation of the effects of volume fraction, aspect ratio and fibre length distribution on the elastic and thermoelastic properties of short fibre composites. *Compos. Sci. Tech.*, 62(10):1445–1453, 2002.

[HRW+19] P. A. Hessman, T. Riedel, F. Welschinger, K. Hornberger, and T. Böhlke. Microstructural analysis of short glass fiber reinforced thermoplastics based on x-ray micro-computed tomography. *Compos. Sci. Tech.*, 183:107752, 2019.

[HS63] Z. Hashin and S. Shtrikman. A variational approach to the theory of the elastic behavior of multiphase materials. *J. Mech. Phys. Solids*, 11:127–140, 1963.

[HS80] C. A. Hieber and S. F. Shen. A finite-element/finite-difference simulation of the injection-molding filling process. *J. Non-Newt. Fluid Mech.*, 7:1–32, 1980.

[HS19] I. Hanhan and M. D. Sangid. ModLayer: A MATLAB GUI drawing segmentation tool for visualizing and classifying 3D data. *Integrating Materials and Manufacturing Innovation*, 8(4):468–475, 2019.

[HS21] I. Hanhan and M. D. Sangid. Damage propagation in short fiber thermoplastic composites analyzed through coupled 3D experiments and simulations. *Composites B*, 218:108931, 2021.

[Hse09] J. Hseih. *Computed Tomography: Principles, Design, Artifacts, and Recent Advances*. SPIE, Bellingham, WA, 2nd edition, 2009.

[Huy01] H. M. Huynh. *Improved Fiber Orientation Predictions for Injection-Molded Composites*. Master's thesis, University of Illinois at Urbana-Champaign, 2001.

[ISO17] ISO. Plastics–Injection moulding of test specimens of thermoplastic materials–Part 5: Preparation of standard specimens for investigating anisotropy. (ISO 294-5:2017), 2017.

[JAS19] J. K. Jørgensen, E. Andreassen, and D. Salaberger. The effect of fiber concentration on fiber orientation in injection molded film gated rectangular plates. *Polym. Compos.*, 40(2):615–629, 2019.

[Jef22] G. B. Jeffery. The motion of ellipsoidal particles immersed in a viscous fluid. *Proc. Roy. Soc.*, A102:161–179, 1922.

[Jon99] R. M. Jones. *Mechanics of Composite Materials.* Taylor & Francis, Philadelphia, 2nd edition, 1999.

[JS04] D. A. Jack and D. E. Smith. Assessing the use of tensor closure methods with orientation distribution reconstruction functions. *J. Compos. Mater.*, 38:1851–1871, 2004.

[JS08] D. A. Jack and D. E. Smith. Elastic properties of short-fiber polymer composites, derivation and demonstration of analytical forms for expectation and variance from orientation tensors. *J. Compos. Mater.*, 42(3):277–308, 2008.

[KBP+21] S. K. Kugler, A. Bechara, H. Perez, C. Cruz, A. Kech, and T. A. Osswald. Data enriched lubrication force modeling for a mechanistic fiber simulation of short fiber-reinforced thermoplastics. *Phys. Fluids*, 33(5):053107, 2021.

[KDA+11] S. Kammoun, I. Doghri, L. Adam, G. Robert, and L. Delannay. First pseudo-grain failure model for inelastic composites with misaligned short fibers. *Composites A*, 42(12):1892–1902, 2011.

[KDBD15] S. Kammoun, I. Doghri, L. Brassart, and L. Delannay. Micromechanical modeling of the progressive failure in short glass–fiber reinforced thermoplastics – First Pseudo-Grain Damage model. *Composites A*, 73:166–175, 2015.

[KDS+20] S. K. Kugler, A. P. Dey, S. Saad, C. Cruz, A. Kech, and T. Osswald. A flow-dependent fiber orientation model. *J. Compos. Sci.*, 4(3):96, 2020.

[Ken95] P. Kennedy. *Flow Analysis of Injection Molds.* Hanser, Munich, 1995.

[KFN+07] V. Kunc, B. J. Frame, B. N. Nguyen, C. Tucker, and G. Vélez-García. Fiber length distribution measurement for long glass and carbon fiber reinforced injection molded thermoplastics. In *Proceedings of the 7th Annual Society of Plastic Engineers Automotive Composites Conference and Exposition*, Troy, MI, 2007.

[KGBF21] T. Karl, D. Gatti, T. Böhlke, and B. Frohnapfel. Coupled simulation of flow-induced viscous and elastic anisotropy of short-fiber reinforced composites. *Acta Mechanica*, 2021.

[KGFB21] T. Karl, D. Gatti, B. Frohnapfel, and T. Böhlke. Asymptotic fiber orientation states of the quadratically closed Folgar–Tucker equation and a subsequent closure improvement. *J. Rheol.*, 65(5):999–1022, 2021.

[KHL94] C. J. Koh, P. Hookham, and L. G. Leal. An experimental investigation of concentrated suspension flows in a rectangular channel. *J. Fluid Mech.*, 266:1–32, 1994.

[KKCO20] S. K. Kugler, A. Kech, C. Cruz, and T. Osswald. Fiber orientation predictions—A review of existing models. *J. Compos. Sci.*, 4(2), 2020.

[KKO20] A. Kech, S. Kugler, and T. Osswald. Significance of model parameter variations in the pARD-RSC model. *J. Compos. Sci.*, 4(3):109, 2020.

[KKT20] C. Kuhn, E. Koerner, and O. Taeger. A simulative overview on fiber predictions models for discontinuous long fiber composites. *Polym. Compos.*, 41(1):73–81, 2020.

[KLC+20a] S. K. Kugler, G. M. Lambert, C. Cruz, A. Kech, T. A. Osswald, and D. G. Baird. Efficient parameter identification for macroscopic fiber orientation models with experimental data and a mechanistic fiber simulation. *AIP Conf. Proc.*, 2205:020050, 2020.

[KLC+20b] S. K. Kugler, G. M. Lambert, C. Cruz, A. Kech, T. A. Osswald, and D. G. Baird. Macroscopic fiber orientation model evaluation for concentrated short fiber reinforced polymers in comparison to experimental data. *Polym. Compos.*, 41(7):2542–2556, 2020.

[KMK+20] R. Karamov, L. M. Martulli, M. Kerschbaum, I. Sergeichev, Y. Swolfs, and S. V. Lomov. Micro-CT based structure tensor analysis of fibre orientation in random fibre composites versus high-fidelity fibre identification methods. *Compos. Struct.*, 235:111818, 2020.

[Kon87] T. S. Konicek. *A Method to Determine Three-Dimensional Fiber Orientation in Fiber Reinforced Polymers.* Master's thesis, University of Illinois at Urbana-Champaign, Urbana, IL, 1987.

[KS01] A. C. Kak and M. Slaney. *Principles of Computerized Tomographic Imaging.* SIAM, 2001.

[KSW+98] A. Kaye, R. F. T. Stepto, W. J. Work, J. V. Aleman, and Y. Y. Malkin. Definition of terms relating to the non-ultimate mechanical properties of polymers. *Pure Appl. Chem.*, 70(3):701–754, 1998.

[KW20] C. Kuhn and S. Wehler. A force-balanced fiber retardation model to predict fiber-matrix-separation during polymer processing. *J. Compos. Sci.*, 4(4), 2020.

[KWTO17] C. Kuhn, I. Walter, O. Taeger, and T. A. Osswald. Experimental and numerical analysis of fiber matrix separation during compression molding of long fiber reinforced thermoplastics. *J. Compos. Sci.*, 1(1), 2017.

[LA87] D. Leighton and A. Acrivos. The shear-induced migration of particles in concentrated suspensions. *J. Fluid Mech.*, 181:415–439, 1987.

[Lar85] R. G. Larson. Flows of constant stretch history for polymeric materials with power-law distributions of relaxation times. *Rheol. Acta*, 24(5):443–449, 1985.

[LB16] G. M. Lambert and D. G. Baird. Evaluating rigid and semiflexible fiber orientation evolution models in simple flows. *J. Manuf. Sci. Eng.*, 139(3), 10 2016. 031012.

[LDHB88] G. G. Lipscomb, M. M. Denn, D. U. Hur, and D. V. Boger. The flow of fiber suspensions in complex geometries. *J. Non-Newt. Fluid Mech.*, 26:297–325, 1988.

[Lev67] V. M. Levin. Thermal expansion coefficients of heterogeneous materials. *Mekhana Tverdogo Tela*, 2:88–94, 1967. English translation in *Mech. Solids*, 2:58–61, 1967.

[LHG02] H. R. Lusti, P. J. Hine, and A. A. Gusev. Direct numerical predictions for the elastic and thermoelastic properties of short fibre composites. *Compos. Sci. Tech.*, 62(15):1927–1934, 2002.

[LL19] T. Li and J.-F. Luyé. Flow-fiber coupled viscosity in injection molding simulations of short fiber reinforced thermoplastics. *Int. Polym. Process.*, 34(2):158–171, 2019.

[LOD+16] T. Laurencin, L. Orgéas, P. Dumont, S. Rolland du Roscoat, P. Laure, S. Le Corre, L. Silva, R. Mokso, and M. Terrien. 3D real-time and *in situ* characterisation of fibre kinematics in dilute non-Newtonian fibre suspensions during confined and lubricated compression flow. *Compos. Sci. Tech.*, 134:258–266, 2016.

[LPC+98] G. Lielens, P. Pirotte, A. Couniot, F. Dupret, and R. Keunings. Prediction of thermo-mechanical properties for compression-moulded composites. *Composites A*, 29:63–70, 1998.

[LU08] S. B. Lindström and T. Uesaka. Simulation of semidilute suspensions of non-Brownian fibers in shear flow. *J. Chem. Phys.*, 128(2):024901, 2008.

[LWB17] G. Lambert, P. Wapperom, and D. G. Baird. Obtaining short-fiber orientation model parameters using non-lubricated squeeze flow. *Phys. Fluids*, 29(12):121608, 2017.

[LWB21] G. M. Lambert, P. Wapperom, and D. Baird. Comparing fiber orientation evolution between startup of shear and nonlubricated squeeze flow. *Polym. Compos.*, 42:98–112, 2021.

[Mac94] C. W. Macosko. *Rheology*. Wiley, New York, 1994.

[Mal69] L. E. Malvern. *Introduction to the Mechanics of a Continuous Medium*. Prentice-Hall, Englewood Cliffs, NJ, 1969.

[MB99] J. F. Morris and F. Boulay. Curvilinear flows of noncolloidal suspensions: The role of normal stresses. *J. Rheol.*, 43(5):1213–1237, 1999.

[MBA+94] L. A. Mondy, H. Brenner, S. A. Altobelli, J. R. Abbott, and A. L. Graham. Shear-induced particle migration in suspensions of rods. *J. Rheol.*, 38(2):444–452, 1994.

[MCH+05] C. Mobuchon, P. J. Carreau, M.-C. Heuzey, M. Sepehr, and G. Ausias. Shear and extensional properties of short glass fiber reinforced polypropylene. *Polym. Compos.*, 26(3):247–264, 2005.

[Mey21] N. Meyer. fiberoripy. *https://github.com/nilsmeyerkit/fiberoripy*, 2021.

[MGB+18] T. Mulholland, S. Goris, J. Boxleitner, T. A. Osswald, and N. Rudolph. Process-induced fiber orientation in fused filament fabrication. *J. Compos. Sci.*, 2(3), 2018.

[MH09] E. Mitsoulis and S. G. Hatzikiriakos. Modelling PTFE paste extrusion: The effect of an objective flow type parameter. *J. Non-Newt. Fluid Mech.*, 159(1):41–49, 2009.

[MHB14] K. J. Meyer, J. T. Hofmann, and D. G. Baird. Prediction of short glass fiber orientation in the filling of an end-gated plaque. *Composites A*, 62:77–86, 2014.

[Mle99] B. Mlekusch. Fibre orientation in short-fibre-reinforced thermoplastics II. Quantitative measurements by image analysis. *Compos. Sci. Tech.*, 59(4):547–560, 1999.

[MLG99] B. Mlekusch, E. Lehner, and W. Geymayer. Fibre orientation in short-fibre-reinforced thermoplastics I. Contrast enhancement for image analysis. *Compos. Sci. Tech.*, 59(4):543–545, 1999.

[MM06] R. M. Miller and J. F. Morris. Normal stress-driven migration and axial development in pressure-driven flow of concentrated suspensions. *J. Non-Newt. Fluid Mech.*, 135(2):149–165, 2006.

[MPS+16] R. Mezher, M. Perez, A. Scheuer, E. Abisset-Chavanne, F. Chinesta, and R. Keunings. Analysis of the Folgar & Tucker model for concentrated fibre suspensions in unconfined and confined shear flows via direct numerical simulation. *Composites A*, 91:388–397, 2016.

[MSB+20] N. Meyer, L. Schöttl, L. Bretz, A. Hrymak, and L. Kärger. Direct bundle simulation approach for the compression molding process of sheet molding compound. *Composites A*, 132:105809, 2020.

[MSHJS11] S. Montgomery-Smith, W. He, D. A. Jack, and D. E. Smith. Exact tensor closures for the three-dimensional Jeffery's equation. *J. Fluid Mech.*, 680:321–335, 2011.

[MSM09] R. M. Miller, J. P. Singh, and J. F. Morris. Suspension flow modeling for general geometries. *Chem. Eng. Sci.*, 64(22):4597–4610, 2009.

[MT73] T. Mori and K. Tanaka. Average stress in matrix and average elastic energy of materials with misfitting inclusions. *Acta Metallurgica*, 21:571–574, 1973.

[Mur82] T. Mura. *Micromechanics of Defects in Solids.* Martinus Nijhoff, The Hague, 1982.

[NB94] P. R. Nott and J. F. Brady. Pressure-driven flow of suspensions: simulation and theory. *J. Fluid Mech.*, 275:157–199, 1994.

[NBH+08] B. N. Nguyen, S. K. Bapanapalli, J. D. Holbery, M. T. Smith, V. Kunc, B. J. Frame, J. H. Phelps, and C. L. Tucker. Fiber length and orientation in long-fiber injection-molded thermoplastics — Part I: Modeling of microstructure and elastic properties. *J. Compos. Mater.*, 42(10):1003–1029, 2008.

[NDK+20] C. Naili, I. Doghri, T. Kanit, M. Sukiman, A. Aissa-Berraies, and A. Imad. Short fiber reinforced composites: Unbiased full-field evaluation of various homogenization methods in elasticity. *Compos. Sci. Tech.*, 187:107942, 2020.

[NF98] J. C. Nadeau and M. Ferrari. Invariant tensor-to-matrix mappings for evaluation of tensorial expressions. *J. Elasticity*, 52(1):43–61, 1998.

[NKF+09] B. N. Nguyen, V. Kunc, B. Frame, J. H. Phelps, C. L. Tucker, S. K. Bapanapalli, J. D. Holbery, and M. T. Smith. Prediction of the elastic-plastic stress/strain response for injection-molded long-fiber thermoplastics. *J. Compos. Mater.*, 43(3):217–246, 2009.

[NS14] R. Nakano and K. Sakaba. Development of CAE software for injection and BMC/SMC molding including short/long fiber reinforcement. In *Proceedings of SAMPE Technical Conference*, Seattle, WA, June 2014.

[OAE+11] K. C. Ortman, N. Agarwal, A. P. R. Eberle, D. G. Baird, P. Wapperom, and A. J. Giacomin. Transient shear flow behavior of concentrated long glass fiber suspensions in a sliding plate rheometer. *J. Non-Newt. Fluid Mech.*, 166(16):884–895, 2011.

[OBW+20] Z. Ouyang, E. Bertevas, D. Wang, B. C. Khoo, J. Férec, G. Ausias, and N. Phan-Thien. A smoothed particle hydrodynamics study of a non-isothermal and thermally anisotropic fused deposition modeling process for a fiber-filled composite. *Phys. Fluids*, 32(5):053106, 2020.

[OBWA12] K. Ortman, D. G. Baird, P. Wapperom, and A. Aning. Prediction of fiber orientation in the injection molding of long fiber suspensions. *Polym. Compos.*, 33(8):1360–1367, 2012.

[OBWW12] K. Ortman, D. Baird, P. Wapperom, and A. Whittington. Using startup of steady shear flow in a sliding plate rheometer to determine material parameters for the purpose of predicting long fiber orientation. *J. Rheol.*, 56(4):955–981, 2012.

[OD91] P. A. O'Connell and R. A. Duckett. Measurements of fibre orientation in short-fibre-reinforced thermoplastics. *Compos. Sci. Tech.*, 42:329–347, 1991.

[OL88] E. T. Onat and F. A. Leckie. Representation of mechanical behavior in the presence of changing internal structure. *ASME J. Appl. Mech.*, 55:1–10, 1988.

[Ona84] E. T. Onat. Effective properties of elastic materials that contain penny shaped voids. *Int. J. Eng. Sci.*, 22:1013–1021, 1984.

[OT03] J. F. O'Gara and C. L. Tucker. Fiber orientation in injection molded parts: Limitations of current predictions. Technical report DRL-R0257, Delphi Research Labs, 2003.

[OWN03] J. F. O'Gara, M. G. Wyzgoski, and G. E. Novak. Development of an ISO standard for determining anisotropic properties of glass-filled thermoplastics. In *ANTEC 2003*, pages 2883–2887. Society of Plastics Engineers, May 2003.

[PAB+92] R. J. Phillips, R. C. Armstrong, R. A. Brown, A. L. Graham, and J. R. Abbott. A constitutive equation for concentrated suspensions that accounts for shear-induced particle migration. *Phys. Fluids A*, 4(1):30–40, 1992.

[PAKT13] J. H. Phelps, A. I. Abd El-Rahman, V. Kunc, and C. L. Tucker. A model for fiber length attrition in injection-molded long-fiber composites. *Composites A*, 51:11–21, 2013.

[PCSŠ94] R. B. Pipes, D. W. Coffin, S. F. Shuler, and P. Šimáček. Non-Newtonian constitutive relationships for hyperconcentrated fiber suspensions. *J. Compos. Mater.*, 28(4):343–351, 1994.

[PDB+18] P. Pinter, S. Dietrich, B. Bertram, L. Kehrer, P. Elsner, and K. A. Weidenmann. Comparison and error estimation of 3D fibre orientation analysis of computed tomography image data for fibre reinforced composites. *NDT & E Intl.*, 95:26–35, 2018.

[Pet99] C. J. S. Petrie. The rheology of fibre suspensions. *J. Non-Newt. Fluid Mech.*, 87:369–402, 1999.

[PFD04] O. Pierard, C. Friebel, and I. Doghri. Mean-field homogenization of multi-phase thermo-elastic composites: a general framework and its validation. *Compos. Sci. Tech.*, 64(10):1587–1603, 2004.

[PGAC+17] M. Perez, S. Guevelou, E. Abisset-Chavanne, F. Chinesta, and R. Keunings. From dilute to entangled fibre suspensions involved in the flow of reinforced polymers: A unified framework. *J. Non-Newt. Fluid Mech.*, 250:8–17, 2017.

[PGBC19] V. Perumal, R. K. Gupta, S. N. Bhattacharya, and F. S. Costa. Fiber migration in shear flow: Model predictions and experimental validation. *Polym. Compos.*, 40(9):3573–3581, 2019.

[PGBC20] V. Perumal, R. K. Gupta, S. N. Bhattacharya, and F. S. Costa. Transient viscosity of fibre-filled composites incorporating evolution of fibre orientation and concentration. *Rheol. Acta*, 59(1):35–46, 2020.

[PGS+07] O. Pierard, C. González, J. Segurado, J. LLorca, and I. Doghri. Micromechanics of elasto-plastic materials reinforced with ellipsoidal inclusions. *Int. J. Solids Struct.*, 44(21):6945–6962, 2007.

[PHB+91] R. B. Pipes, J. W. S. Hearle, A. J. Beaussart, A. M. Sastry, and R. K. Okine. A constitutive relation for the viscous flow of an oriented fiber assembly. *J. Compos. Mater.*, 25(9):1204–1217, 1991.

[Phe09] J. H. Phelps. *Processing-Microstructure Models for Short- and Long-Fiber Thermoplastic Composites.* PhD thesis, University of Illinois at Urbana-Champaign, 2009.

[PPG+20] M. Perez, D. Prono, C. Ghnatios, E. Abisset, J. L. Duval, and F. Chinesta. Advanced modeling and simulation of sheet moulding compound (SMC) processes. *Int. J. Mater. Forming*, 13(5):675–685, 2020.

[PSAC+16] M. Perez, A. Scheuer, E. Abisset-Chavanne, F. Chinesta, and R. Keunings. A multi-scale description of orientation in simple shear flows of confined rod suspensions. *J. Non-Newt. Fluid Mech.*, 233:61–74, 2016.

[PT09] J. H. Phelps and C. L. Tucker. An anisotropic rotary diffusion model for fiber orientation in short- and long-fiber thermoplastics. *J. Non-Newt. Fluid Mech.*, 156:165–176, 2009.

[PTFTZ02] N. Phan-Thien, X.-J. Fan, R. Tanner, and R. Zheng. Folgar–Tucker constant for a fibre suspension in a Newtonian fluid. *J. Non-Newt. Fluid Mech.*, 103(2):251–260, 2002.

[PTVF92] W. H. Press, S. A. Teukolsky, W. T. Vetterling, and B. P. Flannery. *Numerical Recipes in FORTRAN.* Cambridge University Press, 2nd edition, 1992.

[Reu71] R. C. Reuter. Concise property transformation relations for an anisotropic lamina. *J. Compos. Mater.*, 5(2):270–272, 1971.

[RHJS18] T. Russell, B. Heller, D. A. Jack, and D. E. Smith. Prediction of the fiber orientation state and the resulting structural and thermal properties of fiber reinforced additive manufactured composites fabricated using the big area additive manufacturing process. *J. Compos. Sci.*, 2(2), 2018.

[RM01] B. D. Reddy and G. P. Mitchell. Finite element analysis of fibre suspension flows. *Comput. Methods Appl. Mech. Engrg.*, 190:2349–2367, 2001.

[ROCM20] A. Rashidi, T. Olfatbakhsh, B. Crawford, and A. S. Milani. A review of current challenges and case study toward optimizing micro-computed X-ray tomography of carbon fabric composites. *Materials*, 13(16), 2020.

[SAAC+18] A. Scheuer, A. Ammar, E. Abisset-Chavanne, E. Cueto, F. Chinesta, R. Keunings, and S. G. Advani. Data-driven upscaling of orientation kinematics in suspensions of rigid fibres. *Comput. Model. Engr. Sci.*, 117(3):367–386, 2018.

[SAC04] M. Sepehr, G. Ausias, and P. J. Carreau. Rheological properties of short fiber filled polypropylene in transient shear flow. *J. Non-Newt. Fluid Mech.*, 123:19–32, 2004.

[SACCK16] A. Scheuer, E. Abisset-Chavanne, F. Chinesta, and R. Keunings. Second-gradient modelling of orientation development and rheology of dilute confined suspensions. *J. Non-Newt. Fluid Mech.*, 237:54–64, 2016.

[SAS20] Z. Sekkate, A. Aboutajeddine, and A. Seddouki. Elastoplastic mean-field homogenization: recent advances review. *Mech. Adv. Mater. Struct.*, 2020. DOI: 10.1080/15376494.2020.1776431.

[SBA04] N. C. Shapley, R. A. Brown, and R. C. Armstrong. Evaluation of particle migration models based on laser Doppler velocimetry measurements in concentrated suspensions. *J. Rheol.*, 48(2):255–279, 2004.

[SBG+17] K. Schladitz, A. Büter, M. Godehardt, O. Wirjadi, J. Fleckenstein, T. Gerster, U. Hassler, K. Jaschek, M. Maisl, U. Maisl, S. Mohr, U. Netzelmann, T. Potyra, and M. O. Steinhauser. Non-destructive characterization of fiber orientation in reinforced SMC as input for simulation based design. *Compos. Struct.*, 160:195–203, 2017.

[SBO20] S. A. Simon, A. Bechara, and T. Osswald. Experimental validation of a direct fiber model for orientation prediction. *J. Compos. Sci.*, 4(2), 2020.

[SCG+04] M. Sepehr, P. Carreau, M. Grmela, G. Ausias, and P. Lafleur. Comparison of rheological properties of fiber suspensions with model predictions. *J. Polym. Eng.*, 24(6):579–610, 2004.

[Sch17] M. Schneider. The sequential addition and migration method to generate representative volume elements for the homogenization of short fiber reinforced plastics. *Computational Mechanics*, 59(2):247–263, 2017.

[SCMA04] M. Sepehr, P. J. Carreau, M. Moan, and G. Ausias. Rheological properties of short fiber model suspensions. *J. Rheol.*, 48:1023–1048, 2004.

[SFP18] D. E. Sommer, A. J. Favaloro, and R. B. Pipes. Coupling anisotropic viscosity and fiber orientation in applications to squeeze flow. *J. Rheol.*, 62(3):669–679, 2018.

[SIS19] T. Sasayama, M. Inagaki, and N. Sato. Direct simulation of glass fiber breakage in simple shear flow considering fiber-fiber interaction. *Composites A*, 124:105514, 2019.

[SJKK15] D. Salaberger, M. Jerabek, T. Koch, and J. Kastner. Consideration of accuracy of quantitative X-ray CT analyses for short-glass-fibre-reinforced polymers. *Mater. Sci. Forum*, 825:907–913, 2015.

[SKF+18] B. N. Sharma, S. A. Kijewski, L. S. Fifield, Y. Shin, C. L. Tucker, and M. D. Sangid. Reliability in the characterization of fiber length distributions of injection molded long carbon fiber composites. *Polym. Compos.*, 39(12):4594–4604, 2018.

[SKK+11] D. Salaberger, K. A. Kannappan, J. Kastner, J. Reussner, and T. Auinger. Evaluation of computed tomography data from fibre reinforced polymers to determine fibre length distribution. *Int. Polym. Process.*, 26(3):283–291, 2011.

[SL96] A. J. Szeri and D. J. Lin. A deformation tensor model of Brownian suspensions of orientable particles–the nonlinear dynamics of closure models. *J. Non-Newt. Fluid Mech.*, 64:43–69, 1996.

[SLW15] I. Straumit, S. V. Lomov, and M. Wevers. Quantification of the internal structure and automatic generation of voxel models of textile composites from X-ray computed tomography data. *Composites A*, 69:150–158, 2015.

[SNN+18] B. N. Sharma, D. Naragani, B. N. Nguyen, C. L. Tucker, and M. D. Sangid. Uncertainty quantification of fiber orientation distribution measurements for long-fiber-reinforced thermoplastic composites. *J. Compos. Mater.*, 52(13):1781–1797, 2018.

[SSBG18] S. Strednak, S. Shaikh, J. E. Butler, and E. Guazzelli. Shear-induced migration and orientation of rigid fibers in an oscillatory pipe flow. *Phys. Rev. Fluids*, 3:091301, Sep 2018.

[SSK00] C. F. Schmid, L. H. Switzer, and D. J. Klingenberg. Simulations of fiber flocculation: Effects of fiber properties and interfiber friction. *J. Rheol.*, 44(4):781–809, 2000.

[SSKM20] T. Sasayama, N. Sato, Y. Katagiri, and Y. Murayama. Particle-level simulation for the prediction of short fiber orientation in injection molding. *Composites A*, 139:106115, 2020.

[SV98] B. Souloumiac and M. Vincent. Steady shear viscosity of short fibre suspensions in thermoplastics. *Rheol. Acta*, 37(3):289–298, 1998.

[TA89] M. Taya and R. J. Arsenault. *Metal Matrix Composites: Thermomechanical Behavior.* Pergamon Press, Oxford, 1989.

[TA93] S. Toll and P.-O. Andersson. Microstructure of long- and short-fiber reinforced injection molded polyamide. *Polym. Compos.*, 14(2):116–125, 1993.

[TCH13] H.-C. Tseng, R.-Y. Chang, and C.-H. Hsu. Phenomenological improvements to predictive models of fiber orientation in concentrated suspensions. *J. Rheol.*, 57:1597–1631, 2013.

[TCH16] H.-C. Tseng, R.-Y. Chang, and C.-H. Hsu. An objective tensor to predict anisotropic fiber orientation in concentrated suspensions. *J. Rheol.*, 60(2):215–224, 2016.

[TCH18a] H.-C. Tseng, R.-Y. Chang, and C.-H. Hsu. Predictions of fiber concentration in injection molding simulation of fiber-reinforced composites. *J. Thermoplast. Compos. Mater.*, 31(11):1529–1544, 2018.

[TCH18b] H.-C. Tseng, R.-Y. Chang, and C.-H. Hsu. The use of principal spatial tensor to predict anisotropic fiber orientation in concentrated fiber suspensions. *J. Rheol.*, 62(1):313–320, 2018.

[TF19] H.-C. Tseng and A. J. Favaloro. The use of informed isotropic constitutive equation to simulate anisotropic rheological behaviors in fiber suspensions. *J. Rheol.*, 63(2):263–274, 2019.

[THO20] J. Teuwsen, S. K. Hohn, and T. A. Osswald. Direct fiber simulation of a compression molded ribbed structure made of a sheet molding compound with randomly oriented carbon/epoxy prepreg strands—A comparison of predicted fiber orientations with computed tomography analyses. *J. Compos. Sci.*, 4(4), 2020.

[Tit58] E. C. Titmarsch. George Barker Jeffery. *Biographical Memoirs of Fellows of the Royal Society*, 4:128–137, 1958.

[TL99] C. L. Tucker and E. Liang. Stiffness predictions for unidirectional short fiber composites: Review and evaluation. *Compos. Sci. Tech.*, 59:655–671, 1999.

[TM51] B. J. Trevelyan and S. G. Mason. Particle motions in sheared suspensions I. Rotations. *J. Colloid Sci.*, 6(4):354–367, 1951.

[TS19] H.-C. Tseng and T.-H. Su. Coupled flow and fiber orientation analysis for 3D injection molding simulations of fiber composites. *AIP Conf. Proc.*, 2065(1):030021, 2019.

[TT85] Y. Takao and M. Taya. Thermal expansion coefficients and thermal stresses in an aligned short fiber composite with application to a short carbon fiber/aluminum. *ASME J. Appl. Mech.*, 52:806–810, 1985.

[Tuc89] C. L. Tucker, editor. *Fundamentals of Computer Modeling for Polymer Processing.* Hanser, Munich, 1989.

[Tuc91] C. L. Tucker. Flow regimes for fiber suspensions in narrow gaps. *J. Non-Newt. Fluid Mech.*, 39:239–268, 1991.

[TW84] G. P. Tandon and G. J. Weng. The effect of aspect ratio of inclusions on the elastic properties of unidirectionally aligned composites. *Polym. Compos.*, 5:327–333, 1984.

[VB99] F. W. J. Van Hattum and C. A. Bernardo. A model to predict the strength of short fiber composites. *Polym. Compos.*, 20:524–533, 1999.

[VCD94] V. Verleye, A. Couniot, and F. Dupret. Numerical prediction of fiber orientation in complex composite injection-molded parts. In K. Himasekhar, V. Prasad, T. A. Osswald, and G. Batch, editors, *Advances in Computer-Aided Engineering (CAE) in Polymer Processing,* MD-Vol. 175/HTD-Vol. 238, pages 265–279. ASME, New York, 1994.

[VD93] V. Verleye and F. Dupret. Prediction of fiber orientation in complex injection molded parts. In D. A. Siginer, W. E. VanArsdale, C. E. Altan, and A. N. Alexandrou, editors, *Developments in Non-Newtonian Flows,* AMD-Vol. 175, pages 139–163. ASME, New York, 1993.

[Ver98] B. E. VerWeyst. *Numerical Predictions of Flow-Induced Fiber Orientation in 3-D Geometries.* PhD thesis, University of Illinois at Urbana-Champaign, Urbana, IL, 1998.

[VG12] G. M. Vélez-García. *Experimental Evaluation and Simulations of Fiber Orientation in Injection Molding of Polymers Containing Short Glass Fibers.* PhD thesis, Virginia Polytechnic Institute and State University, Blacksburg, VA, 2012.

[VGWB$^+$12] G. M. Vélez-García, P. Wapperom, D. G. Baird, A. O. Aning, and V. Kunc. Unambiguous orientation in short fiber composites over small sampling area in a center-gated disk. *Composites A,* 43(1):104–113, 2012.

[VGWK$^+$12] G. Vélez-García, P. Wapperom, V. Kunc, D. Baird, and A. Zink-Sharp. Sample preparation and image acquisition using optical-reflective microscopy in the measurement of fiber orientation in thermoplastic composites. *J. Microscopy,* 248(1):23–33, 2012.

[VT02] B. E. VerWeyst and C. L. Tucker. Fiber suspensions in complex geometries: Flow/orientation coupling. *Can. J. Chem. Eng.*, 80:1093–1106, 2002.

[VTFO99] B. E. VerWeyst, C. L. Tucker, P. H. Foss, and J. F. O'Gara. Fiber orientation in 3-D injection molded features: Prediction and experiment. *Int. Polym. Proc.*, 14:409–420, 1999.

[Wan07] J. Wang. *Improved Fiber Orientation Predictions for Injection Molded Composites.* PhD thesis, University of Illinois at Urbana-Champaign, Urbana, IL, 2007.

[Wan18] J. Wang. Moldflow R & D update. *https://damassets.autodesk.net/content/dam/autodesk/www/campaigns/autodesk-moldflow-summit-recording/7_R&D-Update-JinWang.pdf*, 2018.

[Wet98] E. D. Wetzel. Personal communication, 1998.

[Wil77] J. R. Willis. Bounds and self-consistent estimates for the overall properties of anisotropic composites. *J. Mech. Phys. Solids*, 25:185–202, 1977.

[WOT08] J. Wang, J. F. O'Gara, and C. L. Tucker. An objective model for slow orientation kinetics in concentrated fiber suspensions: Theory and rheological evidence. *J. Rheol.*, 52:1179–1200, 2008.

[WRB20] F. Willems, P. Reitinger, and C. Bonten. Calibration of fiber orientation simulations for LFT–A new approach. *J. Compos. Sci.*, 4:163, 2020.

[WSEO08] O. Wirjadi, K. Schladitz, P. Easwaran, and J. Ohser. Estimating fibre direction distributions of reinforced composites from tomographic images. *Image Analysis & Stereology*, 35(3):167–179, 2016-12-08.

[WT01] E. D. Wetzel and C. L. Tucker. Droplet deformation in dispersions with unequal viscosities and zero interfacial tension. *J. Fluid Mech.*, 426:199–228, 2001.

[YAMG+20] M. Yun, C. Argerich Martin, P. Giormini, F. Chinesta, and S. Advani. Learning the macroscopic flow model of short fiber suspensions from fine-scale simulated data. *Entropy*, 22(1), 2020.

[YKD94] Y. Yamane, Y. Kaneda, and M. Doi. Numerical simulation of semi-dilute suspensions of rodlike particles in shear flow. *J. Non-Newt. Fluid Mech.*, 54:405–421, 1994.

[YM95] S. Yamamoto and T. Matsuoka. Dynamic simulation of fiber suspensions in shear flow. *J. Chem. Phys.*, 102(5):2254–2260, 1995.

[Yur87] S. Yurgartis. Measurement of small angle fiber misalignments in continuous fiber composites. *Compos. Sci. Tech.*, 30(4):279–293, 1987.

[ZBL97] Y. T. Zhu, W. R. Blumenthal, and T. C. Lowe. Determination of non-symmetric 3-D fiber-orientation distribution and average fiber length in short-fiber composites. *J. Compos. Mater.*, 31(13):1287–1301, 1997.

[ZESAJMS11] D. Zhang, D. E. Smith, D. A. Jack, and S. Montgomery-Smith. Numerical evaluation of single fiber motion for short-fiber-reinforced composite materials processing. *J. Manuf. Sci. Eng.*, 133(5), 09 2011. 051002.

[ZPB01] G. Zak, C. B. Park, and B. Benhabib. Estimation of three-dimensional fibre-orientation distribution in short-fibre composites by a two-section method. *J. Compos. Mater.*, 35(4):316–339, 2001.

[ZTF11] R. Zheng, R. I. Tanner, and X. Fan. *Injection Molding: Integration of Theory and Modeling Methods.* Springer, Berlin, 2011.

Index

D

S

U

V

W

X

Y